NMR of Polymers

NMR OF POLYMERS

FRANK A. BOVEY
PETER A. MIRAU
AT & T Bell Laboratories
Murray Hill, New Jersey

ACADEMIC PRESS
San Diego   London   Boston   New York   Sydney   Tokyo   Toronto

Cover art adapted from Figs. 3.62 and 3.67. Reprinted with permission of the American Chemical Society.

This book is printed on acid-free paper. ∞

Copyright © 1996, 1972 by AT&T

All Rights Reserved.
No part of this publication may be reproduced or transmitted in any form or by any means, electronic or mechanical, including photocopy, recording, or any information storage and retrieval system, without permission in writing from the publisher.

Academic Press, Inc.
525 B Street, Suite 1900, San Diego, California 92101-4495, USA
http://www.apnet.com

Academic Press Limited
24-28 Oval Road, London NW1 7DX, UK
http://www.hbuk.co.uk/ap/

Library of Congress Cataloging-in-Publication Data

Bovey, Frank Alden, date.
    NMR of Polymers / by Frank A. Bovey, Peter A. Mirau.
        p.   cm.
    Includes index.
    ISBN 0-12-119765-4 (alk. paper)
    1. Nuclear magnetic resonance spectroscopy.   2. Macromolecules-
-Analysis.   I. Mirau, Peter A.   II. Title.
QP519.9.N83B68   1996
574.19'285--dc20   96-28240
                                                                              CIP

PRINTED IN THE UNITED STATES OF AMERICA
96 97 98 99 00 01 BC 9 8 7 6 5 4 3 2 1

Contents

Preface — ix

1

FUNDAMENTALS OF NUCLEAR MAGNETIC RESONANCE

1.1 Introduction	1
1.2 Nuclear Spin and the NMR Phenomenon	2
1.3 The Detection of the Resonance Phenomenon	7
1.4 Nuclear Relaxation	11
1.4.1 Spin–Lattice Relaxation	11
1.4.2 Spin–Spin Relaxation and Dipolar Broadening	19
1.4.3 The Bloch Equations	23
1.4.4 Nuclear Electric Quadrupole Relaxation	29
1.5 Magnetic Shielding and the Chemical Shift	30
1.5.1 The Basis of Molecular Shielding: Proton Chemical Shifts	30
1.5.2 Carbon-13 Chemical Shifts	42
1.5.3 Other Nuclei (^{19}F, ^{31}P, ^{14}N, ^{15}N)	46
1.6 Indirect Coupling of Nuclear Spins	47
1.6.1 Introduction	47
1.6.2 Proton–Proton Couplings and Spectral Analysis	49

	1.6.3 Carbon–Proton Couplings and Carbon–Carbon Couplings	53
	1.6.4 Other Nuclei (^{19}F, ^{31}P, ^{14}N, ^{15}N)	53
1.7	Rate Phenomena: Averaging of Chemical Shifts and J Couplings	55
1.8	Experimental Methods	57
	1.8.1 Introduction	57
	1.8.2 The Magnetic Field	57
	1.8.3 Referencing	59
	1.8.4 Decoupling	61
	1.8.5 Solid State Methods	63
	1.8.6 The Observation of Nuclear Relaxation	72
	1.8.7 Two-Dimensional NMR	87
	1.8.8 Spectral Editing and Simplification	105

2

THE MICROSTRUCTURE OF POLYMER CHAINS

2.1	Introduction	117
2.2	Polymer Microstructure	118
	2.2.1 Head-to-Tail versus Head-to-Head–Tail-to-Tail Isomerism: Regioisomerism	118
	2.2.2 Stereochemical Configuration	118
	2.2.3 Isomerism in Diene Polymer Chains	122
	2.2.4 Vinyl Polymers with Optically Active Sidechains	128
	2.2.5 Polymers with Asymmetric Centers in the Main Chain	129
	2.2.6 Branching and Cross Linking	129
	2.2.7 Copolymer Sequences	131
	2.2.8 Other Types of Isomerism	133
2.3	Configurational Statistics and Propagation Mechanism	133
	2.3.1 Bernoulli and First-Order Propagation Models	133
	2.3.2 Fitting of Longer Sequences of the First-Order Markov Model	139
	2.3.3 More Complex Propagation Models	143
2.4	Propagation Errors	151
	2.4.1 Copolymer Propagation	153

THE SOLUTION CHARACTERIZATION OF POLYMERS

3.1 Introduction	155
3.2 Resonance Assignments	158
3.2.1 Resonance Assignments from Model Compounds	158
3.2.2 Assignment by Intensity and Chemical Shift Calculations	160
3.2.3 Two-Dimensional NMR	165
3.3 Microstructural Characterization of Polymers in Solution	167
3.3.1 Regioisomerism	168
3.3.2 Stereochemical Isomerism	181
3.3.3 Geometric Isomerism	196
3.3.4 Branches and End Groups	199
3.3.5 Copolymer Characterization	212
3.4 The Solution Structure of Polymers	220
3.4.1 Chain Conformation	220
3.4.2 Intermolecular Association of Polymers	232

THE SOLID-STATE NMR OF POLYMERS

4.1 Introduction	243
4.2 Chain Conformation in the Solid State	245
4.2.1 Semicrystalline Polymers	245
4.2.2 Amorphous Polymers	255
4.2.3 Solid–Solid Phase Transitions	261
4.2.4 Other Studies	267
4.3 The Organization of Polymers in the Solid State	274
4.3.1 The Organization of Semicrystalline Polymers	279
4.3.2 Multiphase Polymers	289
4.3.3 Polymer Blends	300
4.3.4 NMR Studies of Oriented Polymers	331
4.4 NMR Imaging of Polymers	337

The Dynamics of Macromolecules

5.1 Introduction	353
5.2 Polymer Dynamics in Solution	358
5.2.1 Introduction	358
5.2.2 Modeling the Molecular Dynamics of Polymers	359
5.2.3 Observation of Polymer Relaxation in Solution	367
5.3 Polymer Dynamics in the Solid State	379
5.3.1 Introduction	379
5.3.2 The Dynamics of Semicrystalline Polymers	394
5.3.3 The Dynamics of Amorphous Polymers	412
5.3.4 The Dynamics of Polymer Blends	436
5.3.5 The Dynamics of Multiphase Polymer Systems	445

Index   455

Preface

Over the past three decades, NMR spectroscopy has emerged as one of the most important methods for polymer characterization. In the early history, high resolution solution NMR spectra played a key role in characterizing polymer microstructure and in understanding polymerization mechanisms. More recently, solid-state NMR has been developed and used to probe the structure, conformation, organization, and dynamics of polymers in their native state. In fact, the solid-state characterization of polymers is such an important topic that it has driven the development of new solid-state NMR methods. A further revolution occurred with the introduction of multidimensional NMR, making it possible to observe polymers with much higher resolution, leading to a more detailed understanding of polymer microstructure and to the molecular level assignments of the dynamics that have long been measured by dielectric and dynamic mechanical spectroscopy. The current importance of NMR in polymer characterization can be judged from the many papers published in polymer science that rely on NMR to elucidate the properties of polymers.

The goal of this book is to provide an overview of the applications of NMR to polymer characterization. The book begins with a review of the fundamental principles in Chapter 1 and polymer structure in Chapter 2. The high resolution solution state NMR of polymers is presented in Chapter 3 and includes multinuclear NMR studies and the two-dimensional NMR methods used to study polymer microstructure, chain conformation, and the structure of associating polymers. The solid-state NMR of polymers is presented in Chapter 4 and includes NMR determination of chain conformation in semicrystalline and amorphous polymers, polymer blends, and multiphase polymer systems, as well as the NMR methods used to study chain organization on longer length scales. The molecular dynamics of polymers is covered in Chapter 5, where the methods used to study polymer dynamics both in solution and in the solid state are presented. Included in this chapter are the wideline and multidimensional NMR methods that have been used to relate the transitions measured by other spectroscopies to the chain dynamics at the atomic level.

This book is intended for academic and industrial researchers and does not reference all the data collected on the NMR of polymers. It does, however, illustrate the types of problems in polymer science that can be solved by NMR spectroscopy. In certain instances it is most instructive to show the data from some of the early, classic experiments in polymer NMR, while in other cases experiments from the most current literature are provided.

1

FUNDAMENTALS OF NUCLEAR MAGNETIC RESONANCE

1.1 INTRODUCTION

NMR spectroscopy is a method of great interest and importance for the observation of every aspect of the structure and properties of macromolecular substances. The first studies were reported by Alpert (1) only about a year after the discovery of nuclear resonance in bulk matter (2,3). It was observed that natural rubber at room temperature gives a proton linewidth more like that of a mobile liquid than of a solid, but that the resonance broadens markedly at temperatures approaching that (ca. −70°C) now known as the glass temperature. These phenomena were recognized as being related to the presence (and cessation) of micro-Brownian motion. We shall relate these matters in more detail when we deal with the vital topic of polymer NMR in the solid state. For the present we note that this field was well developed before high-resolution spectra of synthetic polymers *in solution* were first reported by Saunder and Wishnia (4), Odajima (5), and Bovey *et al.* (6). From today's perspective it is difficult to understand why the observation of such spectra was so long delayed, especially since the NMR spectroscopy of small molecules was in an advanced state. It appears that solution spectra were generally expected to be very complex and to exhibit resonances too broad to be usefully interpreted. As we shall see, such fears were in general not well founded. Modern instrumental methods have made it possible to obtain dynamic information from polymer solutions as well as solids and to obtain structural information from solid polymers as well as solutions. Thus, the barrier between two once quite separate modes of study of macromolecules—solid- and liquid-state NMR spectroscopy—has all but disappeared.

1.2 NUCLEAR SPIN AND THE NMR PHENOMENON

In addition to charge and mass, which all nuclei have, many isotopes possess spin, or angular momentum. Since a spinning charge generates a magnetic field, there is a magnetic moment associated with this angular momentum. The magnetic properties of nuclei were first postulated by Pauli (7) to explain certain hyperfine structural features of atomic spectra.

According to a basic principle of quantum mechanics, the maximum experimentally observable component of the angular momentum of a nucleus possessing a spin (or of any particle or system having angular momentum) is a half-integral or integral multiple of $h/2\pi$, where h is Planck's constant. This maximum component is I, which is called the spin quantum number or simply "the spin." Each nuclear ground state is characterized by just one value of I. If $I = 0$, the nucleus has no magnetic moment. If I is not zero, the nucleus will possess a magnetic moment μ, which is always taken as parallel to the angular momentum vector. The permitted values of the vector moment along any chosen axis are described by means of a set of magnetic quantum numbers m given by the series

$$m = I, (I - 1), (I - 2), \ldots, -I. \quad (1.1)$$

Thus, if I is 1/2, the possible magnetic quantum numbers are $+1/2$ and $-1/2$. If I is 1, m may take on the values 1, 0, -1, and so on. In general, then, there are $2I + 1$ possible orientations or states of the nucleus. In the absence of a magnetic field, these states all have the same energy. In the presence of a uniform magnetic field B_0, they correspond to states of different potential energy. For nuclei for which I is 1/2, the two possible values of m, $+1/2$ and $-1/2$, describe states in which the nuclear moment is aligned with and against the field B_0, respectively, the latter state being of higher energy. The detection of transitions of magnetic nuclei (often themselves referred to as "spins") between the states is made possible by the nuclear magnetic resonance phenomenon.

The magnitudes of nuclear magnetic moments are often specified in terms of the ratio of magnetic moment and angular momentum, or

1.2. Nuclear Spin and the NMR Phenomenon

magnetogyric ratio γ, defined as

$$\gamma = \frac{2\pi\mu}{Ih}. \tag{1.2}$$

A spinning spherical particle with mass M and charge e uniformly spread over its surface can be shown to give rise to a magnetic moment $eh/4\pi Mc$, where c is the velocity of light. For a particle with the charge, mass, and spin of the proton, the moment should be 5.0505×10^{-27} joules/tesla (J T^{-1})[1] (5.0505×10^{-27} erg-G^{-1} in cgs units) in this model. Actually, this approximation is not a good one even for the proton, which is observed to have a magnetic moment about 2.79 times as great as that the oversimplified model predicts. No simple model can predict or explain the actual magnetic moments of nuclei. However, the predicted moment for the proton serves as a useful unit for expressing nuclear moments and is known as the *nuclear magneton*; it is the analogue of the Bohr magneton for electron spin. Observed nuclear moments can be specified in terms of the nuclear magneton by

$$\mu = g \frac{ehI}{4\pi M_p c}, \tag{1.3}$$

where M_p is the proton mass and g is an empirical parameter called the nuclear g factor. In units of nuclear magnetons, then,

$$u = gI. \tag{1.4}$$

In Table 1.1 nuclear moments are expressed in these units. It will be noted that some nuclei have negative moments. This is of some practical significance to NMR spectroscopists, as we shall see in later sections. Its theoretical meaning will be evident a little further on. It should also be noted that the neutron, with no net charge, has a substantial magnetic moment. This is a particularly striking illustration of the failure of simple models to predict μ. Clearly, the neutron must contain separated charges (at least a part of the time) even though its total charge is zero.

Although magnetic moments cannot be predicted exactly, there are useful empirical rules relating the mass number A and atomic number Z to the nuclear spin properties:

1. If both the mass number A and the atomic number Z are even, $I = 0$.

---

[1] Magnetic field strengths are expressed in kilogauss or tesla. T; 10 kG = 1 T.

TABLE 1.1
The NMR Properties of Nuclei of Interest to NMR Spectroscopists

Isotope	Abundance (%)	Spin	$\gamma \times 10^{-8a}$	Relative sensitivity[b]	v_0 at 11.7 T (MHz)
^{1}H	99.98	1/2	2.6752	1.0	500.
^{19}F	100.	1/2	2.5167	0.83	470.2
^{29}Si	4.7	1/2	−0.5316	0.078	99.3
^{31}P	100.	1/2	1.0829	0.066	202.3
^{13}C	1.1	1/2	0.6726	0.0159	125.6
^{2}H	0.015	1	0.4107	0.00964	76.7
^{15}N	0.365	1/2	−0.2711	0.0010	50.6

[a] Magnetogyric ratio in SI units.
[b] For an equal number of nuclei at constant field.

2. If A is odd and Z is odd or even, I will have half-integral values 1/2, 3/2, 5/2, etc.
3. If A is even and Z is odd, I will have integral values 1, 2, 3, etc.

Thus, some very common isotopes, such as ^{12}C, ^{16}O, and ^{32}S, have no magnetic moment and cannot be observed by NMR.

When placed in a magnetic field, nuclei having spin undergo precession about the field direction, as shown in Fig. 1.1. The frequency of this so-called Larmor precession is designated ω_0 in radians per second or v_0 in hertz (Hz), cycles per second ($\omega_0 = 2\pi v_0$). The nuclei can be made to flip over, i.e., reverse the direction of the spin axis, by applying a second magnetic field, designated B_1, at right

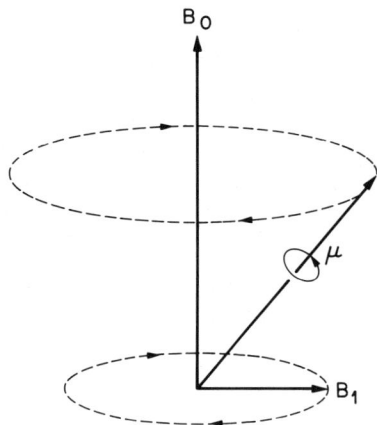

FIGURE 1.1 Nuclear moment in a magnetic field.

1.2. Nuclear Spin and the NMR Phenomenon

angles to B_0 and causing this second field to rotate at the precession frequency v_0. This second field is represented by the horizontal vector in Fig. 1.1, although in practice it is actually very much smaller in relation to B_0 than this figure suggests. It can be seen that if B_1 rotates at a frequency close to but not exactly at the precession frequency, it will cause at most only some wobbling or nutation of the magnetic moment μ. If, however, B_1 is made to rotate exactly at the precession frequency, it will cause large oscillations in the angle between μ and B_0. If we vary the rate of rotation of B_1 through this value, we will observe a resonance phenomenon as we pass through v_0.

One might suppose that the Larmor precession of the nuclear moments could itself be detected by some means without the need to invoke a resonance phenomenon. This, however, is not possible because each nucleus precesses with a completely random phase with respect to that of its neighbors and there is therefore no macroscopic property of the system that changes at the Larmor frequency. By a well-known relationship, the Larmor precession frequency is given by

$$\omega_0 = \gamma B_0, \tag{1.5}$$

or, from Eq. (1.2),

$$h v_0 = \frac{\mu B_0}{I}. \tag{1.6}$$

The result of this classical treatment can also be obtained by a quantum mechanical description, which is in some ways a more convenient way of regarding the resonance phenomenon. It is best, however, not to try to adopt either approach to the exclusion of the other, since each provides valuable insights. From the quantum mechanical viewpoint, the quantity $h v_0$ is the energy separation ΔE between the magnetic energy levels (often termed Zeeman levels, after the investigator who first observed the corresponding splitting in atomic spectra) in a magnetic field B_0, as shown in Fig. 1.2. For a nucleus of spin $1/2$, ΔE will be $2\mu B_0$, and as we have seen, only two energy levels are possible. For a nucleus of spin 1, there are three energy levels, as illustrated in Fig. 1.2. The quantum mechanical treatment gives us an additional result for such systems with $I > 1$, which the classical treatment does not: it tells us that only transitions between adjacent energy levels are allowed, i.e., that the magnetic quantum number can only change by ± 1. Thus, transitions between the $m = -1$ and the $m = 0$ levels and between the $m = +1$ and the $m = 0$ levels are possible, but transitions between the $m = -1$ and the $m = +1$ levels are not possible.

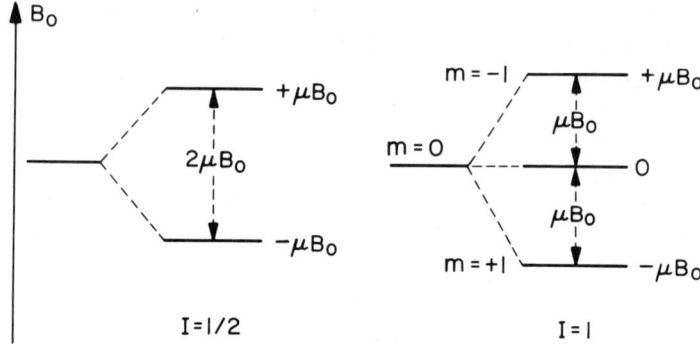

FIGURE 1.2 Magnetic energy levels for nuclei of spin 1/2 and 1.

In Fig. 1.3 the separation of proton magnetic energy levels is shown as a function of magnetic field strength for a number of values of the latter employed in current spectrometers. The resonant proton RF field frequency is indicated (in megahertz, MHz, 10^6 Hz) and is a common way of designating the magnetic field strength. Fields above

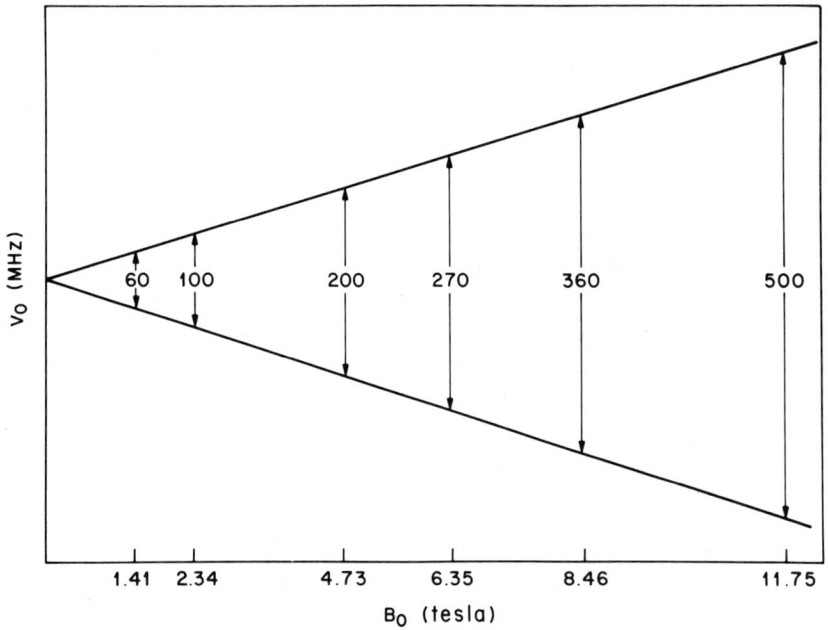

FIGURE 1.3 The splitting of the magnetic energy levels of protons expressed as resonance frequency v_0 as a function of the magnetic field strength in tesla (T).

ca. 2.5 T require superconducting solenoid magnets. For the ^{13}C nucleus, which has a magnetogyric ratio one-fourth that of the proton (Table 1.1), the resonant frequencies will be one-fourth of those shown in Fig. 1.3.

1.3 THE DETECTION OF THE RESONANCE PHENOMENON

We have not yet said how the rotating magnetic field B_1 is to be provided or how the resonance phenomenon is to be detected. In the semischematic diagram of Fig. 1.4, the sample tube is placed in the field of the magnet, which may be a superconducting solenoid (as shown here) or may also be a permanent magnet or an electromagnet. The radiofrequency transmitter applies an RF field of appropriate frequency (Fig. 1.3) by means of the exciting coil indicated, which is of Helmholtz design. The magnetic field direction z is along the axis of the sample tube. The magnetic vector of the RF field oscillates along

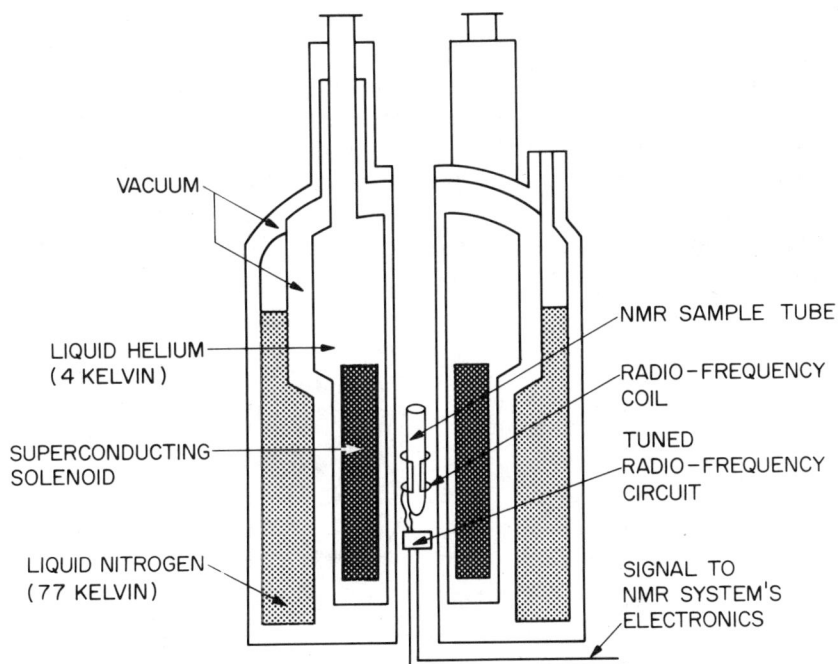

FIGURE 1.4 Cross section of a superconducting magnet for NMR spectroscopy. The magnet and sample tube are not drawn to the same scale. The diameter of the magnet assembly is approximately 70 cm, while that of the tube is approximately 1 cm.

the y direction [i.e., in a direction perpendicular to the axis of the sample tube and to B_0 (see Fig. 1.5)]. This oscillating magnetic field provides what is required for flipping over the nuclear spins, for it may be thought of as being composed of two equal magnetic vectors rotating in phase with equal angular velocities, but in opposite directions. (An exact analogue is the decomposition of plane polarized light into two equal and opposite circularly polarized vectors.) The precessing magnetic moments will pick out the appropriate rotating component of B_1 in accordance with Eq. (1.6) and the sign of μ (see Table 1.1). The other rotating component is so far off resonance that it has no observable effect.

To display the resonance signal, the magnetic field may be increased slowly until Eq. (1.6) is satisfied and resonance occurs. At resonance, the nuclear magnetic dipoles in the lower energy state flip over and in so doing induce a voltage in the RF coil. This induced voltage is amplified and recorded. One may regard the rotating field B_1 as having given the precessing spins a degree of coherence, so that now there is a detectable *macroscopic* magnetic moment, precessing at a rate v_0. The resonance phenomenon observed in this way is termed *nuclear induction* and is the method originally used by the Bloch group. Alternatively, the nuclear flipping may be detected as an *absorption* of energy from the RF field.

The continuous wave (CW) method of detection just described is now employed only in old or very simple NMR spectrometers. Much greater sensitivity and flexibility are achieved in modern instruments by use of *pulse methods* with Fourier transformation of the resulting time-domain signal. In principle, the signal-to-noise ratio of any spectrum can be improved by extending the observing time, t. Background noise accumulates in proportion to $\sqrt{t}$, whereas the strength of

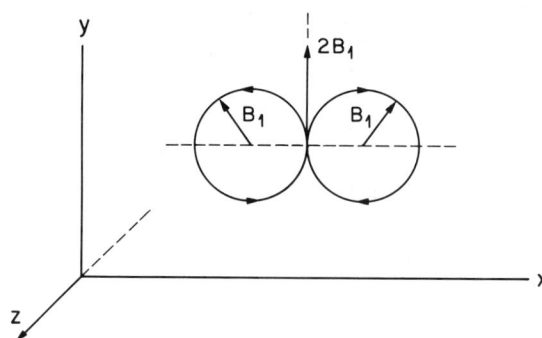

FIGURE 1.5 Decomposition of the magnetic vector B_1 into two counterrotating vectors.

1.3. The Detection of the Resonance Phenomenon

a coherent signal increases as t. Thus, one might obtain increased sensitivity by sweeping through the spectrum over a period of several hours instead of a few minutes, as is usual. This approach is of limited utility, however, as saturation can occur. It is somewhat more practical to sweep through the spectrum rapidly many times in succession, beginning each new sweep at the same point in the spectrum and then summing up the traces. This summation can be performed by a multichannel pulse height analyzer and a small computer that can store several thousand traces. The signal-to-noise ratio increases in proportion to the square root of the number of traces.

This method is still wasteful of time, however, as many spectra, particularly ^{13}C spectra, consists mostly of baseline. The time intervals spent between resonances are not profitable. It is much more efficient to excite all the nuclei of interest at the same time and to avoid sweeping entirely. This can be done by supplying the RF field in the form of *pulses*. These are commonly orders of magnitude stronger than the RF fields used in CW spectroscopy. Pulse spectroscopy utilizes the fact that a burst of resonant RF energy, although nominally monochromatic, with a frequency v_0, actually contains a band of frequencies corresponding to the Fourier series of sine and cosine functions necessary to approximate a square wave. The bandwidth may be estimated approximately as $v_0 \pm 1/t_p$, where t_p is the duration (or "width") of the pulse, generally a few microseconds. The use of pulsed RF fields in NMR spectroscopy was first suggested by Bloch et al. (3). It has long been standard for the measurement of proton relaxation in the solid state, particularly for polymers (8,9). It use with spectrum accumulation for observing high-resolution spectra was first proposed by Ernst and Anderson (10).

In CW spectroscopy, the RF field tips the macroscopic magnetic moment M_0 only very slightly. In pulsed spectroscopy, the angle of tipping, α, is much greater and is proportional to the magnetogyric ratio of the nucleus and to the strength and duration of the pulse, as given by Eq. (1.7), which is a variation of Eq. (1.5):

$$\alpha = \gamma B_1 t_p. \tag{1.7}$$

It is common to employ a pulse that rotates the magnetic moment by $\pi/2$ rad or 90°, i.e., just into the $x'y'$ plane, as shown in Figs. 1.6(b) and 1.6(c). Such a pulse is termed a "90° pulse." In Fig. 1.6, the magnetization M_0 is represented in a coordinate frame that rotates with B_1 and, at resonance, with M_0, and in which both are therefore static. In Fig. 1.6(d) the magnetization vectors are shown fanning out, i.e., precessing at unequal rates, as a result of inhomogeneities in B_0

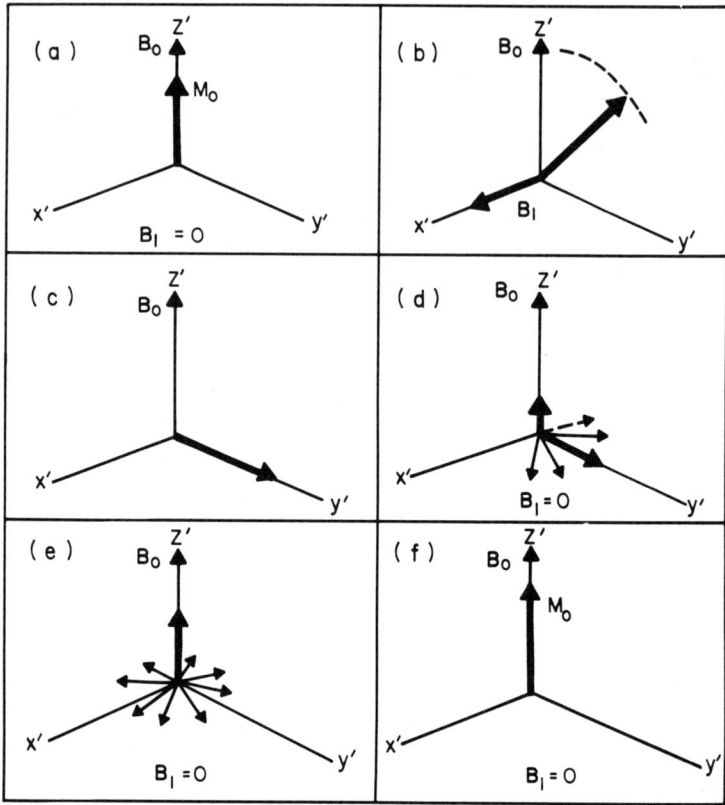

FIGURE 1.6 Rotating frame diagrams describing the pulsed NMR experiment. (The "primed" axes are used to indicate a rotating coordinate system.) (a) The net magnetization M_0 is aligned along the magnetic field direction. In (b) and (c) a 90° pulse is applied along the x' axis to rotate the magnetization into the $x'y'$ plane. In (d) and (e) the spins begin to relax (or dephase) in the $x'y'$ plane due to spin–spin (T_2) relaxation and finally return to their equilibrium state (f) via spin–lattice (T_1) relaxation.

and nuclear relaxation. The resulting loss of phase coherence is manifested as a decay in the net magnetization and in the observed signal, and occurs with a time constant T_2. At the same time, M_z, the longitudinal component of M_0, *is* growing back with a time constant T_1. The decay of the transverse magnetization (which provides the observed signal) is termed a *free induction decay* (FID), referring to the absence of the RF field. It may also be called the *time-domain spectrum*. The FID following a single 90° pulse is shown in a schematic manner in Fig. 1.7(a). The FID is an interferogram with a simple exponential decay envelope; the beat pattern corresponds to the

1.4. Nuclear Relaxation

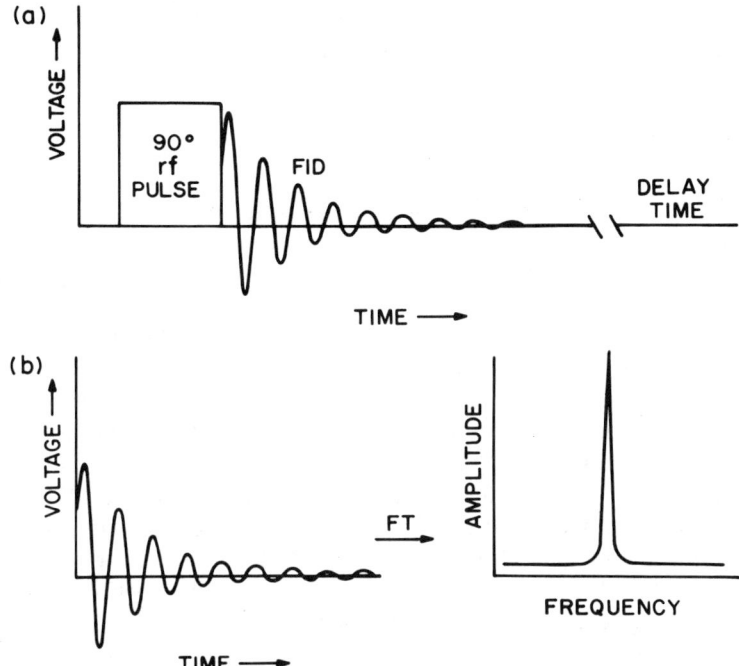

FIGURE 1.7 (a) Representation of a 90° RF pulse and the ensuing free induction decay (FID). (b) The Fourier transformation of the time-domain FID into the frequency-domain signal.

difference between the pulse carrier frequency v_0 and the precession frequency of the nuclei v_c. In Fig. 1.7(b) the FID is shown undergoing transformation to its Fourier inverse, appearing as a conventional or *frequency-domain spectrum*.

1.4 NUCLEAR RELAXATION

1.4.1 SPIN–LATTICE RELAXATION

In the magnetic fields employed in NMR instruments, the separation of nuclear magnetic energy levels is very small. For example, for protons in a field of 14.1 T (near the largest value employed in commercial instruments) it is only about 0.5 J. Even in the absence of the RF field there is usually a sufficiently rapid transfer of spins from the lower to the upper state and vice versa (for reasons to be explained shortly), so that an equilibrium population distribution is attained within a few seconds after B_0 is applied. If this distribution of spins is given by the Boltzmann factor $e^{2\mu B_0/kT}$, this can be expressed with sufficient accuracy by

$$\frac{N_+}{N_-} = 1 + \frac{2\mu B_0}{kT}, \qquad (1.8)$$

where N_+ and N_- represent the spin populations of the lower and upper states, respectively, and T is the Boltzmann *spin temperature*. It can thus be seen that even for protons, the net degree of polarization of the nuclear moments in the magnetic field direction is small; for example, only about 100 ppm, in an 11.75-T (500 MHz for protons; see Fig. 1.3) field; for all other nuclei (with the exception of tritium), it is even smaller. It is also clear that the nuclear magnetic energy cannot be expected to perturb the thermal energies of the molecules to an observable degree, except possibly at temperatures close to 0 K.

For nuclei with spins greater than $1/2$, the $2I + 1$ equally spaced magnetic energy levels will be separated in energy by μB_0, and the relative populations of adjacent levels will be given by expressions analogous to Eq. (1.8). To simplify the subsequent discussion, we shall confine our attention to nuclei with spins of $1/2$, recognizing that what is said for such two-level systems will apply also to any pair of adjacent levels in systems of spin greater than $1/2$.

We have seen that in the presence of the field B_1, there will be a net transfer of spins from the lower energy state to the upper energy state. In time, such a process would cause the populations of the levels to become equal, or nearly so, corresponding to a condition of *saturation* and to a very high Boltzmann spin temperature, unless there were some means by which the upper level spins could relax to the lower level. Such a process transfers energy from or *cools* the spin system. Similarly, when a system of spins is first thrust into a magnetic field, the populations of spins in the upper and lower energy states are equal, and the same relaxation must occur in order to establish an equilibrium spin population distribution and permit a resonance signal to be observed. It should be realized that since the energies involved are very small and the nuclei are, as we shall see, usually rather weakly coupled thermally to their surroundings, i.e., the thermal relaxation is a slow process, the spin temperature may readily be made very high with little or no effect on the actual temperature of the sample as ordinarily observed. We may say that the heat capacity of the nuclear spin system is very small.

The required relaxation can occur because each spin is not entirely isolated from the rest of the assembly of molecules, commonly referred to as the "lattice," a term which originated in dealing with solids but which has long been employed in discussing both liquids

1.4. Nuclear Relaxation

and solids. The spins and the lattice may be considered to be essentially separate coexisting systems with a very inefficient but nevertheless very important link by which thermal energy may be exchanged. This link is provided by *molecular motion*. Each nucleus sees a number of other nearby magnetic nuclei, both in the same molecule and in other molecules. These neighboring nuclei are in motion with respect to the observed nucleus, and this motion gives rise to fluctuating magnetic fields. The observed nuclear magnetic moment will be precessing about the direction of the applied field B_0, and will also be experiencing the fluctuating fields of its neighbors. Since the motions of each molecule and of its neighbors are random or nearly random, there will be a broad range of frequencies describing them. To the degree that the fluctuating local fields have components in the direction of B_1 and at the precession frequency ω_0 (expressed in rad-s^{-1} or $2\pi v_0$), they will induce transitions between energy levels because they provide fields equivalent to B_1. In solids or very viscous liquids, the molecular motions are relatively slow and so the component at ω_0 will be small. The frequency spectrum will resemble curve (a) in Fig. 1.8. At the other extreme, in liquids of very low viscosity, the motional frequency spectrum may be very broad, and so no one component, in particular that at ω_0, can be very intense [curve (c) in Fig. 1.8]. We are then led to expect that at some intermediate condition, probably that of a moderately viscous liquid [curve (b)], the

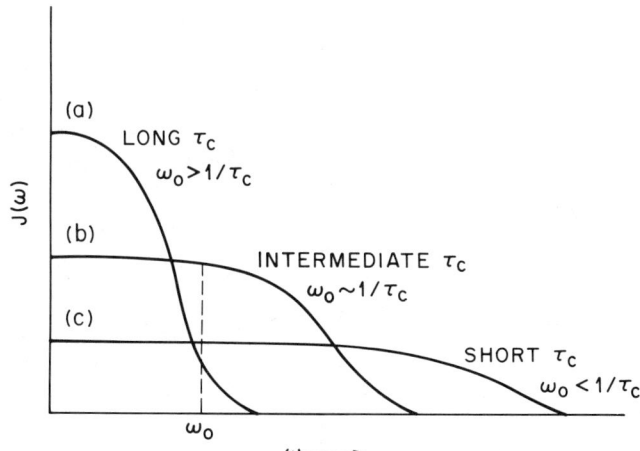

FIGURE 1.8 Motional frequency spectrum at (a) high viscosity, (b) moderate viscosity, and (c) low viscosity. The vertical ordinate $J(\omega)$ represents the relative intensity of the motional frequency ω. The observing frequency is ω_0.

component at ω_0 will be at a maximum, and thermal relaxation of the spin system can occur with optimum efficiency.

The probability per unit time of a downward transition of a spin from the higher to the lower magnetic energy level, $W\uparrow$, exceeds that of the reverse transition $W\downarrow$ by the same factor as the *equilibrium* lower state population exceeds the upper state population [Eq. (1.8)]. We may then write

$$\frac{W\downarrow}{W\uparrow} = 1 + \frac{2\mu B_0}{kT}. \qquad (1.9)$$

This must be the case, or equilibrium could not be maintained. (For isolated spins, i.e., those not in contact with the thermal reservoir provided by the lattice, the transition probabilities are exactly the same in both directions.) Let the spin population difference at any time t be given by n; let the equilibrium population difference (i.e., in the presence of B_0, but in the absence of the RF field) be given by n_{eq}. We shall also define more completely the longitudinal relaxation time T_1, in units of seconds, as the reciprocal of the sum of the probabilities (per second) of upward and downward transitions, i.e., the total rate of spin transfer in both directions:

$$\frac{1}{T_1} = W\downarrow + W\uparrow. \qquad (1.10)$$

The approach to equilibrium will be described by

$$\frac{dn}{dt} = 2N_-W\downarrow - 2N_+W\uparrow, \qquad (1.11)$$

N_- and N_+ being the populations of the upper and lower levels, respectively, as in Eq. (1.8). This represents the *net* transfer of spins from the upper state. (The factors of 2 arise because each transfer of one spin changes the population difference by 2.) If we designate the total spin population $(N_+ + N_-)$ by N, we can express this rate in terms of N and n as

$$\begin{aligned}\frac{dn}{dt} &= (N-n)W\downarrow - (N+n)W\uparrow \\ &= N(W\downarrow - W\uparrow) - n(W\downarrow + W\uparrow). \end{aligned} \qquad (1.12)$$

At equilibrium, we find from Eq. (1.12) that

$$\frac{n_{eq}}{N} = \frac{W\downarrow - W\uparrow}{W\downarrow + W\uparrow}, \qquad (1.13)$$

1.4. Nuclear Relaxation

and so, using Eq. (1.10), we may rewrite Eq. (1.12) as

$$\frac{dn}{dt} = \frac{1}{T_1}(n_{eq} - n), \quad (1.14)$$

which gives upon integration

$$n = n_{eq}(1 - e^{-t/T_1}). \quad (1.15)$$

We see then that, as might perhaps have been intuitively expected, the rate of approach to thermal equilibrium depends upon how far away from equilibrium we are. Here $1/T_1$ is the first-order constant (in the language of chemical kinetics) for this process. T_1 is called the *spin–lattice relaxation time*; it is the time required for the difference between the actual spin population and its equilibrium value to be reduced by the factor e.

Except in a static collection of nuclei, i.e., a solid near 0 K, spin–lattice relaxation through molecular motion always occurs, although it may be very slow. There are two other possible contributing causes to spin–lattice relaxation which may occur under some conditions and are of importance to the chemist.

1. Spin–lattice relaxation by interaction with unpaired electrons in paramagnetic substances. Fundamentally the mechanism is exactly the same as the one just discussed, but is much more effective because the magnetic moment of an unpaired electron is close to the magnitude of the Bohr magneton, $eh/4\pi M_e c$, and on the order of 10^3 times greater than the nuclear magneton owing to the small mass of the electron, M_e.

2. Spin–lattice relaxation by interaction of the *electric quadruple moments* of nuclei of spin 1 or greater with electric fields within the tumbling molecule.

Spin–lattice relaxation is often termed *longitudinal relaxation* because it involves changes of energy and therefore involves the component of the nuclear moment along the direction of the applied magnetic field. It also, of course, shortens the lifetimes of the spins in both the upper and the lower energy states. This leads to an uncertainty in the energies of these states and to a broadening of the resonance lines. This can be estimated from the Heisenberg relation, $\delta E \cdot \delta t \approx h/2\pi$, from which the uncertainty in the frequency of absorption is given by

$$\delta v \approx \frac{1}{2\pi \cdot \delta t}. \quad (1.16)$$

Under the present-day operating conditions, linewidths are about 0.2 Hz, which, if arising from uncertainty broadening, would correspond to a spin lifetime of about 0.3 s. Since T_1 for protons and ^{13}C is usually on the order of 2–20 s in mobile liquids (somewhat shorter for polymers in solution), spin–lattice relaxation ordinarily does not contribute observably to line broadening. As the temperature is lowered and viscosity increases, the component of the local magnetic noise spectrum at the Larmor frequency will increase, pass through a maximum, and decrease again. T_1 will correspondingly decrease, pass through a minimum, and increase again. It is useful to define a *correlation time* τ_c, which we shall understand to be the average time between molecular collisions. The correlation time defines the length of time that the molecule can be considered to be in a particular state of motion. This time is sometimes referred to as the "phase memory" of the system. Thus, molecules with short correlation times are in rapid molecular motion. For rigid, approximately spherical molecules, τ_c is given to a good degree of approximation by

$$\tau_c = \frac{\tau_D}{3} = \frac{4\pi\eta a^3}{3kT}, \qquad (1.17)$$

τ_D being the correlation time used in the Debye theory of dielectric dispersion, η the viscosity of the liquid, and a the effective radius of the molecule. For most nonassociated molecules, T_1 is governed mainly by interactions of nuclei within the same molecule, rather than by motions of neighboring molecules. Under these conditions, it is found that the rate of spin–lattice relaxation for a spin-1/2 nucleus interacting with another nucleus of the same species is given by (11)

$$\frac{1}{T_1} = \frac{3}{10}\frac{\gamma^4\hbar^2}{r^6}\left[\frac{\tau_c}{1+\omega_0^2\tau_c^2} + \frac{4\tau_c}{1+4\omega_0^2\tau_c^2}\right], \qquad (1.18)$$

where γ is the magnetogyric ratio of the interacting nuclei, $\hbar = h/2\pi$, r is the internuclear distance, and ω_0 is the resonant frequency in rad-s^{-1}. For mobile liquids near or above room temperature (or for moderate-sized molecules dissolved in mobile solvents), the correlation time is very short compared to $1/\omega_0$, and consequently Eq. (1.18) may be reduced to

$$\frac{1}{T_1} = \frac{3}{2}\frac{\gamma^4\hbar^2\tau_c}{r^6}. \qquad (1.19)$$

These circumstances are referred to as the "extreme narrowing condition." Combining Eqs. (1.17) and (1.19) and replacing $4\pi\eta a^3/3$ by the

1.4. Nuclear Relaxation

molecular volume V we have

$$\frac{1}{T_1} = \frac{3}{2} \frac{\gamma^4 \hbar^2}{r^6} \frac{V\eta}{kT}. \qquad (1.20)$$

For proton relaxation, one will in general have to take account of more than one interaction with the observed nucleus and employ a summation over the inverse sixth powers of all the protons concerned. The spin–lattice relaxation of carbon-13 nuclei in organic molecules is more commonly studied than that of protons. It is somewhat easier to interpret because for the dipole–dipole mechanism just described the relaxation of the ^{13}C nucleus by directly bonded nuclei (if any) dominates, a consequence of the r^6 dependence, and we find

$$\frac{1}{T_1} = \frac{3}{2} \frac{N\gamma_C^2 \gamma_H^2 \hbar^2}{r^6} \cdot \tau_c. \qquad (1.21)$$

Here, N is the number of directly bonded protons and the other terms are self-explanatory. For mobile liquids, mechanisms other than the dipole–dipole interactions described here may make important contributions, in particular *spin rotation* and *chemical shift anisotropy* (12). For polymer molecules, these other processes often play no appreciable part even in solution, dipole–dipole relaxation being entirely dominant. In certain circumstances, *nuclear electric quadrupolar relaxation* can be important (as we have seen above), e.g., for protons bonded to ^{14}N and for deuterium.

In Fig. 1.9 is shown the expected dependence of T_1 on the correlation time for a carbon-13 nucleus having a single directly bonded proton and located in a molecule or group, the motion of which may be adequately described by a single correlation time, τ_c. (We shall discuss this relaxation process and its observation more fully in Section 1.8.6.) The dependence of T_1 on τ_c is shown for two values of the magnetic field B_0 corresponding to $v_0 (= \omega_0/2\pi)$ of 50.3 and 125 MHz. It will be observed that in the regime of short correlation times at the left of the plot, T_1 is independent of the magnetic field strength, whereas when molecular motion is slower T_1 becomes longer as B_0 is increased, and the T_1 minimum moves to shorter τ_c values.

Figure 1.9 and Eqs. (1.17)–(1.19) tell us at least three things that have practical implications:

1. For a given liquid, T_1 will vary inversely with η/T over the usual temperature ranges employed in NMR spectroscopy, i.e., those over which the sample remains reasonably mobile. However, it is of

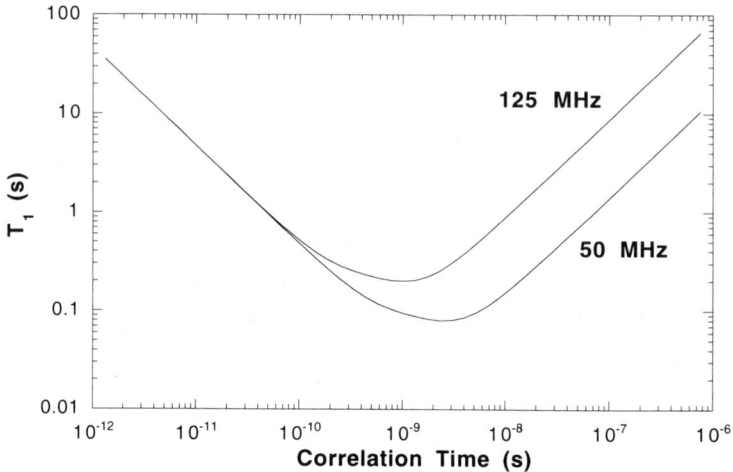

FIGURE 1.9 Spin–lattice relaxation time T_1 for a carbon-13 nucleus with one directly bonded proton as a function of the correlation time τ_c at 50.3 and 125 MHz.

great importance to note that this relationship does *not* hold for polymer solutions, either for the spectral lines of the polymer or for those of the solvent. Polymers confer a high viscosity on their solutions through entanglement and through entrapment of solvent molecules. Despite this very high viscosity—which may readily be such that the solution cannot be poured from the NMR tube—the linewidths and T_1 values for the polymer and solvent may remain nearly unaffected. As we shall see, this observation is very fortunate for the success of high-resolution NMR studies of polymers. It is a consequence of the fact that local segmental motion of polymer chains, as well as that of solvent molecules, may remain very fast in solutions of high viscosity or even in gels.

2. Other things being equal, T_1 will vary inversely as a^3 or V. Again, however, this will apply strictly only to molecules which are not articulated and have "no moving parts." For polymers in a random coil state, T_1 may be nearly independent of V. However, for large rigid molecules and for biological polymers in the native folded or helical state (e.g., proteins and polynucleotides), T_1 may be very short.

3. Substances in which the magnetic nuclei are relatively far apart or have small moments will exhibit longer relaxation times, since T_1 is a sensitive function of both μ and r. We may also note that if τ_c is very long, as, for example, in a crystalline solid, T_1 may become very long—on the order of hours or even days.

1.4.2 SPIN–SPIN RELAXATION AND DIPOLAR BROADENING

Although limiting the lifetime of a spin in a given energy state might in principle lead to line broadening, a more important source is the so-called "dipolar" broadening. To understand this effect, let us first consider not a liquid sample but rather a proton-containing solid. If the protons are sufficiently removed from each other that they do not feel the effects of each other's magnetic fields, the resonant magnetic field at the nucleus will be essentially equal to B_0 (actually very slightly smaller because of the shielding effects of local electrons). Therefore, if by careful design of the magnet the field B_0 can be made very homogeneous over the volume occupied by the sample, the width of the absorption peak may be less than 10^{-4} G (10^{-8} T), i.e., on the order of a few tenths of a hertz, or 1 part in 10^9. In most such substances, the protons are actually near enough to each other so that each is appreciably influenced by the magnetic fields of its neighbors. Let us first imagine that we are dealing with a static system of isolated pairs of protons; i.e., each member of a pair experiences the field of the other member but not those of the other pairs. The field felt by each proton will be made up of the applied field B_0, plus this small additional field, B_{loc}. The sign and magnitude of this increment will depend on the distance between the nuclei, r, and on θ, the angle between the line joining the nuclei and the vector representing the direction of B_0. We suppose that all the pairs have the same values of these parameters. The equation expressing the functional dependence of the separation of the protons' magnetic energy levels upon r and θ is given by

$$\Delta E = 2\mu[B_0 \pm B_{\text{loc}}] = [B_0 \pm \tfrac{3}{2}\mu r^{-3}(3\cos^2\theta - 1)]. \quad (1.22)$$

The $\pm$ sign corresponds to the fact that the local field may add to or subtract from B_0 depending upon whether the neighboring dipole is aligned with or against the direction of B_0. We recall that the net polarization of nuclear moments along B_0 is ordinarily only a few parts per million, and so in a large collection of such isolated pairs the probabilities of a neighboring dipole being aligned with or against B_0 are almost equal. Equation (1.22) then expresses the fact that the spectrum of this proton-pair system will consist of two equal lines [Fig. 1.10(a)] whose separation, at a fixed value of θ, will vary inversely as r^3. Only when the orientation of the pairs is such that $\cos^2\theta = 1/3$ ($\theta = 54.7$) will the lines coincide to produce a single line. We shall see this angle reintroduced and discussed further in Chapter 4 and in subsequent treatments of solid-state spectroscopy.

In many solids, for example, certain hydrated crystalline salts, we actually have pairs of protons in definite orientations and may expect

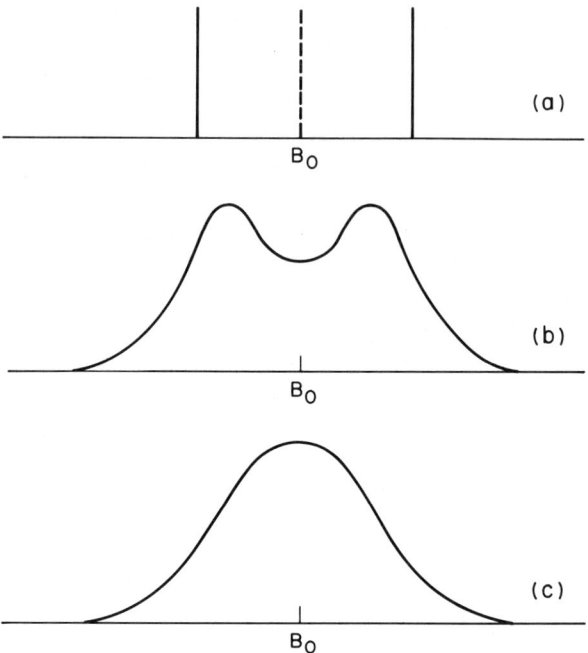

FIGURE 1.10 Schematic NMR spectra of static arrays of protons: (a) an isolated pair of protons; (b) a semi-isolated pair of protons; (c) a random or nearly random array of protons. In actual proton-containing solids, the width of spectra (b) and (c) may be on the order of 50 kHz.

to find a twofold NMR resonance. This is indeed observed (13). The lines, however, are not narrow and isolated as in Fig. 1.10(a), but are broadened and partially overlapping as in Fig. 1.10(b). The separation of the maxima is found to depend on the orientation of the crystal in the magnetic field. In such substances, the magnetic interactions of the members of each pair dominate, but the interactions between pairs are not negligible. These many smaller interactions, varying with both r and θ, give rise in effect to a multiplicity of lines whose envelope is seen as the continuous curve of Fig. 1.10(b). We should perhaps state that the simple picture suggested here, while qualitatively useful, does not correctly give the numbers and intensities of the lines composing this envelope in a multispin system. The solution of the problem actually involves quantum mechanical considerations; in fact, the factor 3/2 in Eq. (1.22) is a result of this more correct treatment (14,15). In most organic solids, pairing of protons does not occur to this degree, and instead we find a nearly Gaussian distribution of magnetic interactions, such as one would expect if the protons

1.4. Nuclear Relaxation

were randomly distributed throughout the substance. For such a complex array of nuclei, the NMR spectrum is a single broad peak as in Fig. 1.10(c). In most rigid-lattice solids, where the protons are on the order of 1 Å apart, the distribution of local field strengths about each proton is such that the half-height width of the peak is on the order of 10 G (10^{-3} T), or ca. 0.5×10^5 Hz. A relationship similar to Eq. (1.22) holds for the broadening of the resonances of a rare nuclear species, e.g., carbon-13 (Table 1.1), by the presence of attached or nearby protons.

We have so far assumed that the nuclei are fixed in position. If *molecular motion* is allowed to take place, i.e., by raising the temperature of the solid, by melting it, or by dissolving it in a mobile solvent, the variables in Eq. (1.22) become functions of time. If we assume that r is constant and only θ varies with time, as would be true for pairs of protons with a rigid molecule, then the time-averaged local field will be given by

$$B_{\text{loc}} = \frac{\mu r^{-3}}{T_2} \int_0^{T_2} (3\cos^2\theta - 1)\, dt, \quad (1.23)$$

where T_2 is the time that the nucleus resides in a given spin state. If θ may vary rapidly over all values, this time average can be replaced by a space average (3)

$$B_{\text{loc}} = \mu r^{-3} \int_0^{\pi} (3\cos^2\theta - 1)\sin\theta\, d\theta = 0. \quad (1.24)$$

Thus, if the correlation time τ_c [Eq. (1.17)] is so short that this space averaging is valid, then the net effect of the neighboring magnetic nuclei is effectively erased, and the line will be drastically narrowed (by a factor of 10^4 to 10^5) compared to its rigid-lattice value. If we begin with a solid at a temperature so low that it is virtually a rigid lattice, and then permit molecular motion (chiefly rotation) to occur with increasing frequency by raising the temperature, a narrowing of the resonance line will be observed when

$$\frac{1}{\tau_c} \geq 2\pi\, \delta v, \quad (1.25)$$

where δv is the static linewidth (hertz). In liquids of ordinary viscosity, molecular motion is so rapid that Eq. (1.24) holds; the local variations in magnetic field strength have become so short lived that *motional averaging* is complete. But for large molecules and for viscous liquids this line narrowing effect may not be complete and broadening may still be observable. For high polymers, and even for molecules of 400–500 molecular weight (particularly if they are rigid), this broadening can remain conspicuous.

There is another aspect of the interaction of neighboring magnetic dipoles that is closely related to the effects we have just considered and must also be considered in relation to line broadening. We must recall that our nuclear spins are not merely small static magnetic dipoles, but that even in a rigid solid they are precessing about the field direction. We may resolve a precessing nuclear moment (Fig. 1.11) into a *static component a* along the direction of B_0, which we have considered so far, and into a *rotating component b*, whose effect we must now also consider. This component constitutes the right type of magnetic field, as we have seen (Section 1.4), to induce a transition in a neighboring nucleus if this is precessing at the same frequency. If this spin exchange or flip-flop occurs, the nuclei will exchange magnetic energy states with no overall change in the energy of the system but with a shortening of the lifetime of each. The magnitude of local field variations may be taken as μ/r^3 [an approximation of Bloch from Eq. (1.22)], and consequently the relative phases of the nuclei will change in a time on the order of hr^3/μ^2, the "phase memory time." From Eq. (1.16), we expect an uncertainty broadening of about μ^2/hr^3, i.e., of the same form and order of magnitude as that produced by the interaction of the static components of the nuclear

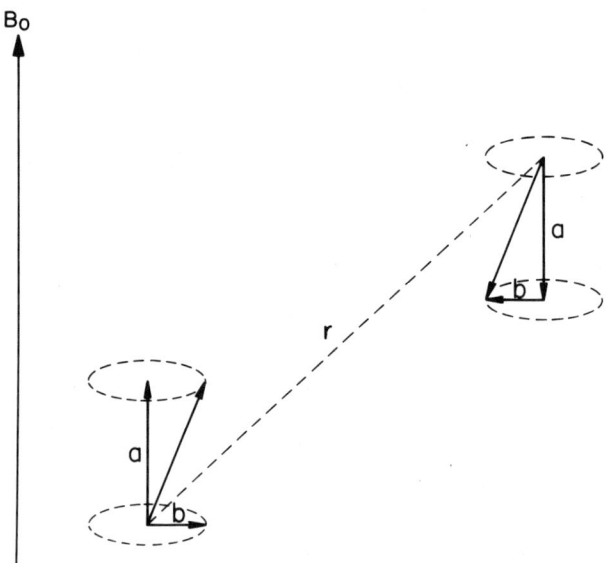

FIGURE 1.11 Static and rotating components of a precessing nuclear moment.

moments. It has become customary to include both effects in the quantity T_2, which we defined as the spin lifetime. It should be recalled that we have already introduced T_2 earlier as the time constant for the decay of the x, y (i.e., observable) component of the nuclear magnetization following a 90° pulse. T_2 is an inverse measure of the broadening of the spectral lines:

$$T_2 = \frac{1}{\pi \delta v}. \qquad (1.26)$$

It is called the *spin–spin relaxation time* or *transverse relaxation time*. A detailed theory of its dependence on molecular correlation time has been given by Bloembergen *et al.* (11). At short correlation times, as in mobile liquids (where "phase memory" is short), it becomes equal to T_1, but after T_1 passes through its minimum, T_2 continues to decrease as molecular motion becomes slower and finally levels out as the system begins to approach a rigid lattice.

Spin–spin exchange and dipolar broadening should not be considered as merely two alternative ways of looking at the same phenomenon, closely interrelated though they are. For example, in a lattice composed of magnetic nuclei α containing a dilute distribution of a different magnetic nuclear species β, spin exchange between nuclei α and β cannot occur since they precess at greatly different frequencies, but, nevertheless, dipolar broadening will be present. An example is provided by solid-state ^{13}C spectra, which will be fully discussed in later chapters.

Spin–spin relaxation is associated with a decay of the macroscopic nuclear moment in the x, y plane; spin–lattice relaxation is associated with a decay of the macroscopic moment along the z direction. We may suspect that these relaxation rates, although they can be quite different in magnitude, are closely associated, and that both T_1 and T_2 must be considered in describing resonance signal shapes and intensities.

1.4.3 THE BLOCH EQUATIONS

In treating the experimental observation of nuclear magnetic resonance, it is convenient to adopt the approach of Bloch *et al.* (3) and to consider the assembly of nuclei in macroscopic terms. We define a total moment M, which is the resultant sum, per unit volume, of all the individual nuclear moments in an assembly of identical nuclei, with magnetogyric ratio γ and $I = 1/2$. We consider that M is not colinear with any of the axes x, y, and z, as in Fig. 1.12. The static field B_0 is in the z direction, and M, like the individual moments,

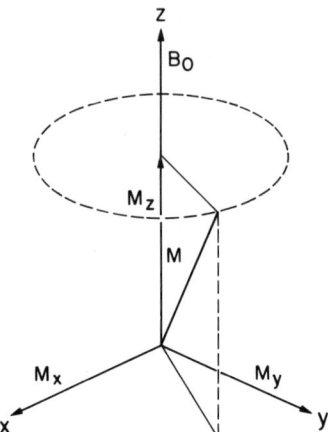

FIGURE 1.12 Precessing nuclear moment M in a magnetic field B_0 directed along the z axis of a fixed coordinate frame.

will precess about z with an angular frequency ω_0. In the absence of relaxation effects and the rotating field B_1, the projection of M on the z axis, M_z, would remain constant:

$$\frac{dM_z}{dt} = 0. \tag{1.27}$$

The magnitudes of the x and y projections, M_x and M_y, will, however, vary with time as M precesses and, as can be seen from Fig. 1.12, will be 90° out of phase, since when the projection of M along the x axis is at a maximum it will be zero along the y axis, and vice versa. This time dependence can be expressed by

$$\frac{dM_x}{dt} = \gamma M_y \cdot B_0 = \omega_0 M_y \tag{1.28}$$

$$\frac{dM_y}{dt} = -\gamma M_x \cdot B_0 = -\omega_0 M_x. \tag{1.29}$$

(Note that the inclusion of y in macroscopic expressions causes no difficulty if dimensions are consistent.) In addition to the fixed field B_0, we must now consider the rotating field B_1, which we recall from earlier discussion will be provided by one of the two counterrotating magnetic vectors of a linearly polarized RF field. This field B_1 rotates in the xy plane, i.e., perpendicularly to B_0, with frequency ω (equal to ω_0 only at exact resonance, i.e., at the center of the absorption peak). We note that B_1 need not be of constant magnitude but may be applied as a brief pulse. Consideration of the effect of B_1 upon the

1.4. Nuclear Relaxation

magnitudes of M_x, M_y, and M_z in accordance with the basic laws describing the tipping of magnetic vectors in magnetic fields leads to modifications of Eqs. (1.27)–(1.29),

$$\frac{dM_x}{dt} = \gamma\left[M_y B_0 - M_z(B_1)_y\right] \quad (1.30)$$

$$\frac{dM_y}{dt} = -\gamma\left[M_x B_0 + M_z(B_1)_x\right] \quad (1.31)$$

$$\frac{dM_z}{dt} = \gamma\left[M_x(B_1)_y - M_y(B_1)_x\right], \quad (1.32)$$

where $(B_1)_x$ and $(B_1)_y$ are the components of B_1 along the x and y axes and are given by

$$(B_1)_x = B_1 \cos \omega t \quad (1.33)$$

$$(B_1)_y = -B_1 \sin \omega t. \quad (1.34)$$

We have so far omitted any consideration of the relaxation of the components of M in the x, y, and z directions. By putting Eq. (1.14) in phenomenological terms, we may express the relaxation of the z component toward its equilibrium value M_0 as

$$\frac{dM_z}{dt} = -\frac{M_z - M_0}{T_1}. \quad (1.35)$$

The transverse relaxations may be expressed similarly, but with T_2 as the time constant:

$$\frac{dM_x}{dt} = -\frac{M_x}{T_2} \quad (1.36)$$

$$\frac{dM_y}{dt} = -\frac{M_y}{T_2}. \quad (1.37)$$

The transverse relaxation processes differ in that they go to zero rather to equilibrium values. Upon adding these terms to Eqs. (1.30)–(1.34) and using Eqs. (1.33) and (1.34), we obtain the complete Bloch equations:

$$\frac{dM_x}{dt} = \gamma(M_y B_0 - M_z B_1 \sin \omega t) - \frac{M_x}{T_2} \quad (1.38)$$

$$\frac{dM_y}{dt} = -\gamma(M_x B_0 - M_z B_1 \cos \omega t) - \frac{M_y}{T_2} \quad (1.39)$$

$$\frac{dM_z}{dt} = \gamma(M_x B_1 \sin \omega t + M_y B_1 \cos \omega t) - \frac{M_z - M_0}{T_1}. \quad (1.40)$$

A clearer insight into the significance of these equations in experimental terms is gained if the frame of reference is changed from fixed axes x, y, and z to the rotating frame (Fig. 1.6; see also Section 1.4), where both B_0 and B_1 are fixed. We may then resolve the projection of M on the x, y plane into components u and v, which are along and perpendicular to B_1, respectively, and will accordingly be in phase and out of phase, respectively, with B_1 (see Fig. 1.13). To transform to this new frame, we note that

$$M_x = u \cos \omega t - v \sin \omega t \tag{1.41}$$

$$M_y = -u \sin \omega t - v \cos \omega t \tag{1.42}$$

and also that $\gamma B_0 = \omega_0$. We then may replace Eqs. (1.38)–(1.40) with

$$\frac{du}{dt} + \frac{u}{T_2} + (\omega_0 - \omega)v = 0 \tag{1.43}$$

$$\frac{dv}{dt} + \frac{v}{T_2} - (\omega_0 - \omega)u + \gamma B_1 M_z = 0 \tag{1.44}$$

$$\frac{dM_z}{dt} + \frac{M_z - M_0}{T_1} - \gamma B_1 v = 0. \tag{1.45}$$

Here, the terms $(\omega_0 - \omega)$ are a measure of how far we are from exact resonance. [Again, we may not in actual fact sweep B_0 or B_1 through resonance, but rather we apply B_1 as a pulse containing in effect all the resonance frequencies of the observed nuclei.] We see from Eq.

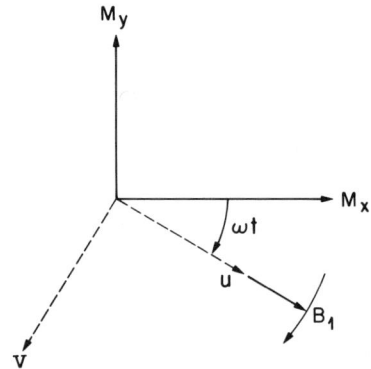

FIGURE 1.13 Components of the transverse macroscopic moment in a fixed plane (M_x, M_y) and in a plane rotating at the Larmor frequency about the static field B_0, taken as normal to the xy plane (and upward); the rotating components u and v are along the directions parallel and perpendicular to B_1, respectively.

1.4. Nuclear Relaxation

(1.45) that changes of M_z, i.e., changes in the energy of the spin system, are associated only with v, the out-of-phase component of the macroscopic moment, and not with u. We may then anticipate that absorption signals will be associated with the measurement of v. The component u will be associated with "dispersion-mode" signals.

Under certain conditions of experimental observation, we are dealing with a steady state in which u, v, and M_z are constant in the rotating frame. We pass through the resonance peak varying ω (or in practice B_0) at a rate so slow that u, v, and M_z always have time to reach these steady values. Under such "slow-passage" conditions we obtain from Eqs. (1.43)–(1.45)

$$u = M_0 \frac{\gamma B_1 T_2^2 (\omega_0 - \omega)}{1 + T_2^2(\omega_0 - \omega)^2 + \gamma^2 B_1^2 T_1 T_2} \quad (1.46)$$

$$v = -M_0 \frac{\gamma B_1 T_2}{1 + T_2^2(\omega_0 - \omega)^2 + \gamma^2 B_1^2 T_1 T_2} \quad (1.47)$$

$$M_z = M_0 \frac{1 + T_2^2(\omega_0 - \omega)^2}{1 + T_2^2(\omega_0 - \omega)^2 + \gamma^2 B_1^2 T_1 T_2}. \quad (1.48)$$

We see from Eq. (1.47) that under conditions when $\gamma^2 B_1^2 T_1 T_2$ is very small, i.e., when B_1 is a few milligauss and T_1 and T_2 are no greater than a few seconds, the absorption or "v-mode" signal should be proportional to $\gamma B_1 T_2 / [1 + T_2^2(\omega_0 - \omega)^2]$. This describes what is known as a Lorentzian lineshape, as shown in Fig. 1.14(a), and is the type of signal normally used with high-resolution spectra. At the center, when the resonance condition is exactly fulfilled, $\omega_0 - \omega$ becomes zero and the signal height is proportional to $\gamma B_1 T_2$. It follows that the width must be inversely proportional to T_2, as we have already seen [Eq. (1.26)]. If we define the peak width $\delta\omega$ or δv in the customary way as its width at half the maximum height, it is easily shown that

$$\delta\omega = \frac{2}{T_2} \quad (1.26a)$$

or

$$\delta v = \frac{1}{\pi T_2}. \quad (1.26b)$$

For some purposes, it is preferable to employ the dispersion or "u-mode" signal, which under the same conditions will be proportional to $\gamma B_1 T_2^2(\omega_0 - \omega)/[1 + T_2^2(\omega_0 - \omega)^2]$. This will give the lineshape shown in Fig. 1.14(b). The maximum and minimum are separated by $1/\pi T_2$ for Lorentzian signals.

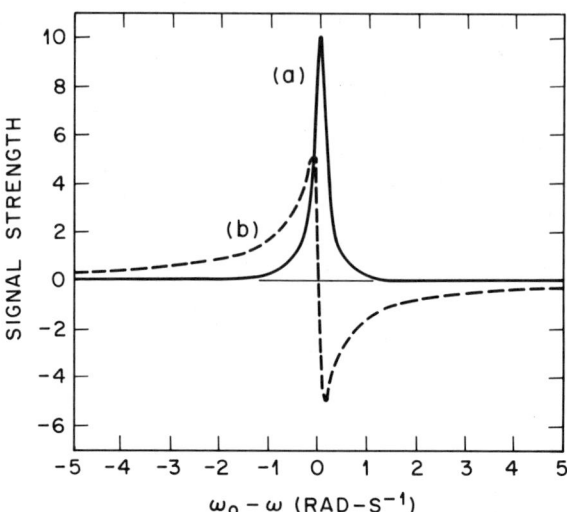

FIGURE 1.14 v-mode [absorption (a)] and u-mode [dispersion (b)] resonance signals.

The preceding discussion contemplates an NMR experiment in which the RF field is of constant strength, but much of it applies (*mutatis mutandis*) to pulsed NMR. The phenomena peculiar to the latter—notably the free induction decay and its Fourier transform—are most commonly used in NMR today.

Equations (1.46)–(1.48) allow us to express the phenomenon of *saturation* in more quantitative terms. In CW spectroscopy, partial saturation will begin to be observed when the term $\gamma^2 B_1^2 T_1 T_2$ is no longer much less than unity, as assumed in the preceding discussion. The field B_1 is now beginning to cause a net transfer of lower state spins to the upper state that cannot be neglected in comparison to that of spin–lattice relaxation. The new equilibrium population difference $n_{eq'}$ will be given by

$$n_{eq'} = \frac{n_{eq}}{1 + \gamma^2 B_1^2 T_1 T_2}. \quad (1.49)$$

The observable signal strength will be correspondingly reduced. Since $T_1 T_2$ will be on the order of 10 and γ^2 approximately 10^8 for protons, observable saturation effects can be expected when B_1 is no greater than 10^{-8}–10^{-9} T.

In pulsed NMR, the condition represented in Fig. 1.6(c) following the 90° pulse might be thought to correspond to saturation since the populations of the energy levels are now equal. However, we must

1.4. Nuclear Relaxation

remember too this is not an equilibrium situation; the magnetization has been rotated from its position along the z' axis in a time so short that no relaxation has had time to occur and the observable magnetization is along y', so that the signal is a maximum. The system in effect retains a memory (for a time T_2) of the conditions under which the macroscopic magnetization was produced. [This is well discussed by Farrar and Becker (16).] In true saturation there is no magnetization in any direction. The occurrence of saturation is not a major concern in pulse NMR. A condition of partial saturation may be induced by a train of pulses separated by intervals small compared to T_1.

1.4.4 Nuclear Electric Quadrupole Relaxation

Nuclei with spins of 1/2 possess a spherical distribution of the nuclear charge and are therefore not affected by the electric environment within the molecule. Nuclei with spins of 1 or greater, however, are found to have *electric quadrupole moments*, and their charge distribution can be regarded as spheroidal in form; the nucleus is to be regarded as spinning about the principal axis of the spheroid. The quadrupole moment may be *positive*, corresponding to a prolate spheroid [Fig. 1.15(a)], or *negative*, corresponding to an oblate spheroid [Fig. 1.15(b)]. The energies of spheroidal charges will depend upon their orientation in the molecular electric field gradient. In certain classes of molecules where an approximately spherical or tetrahedral charge distribution prevails, as, for example, in the ammonium ion NH_4^+, little or no electric field gradient will be present, and the nuclear quadrupole moment will not be disturbed by the tumbling of the molecule. But in most molecules, substantial electric field gradi-

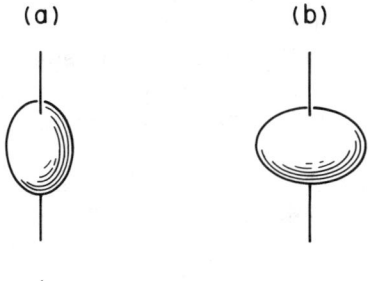

FIGURE 1.15 Nuclear charge distribution corresponding to a (a) positive electric quadrupole (prolate spheroid) and a (b) negative electric quadrupole moment (oblate spheroid).

ents are present and can interact with the nuclear quadrupoles so that the spin states of such nuclei may be rapidly changed by the tumbling of the molecular framework. This furnishes an additional pathway for energy exchange between the spins and the lattice, i.e., an important contribution to spin–lattice relaxation, and can broaden the resonance peaks very markedly. The resonance of nuclei such as ^{2}H or ^{14}N (Q positive) or ^{17}O, ^{35}Cl, and ^{36}Cl (Q negative) may be so broad as to be difficult or impossible to observe. Nuclear quadrupole relaxation can also affect the resonance even of nuclei of spin 1/2 when these are in sufficiently close proximity to nuclei of spin 1 or greater.

1.5 MAGNETIC SHIELDING AND THE CHEMICAL SHIFT

1.5.1 THE BASIS OF MOLECULAR SHIELDING: PROTON CHEMICAL SHIFTS

From what we have said so far, it might be supposed that at any particular radiofrequency v_0 all nuclei of a given species, for example, all protons, would resonate at the same value of B_0. Indeed this is true within the limits of the significant figures with which nuclear properties are expressed in Table 1.1. If this were *strictly* true, NMR would be of little interest to chemists. But about four years after the first demonstration of NMR in condensed phases, it was found that the characteristic resonant frequency of a nucleus depends to a very small, but measurable, extent upon its chemical environment. It was found that the protons of water do not absorb at quite the same frequency as those of mineral oil (17), the difference being only a few parts per million. For heavier nuclei, much larger effects are noted—up to 2% for certain metals (18). In 1950, Gutowsky and students observed different resonance frequencies for ^{19}F nuclei in an aromatic ring and on a CF_3 group in the same molecule. Similarly, Arnold et al. (19) reported separate spectral lines for chemically different nuclei in alcohols. These discoveries opened a new era for the organic chemists.

The total range of variation of B_0 for protons is about 13 ppm or ca. 2600 Hz in a 4.73-T field (200 MHz). The proton spectrum of ethyl orthoformate, $CH(OCH_2CH_3)_3$, shown in Fig. 1.16(a), exhibits successive peaks (or groups of peaks; the cause of the splittings within the groups will be discussed shortly) for CH, CH_2, and CH_3 protons as we in effect increase the resonant magnetic field from left to right. The corresponding ^{13}C spectrum is shown in Fig. 1.16(b) (the much

1.5. Magnetic Shielding and the Chemical Shift

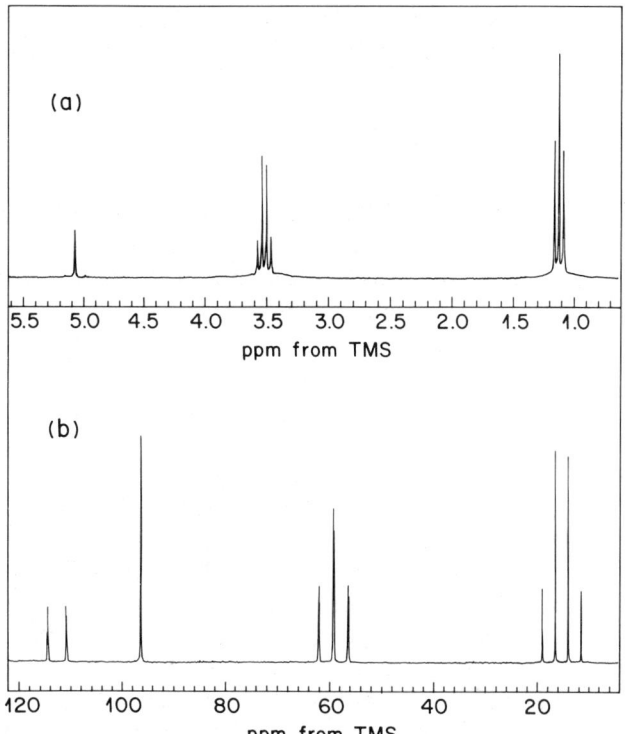

FIGURE 1.16 (a) The 200-MHz proton spectrum of ethyl orthoformate [CH(OCH$_2$CH$_3$)$_3$] as a 15% solution in carbon tetrachloride. (b) The 50-MHz carbon spectrum of the same solution. The chemical shifts in both spectra are reported relative to TMS at 0 ppm.

larger spacing of resonances is noteworthy). In both spectra the intensities of the peaks are proportional to the number of nuclei of each type.

The origin of this variation in resonant field strength is the cloud of electrons about each of the nuclei. When a molecule is placed in a magnetic field B_0, orbital currents are induced in the electron clouds, and these give rise to small, local magnetic fields, which, by a well-known physical principle, are always proportional to B_0 but opposite in direction.[2] Such behavior is common to all molecules and gives rise to the universally observed diamagnetic properties of mat-

---

[2] It is assumed that all the electrons in the molecule are paired off, so that they have no net magnetic moment per se. If this is not the case, i.e., if the molecule is a free radical, the much stronger magnetic field arising from the unpaired electron will be *independent* of B_0.

ter. Each nucleus is, in effect, partially shielded from B_0 by the electrons and requires a slightly higher value of B_0 to achieve resonance. This can be expressed as

$$B_{\text{loc}} = B_0(1 - \sigma), \qquad (1.50)$$

where B_{loc} is the actual local field experienced by the nucleus and σ is the *screening constant*, expressing the reduction in effective field; σ is independent of B_0 but is highly dependent on chemical structure. Equation (1.6) can then be modified to

$$h v_0 = \frac{\mu B_0(1 - \sigma)}{I}. \qquad (1.51)$$

The effect of nuclear screening is to decrease the spacing of the nuclear magnetic energy levels. It will be seen that at constant RF field frequency v_0, an increase in σ, i.e., an increase in the magnetic shielding of the nucleus, means that B_0 will have to be *increased* to achieve resonance. Thus, if resonance peak positions are expressed on a scale of magnetic field strength increasing from left to right, as is now almost universally done, the peaks for the more shielded nuclei will appear on the right-hand side of the spectrum.

The causes of variations in nuclear shielding may be thought of as arising from:

a. Variations in local electron density. Nuclei attached to or near electronegative groups or atoms such as O, OH, halogens, CO_2H, NH_3, and NO_2 experience a lower density of shielding electrons and resonate at lower values of B_0. Nuclei removed from such groups appear at higher values. Such inductive effects are the principal cause of shielding variations. The spectra of ethyl orthoformate (Fig. 1.16) furnish an illustration of these effects. The methyl group carbons and protons, being furthest removed from the oxygen atoms, appear on the right-hand side of the spectrum; the methylene resonances are further "downfield"; and the formyl carbon and proton, because of the three attached oxygens, are the least shielded.

b. Special shielding effects produced by certain groups and structures that allow circulation of electrons only in certain preferred directions within the molecule, i.e., which exhibit diamagnetic anisotropy. Benezene rings, for example, show this behavior very strongly, and many other groups do so in varying degrees, including even carbon–carbon single bonds. These effects are in general smaller than inductive effects, but nonetheless can be very marked and can provide valuable structural clues.

1.5. Magnetic Shielding and the Chemical Shift

Protons, because of their low density of screening electrons, show smaller variations due to (a) than do other nuclei, but for the same reason exhibit relatively greater effects from (b) than do heavier nuclei.

Because nuclear shielding is proportional to the applied magnetic field, it of course follows that the spacing between peaks (or groups of peaks) corresponding to different types of nuclei is also proportional to the magnetic field. Thus, in NMR spectroscopy, in contrast to optical spectroscopy, there is no natural fundamental scale unit; the energies of transitions between quantum levels are proportional to the laboratory field. There is also no natural zero of reference. For practical purposes, these difficulties are evaded by these devices:

a. Using parts per million relative change in B_0 or v_0 as the scale unit. (This convention is retained in Fourier transform spectra.)
b. Using an arbitrary reference substance dissolved in the sample and referring all displacements in resonance, called *chemical shifts*, to this "internal" reference.

The use of a dimensionless scale unit has the great advantage that the chemical shift values so expressed are independent of the value of B_0 of any particular spectrometer, and so a statement of chemical shift does not have to be accompanied by a statement of the frequency (or field) employed, as is the case when actual magnetic field strength or frequency are used as the scale unit. Tetramethylsilane, $(CH_3)_4Si$, was proposed many years ago by Tiers (20) as a reference for proton spectra because its resonance appears in a more shielded position than those of most organic compounds. For a similar reason, it is now also accepted as the reference for ^{13}C spectra. It is soluble in most solvents and is normally unreactive. For both protons and ^{13}C, tetramethylsilane, commonly abbreviated TMS, is assigned a chemical shift of zero. (In older proton spectra, TMS appears at 10.0 ppm, as originally proposed by Tiers. This scale and the shielding both increase from left to right, but it has now been superseded.)

In an atom, the electron cloud may be regarded as spherical; i.e., the atom is in a ^1S state. For the hydrogen atom, the local field at the nucleus is given by (21)

$$B_{\text{loc}} = B_0 - \frac{4\pi e^2 B_0}{3 M_e c^2} \int_0^\infty r\rho(r)\, dr, \qquad (1.52)$$

where $\rho(r)$ describes the electron density as a function of the distance r from the nucleus. For ^1S atoms of nuclear charge Z, we find from

Eqs. (1.51) and (1.52) that the nuclear *screening constant* σ is given by

$$\sigma = \frac{4\pi Z e^2}{3 M_e c^2} \int_0^\infty r\rho(r)\,dr. \tag{1.53}$$

Calculation of σ for atoms from $Z = 1$ through $Z = 92$ (22), using appropriate functions for the electron density, gave values of σ from 18×10^{-6} for hydrogen to values of the order of 10^{-2} for the heaviest atoms ($Z > 80$).

When the atoms are incorporated into molecules the electron distributions are no longer spherical, and the electrons are not free to move in the circular path contemplated by Eqs. (1.52) and (1.53). The Larmor precession of the electron system about the field direction z is now impeded by the presence of other nuclei. The simple model of circular electron paths is applicable only if the molecule can be so oriented that the nuclei and electrons have an arrangement that is axially symmetric along a line parallel to the z axis. For all other arrangements and for axially symmetric molecules that tumble freely —in short, for all actual liquid systems—a Lamb-type term does not describe the shielding correctly, even if averaged over all orientations, but instead predicts a screening that is too large. Another term is necessary to describe the hindering of electron circulation. Because this second term corresponds to a magnetic moment opposing that of the diamagnetic Lamb term, it is referred to as the paramagnetic term. The shielding may thus be expressed as

$$\sigma = \sigma_D + \sigma_P. \tag{1.54}$$

For molecular hydrogen, an averaged Lamb term of 32.1 is calculated; this is reduced to 26.7 when the paramagnetic term is included, in good agreement with the experimental value of 26.6 ppm. For larger molecules similar success has not been attained.

The molecular screening constant is actually anisotropic or *directional*: it depends on the orientation of the molecule with respect to magnetic field direction. It is expressed as a tensor, a mathematical quantity having both direction and magnitude, and is composed of three principal components, σ_{ii},

$$\sigma = \lambda_{11}^2 \sigma_{11} + \lambda_{22}^2 \sigma_{22} + \lambda_{33}^2 \sigma_{33}, \tag{1.55}$$

where λ_{ii} are the direction cosines of the principal axes of the screening constant with respect to the magnetic field. (Direction cosines represent the values of the cosines of the angles describing the orientation of the axes). In Fig. 1.17 these mutually perpendicular axes are represented in the laboratory frame for a molecule contain-

1.5. Magnetic Shielding and the Chemical Shift

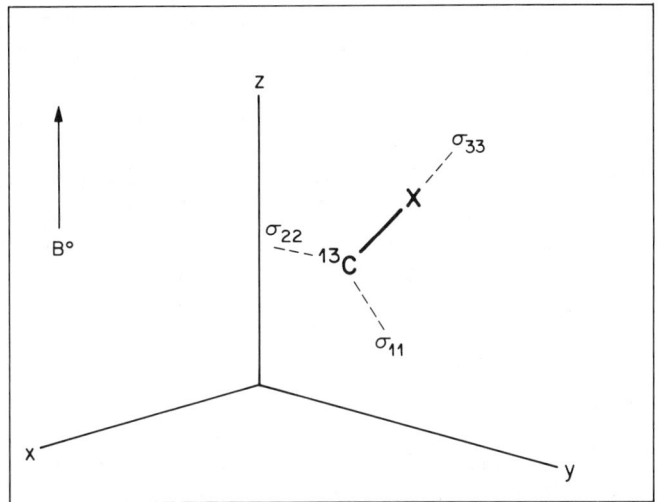

FIGURE 1.17 The principal values of the carbon-13 screening tensor σ_{11}, σ_{22}, and σ_{33}, represented along three mutually perpendicular axes in a laboratory frame for a molecule containing a ^{13}C–X bond. One of these axes (σ_{33}) is shown as coinciding with the ^{13}C–X bond, although in general it is not necessary that the axes coincide with the bonds.

ing a ^{13}C–X bond. One of these axes, corresponding to σ_{33}, is shown as coinciding with the ^{13}C–X bond although it is not necessary that any of the axes coincide with chemical bonds. If the molecule is oriented so that the ^{13}C–X bond is along the field direction, the observed chemical shift will correspond to the screening constant σ_{33}; similar statements apply to σ_{11} and σ_{22}. For any arbitrary orientation the chemical shift is prescribed by Eq. (1.55). In solution, the screening constant is given by the isotropic average of this equation. Since this is 1/3 for each λ_i^2 we have

$$\sigma = \tfrac{1}{3}(\sigma_{11} + \sigma_{22} + \sigma_{33}). \tag{1.56}$$

(The quantity in parentheses is the *trace* of the tensor, i.e., the sum of the diagonal elements of the tensor representing the screening.) Such averaging also occurs when the solid is rotated at the magic angle.

Chemical shift anisotropy—and the elimination of its broadening effects—is of greatest significance for the observation of nuclei other than the proton, principally carbon-13 (Section 1.8.5.3).

Most organic solids are polycrystalline, the crystals and their constituent molecules being oriented at all angles with equal probability. In these circumstances, the screening constant—or chemical shift—takes on a continuum of values, forming the lineshapes shown in Fig. 1.18. The sharp-edged curves represent the theoretical lineshape and the smooth curves represent the experimental lineshape, in which a Lorentzian broadening function has been incorporated. Figure 1.18(a) shows a general, asymmetric powder lineshape where the three principal values of the screening tensor all differ. In Figure 1.18(b) the chemical shift is axially symmetric; the screening constants σ_{11} and σ_{22} are equal and may be designated as $\sigma_\perp$, σ_{33} being then designated as $\sigma_\parallel$. If instead $\sigma_{22} = \sigma_{33}$, as may also happen, the pattern will be reversed left to right. σ_{33} is customarily taken as the largest value.

The observation of such patterns does not actually tell us the orientations of the principal axis of the tensor with respect to the molecular framework. To determine this, one must study single crystals or consider symmetry questions and relationships to other known molecules. The remainder of the discussion in this section will deal only with isotropic, motionally averaged chemical shifts, as observed in the molten or dissolved state. We return to the solid state in Chapter 4 and in subsequent chapters.

Of the two principal contributions to variations in molecular shielding described earlier, let us first consider *inductive effects*. The most obvious correlation of chemical shifts with molecular constitu-

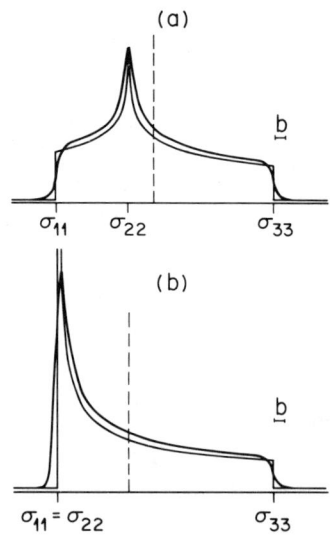

FIGURE 1.18 Powder lineshapes for the shielding tensor: (a) general $\sigma_{11} \neq \sigma_{22} \neq \sigma_{33}$; (b) having axial symmetry, $\sigma_{11} = \sigma_{22}$. The theoretical lineshape is given by spectra with sharp edges. In reality the spectra are more likely to be slightly broadened. The dashed lines shows the position of the isotropic chemical shift.

tion is the observation that electronegative atoms and groups cause deshielding of nuclei attached to or near them, and that this deshielding effect falls off rapidly as the number of intervening bonds is increased. This correlation is not exact and can be perturbed by other effects but is nevertheless one of the basic rules that the chemist relies on when interpreting an unfamiliar spectrum.

The influence of substituent negativity is illustrated by the general trends summarized for ^{1}H in Fig. 1.19 and for ^{13}C in Fig. 1.20; these are the principal nuclei which will concern us. Electron attracting groups such as $-O-$, $-OH$, $-NO_2$, $-CO_2R$, and halogen atoms reduce the electron density around nearby nuclei and thereby reduce their shielding. The organic chemist seeking to interpret the spectrum of a complex compound is concerned with the propagation of such induc-

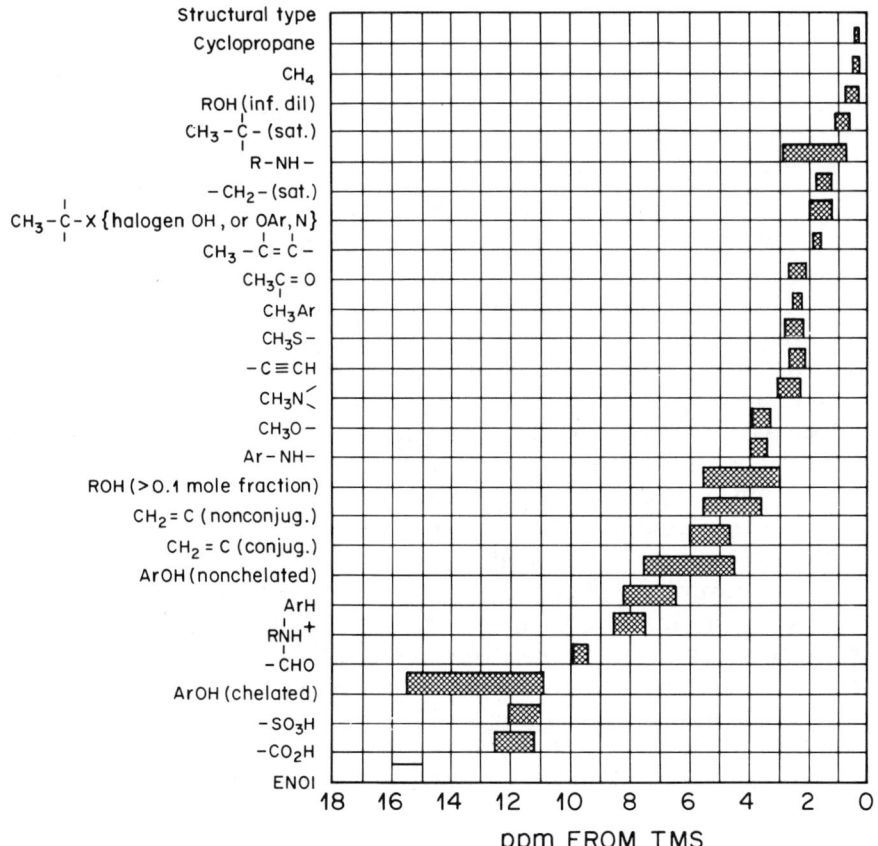

FIGURE 1.19 Chemical shifts from protons in various structures and function groups, expressed in ppm from tetramethysilane.

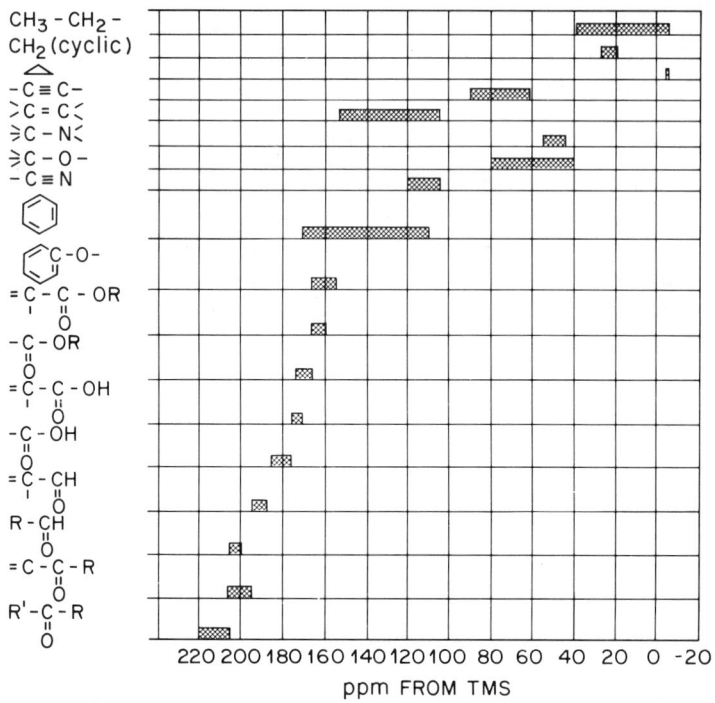

FIGURE 1.20 Chemical shifts for ^{13}C in various structures and functional groups, expressed in ppm from tetramethylsilane.

tive effects through a system of bonds. In Fig. 1.21 are shown the proton spectra of a number of aliphatic compounds that exhibit the transmission of the inductive effect of electronegative groups in straight chains. Multiple substitutions of groups on the same carbon atom produce at least approximately additive effects on the shielding of protons remaining on that carbon atom.

We now consider the special shielding effects produced by magnetically anisotropic neighboring groups. In terms of relative effects, these are more important for protons than for any other nucleus. Let us suppose first that the electron cloud in this neighboring group is spherically symmetrical, as might be nearly the case if it were simply a single atom. If a line joining such a group to the observed nucleus tumbles rapidly and randomly in all directions, as would be the case for a molecule in a liquid, it can be shown that the net average field produced by the group at the observed nucleus will be zero. Thus, remote shielding effects can be produced only by electron groups that are *not* spherically symmetrical and consequently are magnetically anisotropic. Such a group is represented in a generalized manner as g

1.5. Magnetic Shielding and the Chemical Shift

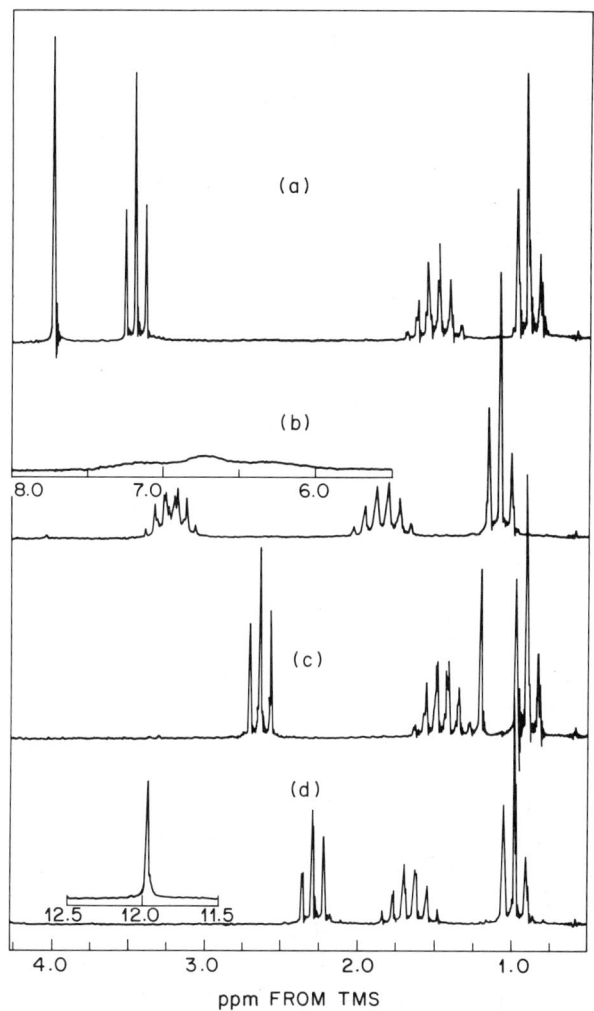

FIGURE 1.21 Proton spectra illustrating the transmission of inductive effects in straight-chain aliphatic compounds, all observed at 100 MHz at 25°C using 10% w/v solutions in CCl_4, except (b), which was observed in trifluoroacetic acid: (a) n-propyl alcohol, $CH_3CH_2CH_2OH$; (b) n-propyammonium ion, $CH_3CH_2CH_2NH_3^+$; (c) n-propylamine, $CH_3CH_2CH_2NH_2$; (d) n-butyric acid, $CH_3CH_2CH_2CO_2H_2$. In all spectra the methyl group appears near 1.0 ppm while the methylene protons adjacent to the methyl appear near 1.5 ppm, but the chemical shift depends on the structure.

in Fig. 1.22. It has a principal axis A, along which the diamagnetic susceptibility is $\chi_{\parallel}$. This means that if all the molecules were aligned so that A were parallel to B_0, the observed magnetic susceptibility would be $\chi_{\parallel}$ and consequently the observed molecular diamagnetic moment would be $B_0 \chi_{\parallel}$. For a group with a threefold or higher

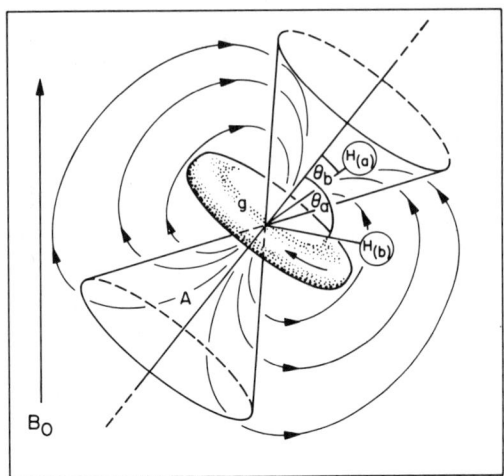

FIGURE 1.22 The shielding region about an electron grouping g, which has cylindrical but not spherical symmetry. Protons $H_{(a)}$ and $H_{(b)}$ are in the shielding and deshielding zones, respectively, the nodal surface of zero shielding being the double cone.

principal symmetry axis, there will be one other susceptibility, $\chi_\perp$, perpendicular to A, giving rise to a moment $B_0 \chi_\perp$; if all the molecules were aligned so that A were perpendicular to B_0, the observed diamagnetic susceptibility would be $\chi_\perp$. (For groups of lower symmetry, a third susceptibility must be defined, but we shall not consider this here.) Such a magnetically anisotropic group will influence the shielding of the "distant" nuclei $H_{(a)}$ or $H_{(b)}$ even when rapid randomization occurs. McConnell (23) has shown that if these observed nuclei are sufficiently distant so that the anisotropic group g can be regarded as a point magnetic dipole, then the mean screening of these nuclei by the group g in the tumbling molecule will be expressed by

$$\sigma_g = \frac{(\chi_\parallel - \chi_\perp)(1 - 3\cos^2\theta)}{3r^3}, \qquad (1.57)$$

where r is the distance from the observed nucleus to the point dipole and θ is the angle between A and the line joining the nucleus to the point dipole. Both θ and r will of course be fixed in a rigid molecule, even though tumbling. Here, $\chi_\parallel - \chi_\perp$ or $\Delta\chi$ is the anisotropy of the diamagnetic susceptibility. Since the susceptibilities are negative numbers, $\Delta\chi$ will be negative when $\chi_\parallel$ is larger (more negative) than $\chi_\perp$.

As Eq. (1.57) and Fig. 1.22 indicate, the shielding of nuclei, protons in particular, in the presence of such an anisotropic group will depend

1.5. Magnetic Shielding and the Chemical Shift

on their geometrical relationship to the symmetry axis. A proton $H_{(a)}$ or other nucleus located in the cone-shaped zones extending along the symmetry axis will experience shielding since the induced field opposes B_0 here, which consequently must be made larger to compensate. Correspondingly, a nucleus located in the region outside these zones will be deshielded. The surface of this cone is a nodal surface in which σ_g is zero.

A particularly thoroughly studied type of anisotropic shielding group is the aromatic ring (24–26). Here the point dipole approximation is replaced by a more realistic model in which for a single aromatic group, e.g., a benzene ring, six π electrons are pictured as circulating under the influence of B_0 in a circular path equal in radius to the distance from the center of the ring to each carbon, i.e., the length of a carbon–carbon bond, 1.39 Å. Actually the electrons are taken as circulating in two doughnut-like π-electron clouds on each face of the ring, separated by 1.28 Å (23) (see Fig. 1.23). The results of calculations based on this model are shown in Fig. 1.24, in which "isoshielding" lines are shown over a region extending outward 5.0 ring radii (6.95 Å) in the plane of the carbon ring and 3.5 ring radii (4.87 Å) along the hexad symmetry axis. (It should be noted that quantum mechanical calculations (25) produce results very close to the classical approach embodied in Fig. 1.22.) Since the shielding zone about an aromatic ring is somewhat more spacious than the deshielding zone, the effect of aromatic solvents and of interactions within polymers having aromatic components is generally a shielding one.

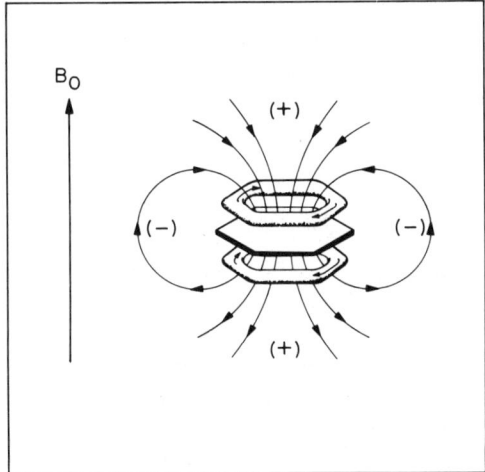

FIGURE 1.23 Electron density, ring currents, and magnetic lines of force about a benzene ring.

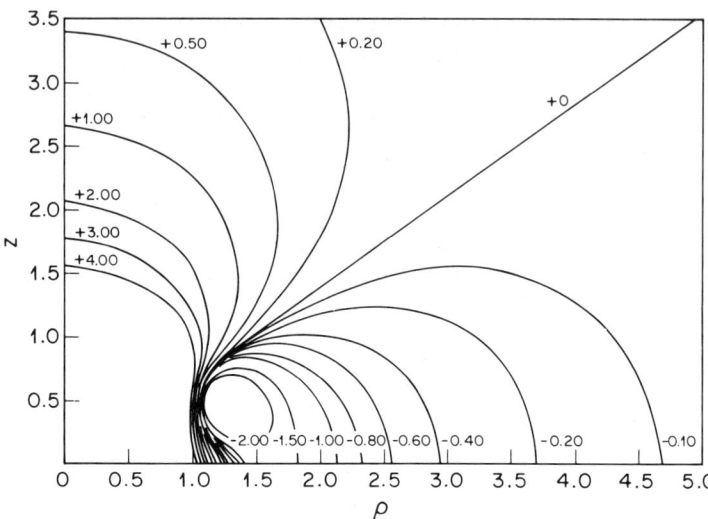

FIGURE 1.24 The shielding zone about a benzene ring. The "isoshielding" lines are calculated from the model represented by Fig. 1.23.

Although there are several other groups which can produce appreciable anisotropic shielding effects, they do not merit special consideration in connection with polymers. A potent influence on proton chemical shifts which we must consider is that of *hydrogen bonding*, always a deshielding effect. A common example is that of the hydroxyl group in alcohols. This usually causes the hydroxyl proton of ethanol to be less shielded than those of the methyl and methyl groups, but upon dilution in carbon tetrachloride to 0.5% or less this proton actually becomes more shielded than the methyl protons. Hydrogen bonding also causes important effects in amino acids, polypeptides and polynucleotides.

1.5.2 Carbon-13 Chemical Shifts

Carbon-13 chemical shifts are very sensitive to molecular structure and geometry, quite apart from the influence of substituent groups. This is particularly clearly revealed in the ^{13}C spectra of paraffinic hydrocarbons, a class to which many important polymers belong. In contrast to paraffinic protons, which embrace a range of only about 2 ppm (Fig. 1.19), carbon chemical shifts are spread over more then 40 ppm (Fig. 1.20). This makes ^{13}C spectroscopy a powerful means for the study of such materials. The empirical ordering of such effects (theoretical understanding lags far behind) may be done in terms of

1.5. Magnetic Shielding and the Chemical Shift

α, β, and γ effects (27,28) (δ and ε effects are further refinements which we omit here). Table 1.2 shows a series of data for simple hydrocarbons that illustrate the α effect. Here, we are to observe $^\circ$C carbon and ask what happens on adding carbons α to this one. We see a regular deshielding of about 9 ± 1 ppm for each added carbon, except in neopentane, where crowding apparently reduces the effect.

In Table 1.3 we see examples of the effect of carbons added β to the observed one, $^\circ$C. The effect is of similar magnitude to that produced by α carbons, a particularly difficult point theoretically. For nonterminal carbons the effect is similar, but reduced in magnitude if $^\circ$C is a branch point.

Finally, we must consider the γ effect (Table 1.4), which, although smaller than α and β effects, is of particular interest because unlike them it is a shielding effect and also is clearly dependent on molecular conformation (29). The γ effect is found to be produced by ele-

TABLE 1.2
The α Effect in ^{13}C Shieldings

Structure	δ_C (ppm)	α Effect (ppm)
$^\circ$**CH$_3$**–H	−1.2	—
$^\circ$**CH$_3$**–CH$_3$	5.9	8.0
$^\circ$**CH$_2$**–(CH$_3$)$_2$	16.1	10.2
$^\circ$**CH**–(CH$_3$)$_3$	25.2	9.1
$^\circ$**C**–(CH$_3$)$_4$	27.9	2.7

TABLE 1.3
The β Effect in ^{13}C Shieldings

Structure	δ_C (ppm)	β Effect (ppm)
$^\circ$**CH$_3$**–$^\alpha$CH$_3$	5.9	—
$^\circ$**CH$_3$**–$^\alpha$CH$_2$–$^\beta$CH$_3$	15.6	9.7
$^\circ$**CH$_3$**–$^\alpha$CH–($^\beta$CH$_3$)$_2$	24.3	8.7
$^\circ$**CH$_3$**–$^\alpha$C–($^\beta$CH$_3$)$_3$	31.5	7.2
$^\alpha$CH$_3$–$^\circ$**CH$_2$**–$^\alpha$CH$_3$	16.1	—
$^\alpha$CH$_3$–$^\circ$**CH$_2$**–$^\alpha$CH$_2$–$^\beta$CH$_3$	25.0	8.9
$^\alpha$CH$_3$–$^\circ$**CH$_2$**–$^\alpha$CH–($^\beta$CH$_3$)$_2$	31.8	6.8
$^\alpha$CH$_3$–$^\circ$**CH$_2$**–$^\alpha$C–($^\beta$CH$_3$)$_3$	36.7	4.9
($^\alpha$CH$_3$)$_2$–$^\circ$**CH$_2$**–$^\alpha$CH$_3$	25.2	—
($^\alpha$CH$_3$)$_2$–$^\circ$**CH$_2$**–$^\alpha$CH$_2$–$^\beta$CH$_3$	29.9	4.7
($^\alpha$CH$_3$)$_2$–$^\circ$**CH$_2$**–$^\alpha$CH–($^\beta$CH$_3$)$_2$	34.1	4.2
($^\alpha$CH$_3$)$_2$–$^\circ$**CH$_2$**–$^\alpha$C–($^\beta$CH$_3$)$_3$	38.1	4.0

TABLE 1.4
The γ Effect in ^{13}C Shieldings

Structure	δ_C (ppm)	γ Effect (ppm)
$^o\mathbf{CH_3}-^{\alpha}CH_2-^{\beta}CH_3$	15.6	—
$^o\mathbf{CH_3}-^{\alpha}CH_2-^{\beta}CH_2-^{\gamma}CH_3$	13.2	−2.4
$^o\mathbf{CH_3}-^{\alpha}CH_2-^{\beta}CH-(^{\gamma}CH_3)_2$	11.3	−1.9
$^o\mathbf{CH_3}-^{\alpha}CH_2-^{\beta}C-(^{\gamma}CH_3)_3$	8.8	−2.5
$^{\alpha}CH_3-^o\mathbf{CH_2}-^{\alpha}CH_2-^{\beta}CH_3$	25.0	—
$^{\alpha}CH_3-^o\mathbf{CH_2}-^{\alpha}CH_2-^{\beta}CH_2-^{\gamma}CH_3$	22.6	−2.4
$^{\alpha}CH_3-^o\mathbf{CH_2}-^{\alpha}CH_2-^{\beta}CH-(^{\gamma}CH_3)_2$	20.7	−1.9
$^{\alpha}CH_3-^o\mathbf{CH_2}-^{\alpha}CH_2-^{\beta}C-(^{\gamma}CH_3)_3$	18.8	−1.9

ments other than carbon which are three bonds removed from the observed carbon and to show a correlation with the electronegativity of this element (*vide infra*).

For a better understanding of the γ effect, consider the staggered conformers of three of the compounds in Table 1.4, represented in Scheme 1. In the case of butane, we divide the observed average γ shift by the *gauche* conformer content (ca. 0.45) and find a shielding value of −5.3 ppm per *gauche* interaction. It is proposed that this shielding occurs when the four-carbon three-bond system is in or near a *gauche* state. We shall see in later discussion that when the dihedral angle decreases from the *gauche* value of 60° toward the eclipse conformation (0°), the shielding increases to about 8 ppm, while as the angle increases the effect decreases, becoming zero beyond about 90°–100°. Shielding differences of this magnitude can be readily observed and measured in the solid state and provide a welcome substantiation of crystal structures derived from X-ray diffraction.

We shall see later that the *gauche* shielding value of −5.3 ppm explains quite accurately, among other observations, the dependence of carbon chemical shifts on stereochemical configuration in many chiral compounds and macromolecules. For the more crowded compounds in Scheme 1, 2-methylbutane and 2,2-dimethylbutane, the effect appears somewhat smaller.

The rules and regularities embodied in Tables 1.2, 1.3, and 1.4 and in the γ-*gauche* effect enable one to *predict* as well as interpret carbon chemical shifts, and thereby to test proposed structures and conformations. A more detailed discussion of the manner of making such prediction is given elsewhere (30) and will not concern us further here.

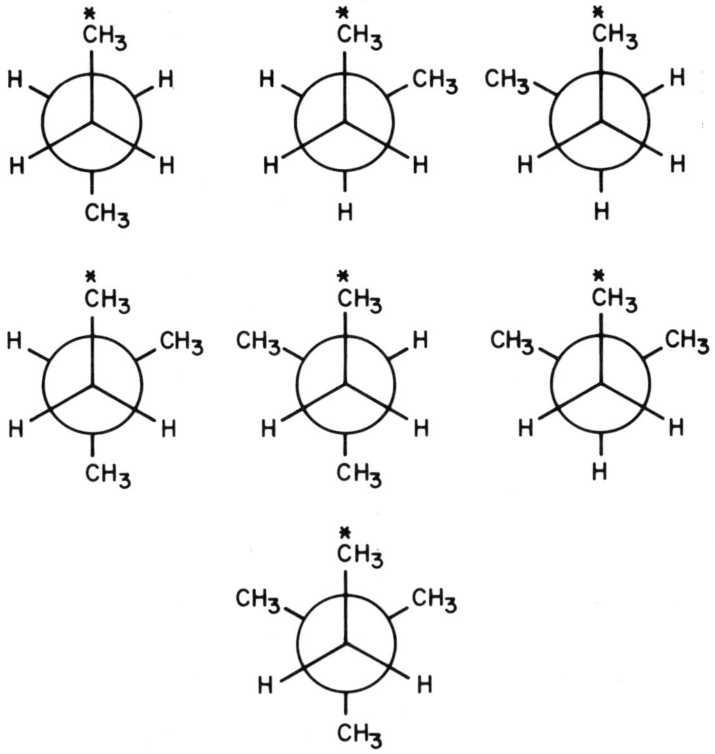

1.5. Magnetic Shielding and the Chemical Shift

SCHEME 1

Inductive effects are of great importance in carbon-13 chemical shifts of non-paraffinic compounds and macromolecules. The transmission of inductive effects is illustrated by the schematic spectra in Fig. 1.25. Theory (31,32) suggests that the deshielding influence of electronegative elements or groups should be propagated down the chain with alternating effect, decreasing as the third power of the distance. This prediction is not borne out. It is true that the shielding of C_1 shows the expected dependence on the electronegativity of the attached atom X. Note particularly the effect of F and O. The relative effects of the halogens are also what one expects. Note that iodine actually causes C_1 to be more shielded than C_5. We observe, however, that the shielding of C_2 is independent of electronegativity. The substitution of any element for the H of pentane produces about the same deshielding effect at C_2, i.e., 8 to 10 ppm. Such substitution causes a *shielding* of C_3 by ca. 2–6 ppm, an effect we have already noted and discussed when the substituting element is carbon but

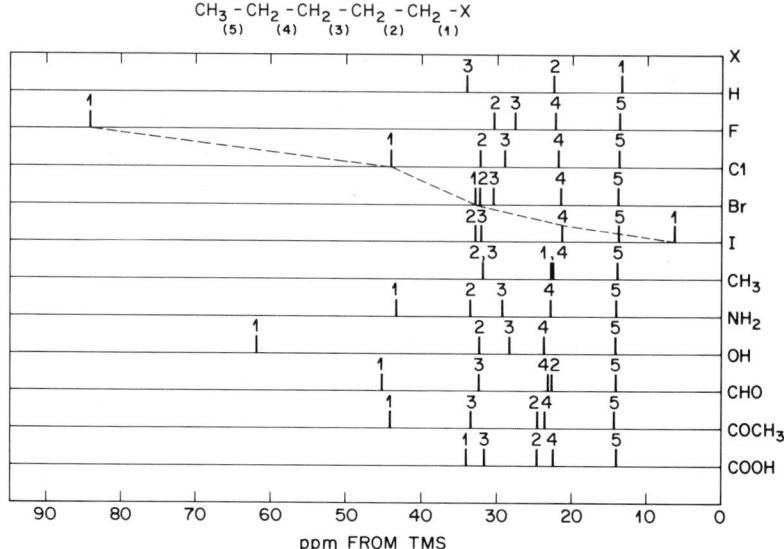

FIGURE 1.25 Schematic carbon-13 spectra illustrating the transmission of inductive effects in straight-chain aliphatic compounds. All spectra represented as proton decoupled so that carbon resonances appear as singlets.

which is caused by other elements as well. In fact, carbon is at the low end of this 2 to 6-ppm range; N and O produce a larger "γ" effect.

In Fig. 1.20 we saw that olefinic and aromatic carbons are strongly deshielded compared to those of saturated carbon chains—an effect parallel to that for protons (Fig. 1.19) but much larger. Theoretical explanations (31,32) involve the paramagnetic screening term rather than diamagnetic shielding or ring current considerations (33–35). Carbonyl carbon atoms show even greater deshielding (Fig. 1.20).

1.5.3 Other Nuclei (^{19}F, ^{31}P, ^{14}N, ^{15}N)

1.5.3.1 Fluorine

The magnetic moment of ^{19}F is only slightly smaller than that of ^{1}H, so its relative sensitivity is high (Table 1.1). (Relative sensitivities are approximately proportional to the cubes of the ratios of magnetic moments at constant field strength.) It has a spin of 1/2 and therefore no quadrupole moment. Because of the greater polarizability of its electron cloud, ^{19}F, like ^{13}C and most other nuclei, exhibits a much greater range of chemical shifts than does ^{1}H. This greater range often makes possible the discrimination of subtler structural and environmental differences than possible by ^{1}H (or even ^{13}C). Fluorine chemical shifts are commonly expressed on the Φ scale (36), on which CCl_3F serves as reference. They have been extensively reviewed (37). We shall be concerned only with fluorine bonded to carbon.

1.5.3.2 Phosphorus

^{31}P is the only naturally occurring isotope of phosphorus. It has a spin of 1/2, a magnetic moment of 1.1305, a magnetogyric ratio of 1.0829×10^8, and a relative sensitivity of 0.0664 compared to the proton (Table 1.1). It is thus easy to observe with modern instrumentation. Relaxation times are longer than for protons but not so long as to cause problems, and the chemical shift range is over 700 ppm. The standard of reference is external 85% aqueous phosphoric acid as zero, with ^{31}P nuclei less shielded than this given positive values and those more shielded given negative values. (In earlier literature this sign convention may be reversed.) ^{31}P NMR spectroscopy has become of great importance in the study of polynucleotide structure and also in the observation of metabolic processes *in vitro* and *in vivo*.

1.5.3.3 ^{14}N and ^{15}N

Next to hydrogen and carbon, nitrogen is the most important of the elements composing organic and biorganic macromolecules, occurring both in skeletons and in functional groups. The abundant species, ^{14}N (99.63%), is inconvenient to observe because of its nuclear electric quadrupole (Table 1.1). Sometimes this is beneficial; for example, it makes possible the observation of protons directly bonded to nitrogen in polymers with good resolution. For the observation of ^{14}N itself, however, it is not helpful since it causes severe broadening, sometimes by as much as 10^3 Hz.

With the introduction of Fourier transform NMR spectroscopy, it becomes possible to observe ^{15}N, which has a spin of 1/2 (Table 1.1) and therefore suffers no quadrupolar broadening. Although its observing sensitivity in natural abundance is only 0.357% of that of ^{14}N, this drawback is in part compensated by a large nuclear Overhauser enhancement.

The NMR of nitrogen-15 has been extensively reviewed. The 1981 monograph of Witanowski *et al.* (37) contains 156 tables of data while that of Stefaniak *et al.* (38) contains 163 tables. The review of von Philipsborn and Muller (39) is recommended as a general survey.

1.6 INDIRECT COUPLING OF NUCLEAR SPINS

1.6.1 INTRODUCTION

We have seen in Section 1.4.2 that nuclear spins may be directly coupled to each other through space, giving rise to the phenomenon of dipolar broadening. In addition to this direct coupling, magnetic nuclei may also transmit information to each other indirectly through the intervening chemical bonds. This interaction, discovered independently by Gutowsky et al. (40) and Hahn and Maxwell (41), occurs by slight polarizations of the spins and orbital motions of the valence electrons and, unlike the direct couplings, is not affected by the tumbling of the molecules and is independent of B_0. If two nuclei of spin 1/2 are so coupled, each will spit the other's resonance to a doublet, for in a collection of many such pairs of nuclei there is an almost exactly equal probability of each finding the other's spins to be oriented with $(+1/2)$ or against $(-1/2)$ the applied field [Fig. 1.26(a)]. If one nucleus is coupled to a second group of two identical nuclei, the possible combinations of orientations of the latter will be as shown in Fig. 1.26(b). It is therefore to be expected that the resonance of the first nucleus will appear as a 1:2:1 triplet, while that of the group of two identical nuclei will be a doublet. Three equivalent neighboring spins would split the single nucleus's resonance to a 1:3:3:1 quartet [Fig. 1.26(c)]. Generalizing, we may say that if a nucleus of spin 1/2 has n equivalently coupled neighbors of spin 1/2, its resonance will be split into $n + 1$ peaks corresponding to the $n + 1$ spin states of the

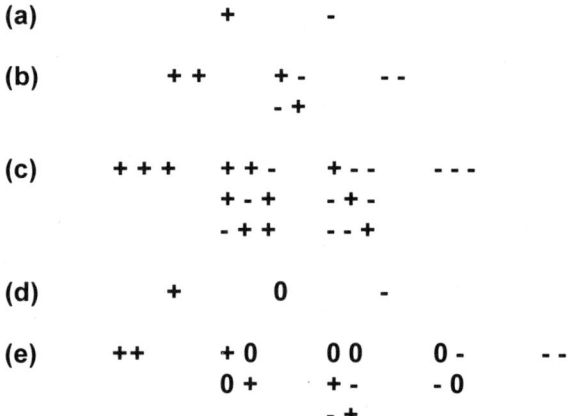

FIGURE 1.26 Spin states and statistical weights for: (a) one spin-1/2 nucleus, (b) two spin-1/2 nuclei, (c) three spin-1/2 nuclei, (d) one spin-1 nucleus, and (e) two spin-1 nuclei.

1.6. Indirect Coupling of Nuclear Spins

neighboring group of spins. The peak intensities are determined by simple statistical considerations and are proportional to the coefficients of the binomial expansion. A convenient mnemonic device for these is the Pascal triangle (Fig. 1.27).

The proton spectrum of the ethyl groups of ethyl orthoformate [Fig. 1.16(a)] illustrates these features of NMR spectra. The methyl resonance at 1.16 ppm is a triplet because of coupling to the CH_2 protons, which appear as a quartet centered at 3.52 ppm. The formyl proton is too distant to experience observable coupling. The strength of the coupling is denoted by J and is given by the spacing of the multiplets expressed in hertz. Couplings through more than three intervening chemical bonds tend to be weak or unobservable, as for the formyl proton in this case.

In the carbon spectrum [Fig. 1.16(b)], the multiplicity arises from the direct (one bond) couplings of the ^{13}C nuclei to the attached protons. These couplings are approximately 125 Hz. The ^{1}H–^{13}C couplings through two or more bonds are much smaller and are not resolved on the scale of Fig. 1.16(b). The numbers and intensities of the resonances are strictly binomial and can often serve to identify the carbon types. (In practice, these couplings are usually removed and the carbon resonances reduced to singlets in order to increase the effective signal-to-noise ratio; see Section 1.6.3.)

The coupling of nuclei with spins of 1 or greater is more rarely dealt with, deuterium (^{2}H) and ^{14}N being the most frequently encountered. For nuclei of spin 1, we have seen that three spin states are possible. One such nucleus will split another nucleus (or group of equivalent nuclei), of whatever spin, to a $1:1:1$ triplet; two will give a $1:2:3:2:1$ pentuplet; and so on [Figs. 1.26(d) and 1.26(e)]. In general, a nucleus coupled to n nuclei with spin I will give $2nI + 1$ lines.

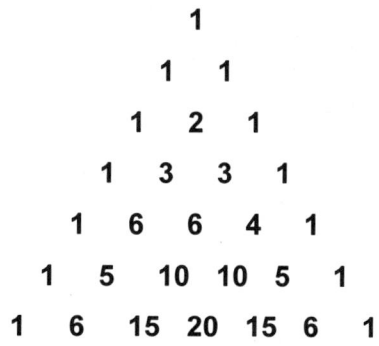

FIGURE 1.27 The Pascal triangle.

1.6.2 PROTON–PROTON COUPLINGS AND SPECTRAL ANALYSIS

It is particularly important to understand that the occurrence of spin–spin coupling does not depend on the presence of the magnetic field B_0, and that the magnitude of J, unlike that of the chemical shift, is independent of the magnitude of B_0. Thus, the ratio $J/\Delta v$, where Δv is the chemical shift difference between two given nuclei or two groups of equivalent nuclei, decreases as B_0 increases. The multiplets move further apart on a field-strength or hertz scale (but not on a parts-per-million scale); however, the spacings within each multiplet (in hertz) remain the same.

It is observed that as $J/\Delta v$ becomes larger (as in a series of suitably chosen molecules or by observing a particular molecule at decreasing B_0), the spectrum begins to deviate noticeably from the simple "first-order" appearance which we have been describing. In the 200-MHz proton spectrum of the ethyl groups in ethyl orthoformate [Fig. 1.16(a)], for which $J/\Delta v$ is only about 0.01, this tendency is not noticeable; but it can be clearly seen at 60 MHz. The members of the multiplets deviate from strict binomial intensities in such a way as to "lean" toward each other. As $J/\Delta v$ approaches and exceeds unity, these deviations become very severe, and chemical shifts and coupling constants can no longer be obtained by simple spacing measurements. The simplest system showing this behavior is the so-called AB spin system consisting of two spins of the same species, usually ^{1}H or ^{19}F, differing in chemical shift and J coupled with a magnitude comparable to the chemical shift difference, expressed in hertz. Calculated spectra in which $J/\Delta v$ is varied from 0.20 to 1.0 are shown in Fig. 1.28. The spacing of the doublets gives J_{AB}, but as $J/\Delta v$ increases the chemical shifts of A and B are no longer given by the centers of the doublets; v_{AB} is given by $[(v_1 - v_4)(v_2 - v_3)]/2$, i.e., the geometric mean of the outer and inner line positions, expressed as frequencies. This will approach the arithmetic mean, $1/2[(v_1 - v_4) + (v_2 - v_3)]$ as $J/\Delta v$ becomes small. This corresponds to what is normally designated as an AX case. The relative intensities of the inner and outer peaks are inversely proportional to their spacings, and so as $J/\Delta v$ is increased without limit the quartet will approach a singlet. The outer peaks are reduced to vanishingly small satellites.

It is observed and can be shown from quantum mechanical principles that in such systems as A_2, A_3, A_2B, A_3B, A_nB, etc., the coupling J_{AA} has no influence on the spectrum and hence cannot be deduced from it. Spin groups such as $A_2, A_3, \ldots, A_n$ are termed "equivalent." The CH_3 and CH_3 groups in the ethyl spectrum are each examples of such equivalent sets of spins. It also must be

1.6. Indirect Coupling of Nuclear Spins

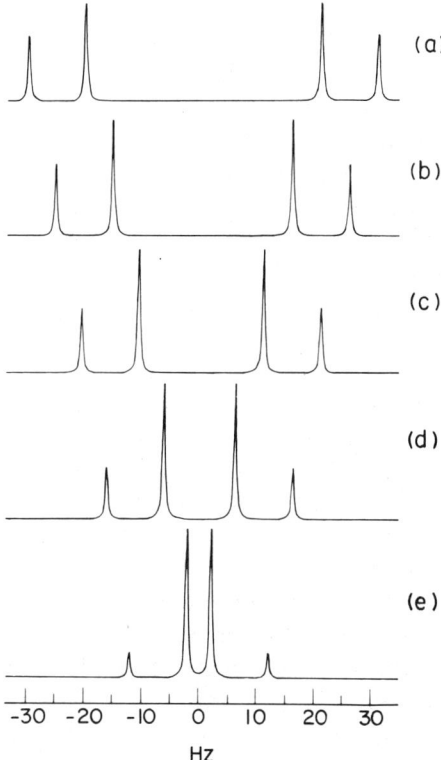

FIGURE 1.28 Calculated AB spectra for $J = 10$ Hz and $J/\Delta v$ equal to (a) 0.20, (b) 0.25, (c) 0.33, (d) 0.50, and (e) 1.00.

recognized that J couplings may be positive or negative in *sign* and that relative signs in systems of three or more spins will affect the spectra.

Although certain patterns become recognizable with experience, proper solutions of strong-coupled spectra involving more than two spins require spectral simulation using quantum mechanical programs which are now commonly included in the software of larger commercial spectrometers. This is not very commonly required, however, and so we will not enlarge on the details here.

The proton–proton couplings with which we shall be mainly concerned in polymer spectra are *geminal*, i.e., between protons on the same atom, usually carbon, and *vicinal*, between protons on neighboring atoms. These are often designated ^2J and ^3J, respectively, the superscript denoting the number of intervening bonds. Of one-bond or direct couplings, ^1J, only that between ^{13}C and directly bonded protons (or ^{19}F) is of importance in polymer spectroscopy.

Geminal couplings exhibit very wide variations. In groups with sp^3 hybridization, they are generally in the range of −12 to −15 Hz. For example, the geminal coupling in methane is −12.4 Hz and in the methylene groups of poly(methyl methacrylate) −14.9 Hz. The value increases algebraically, becoming less negative or actually positive as the H–C–H angle increases. Thus, the coupling in ethylene is +2.5 Hz (sp^2 hybridized) and −4 to −6 Hz in cyclopropanes (intermediate between sp^2 and sp^3). Substituent effects are often quite large and can overwhelm geometrical effects. An electronegative substituent on a vinyl group decreases the geminal coupling: in vinyl chloride, ^2J is −1.3 Hz. Other more marked substituent effects have also been observed, but they are seldom important in polymer spectra.

Vicinal couplings are markedly dependent on geometry, in both unsaturated and saturated systems, and likewise show substituent effects. They do not, however, vary over so large a range as geminal couplings and are rarely—probably never—negative. *Trans* couplings across double bonds are always about twice as large as *cis* couplings in the same structure, but both decrease with increasing electronegativity of the substituent. Thus, in ethylene, J_{cis} = 11.5 Hz and J_{trans} = 19.0 Hz, while in vinyl chloride J_{cis} = 7.4 Hz and J_{trans} = 14.8 Hz. In saturated systems, *trans* couplings are substantially larger than *gauche* couplings. Both in ring compounds like cyclohexanes and in open-chain compounds, J_{gauche} (commonly denoted J_g) is typically 2–4 Hz and J_{trans} (J_t) ranges from 8 to 13 Hz. These considerations are of great importance in studying the conformation of polymer chains, and will be considered in greater detail in Chapter 3. Karplus (42,43) has given a valence-bond interpretation of these observations and concludes that vicinal coupling of protons on adjacent sp^3 carbon atoms should depend on ϕ, the dihedral angle, as follows:

$$J = \begin{cases} 8.5 \cos^2 \phi - 0.28 & \text{for } 0° \leq \phi \leq 90° \\ 9.5 \cos^2 \phi - 0.28 \text{ (Hz)} & \text{for } 90° \leq \phi \leq 180°. \end{cases} \quad (1.58)$$

A plot of these functions (Fig. 1.29) shows that J should be 1.8 Hz when $\phi = 60°$ (J_g) and 9.2 Hz when $\phi = 180°$ (J_t). Actual couplings tend to be somewhat larger than these values, particularly in open-chain systems. Proton couplings through more than three bonds are well recognized in small molecules but are not commonly resolvable in polymer spectra.

1.6. Indirect Coupling of Nuclear Spins

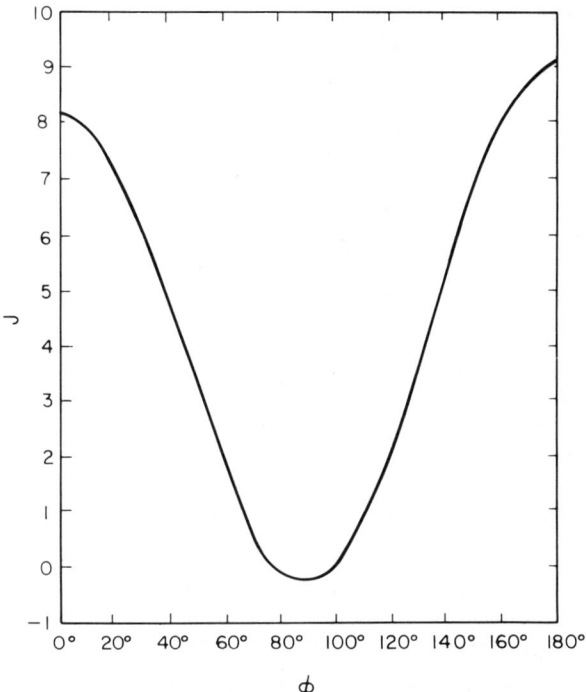

FIGURE 1.29 The Karplus function describing the magnitude of vicinal proton–proton J coupling as a function of the dihedral angle ϕ in the H–C–C–H bond system.

1.6.3 CARBON–PROTON COUPLINGS AND CARBON–CARBON COUPLINGS

Carbon–proton J couplings do not have the same significance in structure determination as proton–proton couplings but are nevertheless of some significance, particularly in identifying carbon types. One-bond ^{13}C–^{1}H couplings depend upon the fractional s character of the bonding orbital (44–46). Thus we find the following values of ^{1}J ^{13}C–^{1}H couplings (in Hz): CH_4, ~ 125; cyclohexane, 123; $(CH_3)_4C$, ~ 124; CH_3C–CCH_3, ~ 131; benzene, 158; cyclobutane, 136; cyclopropane, 161; CH_3C ^{13}C–H, 251. This relationship may be summarized as,

$$^{1}J_{CH} = 5\rho_{CH}, \tag{1.59}$$

where ρ_{CH} is the percentage of s character in the bonding orbitals.

Carbon-13–proton couplings through two or more bonds are very much smaller than directly bonded couplings: commonly 4–8 Hz for

$^2J_{CH}$, and 5–11 Hz for $3J_{CH}$. They do not fall off regularly with the number of intervening bonds.

The observation of $^{13}C-^{13}C$ couplings in natural abundance requires that the molecule have the rare isotope in two positions, which will be the case for approximately 1 molecule in 8300. Thus, sensitivity is a problem for all but the simplest molecules, and for macromolecules, isotopic enrichment is a necessity. Up to the present little use has been made of the structural—including conformational—information inherent in such measurements on polymers. Elimination of carbon-13 multiplicity of proton irradiation is discussed (for both solution and solid state) in Section 1.8.4.

1.6.4 OTHER NUCLEI (^{19}F, ^{31}P, ^{14}N, ^{15}N)

1.6.4.1 ^{19}F

$^{19}F-^{19}F$ couplings show unexpected behavior, particularly in highly fluorinated groups. Vicinal couplings in perfluoroalkyl groups are often nearly zero, but couplings through four or five bonds may be 5 to 15 Hz. Since both ^{19}F and $^{19}F-^{1}H$ couplings are transmitted to an observable degree through a larger number of bonds than $^{1}H-^{1}H$ couplings, ^{19}F multiplets are frequently quite complex and may appear poorly resolved and broad because of the many closely spaced transitions. The geometrical dependence of $^{19}F-^{19}F$ couplings is not so well worked out as that of $^{1}H-^{1}H$ couplings but it is clear that strong effects occur. For example, in $CBrF_2 \cdot ^{13}CBrF_2$, Harris and Sheppard (47) deduced from ^{13}C sideband spectra that the *gauche* coupling is about 12 Hz and the *trans* coupling is about 1.5 Hz, a reversal of the $^{1}H-^{1}H$ magnitudes. However, on increasing substitution by heavier halogens, J_t increases rapidly and may overtake J_g. Geminal $^{19}F-^{19}F$ couplings are much larger than vicinal couplings or geminal proton couplings. In saturated systems they range from about 150 Hz to nearly 300 Hz; on double bonds they are generally less then 100 Hz. Vicinal $^{19}F-F^{19}$ couplings are negative, whereas geminal couplings are positive.

In that they tend to fall off in magnitude monotonically with the number of intervening bonds, $^{19}F-^{1}H$ couplings are more "normal" than $^{19}F-^{19}F$ couplings. Geminal $^{19}F-^{1}H$ couplings are in the range of 45–55 Hz. Vicinal couplings usually have less than half the magnitude of geminal couplings. Both are positive and sensitive to substituent effects.

Direct, one-bond $^{19}F-^{13}C$ couplings range from ca. 250 to ca. 350 Hz with occasional higher values. Accumulation of fluorines or protons tends to decrease the magnitude of the couplings. Couplings to

1.7. Rate Phenomena: Averaging of Chemical Shifts and J Couplings

trigonal carbons are comparable to tetrahedral couplings, while those of aromatic carbons are somewhat reduced.

1.6.4.2 ^{31}P

^{31}P–^{31}P couplings are mainly of interest in nucleotides and polynucleotides, where they are in the range of 19–23 Hz. ^{31}P chemical shifts and couplings have been reviewed by Gorenstein (48).

1.6.4.3 ^{14}N and ^{15}N

We have seen that direct (one-bond) ^{14}N–^{1}H J couplings are frequently not observed because the ^{14}N quadrupole causes decoupling. When observable, they are in the range 50–60 Hz. The corresponding ^{15}N couplings are readily observed as proton or nitrogen multiplicity, the magnitude of which is calculable from the relationship

$$J(^{15}\text{N} - X) = J(^{14}\text{N} - X) \frac{\gamma_{^{15}\text{N}}}{\gamma_{^{14}\text{N}}}$$

$$= -1.40 \cdot J(^{14}\text{N} - X). \quad (1.60)$$

The negative sign arises from the negative sign of the ^{15}N magnetogyric ratio (Table 1.1); as expected, a magnitude of ca. 70–85 Hz is commonly observed for the ^{15}N–^{1}H coupling. Two-bond ^{15}N–^{1}H couplings are smaller, usually by an order of magnitude, but are more sensitive to structure and may be of either sign. Three-bond ^{15}N–^{1}H couplings appear to offer the possibility of measurement of torsional angles in the same manner as ^{1}H–^{1}H couplings, but as they exhibit a narrow range of absolute values—ca. 1–5 Hz—and may be of either sign this application has been limited.

The observation of ^{15}N–^{13}C couplings generally requires enrichment in at least one of these rare nuclei. One-bond ^{15}N–^{13}C couplings are usually of 2 to 20-Hz magnitude and negative. Two-bond ^{15}N–^{13}C couplings are weaker and may be of either sign. Three-bond couplings are negative and weak, seldom exceeding 5 Hz; they have not proved broadly useful for establishing torsional angles.

1.7 RATE PHENOMENA: AVERAGING OF CHEMICAL SHIFTS AND J COUPLINGS

In the ultraviolet or infrared spectrum of a molecule having two or more coexisting subspecies—for example, two different conformations or ionized states—one may expect to see absorption peaks corresponding to both forms even if they are in very rapid equilibrium with each other. If $\delta\tau$ is the lifetime of each form (we shall suppose for the present that these exist in equal amounts and therefore that $\delta\tau$ is the same for both), then from the uncertainty principle one may expect a broadening of the absorption peaks given by

$$\delta v = \frac{1}{2\pi\,\delta\tau}. \tag{1.61}$$

At the high frequencies associated with optical spectra (10^{13}–10^{15} s^{-1}) this broadening cannot be appreciable compared to the resolving power of optical spectrometers unless $\delta\tau$ is very small: 10^{-12}–10^{-13} s or less. Conformations, complexed states, and ionized forms of organic molecules usually have lifetimes considerably longer than this. However, in NMR spectroscopy the observing frequency is one-millionth as great, spectral lines may be less than 1 Hz in width, and those corresponding to interchanging subspecies may be only 1 Hz apart. Very marked broadening, sufficient to merge the lines for each subspecies, may occur even when lifetimes are as long as 10^{-1}–10^{-3} s. If their lifetimes are much shorter than this, the two forms will give a single narrow peak and be entirely indistinguishable. These lifetimes (10^{-1}–10^{-5} s) correspond to a range of rates of reaction and isomerization that is of great importance in chemistry. By suitable modification of the Bloch equations (Section 1.4.3) the spectra lineshapes corresponding to the broadening and merging of peaks can be given an analytical form, and in this way the rates of a wide variety of processes have been measured.

A process of importance in some polymer spectra, particularly those of biopolymers, is *proton exchange*. The exchange rates of protons between organic molecules, in particular those of OH_2, NH_2, and CONH groups, are often in the range amenable to NMR measurement and have been the object of many studies. In such systems, the rate information is contained in the collapse of spin multiplets rather than in the merging of chemically shifted peaks. For example, in pure ethanol, the hydroxyl proton is a triplet because of coupling to the two methylene protons. If we add water and particularly if we add an acid catalyst this triplet will collapse to a singlet, passing through various stages, as shown in Fig. 1.30, as the rate increases due to an increase of temperature or catalyst. The reason for this behavior is that the hydroxyl protons undergo intermolecular exchange, and if the rate of this process is sufficient, their residence time on any one molecule is so short compared to $1/J$ that the coupling is in effect "turned off." Following the departure of a hydroxyl proton, the chances are nearly even that the next proton to arrive at that site will be in

1.8. Experimental Methods

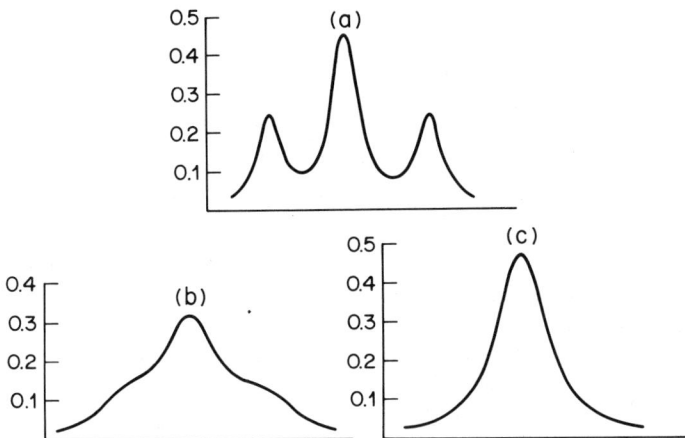

FIGURE 1.30 Collapse of a 1:2:1 triplet as a function of τJ: (a) $\tau J = 100$; (b) $\tau J = 0.25$; (c) $\tau J = 0.0625$.

the other spin state. When this happens, decoupling must occur. The spectra in Fig. 1.30 are shown as a function of τJ, where τ is the residence time of the hydroxyl proton.

A close relationship exists between exchange of magnetization occurring by through-space dipole–dipole interaction (Section 1.4.2) and that occurring by actual chemical exchange. In fact, the mathematical forms of the two exchanges are identical. This parallel can be very effectively exploited by two-dimensional methods; the 2D technique known as NOESY can be used to measure chemical exchange as well as to detect and measure close nuclear approaches. This will be discussed in Section 1.8.7 and subsequent sections.

1.8 EXPERIMENTAL METHODS

1.8.1 INTRODUCTION

We will not describe experimental techniques and procedures in detail since such a discussion is not very useful in this rapidly evolving field. This is in no sense a laboratory handbook. We emphasize principles and basic requirements, especially but not exclusively those that concern polymers and the means of obtaining structural and dynamic information about them.

1.8.2 THE MAGNETIC FIELD

We have already seen (Fig. 1.4) a diagram of a cross section of a superconducting magnet assembly. (Magnets of other types are not likely to be used in polymer studies.) The solenoid, wound from superconducting wire, is bathed in liquid helium. The helium dewar is thermally isolated by means of a vacuum chamber and thermal baffles (not shown) from the outer dewar, which contains liquid nitrogen at 77 K. Once energized, a persistent superconducting magnet requires no additional current so long as it is held at superconducting temperature. (Recent developments in superconducting materials may make possible superconducting magnets that need only liquid nitrogen cooling.)

The magnetic field must be *locked* to ensure stability. Field stability is especially important for repeated signal accumulations. *Internal* lock systems are most widely used as they afford the most exact control of field drift. In this method, the lock material is placed in the sample tube itself. The lock material is most commonly deuterated and serves as the solvent (*vide infra*). The actual manner in which the lock signal is detected depends upon the detection method used by the spectrometer. In Fourier transform instruments, the lock circuit is very similar to the observe circuit. A lock transmitter and receiver coil is wound so that it surrounds the sample. The sample, through this circuit, is pulsed repeatedly (several hundred times a minute) with pulses having a frequency corresponding to the deuterium frequency of the solvent. For example, the deuterium frequency would be 30.7 MHz when the proton frequency is 200 MHz. The free induction decay from the lock signal is amplified and detected. The operator makes small adjustments to the magnetic field so that the lock signal and the lock transmitter occur at identical frequencies. At this point, the *feedback circuit* is engaged. The frequency difference between the lock signal and the lock transmitter is monitored and kept as small as possible by applying a compensating current to the magnetic field. In Table 1.5 are listed the commonly used deuterated solvents together with their molecular formulas, melting and boiling points, and proton and carbon chemical shifts and multiplicities. The latter express the position and appearance of the solvent (and lock) resonances which appear in the spectrum (*vide infra*).

For solid samples one must necessarily employ *external* references (*vide infra*) and locks. However, the drift of the field of a superconducting magnet is so slow that it is usually feasible to forego locking and run solid-state experiments in the unlocked mode. Superconducting magnets are also employed in the unlocked mode in studies of

1.8. Experimental Methods

TABLE 1.5
The Solution Properties and NMR Characteristics of Commonly Used Deuterated Solvents

Solvent	Density	mp (°C)	bp (°C)	$\delta_H{}^a$ (ppm)	$\delta_C{}^a$ (ppm)
Acetone-d_6	0.87	−94	57	2.04 (br)	206 (13) 29.8 (7)
Acetonitrile-d_3	0.84	−45	82	1.93 (br)	118.2 (br) 1.3 (7)
Benzene-d_6	0.95	5	80	7.15 (br)	128.0 (3)
Chloroform-d	1.50	−64	62	7.24 (1)	77.0 (3)
Deuterium oxide	1.11	3.8	101.4	4.63	
Dimethyl formamide-d_7	1.04	−61	153	8.01 (br) 2.91 (5) 2.74 (5)	162.7 (3) 35.2 (7) 30.1 (7)
Dimethyl sulfoxide-d_6	1.18	18	189	2.49 (5)	39.5 (7)
p-Dioxane-d_8	1.13	12	101	3.53 (br)	66.5 (5)
Methylene chloride-d_3	1.35	−95	40	5.32 (3)	53.8 (5)
Pyridine-d_5	1.05	−42	116	8.71 (br) 7.55 (br) 7.19 (br)	149.9 (3) 135.5 (3) 123.5 (3)
Tetrahydrofuran-d_8	0.99	−109	66	3.58 (br) 1.73 (br)	67.4 (br) 25.3 (br)
Toluene-d_8	0.94	−95	111	7.09 (m) 7.00 (br) 2.09 (5)	137.5 (1) 128.9 (3) 128.0 (3) 125.2 (3) 20.4 (7)
Trifluoroacetic acid-d	1.50	−15	72	11.50 (br)	164.2 (4) 116.6 (4)

[a] The multiplicity is given in parentheses.

biological tissues where the lines are broad or it is inconvenient to provide a lock signal.

The shimming and shaping of the magnetic field have been discussed elsewhere (34) and will not be dealt with here. When observing the high-resolution spectra of macromolecules, either in the solution or in solid state, it is commonly found that the lines are broader than for small molecules. The reasons for this will become apparent a little later. This observation has led some workers to conclude that in polymer studies it is not important to shim the field for optimum resolution. Actually, one should always work with the best field, as nothing is thereby lost and much may be gained in some cases.

1.8.3 Referencing

We have seen (Section 1.5) that chemical shifts are almost always reported with respect to a reference material. A small amount of this material is placed in the sample tube along with the sample and (like the lock substance) is termed *internal* if dissolved in the sample. The almost universal substance for this purpose is tetramethylsilane, which serves for ^{1}H, ^{13}C, and ^{29}Si. DSS (4,4-dimethyl-4-silapentane sodium carboxylate) and similar compounds are used for aqueous solutions. Table 1.6 lists some common reference materials for NMR measurements.

Solutions, particularly those of small molecules for relaxation studies, should be filtered through a plug of glass wool or a millipore filter and degassed prior to relaxation measurements. Dissolved oxygen should be removed because it is paramagnetic and competes with the nuclear dipole–dipole mechanism (Section 1.4.1) in inducing relaxation. (But again, because T_1 values of macromolecules are relatively short, this is sometimes not done.)

It is sometimes desirable to add *relaxation reagents* to solution NMR samples to decrease the relaxation time and thereby decrease the accumulation time necessary to obtain a spectrum. Such relaxation agents are composed of paramagnetic substances that reduce relaxation times but do not greatly increase linewidths. Examples include chromium acetylacetonate [CR(acac)$_3$] and the analogous iron complex. Nontoxic, site-selective relaxation reagents are useful as contrast enhancers for biomedical NMR imaging (12).

Polymer samples are generally dissolved in deuterated solvents, the deuterium providing the necessary signal for the lock channel, as we have seen (Table 1.5). The most common NMR solvents are chloroform, benzene, and (for biological samples) water, but other solvents are necessary if the spectroscopy involves poorly soluble materials or measurements are made at very low or high temperatures.

Most deuterated solvents are hygroscopic and pick up water from the atmosphere. This becomes a problem in proton NMR, particularly when recording spectra of material in low concentrations. Deuterated solvents are best stored in a dry box under a nitrogen atmosphere and opened and transferred only in the dry box. Contamination from residual water on glass surfaces can also be avoided if the sample tubes and pipetters are stored in the dry box.

1.8. Experimental Methods

TABLE 1.6
Commonly Used Reference Materials

Compound	Nuclei	δ (ppm)
$(CH_3)_4Si$ Tetramethylsilane (TMS)	^1H	0.0
$(CH_3)_3Si(CH_2)_5SO_3Na$ 4,4-Dimethyl-4-silapentane sodium sulfonate (DSS)	^1H	0.015
$(CH_3)_4Si$ Tetramethylsilane (TMS)	^{13}C	0.0
$(CH_3)_3Si-Si(CH_3)$ Hexamethyldisilane (HMDS)	^{13}C	2.0
CS_2 Carbon disulfide	^{13}C	192.8
CH_3NO_2 Nitromethane	^{15}N	0.0
85% PO_4H_3 Phosphoric acid	^{31}P	0.0
$(CH_3)_4Si$ Tetramethylsilane (TMS)	^{29}Si	0.0
$CFCl_3$ Fluoroform	^{19}F	0.0

The ideal sample for solution-state NMR of polymers is a fairly concentrated solution (2–5%) of a volume barely exceeding the dimensions of the NMR coil, usually 3–4 cm. The most commonly used sample cells (or NMR tubes as they are usually called) have outer diameters of 5 mm for proton spectroscopy and 10 mm for ^{13}C spectroscopy and require approximately 0.5 and 4 ml of solution, respectively. If the amount of sample is limited it is advantageous to use microcells, which require between 10 and 30 μl of solution. The rest of the volume in the coil, i.e., that around the microcell, is filled with an index-matching solvent. On the other hand, if the material is of limited solubility or is composed of biological cells (in which the substances of interest are dilute), it is desirable to use sample cells that are larger than 10 mm in outer diameter. Finally, it is possible to obtain NMR spectra of air-sensitive compounds by preparing the sample on a vacuum line and sealing the sample tube under either vacuum or an inert atmosphere. Normally they are closed with plastic caps. It is essential to use tubes of the highest quality for high-resolution, high-field work. They are usually made of borosilicate glass but also can be made of plastic for corrosive fluorinated compounds, quartz for photochemical applications, and amber glass for photosensitive materials. Vortex plugs are useful, particularly in the case of large sample tubes, to keep the liquid in the bottom of the tube while it is being spun in the probe.

1.8.4 Decoupling

Double resonance or spin decoupling can provide an effective means for simplification of complex spectra and for determining which nuclei are coupled in such spectra. For the second purpose, the use of the form of two-dimensional spectroscopy known as COSY is particularly effective. This is discussed in Section 1.8.7 and will be illustrated by numerous examples in later chapters.

Decoupling may be considered under the general headings of *homonuclear* or *heteronuclear*, depending on whether the two spins or groups of spins are of the same or different species. Homonuclear proton decoupling has not been much employed in polymer studies because proton multiplet structure is usually not well resolved for selective irradiation in polymer spectra. The simplest case of its use is that of an AX system (two doublets; see Section 1.6), in which, while observing nucleus A with the usual weak RF field B_1, we simultaneously provide a second stronger RF field B_2 at the resonance frequency of X. We find that the A doublet collapses to a singlet, which appears at its midpoint. The field B_2 has caused the nucleus X to jump back and forth so rapidly between its spin states that it no longer perturbs the spin states of A. The experiment may of course be reversed, and the X multiplet collapsed by irradiating A. Double resonance is often discussed in terms of the saturation of X, but the phenomenon is really not this simple, as is clearly shown by the fact that, at levels of B_2 below that corresponding to complete collapse of a multiplet, additional lines appear.

The most prevalent form of heteronuclear decoupling is the irradiation of protons to collapse the spin multiplets in carbon-13 spectra. For this purpose one usually wishes to decouple all protons simultaneously regardless of their chemical shift and to generate spectra in which the intensities of the carbon resonances are approximately proportional to the numbers of each type. Covering the chemical shift range of all the protons at superconducting frequencies may require supplying several watts of power to the decoupler coil in the probe. This may generate considerable heat, particularly when biological

1.8. Experimental Methods

samples in aqueous solvents are employed. There is therefore a need to provide effective decoupling with the least expenditure of power. The use of a random noise modulated proton decoupling field (49,50) is not highly efficient in this regard because any particular proton is completely decoupled during only a small fraction of the random noise cycle time. This method continues to be used, but other more efficient schemes are now available. Among these is the method of Grutzner and Santini (51), in which the phase of the CW proton irradiation is periodically reversed by square wave frequency modulation with a 50% duty cycle having a frequency of 20 to 100 Hz. Other decoupling schemes involve sequences of pulses that rotate the components of $^1H-^{13}C$ multiplets through large precession angles—e.g., 180°—repetitively (52,53). To be most effective such inversion pulses must be applied in a specific sequence, for example, 90° ($+x$), 90° ($+x$), 90° ($-x$), 90° ($-x$) (i.e., application of the decoupling field along the direction indicated, with appropriate intervals). This cycle may be repeated up to 16 times, a "supercycle" known as MLEV-64. A further refinement, useful for attaining optimum resolution at the highest superconducting frequencies, is attained by combining 90°, 180°, and 270° pulses along the $\pm x$ and $\pm y$ axes in varying combinations to produce the highest cycling precision (52–55). The WALTZ-16 supercycle, involving 36 pulses before repeating, is commonly used. These broadband proton decoupling methods may of course also be used for the observation of other nuclei, such as ^{14}N, ^{15}N, and ^{19}F.

The employment of proton decoupling discards information concerning carbon types—whether CH_3, CH_2, CH, or quaternary—in order to simplify the spectrum and improve sensitivity. The use of heteronuclear J-resolved 2D NMR (Section 1.8.7.2), which segregates chemical shift and multiplicity information onto different axes of the spectrum, allows one to retain both, although at considerable cost in sensitivity and spectrum accumulation time. Spectral editing methods (Section 1.8) may be more practical for examining minor features of polymer microstructure.

1.8.5 SOLID STATE METHODS

1.8.5.1 Introduction

We have seen (Section 1.4.2) that in solids the direct magnetic interaction of protons results in kilohertz broadening of their resonances, effectively masking most structural information. Although this broadening can be dealt with, we first consider the corresponding effects in carbon-13 spectra, which can be more readily minimized and with greater rewards because of the large range of carbon chemical shifts.

The local magnetic field at a carbon-13 nucleus in an organic solid is given by [see Eq. (1.22)]

$$B_{\text{loc}} = \pm \frac{h}{4\pi} \gamma_H \frac{(3\cos^2 \theta - 1)}{r_{\text{CH}}^3}, \qquad (1.62)$$

where γ_H is the magnetogyric ratio of the proton [Eq. (1.2)] and r_{CH} is the internuclear C–H distance to a nearby proton, assumed to be bonded ($r_{\text{CH}} = 1.09$ Å). As we have seen [Eq. (1.22)], the $\pm$ sign corresponds to the fact that the local field may add to or subtract from the laboratory field B_0 depending on whether the neighboring proton dipole is aligned with or against the direction of B_0, an almost equal probability. The angle θ_{CH} is defined in Fig. 1.31. If r_{CH} and θ_{CH} were fixed throughout the sample, as for isolated ^{13}C–^{1}H pairs in a single crystal, the interaction would result in a splitting of the ^{13}C resonance into equal components, the separation of which would depend on the orientation of the crystal in the magnetic field. This is a large effect; it is found from Eq. (1.60) that for ^{13}C–^{1}H bonded pairs oriented parallel to B_0, the splitting is on the order of 40 kHz.

Polymers in the solid state are glassy or partially microcystalline and one observes a summation over many values of θ_{CH} and r_{CH}, resulting in a proton dipolar broadening of many kilohertz. We have seen (Section 1.4) that rapid reorientation of internuclear vectors, especially if isotropic, will result in narrowing of such broadened lines if the rate of reorientation exceeds the linewidth, expressed as a frequency. When reorienting ^{13}C–^{1}H vectors, if all angles θ_{CH} are

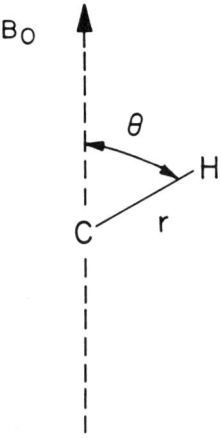

FIGURE 1.31 An internuclear ^{13}C–^{1}H vector of length r making an angle θ with the magnetic field B_0.

sampled in a time which is short compared to the reciprocal of the dipolar coupling, i.e., $\ll 2 \times 10^{-5}$ s, dipolar broadening is reduced to a small value, as in liquids. In solids, such rapid isotropic tumbling is not possible, but since the term $(3\cos^2 \theta_{CH} - 1)$ in Eq. (1.60) is zero when θ_{CH} equals $\cos^{-1} 3^{-1/2}$ or ca. 54.7°, it was long ago realized that spinning solids at this angle might be effective in reducing the linewidth. These early studies were devoted to proton–proton broadening rather than proton–carbon broadening. Lowe (56) and Andrew (57) independently and simultaneously first attempted such "magic angle" spinning (MAS) experiments but they were not really successful for this purpose because it is not mechanically feasible to design rotors that can spin faster than about 10^4 Hz, well short of the 40–50 kHz required. At high speeds all suitable materials (metals being excluded) disintegrate under the extreme g forces generated. We shall see, however, that while MAS is not generally useful for removing proton dipolar broadening, it is very effective for collapsing the lesser broadening arising from chemical shift anisotropy.

1.8.5.2 Dipolar Decoupling

A more practical method for the removal of proton dipolar broadening is to employ a high-power proton decoupling field, i.e., *dipolar decoupling*. This is analogous to the decoupling experiment discussed in Section 1.8.4 for the removal of $^{13}C-^{1}H$ scalar couplings except that much greater power is required. Instead of the ca. 1-G decoupling field used for solutions, a field of ca. 10^{-3} T (ca. 10 G) must be used for the removal of direct dipole–dipole $^{13}C-^{1}H$ coupling in solids, corresponding to ca. 40 kHz expressed as a frequency. This approach is now standard for the observation of solid-state spectra of carbon-13 and other rare or dilute spins. We shall see several examples of its application in the subsequent discussion. It is obvious that it will not serve for the observation of high-resolution proton spectra of solids since one cannot observe the same nucleus that one is simultaneously irradiating with high power. There is a way, however, of observing such proton spectra of solids by combining magic angle spinning and multipulse proton decoupling (58).

1.8.5.3 Chemical Shift Anisotropy and Magic Angle Spinning

Polymers are amorphous or microcrystalline; their carbon chemical shifts take on a continuum of values and form the lineshapes shown in Fig. 1.18. The top pattern corresponds to a chemical shift anisotropy that lacks axial symmetry, while the bottom pattern corresponds to an axially symmetric anisotropy. In the axially symmetric case, $\sigma_{22} = \sigma_{33}$ and both can be designated as $\sigma_\perp$, while $\sigma_{11} = \theta_\parallel$. The principal values σ_{22} and σ_{33} are usually expressed in ppm, as we have seen, and can be measured directly from the singularities or discontinuities in the spectrum, as shown in Fig. 1.18. The isotropic chemical shifts are indicated by dashed lines. For axially symmetric patterns, the position of $\sigma_\parallel$ represents the observed frequency when the principal axis system is parallel with the field direction, while $\sigma_\perp$ represents the observed frequency when the orientation is perpendicular to the field.

Table 1.7 shows the principal values of the chemical shift tensors for a number of carbon types in common polymers. These values are taken from the extensive compilation of Duncan (59), which also

TABLE 1.7
The Chemical Shift Anisotropies for Groups Commonly Occurring in Polymers

	σ_{11}	σ_{22}	σ_{33}	σ_{iso}
a	79	48	29	52
b	53	43	37	44
c	268	150	112	177
d	80	63	10	51
	52	39	14	35
	91	83	33	69
a	226	153	15	137
b	250	122	122	165
c	80	80	28	63

1.8. Experimental Methods

includes data and literature references for over 350 specific compounds. A smaller compilation will also be found in the book by Mehring (60). Such anisotropies are of much interest and repay careful experimental and theoretical study, particularly in connection with solid-state chain dynamics. However, in molecules of any degree of complexity there will be several such patterns that may strongly overlap, producing a broad, uninterpretable spectrum. In these circumstances it usually becomes necessary to sacrifice the anisotropy information in order to retain the isotropic chemical shifts with a reasonable degree of resolution. The width of the carbon-13 anisotropy line-shapes in Table 1.7 correspond to linewidths of about 800–5000 Hz at an observing frequency of 100 MHz (proportionately higher at higher frequencies.) These frequencies are within the range attainable by mechanical spinning with air-driven rotors, such as those shown in Fig. 1.32. Under such rotation the direction cosines in Eq. (1.55) become time dependent in the rotor period. Taking the time average under rapid rotation, Eq. (1.55) becomes

$$\sigma = \left[\tfrac{1}{2}\sin^2\beta(\sigma_{11} + \sigma_{22} + \sigma_{33}) + \tfrac{1}{2}(3\cos^2\beta - 1)\right]$$

$$\times (\text{functions of directions of cosines}). \quad (1.63)$$

The angle β is now the angle between the rotation axis and the magnetic field direction. When β is the magic angle, $\sin^2\beta$ is $2/3$ and the first term becomes equal to one-third of the trace of the tensor, i.e., the isotropic chemical shift [see Eq. (1.55)]. The $(3\cos^2\beta$

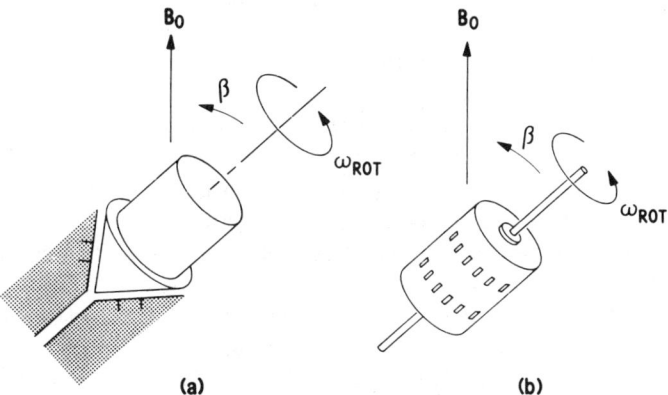

FIGURE 1.32 Two designs for magic angle sample spinners. (a) An Andrew design sample holder, rotating on air bearings within the stator (shaded), making an angle β to the magnetic field B_0. The high-pressure nitrogen jets are indicated by solid lines on the stator. (b) Lowe design rotor, supported by phosphor–bronze axles. High-pressure nitrogen impinges on the flutes.

− 1) term becomes zero, as we have seen. Thus, under magic angle rotation the chemical shift pattern collapses to the isotropic average.

To achieve minimum linewidth in magic angle spectra, it is necessary that the spinning axis be within 0.5° of the magic value in order to reduce the anisotropy pattern to 1% of its static value (61). At high magnetic field strengths, the anisotropy pattern may be as much as 10 kHz in width. Since this spinning rate is difficult to achieve, one must accept the presence of spinning sidebands located on each side of the isotropic chemical shift position by a distance equal to the spinning frequency. Figure 1.33 shows an example of how the position and intensity of sidebands depend on the spinning speed. The sidebands are not necessarily a nuisance and may in fact be put to good use since at relatively low spinning rates they can be used to trace out complex anisotropy patterns that are difficult to disentangle in the static spectra (62).

1.8.5.4 Cross Polarization

We have seen that a single 90° pulse, applied at the resonant frequency, produces a free induction decay signal. Our earlier discussion has implied that the resonant system of nuclei is a liquid, but this situation holds true for a solid as well. In solids, however, molecular motion is limited and this leads to very inefficient spin–lattice relaxation. For carbon-13 nuclei one may see this by inspection of Fig. 1.9; in solids we are on the right-hand branches of the curves, since correlation times will usually be well in excess of 10^{-6} s and may readily be much longer. Even in solids that contain methyl groups (methyl groups generally rotate rapidly), the T_1 values for the other carbons are prohibitively long. The carbon of the methyl group cannot communicate its short T_1 value to other carbons by spin "flip–flops," i.e., *spin diffusion*, as its nearest carbon-13 neighbor is on average 7 Å distant. Even though the protons on the methyl carbons do communicate their short T_1 to other protons in the sample via diffusion to nearby spins, these protons cannot communicate their short T_1 back to other carbons in the sample because the carbons and protons resonate at very different frequencies.

Cross polarization takes advantage of the fact that proton spin diffusion generally causes all of the protons in a solid to have the same T_1 value, and that the proton T_1 is usually short compared to the carbon T_1 values. (For the sake of concreteness and because it is the commonest case, we shall discuss cross polarization from protons to carbons; however, these concepts apply to polarization between any abundant spin, ^{19}F, for example, and a rare spin.) Cross polarization works by effectively forcing an overlap of proton and carbon energies

1.8. Experimental Methods

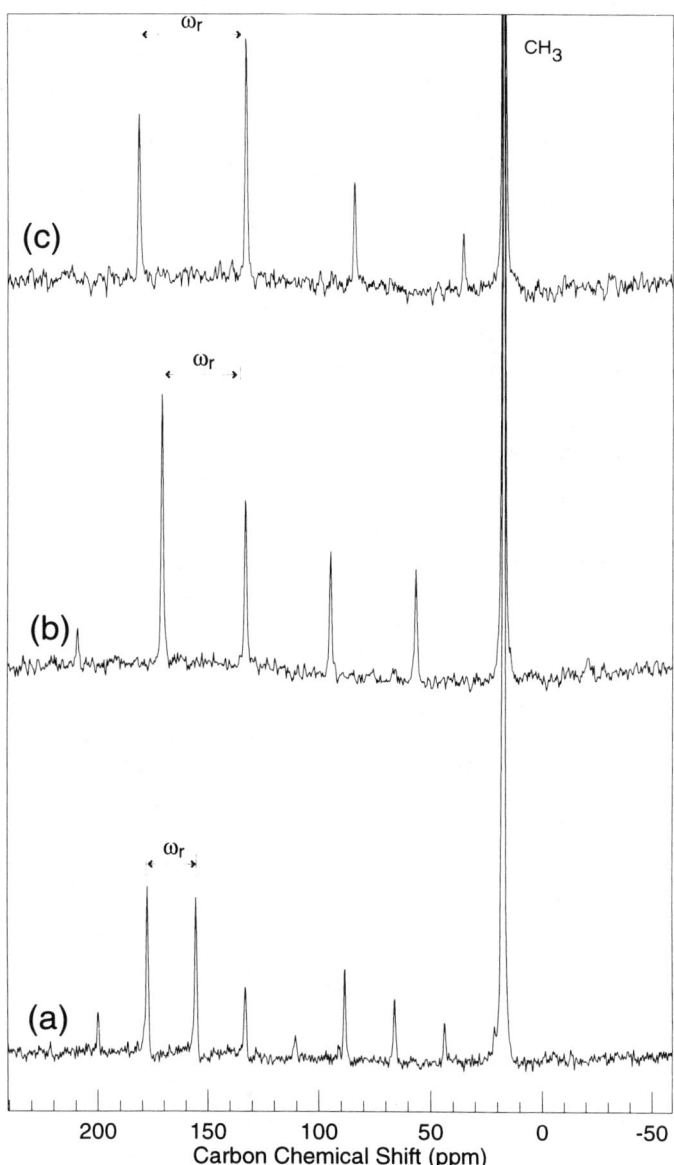

FIGURE 1.33 The effect of sample spinning speed on the sideband pattern for aromatic signals in hexamethyl benzene.

in the rotating frame despite the absence of such an overlap in the laboratory frame. The means of doing this was demonstrated by Hartmann and Hahn in 1962 (63). Energy transfer between nuclei with widely differing Larmor frequencies can be made to occur when

$$\gamma_C B_{1C} = \gamma_H B_{1H}. \qquad (1.64)$$

This equation expresses the *Hartmann–Hahn condition*. Since γ_H is four times γ_C, the Hartmann–Hahn match occurs when the strength of the applied carbon field B_{1C} is four times the strength of the applied proton field B_{1H}. When the proton and carbon rotating frame energy levels match, polarization is transferred from the abundant protons to the rare carbon-13 nuclei. Because polarization is being transferred from the protons to the carbons, it is the shorter T_1 of the protons that dictates the repetition rate for signal averaging.

In Figs. 1.34 and 1.35 we show the pulse sequences for ^{1}H and ^{13}C nuclei and the behavior of the magnetization vectors in the cross-polarization experiment, the goal of which is to create a time dependence of the spins that is common to the two rotating frame systems of ^{1}H and ^{13}C. We first apply a 90_x pulse to the protons [Fig. 1.34(a); Fig. 1.35, top] so that they are now along the y' axis. The phase of the

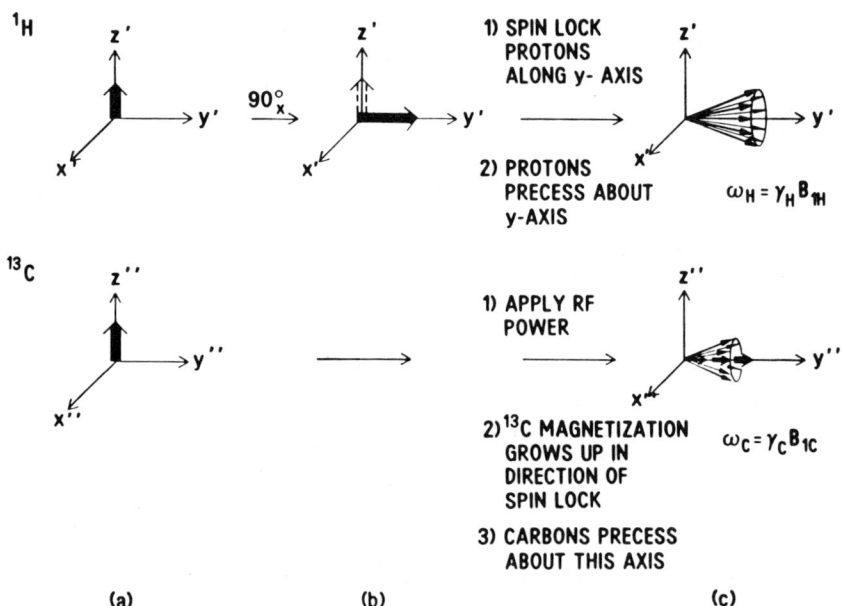

FIGURE 1.34 Vector diagram for a ^{1}H and ^{13}C double rotating frame cross-polarization experiment. The carbon and proton frames are rotating at different frequencies (see text).

1.8. Experimental Methods

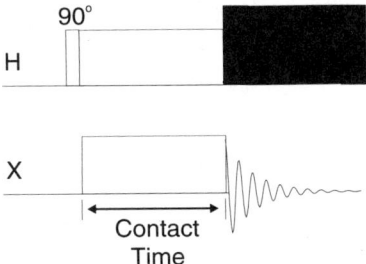

FIGURE 1.35 The ^{1}H–^{13}C cross-polarization pulse sequence for solid-state observations.

proton B_1 field is then shifted by 90° [Fig. 1.34(b)]; i.e., the RF signal is shifted by one-quarter of a wavelength. In this way, the protons are *spin locked* along the y' axis, and for the duration of the spin-locking pulse [Fig. 1.34(c); Fig. 1.35, top] they are forced to precess about the y' axis of their rotating frame with a frequency $\omega_H = \gamma H B_{1H}$. Meanwhile the carbons are put into contact with the protons. This is accomplished by turning the carbon field B_{1C} on during the spin-lock time (Fig. 1.35, bottom), causing the carbon magnetization to grow up in the direction of the spin-lock field [Fig. 1.34(c)]. The carbons are now precessing about their y'' axis with a frequency $\omega_C = \gamma_C B_{1C}$.

Figure 1.36 is intended to show in more detail the mechanism by which the polarization transfer occurs. The protons are now precessing about the B_{1H} field with frequency $\gamma_H B_{1H}$, and the carbons are simultaneously precessing with frequency B_{1C} about the B_{1C} field. Since the Hartmann–Hahn match has been established, these frequencies are equal, and therefore the z components of both the proton and the carbon magnetization must have the same time dependence, as Fig. 1.36 is meant to convey. Because of this common time dependence, mutual spin flip-flops can occur between protons and carbons. This process can be visualized as a flow of polarization from the abundant proton spins to the rare carbon spins. It can be shown that in these circumstances there is an enhancement of the rare spin signal intensity by as much as the ratio of magnetogyric ratios of the abundant and rare spins (64). In the ^{1}H–^{13}C case this factor is γ_H/γ_C or 4, greater than the maximum Overhauser enhancement (Section 1.8.6.4) of 3. This is in addition to the great advantage of not having to deal with long carbon-13 spin–lattice relaxation times in spectrum accumulation.

The signal intensity during cross polarization represents a compromise between several competing processes, including the build-up of intensity from carbon–proton cross polarization and the decay of the

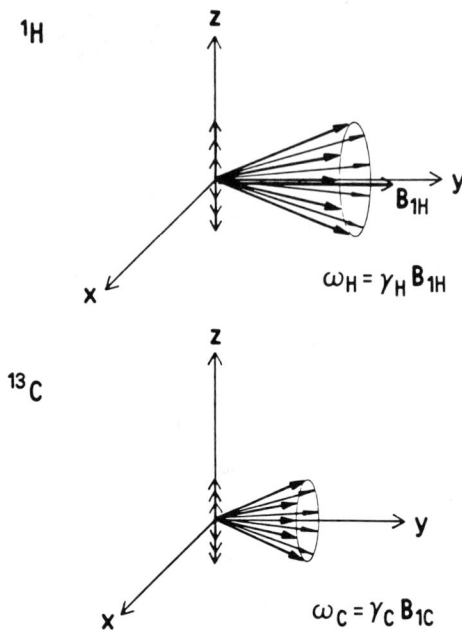

FIGURE 1.36 A more detailed representation of Fig. 1.34. The carbons are precessing with frequency $\omega_C = \gamma_C B_{1C}$ and the protons with frequency $\omega_H = \gamma_H B_{1H}$. When the Hartmann–Hahn match is achieved ($\gamma_C B_{1C} = \gamma_H B_{1H}$), and mutual spin flip-flops can occur.

proton and carbons signals under the spin-locking fields. The intensity at any given cross-polarization contact time t is given by

$$M(t) = M_0 \lambda^{-1}[1 - e^{-\lambda t/T_{CH}}]e^{-t/T_{1\rho}^H} \qquad (1.65)$$

where $\lambda = 1 + T_{CH}/T_{1\rho}^C - T_{CH}/T_{1\rho}^H$ and T_{CH} is the carbon–proton cross polarization time constant, and $T_{1\rho}^C$ and $T_{1\rho}^H$ are the carbon and proton rotating-frame spin–lattice relaxation time constants. In most cases $T_{1\rho}^H$ and $T_{1\rho}^H$ are much greater than T_{CH} so Eq. (1.65) simplifies to

$$M(t) = M_0[1 - e^{-t/T_{CH}}]e^{-t/T_{1\rho}^H}. \qquad (1.66)$$

1.8.6 The Observation of Nuclear Relaxation

1.8.6.1 Spin–Lattice Relaxation

We have seen (Section 1.4.1) that the most important and general mechanism for spin–lattice relaxation is *dipole–dipole* relaxation, and that this is particularly the case for polymers because of their relatively slow motions. In order to understand the quantitative

1.8. Experimental Methods

dependence of T_1 on molecular motion we must consider a system of two spin-1/2 nuclei I and S which differ in chemical shift but do not experience scalar coupling. We show such a system in Fig. 1.37. Our attention centers on the transition probabilities W_0, W_1, and W_2. W_0 is the probability of a simultaneous inversion of both I and S, often termed a "zero-quantum" transition or spin flip-flop; this is the basis of the phenomenon of *spin diffusion*, particularly important for protons when molecular motion is relatively slow, as in solids. W_I is the probability of a one-quantum transition that inverts spin I, while W_S is the corresponding probability for spin S. W_2 is the probability of simultaneous transitions of both spins in the same direction. Only the transitions represented by W_I and W_S are ordinarily directly observable as spectral lines.

Let us assume that we are observing spin I, the Z magnetization of which is M_I^Z. The Z magnetization of spin S is M_S^Z. The rate of change of the Z magnetization of spin I is given by

$$\frac{dM_I^Z}{dt} = -\rho(M_I^Z - M_I^0) - \sigma(M_S^Z - M_S^0), \tag{1.67}$$

as shown by Solomon (65). Here, M_I^0 and M_S^0 are the equilibrium magnetizations and

$$\rho = W_0 + 2W_1^I + W_2 \tag{1.68}$$

$$\sigma = W_0 - W_2, \tag{1.69}$$

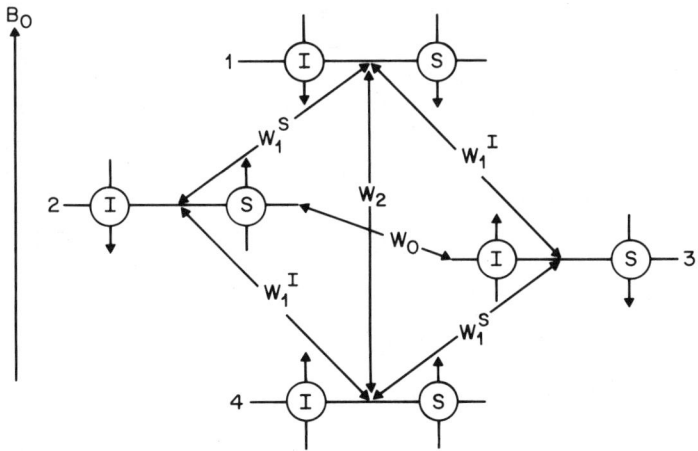

FIGURE 1.37 Energy levels of an AX spin system. Both nuclei have a spin I of 1/2 and no scalar coupling.

where the transition probabilities are those shown in Fig. 1.37. It may be shown (66) that these probabilities are related to the motional frequencies, expressed as spectral densities J_n, as

$$W_0 = \tfrac{1}{20}K^2 J_0(\omega_S - \omega_I) \tag{1.70}$$

$$W_1^I = \tfrac{3}{40}K^2 J_1(\omega_I) \tag{1.71}$$

$$W_1^S = \tfrac{3}{20}K^2 J_1(\omega_S) \tag{1.72}$$

$$W_2 = \tfrac{3}{10}K^2 J_2(\omega_S + \omega_I), \tag{1.73}$$

where ω_S and ω_I are the respective resonant frequencies of spins S and I, and K is given by

$$K = \frac{\gamma_I \gamma_S \hbar}{r_{IS}^3}. \tag{1.74}$$

Here, r_{IS} is the internuclear distance between I and S. The spectral density functions have the form

$$J_n(\omega) = \frac{2\tau_c}{1 + \omega^2 \tau_c^2}. \tag{1.75}$$

This spectral density function corresponds to the simple motional model discussed in Section 1.4.1, i.e., a sphere rotating by small random diffusive steps in a viscous continuum. We have seen that the correlation time τ_c describes the time required for the molecule to lose memory of its previous state of motion. In practice, it may be regarded as the time required to rotate approximately 1 rad.

We consider a homonuclear system for which $\omega_S \approx \omega_I = \omega$ but for which the chemical shift difference is still experimentally observable. Protons are the most common case. We may imagine experiments in which we excite individual spins in a *selective* manner (Fig. 1.38). In the present case, we suppose that we are inverting the magnetization only of spin I, leaving spin S undisturbed. We then find that at time zero

$$M_I^Z - M_I^0 = -2M_I^Z \tag{1.76}$$

and

$$M_S^Z - M_S^0 = 0, \tag{1.77}$$

and therefore from Eq. (1.67) and Eqs. (1.70)–(1.73) we find

$$\frac{dM_I^Z}{dt} = \frac{K^2}{20}(M_I^Z - M_I^0)[J_0(0) + 3J_1(\omega) + 6J_2(2\omega)]. \tag{1.78}$$

1.8. Experimental Methods

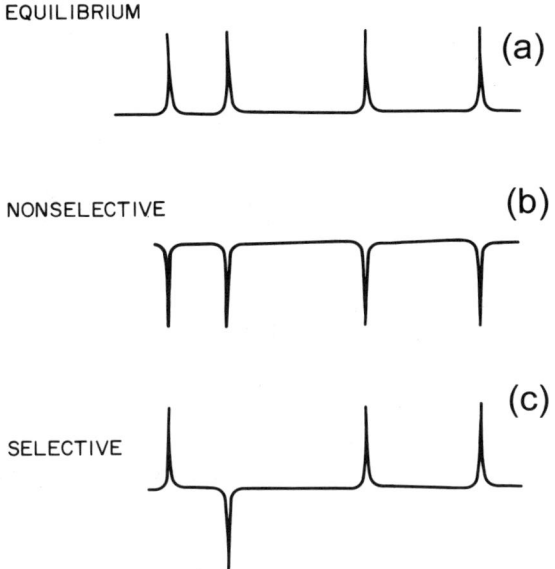

FIGURE 1.38 Resonance signals of a system of spins: (a) at equilibrium; (b) all spins inverted; (c) only one spin selectively inverted.

Therefore since

$$\frac{dM_I^Z}{dt} = \frac{1}{T_1(I)}(M_I^Z - M_I^0), \tag{1.79}$$

$$\frac{1}{T_1} = \frac{K^2}{20}[J_0(0) + 3J_1(\omega) + 6J_2(2\omega)]. \tag{1.80}$$

We note that in these circumstances the spin flip-flop process participates. Under extreme narrowing conditions

$$\frac{1}{T_1(I)} = \frac{\gamma^4 \hbar^2}{r_{IS}^6} \tau_c. \tag{1.81}$$

For *nonselective* excitation, we have again

$$M_I^Z - M_I^0 = -2M_I^0, \tag{1.82}$$

but also that

$$M_S^Z - M_S^0 = -2M_S^0. \tag{1.83}$$

For a homonuclear system M_I^0 and M_S^0 are very nearly equal and therefore we may write

$$\frac{dM_I^Z}{dt} = -(\rho + \sigma)[M_I^Z - M_I^0]$$

$$= \frac{K^2}{20}(M_I^Z - M_I^0)[3J_1(\omega) + 12J_2(2\omega)] \quad (1.84)$$

or

$$\frac{1}{T_1} = \frac{K^2}{20}[3J_1(\omega) + 12J_2(2\omega)]. \quad (1.85)$$

The flip-flop process does not contribute since both spins—or *all* spins in a multispin system—are pointing in the same direction, which from Eq. (1.77) leads to Eq. (1.81). In the extreme narrowing limit

$$\frac{1}{T_1} = \frac{3}{2}\frac{\gamma^4\hbar^2}{r_{IS}^6}\cdot\tau_c, \quad (1.86)$$

as may be seen in Section 1.4.1, Eq. (1.18).

We see that selective and nonselective relaxation rates [Eqs. (1.80) and (1.85)] depend on the internuclear distance r_{IS} in the same way, but depend on molecular motion in different ways. This has important consequences for the observation of molecular motion—particularly of larger molecules—through proton relaxation.

We find for *cross relaxation*—simultaneous transitions of both species—that

$$\frac{dM_I^Z}{dt} = \frac{K^2}{20}(M_I^Z - M_S^z)[6J_2(2\omega) - J_0(0)] \quad (1.87)$$

or

$$\frac{1}{T_1(I)} = \frac{\gamma^4\hbar^2}{20r_{IS}^6}[6J_2(2\omega) - J_0(0)]. \quad (1.88)$$

Cross relaxation dominates proton relaxation under conditions of relatively slow motion. In solids, the zero-quantum term entirely dominates and this mechanism called *spin diffusion*.

The measurement of the relaxation of carbon-13 nuclei in natural abundance (1.1%) has received extensive application for the study of molecular dynamics. There are several reasons for this, including:

a. Carbon-13 resonances are generally well resolved in the chemical shift so that separate T_1 measurements are feasible

1.8. Experimental Methods

  b. Carbon-13 nuclei are spatially isolated so that mutual dipole–dipole influence is negligible. Only nearby protons are effective for relaxation and for other than quaternary carbons only directly bonded protons need be considered, thus simplifying the r_{IS} term in the relaxation equations
  c. Carbon atoms are generally not at the periphery of the molecule and so only intramolecular influences need be considered

For reasons we have already discussed (Section 1.8.4), it is customary to decouple all protons by broadband irradiation when observing carbon-13 spectra. This is the practice also in T_1 measurements and in this case has the further advantage of simplifying the relaxation equation. Let us suppose that spin I is ^{13}C and spin S is a proton. The proton irradiation causes M_s to become zero, and it can be shown (34) that under these conditions, in effect

$$\frac{1}{T_1(C)} = -\rho_{CH} \tag{1.89}$$

$$= \frac{K^2}{20}[J_0(0) + 3J_1(\omega_c) + 6J_2(\omega_H + \omega_C)], \tag{1.90}$$

where J_n is of the form shown in Eq. (1.70) and $K = \gamma_H \gamma_C \hbar / r_{IS}^3$. A plot of Eq. (1.89) is shown as a function of τ_c in Fig. 1.9 for two values of the magnetic field corresponding to carbon resonant frequencies $v_0(= \omega_0/2\pi)$ of 50.3 and 125 MHz. It shows the relaxation of a carbon-13 bonded directly to a single proton with r_{CH} taken as 1.09 Å.

Under extreme narrowing conditions (where $\tau_c \ll \omega_c^{-1}$) for N_H directly bonded protons,

$$\frac{1}{T_1(C)} = \frac{N_H \gamma_H^2 \gamma_C^2 \hbar^2}{r_{CH}^6} \cdot \tau_c$$

$$= 2.11 \times 10^{10} \cdot \tau_c \cdot N_H \tag{1.91}$$

or

$$\tau_c = \frac{4.7 \times 10^{-11}}{N_H \cdot T_1(C)}. \tag{1.92}$$

Equation (1.92) provides a ready estimate of the correlation times for small molecules in mobile solvents, but it is in general best to employ the full equation.

When the proton is irradiated there is an enhancement in the value of the carbon-13 magnetization due to the nuclear Overhauser effect (Section 1.8.6.4).

Although dipole–dipole interactions generally dominate spin-lattice relaxation in polymers, there is another potent source of relaxation which must be taken into consideration when quadrupolar nuclei such as ^{14}N and ^{2}H are present; we have already discussed this briefly in Section 1.4.4.

The most commonly used method for the measurement of spin-lattice relaxation is *inversion recovery*, and the pulse sequence and vector diagrams for this experiment are shown in Fig. 1.39. Most modern spectrometers come equipped with this pulse sequence and the software for fitting the data to obtain the relaxation times. The Z magnetization M^Z [Fig. 1.39(a)] is inverted by a 180° pulse (b), and the magnetization begins to grow back through zero toward its equilibrium value M^0, as shown at (c). To sample this regrowth, the magnetization vector is turned back along the y' axis (d) at time τ, where it is observed and recorded (after transformation) as $M^Z(t)$. Following the acquisition of the signal at time τ, a delay τ_D is allowed. This interval should be long enough compared to T_1 to permit attainment of the equilibrium value before the application of the next 180° pulse; the appropriate interval may be judged from the value of the quantity $(-e^{-\tau/T_1})$, which is 0.993 if $\tau_D = 5T_1$, 0.982 if $\tau_D = 4T_1$, and 0.950 if $\tau_D = 3T_1$. This sequence is repeated for six to eight values of τ, based on an estimated value of T_1. The longest value of τ must be such as to provide a value of the equilibrium magnetization, M_0. This may be judged by the same criterion as for

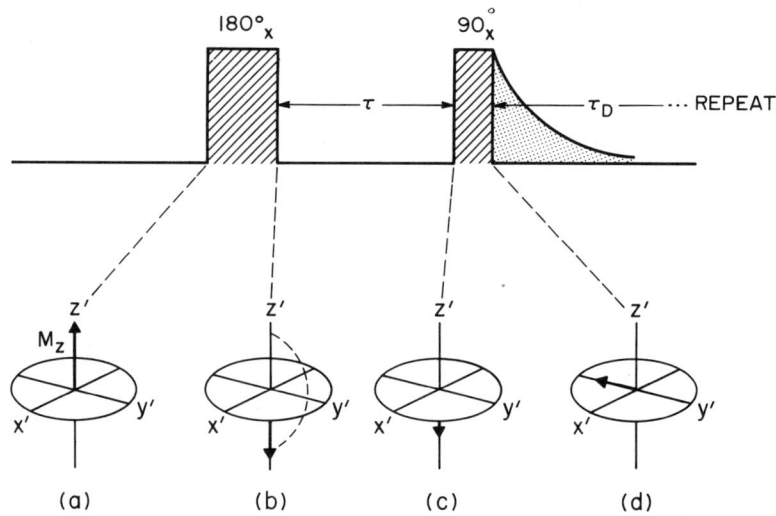

FIGURE 1.39 Inversion-recovery pulse sequence for measurement of T_1. (Pulse widths are greatly exaggerated in comparison to the pulse intervals.)

1.8. Experimental Methods

τ_D. The recovery of the Z magnetization is given by

$$\frac{dM^Z}{dt} = \frac{M^0 - M^Z}{T_1}. \tag{1.93}$$

Replacing M^Z by $M^Z(t)$ and integrating we have

$$M^0 - M^Z = Ae^{-\tau/T_1}. \tag{1.94}$$

The constant A depends on the initial conditions. For an inversion-recovery experiment $M^Z(t)$ at $t = 0$ equals $-M^0$ and so $A = 2M^0$. To determine T_1, one thus plots $M^Z(t)$ versus t according to the relationship

$$\ln[M^0 - M^Z(t)] = \ln(2) + \ln(M^0) - \tau/T_1. \tag{1.95}$$

T_1 is given by the reciprocal of the slope of this plot. Alternatively, the T_1 can be obtained from a nonlinear fit to Eq. (1.94).

The inversion-recovery method can require many hours of spectrum accumulation because of the long delays needed to achieve spin equilibrium. The method of *progressive saturation*, although less accurate, is faster. It employs a series of 90_x pulses to achieve a dynamic steady state between RF absorption and spin–lattice relaxation. The magnitude of the steady-state signal (early signals are discarded) depends on the interval t between pulses according to

$$M^0 - M^Z(t) = M^0 e^{-t/T_1}, \tag{1.96}$$

and therefore a plot of $\ln[M^0 - M^Z(t)]$ versus t yields T_1 as the reciprocal of the slope in much the same manner as for the inversion-recovery method.

It is important for the successful application of this method that no y' magnetization remain prior to each pulse, since this would be rotated to the $-z'$ axis and invalidate Eq. (1.95). One must not, in other words, allow this to be a spin echo experiment (Section 1.8.6.3) but rather must confine it to longitudinal relaxation. This may be ensured by various means that we shall not detail here. The method is well adapted to carbon-13 relaxation measurement since in this case broadband proton irradiation ensures rapid ^{13}C spin–spin relaxation.

For the measurement of T_1 of carbon-13 (or other dilute spins) in solid polymers, where spin–lattice relaxation may be very slow, the method of choice is that of Torchia (67). It involves proton cross polarization (Section 1.8.5.4), with a resulting four-fold enhancement of the carbon resonances, and is terms the CPT_1 experiment. The waiting time between pulse cycles is dependent not on the ^{13}C T_1 but on the much shorter proton T_1.

The pulse sequence for CPT_1 begins with a 90° proton pulse followed by cross polarization, as shown in Fig. 1.40. After the desired ^{13}C polarization is established, the ^{13}C magnetization is rotated to the $-z$ axis with a phase shifted 90° carbon pulse. During the following relaxation delay the four-fold enhance ^{13}C magnetization decays exponentially toward its equilibrium with the time constant T_1, and the resulting magnetization is sampled with the final carbon pulse. On alternate scans the phase of the first proton pulse is reversed and the ^{13}C magnetization is rotated to the $+z$ axis after the carbon pulse following the cross polarization, and the magnetization is recorded in the same manner. This signal is subtracted from the previous one and the T_1 is obtained from the exponential decay in signal intensity.

The advantage of this procedure is that it suppresses two kinds of artifacts which may occur in proton-enhanced spectra: (a) signals from carbons whose polarization arises from normal spin–lattice relaxation rather than cross polarization, and (b) voltage transients generated by the pulsed ^{13}C RF field—for example, magnetoacoustic ringing of the receiver coil. These contribute equally in both sequences and thus are eliminated in the different spectrum. Examples of the use of the CPT_1 procedure will be given in subsequence chapters.

The time constant T_{CH} for cross polarization is also of interest in the study of the molecular dynamics of polymers. This can sometimes be measured by fitting the cross-polarization build-up to Eq. (1.66), but least-squares fits to a difference of exponentials may not always yield a unique result if the T_{CH} and $T_{1\rho}^H$ time constants are not sufficiently different from each other. A better way to measure T_{CH} is shown in Fig. 1.41 (68). This experiment begins like cross polarization, but after a short time the phase of the carbon spin-locking pulse is inverted and the signal intensity changes as the carbon magnetization is cross polarized along the $-y'$ axis instead of the y' axis in the

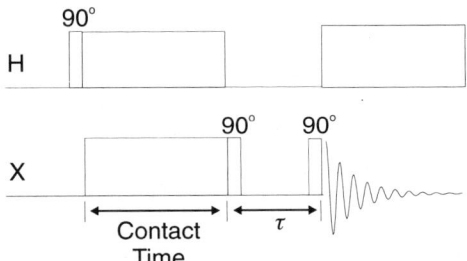

FIGURE 1.40 Pulse sequence for measuring T_1^C using cross polarization and inversion recovery.

1.8. Experimental Methods

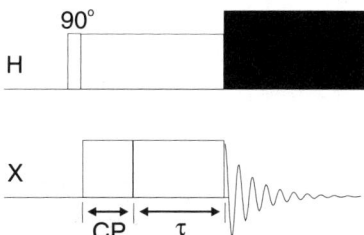

FIGURE 1.41 The pulse sequence for inversion-recovery cross polarization. After a constant contact time the phase of the carbon spin-locking pulse is reversed for a variable period of time τ and the signal is recorded with magic angle spinning and dipolar decoupling.

normal cross polarization. The time constant can be obtained from an exponential fit to the change in signal intensity.

The proton spin–lattice relaxation time can be measured by inserting a 180° pulses and relaxation delay time before the cross-polarization pulse sequences, as shown in Fig. 1.42.

1.8.6.2 Spin–Lattice Relaxation in the Rotating Frame

A related but quite distinct experiment enables one to measure the lifetime of the carbon and proton magnetization in its spin-locking field, a quantity characterized by $T_{1\rho}$, the spin–lattice relaxation time in the rotating field. We will outline the procedure here and discuss its use and value in subsequent chapters. As with T_1 measured by the CPT_1 procedure, it is primarily a technique intended for polymers in the solid state.

Figure 1.43 shows the pulse sequence diagrams for the measurement of the proton and carbon $T_{1\rho}$'s. There are two very closely

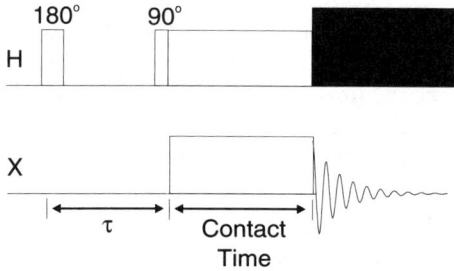

FIGURE 1.42 Pulse sequence for measuring T_1^H using cross polarization and inversion recovery.

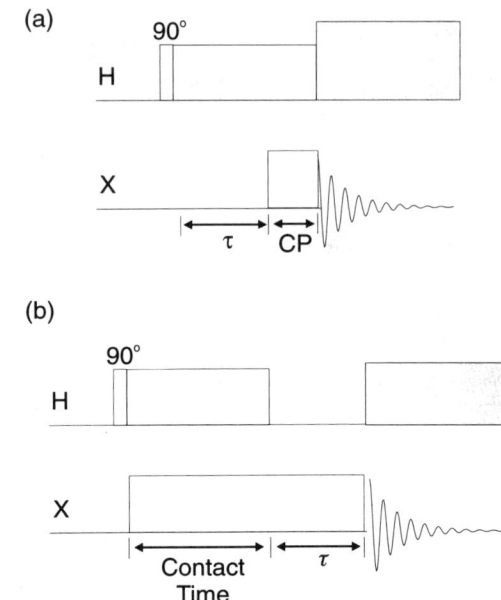

FIGURE 1.43 The pulse sequences for measuring (a) $T_{1\rho}^H$ and (b) $T_{1\rho}^C$.

related pulse sequences that can be used to measure the proton $T_{1\rho}$'s. One approach is to measure carbon intensity as a function of spin-locking time using the normal cross-polarization pulse sequence (Fig. 1.35). The magnetization initially builds up to a maximum due to the carbon–proton dipolar interactions, and the signal decays exponentially to equilibrium with the time constant $T_{1\rho}^H$. Thus the $T_{1\rho}^H$ can be obtained from a semilog plot of the intensity as a function of spin-locking time after the signals have reached their maximum intensity. An alternative method is shown in Fig. 1.43 (a) (64). This pulse sequence differs from the normal cross polarization in that after the 90° proton pulse the proton spin-locking field is turned on, but the carbon field is not. During this variable period τ the proton magnetization decays with the time constant $T_{1\rho}^H$. This decay is monitored by turning on the carbon spin-locking field to initiate cross polarization followed by signal acquisition, and the $T_{1\rho}^H$ is obtained from a semilog plot of the signal decay. The values of $T_{1\rho}^H$ are used both to measure the molecular dynamics of polymers in the solid state and to measure the length scale of polymer mixing.

The carbon $T_{1\rho}$'s are obtained in an analogous manner, as shown in the pulse sequence diagram in Fig. 1.43 (b). This pulse sequence begins as the normal cross polarization with the 90° proton pulse and the spin-locking of the carbons and protons under the Hartmann–Hann condition. After this cross-polarization period the proton spin-locking field is turned off and the magnetization decays exponentially with the time constant $T_{1\rho}^C$. The values of $T_{1\rho}^C$ are useful for measuring the molecular dynamics of polymers (69), although the interpretation of the results is often not as clear as for other relaxation parameters.

1.8.6.3 Spin–Spin Relaxation

We have seen in Section 1.4.2 that by definition T_2 is a measure of the linewidth in solids and viscous liquids:

$$T_2 = \frac{1}{\pi \delta v}. \qquad (1.26b)$$

Another contribution to line broadening is inhomogeneity δB_0 in the laboratory magnetic field. This may be minimized by careful shimming of the field (see Section 1.8.2 and references therein). It is usually included in the experimental values of the spin–spin relaxation time, commonly termed T_2^*:

$$\frac{1}{T_2^*} = \frac{1}{T_2} + \frac{\gamma \delta B_0}{2}. \qquad (1.97)$$

The "true" T_2 may be measured by the method of *spin echoes*, which was first proposed and named by Hahn (70), and is illustrated in Fig. 1.44. At time zero (a), a 90° pulse is applied along the positive x' axis of the rotating frame, causing the magnetization M to precess into the positive y' axis. The magnetization is the vector sum of individual spin vectors arising from nuclei in different parts of the sample and therefore experiencing slightly different values of the magnetic field B_0, which is never perfectly homogeneous. The individual vectors or *isochromats* (each of which represent many spins) will begin to fan out since some will be precessing slightly faster and some slightly slower than the RF field frequency v_0; this is shown at (b). After time τ—usually a few milliseconds—we apply a 180° pulse. This rotates the spin vectors into the $-y'$ axis as shown at (c). Since they continue to precess in the same sense, they will now be rotating together rather than fanning out (d). After a second interval τ they will "refocus" as shown at (e). In this context, refocusing means that all vectors simultaneously attain the same negative (because they are along the $-y'$ axis) emission phase signal form corresponding to an inverted spin population. Beyond the signal buildup at 2τ, the vectors again dephase (f) and become virtually unobservable. It should be

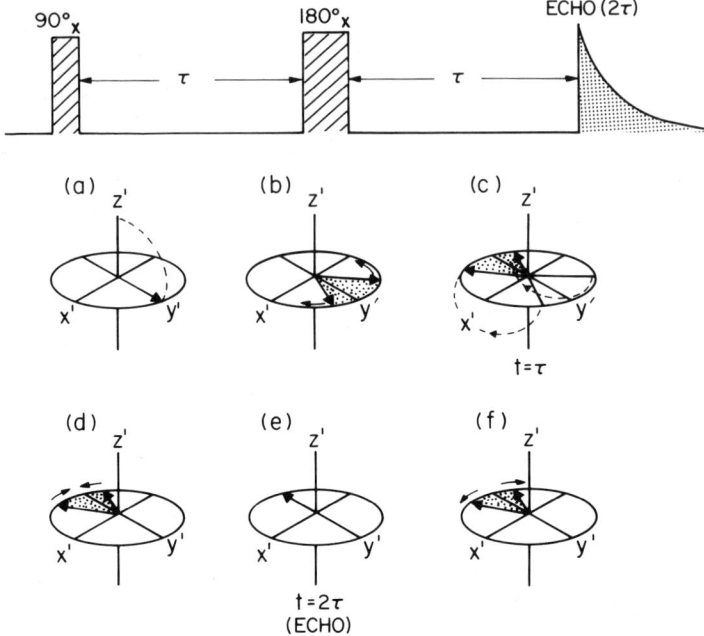

FIGURE 1.44 The spin echo experiment. See text for details.

remarked that refocusing of the vectors to an apparently "single" vector at (e) does not imply any narrowing of the observed resonance over that which would be normally observed. Upon detecting the signal, the varied isochromats contained within this apparently single vector will again become evident as signal broadening just as would have been the case if detection had followed the 90_x pulse. The signal at 2τ is reduced somewhat in intensity, however, because the effects of the true spin–spin relaxation processes are not refocused in the spin echo experiment.

In the original Hahn experiment, it is necessary to repeat the 90_x—τ—180_x sequence several times with varied values of τ and then plot the decline of the echo magnitude as a function of τ to obtain T_2. This is somewhat troublesome, and it may also be shown that the echo will decay not only as a function of "true" T_2 but also as a function of diffusion of the nuclei between portions of the sample experiencing slightly different magnetic fields. The effect of diffusion is more serious as τ is extended. Carr and Purcell (71) showed that the Hahn experiment may be modified to increase its convenience and at the same time greatly reduce the effect of diffusion. This method

1.8. Experimental Methods

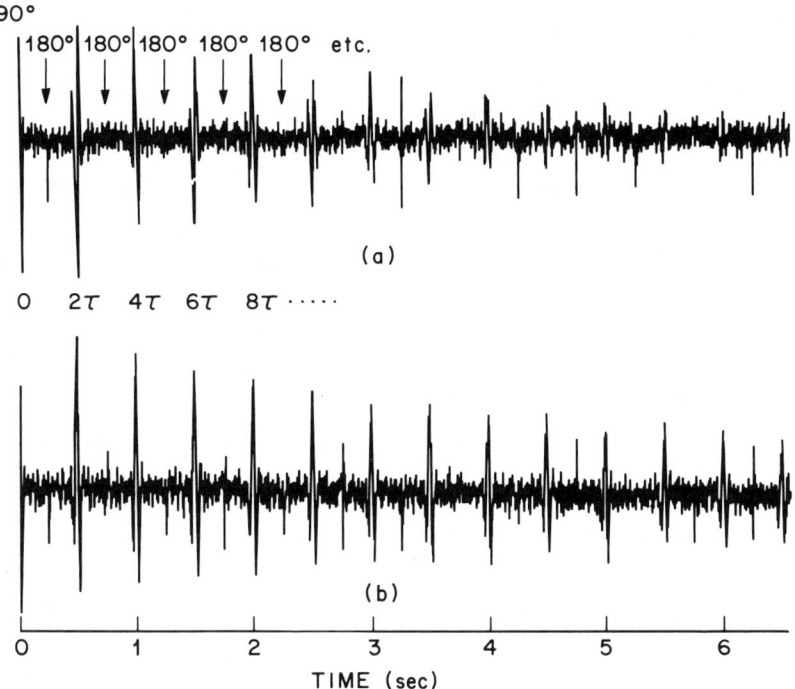

FIGURE 1.45 (a) Carr–Purcell echoes for 60% enriched CH_3 $^{13}COOH$. The sign of the echoes alternated. The pulse sequence begins with a 90° pulse, and 180° pulses are applied as indicated. (b) Echoes from the Meiboom–Gill modification of the Carr–Purcell experiment. Adapted with permission (16).

employs the sequence

$$90_x - \tau - 180_x - 2\tau - 180_x - 2\tau - 180_x - 2\tau \cdots ,$$

where the 180_x pulses (after the first) are applied at 3τ, 5τ, 7τ, etc., giving rise to a train of echoes at $2\tau, 4\tau, 6\tau, 8\tau \ldots$, as may be seen from Fig. 1.45(a)(16). The effect of diffusion is proportional to τ rather than to the accumulated interval and so may be made as small as desired by making τ short. It may also be seen from Fig. 1.45(a) that the echoes will alternate in sign.

A further improvement in the Hahn echo sequence is that made by Meiboom and Gill (72). This is similar to the Carr–Purcell sequence except that the 180° pulses are applied along the *positive y'* axis. As shown in Fig. 1.45(b) this results in positive echoes. It may be further shown [see Martin *et al.* (73) for a good discussion] that there is also an automatic correction for imperfections in the effective flip angle of the 180^0_{+y} pulses. It will be seen that in the latter the echoes persist longer, giving a more correct and accurate measure of T_2.

1.8.6.4 Nuclear Overhauser Effect

This effect has many ramifications but, as we shall see, becomes less important in the solid state. As in the treatment of carbon-13 relaxation (Section 1.8.6.1), let us take spin I as ^{13}C and spin S as ^{1}H. We irradiate the protons so that spin S is saturated and $M_s^t = 0$. Then under steady-state conditions we can show from Eqs. (1.68) and (1.69) that

$$\frac{M_I(S \text{ irrad.})}{M_I^0} = 1 + \frac{\sigma}{\rho} \cdot \frac{M_S^0}{M_I^0}, \qquad (1.98)$$

or, since equilibrium spin populations and magnetizations are proportional to magnetogyric ratios, we may write

$$\frac{M_I(S \text{ irrad.})}{M_I^0} = 1 + \frac{\sigma}{\rho} \cdot \frac{\gamma_S}{\gamma_I}. \qquad (1.99)$$

For carbon-13 enhancement under proton irradiation

$$\frac{M_C(^1\text{H irrad.})}{M_C^0} = 1 + \left[\frac{6J_2(\omega_C + \omega_H) - J_0(\omega_H - \omega_C)}{J_0(\omega_H - \omega_C) + J_1(\omega_C) + J_2(\omega_H + \omega_C)} \right] \cdot \frac{\gamma_H}{\gamma_C} \qquad (1.100)$$

$$\frac{M_C(^1\text{H irrad.})}{M_C^0} = 1 + \eta_{CH}, \qquad (1.101)$$

where η_{CH} is termed the ^{13}C-{^{1}H} NOE enhancement factor. The ratio γ_H/γ_C equals 3.977. Under extreme motional narrowing, Eq. (1.96) becomes

$$\frac{M_C(^1\text{H irrad.})}{M_C^0} = 1 + 0.5 \cdot \frac{\gamma_H}{\gamma_C} = 2.988 \qquad (1.102)$$

or

$$\eta = 1.988. \qquad (1.103)$$

The nuclear Overhauser effect depends upon dipole–dipole relaxation. Other competing relaxation pathways detract from it, and since such pathways are common, especially in smaller molecules, the carbon–proton NOE is often less than maximal even under fast-motion conditions. It may also be less than maximal when the simple isotropic motion contemplated in these developments does not hold. Isotropic motion is particularly unlikely to characterize polymer molecules, especially rigid ones. The NOE will also be less than maximal if the motional narrowing limit does not apply. In Fig. 1.46

1.8. Experimental Methods

are shown plots of Eq. (1.96) for the ^{13}C-{^{1}H} Overhauser enhancement at carbon-13 frequencies of 50.3 and 125 MHz. Corresponding plots of T_1 for a carbon with a single-bonded proton are shown in Fig. 1.9. The inflection points of the NOE occur at the minima of the T_1 plots.

The *proton–proton* Overhauser effect was the first to be used for making resonance assignments and has much potential for dynamic and structural measurements. Since the magnetogyric ratio term is now unity, Eq. (1.100) may be rewritten as

$$\frac{M_H(^1\text{H irrad.})}{M_H^0} = 1 + \left[\frac{6J_2(2\omega_H) - J_0(0)}{J_0(0) + 3J_1(\omega_H) + 6J_2(2\omega_H)}\right]. \quad (1.104)$$

The proton–proton NOE can be very effectively employed in the 2D mode (74); further discussion will be postponed until Section 1.8.7.

The simplest experimental procedure for the measurement of ^{13}C-{^{1}H} nuclear Overhauser enhancements might appear to consist of obtaining the integrated carbon-13 spectra with and without broad-band proton irradiation. In practice, this is found to be undesirable because of the overlapping multiplets of the coupled spectrum, which make it difficult to measure enhancement ratios for individual carbon resonances. A better method is to gate the proton decoupler on only during data acquisition. The multiplets collapse instantaneously, but the NOE requires a time on the order of T_1 to build up. Figure 1.47(a)

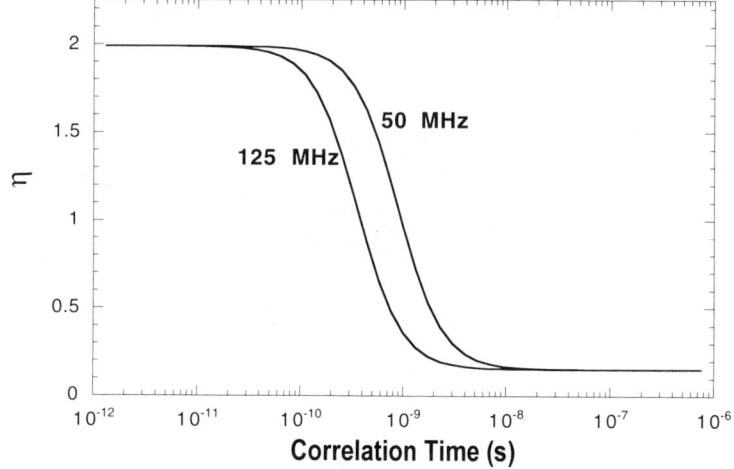

FIGURE 1.46 Nuclear Overhauser enhancement factor η for a ^{13}C nucleus produced by irradiation of neighboring protons as a function of the correlation time τ_c, calculated at carbon frequencies of 125 and 50.3 MHz.

shows the pulse sequence employed. If it is desired to retain both the coupling and the NOE, this may be done by inverse gating of the proton decoupler, i.e., gating it on *except* during data acquisition. In this way carbon–proton couplings may be measured with greater effective sensitivity [Fig. 1.47(c)].

1.8.7 Two-Dimensional NMR

1.8.7.1 Introduction

One of the fundamental limitations in NMR studies is the overlap of spectral lines. This is particularly troublesome in the spectra of both synthetic and biopolymers because of the presence of similar repeating units that may differ only in stereochemistry, side chain, etc. One way to overcome these limitations is to perform NMR

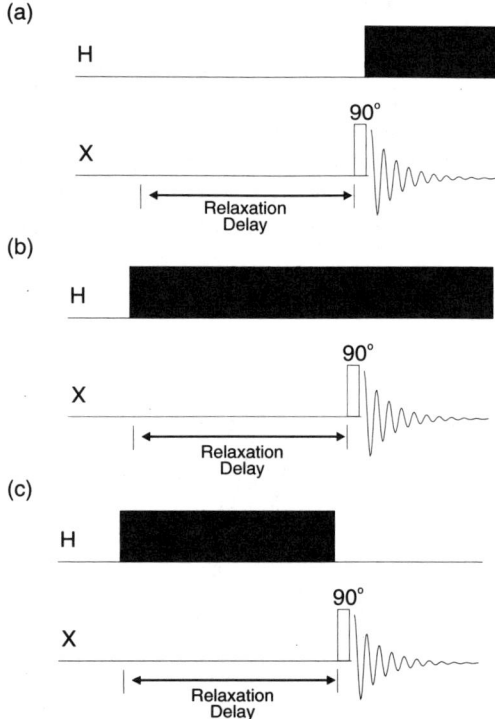

FIGURE 1.47 Gated decoupling pulse sequences for observation of ^{13}C nuclei. (a) The pulse sequence for eliminating the NOE while collapsing the multiplicity arising from carbon–proton J coupling. (b) The pulse sequence with NOE and collapse of the multiplet structure, and (c) the pulse sequence for NOE-enhanced spectra with the carbon–proton multiplicities.

1.8. Experimental Methods

experiments that have more than one frequency dimension. Two-dimensional NMR spectra have two frequency coordinates and exhibit greatly improved resolution. These experiments utilize principles discussed previously in this chapter—i.e., through-bond J coupling and dipolar coupling, etc.—and their primary advantage is that they allow us to study resonances that are partially or completely overlapped in the normal spectrum. There are hundreds of 2D NMR experiments that can be categorized as *correlated* or *resolved* 2D NMR experiments. In *correlated* experiments the resonance frequency of one peak is related to those of its neighbors and thus molecular connectivities or distances between atoms can be determined. In *resolved* experiments the frequency axes show two different interactions; for example, in J-resolved spectroscopy the chemical shift appears along one axis and the J coupling appears along the other. Two-dimensional NMR has been used mostly for solution studies, but many of the same principles can be applied to the study of solids. Extension of the same principles leads to 3D NMR experiments in which there are three independent frequency axes. This has the potential of providing still greater resolution for those cases where the 2D spectrum exhibits overlaps.

Two-dimensional NMR experiments consist of a series of pulses and delays as in other multipulse NMR experiments. One of the differences between 1D and 2D NMR experiments is that we allow the spin system to *evolve* instead of immediately transforming the free induction decay following the pulse. Two-dimensional NMR experiments can be schematically divided into four time periods.

During the *preparation* period, the spins are allowed to come to equilibrium; i.e., the populations of the Zeeman levels are allowed to equilibrate with their surroundings. This interval allows the establishment of reproducible starting conditions for the remainder of the experiment. During the *evolution period* t_1, the x, y, and z components evolve under all the forces acting on the nuclei, including the interactions between them. The interactions may occur through indirect nuclear–nuclear effects (J coupling) or by direct through-space dipole–dipole interactions, as we have seen. One of the things that commonly happens during the t_1 period is *frequency labeling*, where the spins are labeled (i.e., prepared in a nonequilibrium state) by their frequency in the NMR spectrum. The mixing period may consist of either pulses or delays and results in the transfer of magnetization between spins that have been frequency labeled in the t_1 period. The final period, t_2, is the signal *acquisition* common to all pulsed NMR experiments. The second frequency is introduced to the 2D NMR experiment by systematically incrementing the delay between pulses, t_1, as shown in Fig. 1.48. Data collection in a 2D experiment consists

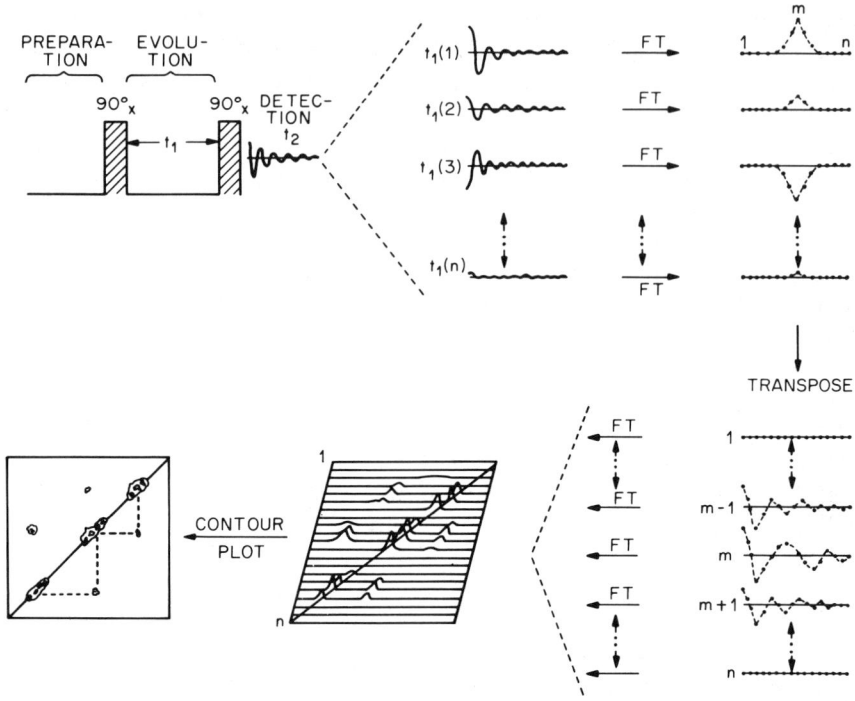

FIGURE 1.48 Schematic representation of the formation of a 2D NMR spectrum. The pulse sequence shown is actually appropriate for the development of a COSY spectrum.

of gathering many free induction decays, each corresponding to a different value of t_1. In a typical experiment, 512 or 1024 FIDs may be acquired for t_1 values incremented between 0.5 and 500 ms. If 1K data points are collected in the t_2 period, the entire 2D data set would be 1K × 1K (because of the large size of the data matrix, high-speed computers will play an increasingly important role in 2D and 3D NMR). After data collection the FIDs are transformed with respect to t_2 to obtain a set of spectra in which the peak intensities or phases are modulated as a function of the t_1 delay. Fourier transformation with respect to t_1 converts the frequency modulation into peaks in the 2D spectrum, which is actually a surface in three-dimensional space and may be represented either as a stacked plot or as a contour plot, as shown. The stacked plot conveys relative peak intensities more vividly than the contour plot. However, it is very time consuming to record and does not clearly show complex relationships. The contour plot is much preferred for most purposes.

In the following discussion, we will describe the most useful forms of 2D NMR, using as illustrations molecules that for the most part are not of highly complex structure, since we believe these relatively simple systems are more effective in showing the principles involved.

Two-dimensional NMR was first proposed in 1971 by Jeener (74), but this proposal was not published. The basic experiment was later described by Aue et al. (75) in terms of density matrix theory. Even before this, in 1975, Ernst (76) described two-dimensional spectra of trichloroethane and a mixture of dioxane and tetrachloroethane obtained by Jeener's technique, but this work also was not published. In 1975, Ernst (76) reviewed the various forms which two-dimensional NMR might take and Hester et al. (77) and Kumar et al. (78) described 2D experiments on solids. The field was properly launched the next year, when several laboratories reported a variety of experiments. The early history has been reviewed by Freeman and Morris (79). General reviews of the technique have been provided by Benn and Gunther (80), Bax and Lerner (81), Bax (82), Kessler et al. (83), and Ernst and co-workers (84).

1.8.7.2 Two-Dimensional Resolved Spectroscopy

Separation of various types of interactions along independent frequency axes is one of the most important uses of 2D NMR experiments. The results are used to aid spectral assignments as well as for the study of polymer chain conformation and dynamics. In all resolved experiments the normal 1D spectrum appears along one of the frequency axes. In solution the most commonly employed type of resolved 2D experiment is 2D *J-resolved* spectroscopy, where the spin–spin coupling of each resonance appears along the second frequency axis. This allows the determination of coupling patterns that can in turn be used for assignment purposes and also for measurement of the magnitude of the J coupling. The latter can be related to the average conformation. Polymer dynamics in solids have been effectively explored with *magic angle spinning* 2D experiments in which the high-resolution spectrum appears along one axis and either the dipolar coupling or the chemical shift anisotropy (CSA) pattern appears along the second dimension. Molecular motion that occurs at frequencies faster than the breadth of the CSA or dipolar patterns leads to averaging that can be interpreted with detailed models of molecular motion. As an introduction to 2D resolved experiments we will discuss 2D J-resolved spectroscopy.

It has long been known that the echo train in a spin echo experiment (Section 1.8.6.3) is modulated by the J coupling when a system of coupled spins is observed. In Fig. 1.49 is shown a series of proton spectra of 1,1,2-trichloroethane, CH_2CHCl_2, in which $J_{AX} = 6.0$ Hz (85). The spectra were obtained by Fourier transformation of the

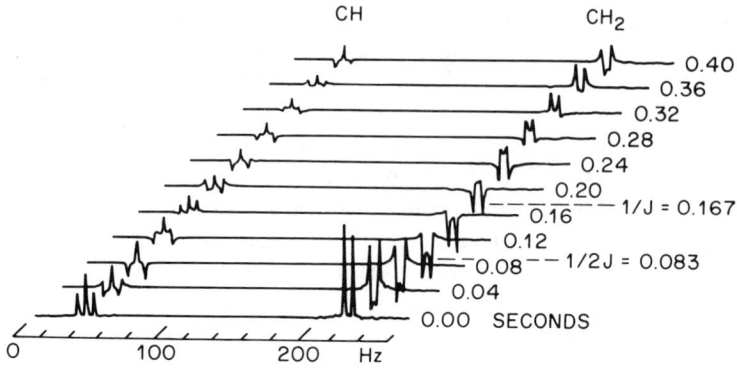

FIGURE 1.49 The 100-MHz proton spectra of 1,1,2-trichloroethane for which $^3J = 6$ Hz obtained with the Carr–Purcell sequence in which $2\tau = 0.02$ s. Adapted with permission (85).

second half of each of the echoes in a Carr–Purcell echo train in which $2\tau = 0.020$ s. It will be observed that the triplet of the CH protons passes through a series of phases as a function of time, expressed in units of $1/J$; the triplet center peak is unaffected and the triplet outer peaks pass through their phases twice as fast as those of the doublet. This J dependence is troublesome with respect to the measurement of T_2, but can be very valuable for the observation of J. Even very weak couplings that are too small to appear as observable splittings in the 1D spectrum can be measured if the system is allowed to evolve long enough. The time scale in this spectrum may be regarded as corresponding to the evolution time t_1 (Fig. 1.48). A second Fourier transform of the intensity oscillations along this scale in Fig. 1.49 will yield a doublet and triplet with spacings of 6.0 Hz, just as we see along the lower axis at time zero. The difference is that now the chemical shift difference between A and X does not appear in this dimension. We have thus achieved a partial separation of chemical shift and J coupling, but we must

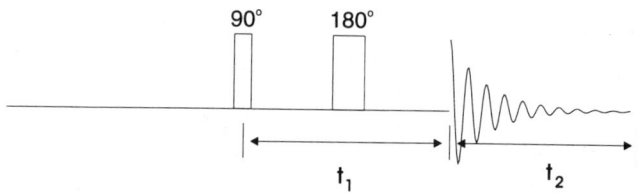

FIGURE 1.50 The pulse sequence for the homonuclear 2D J-resolved spectroscopy. The value of t_1 is systematically incremented throughout the experiment.

1.8. Experimental Methods

TABLE 1.8
The Coordinates of Peaks in the 2D J-Resolved Spectrum

Multiplicity	F_2 axis	J_{axis}
Doublet	$v \pm J/2$	$\pm J/2$
Triplet	$v \pm J$	$\pm J$
	v	0
Quartet	$v \pm 3J/2$	$\pm 3J/2$
	$b \pm J/2$	$\pm J/2$

consider further the means of generating a truly J-resolved homonuclear 2D spectrum.

The pulse sequence employed to do this is shown in Fig. 1.50, and the coordinates of the resonances will be those listed in Table 1.8. As shown schematically in Fig. 1.51 for a doublet and a triplet, this means that (assuming that the scales of the F_2 and J axes are the same) the component peaks of each multiplet will lie on a line making a 45° angle with the F_2 axis. If the spectra are rotated counterclockwise by 45° about their centers, which may be done by a computer routine, then projections on the F_2 axis contain only chemical shift

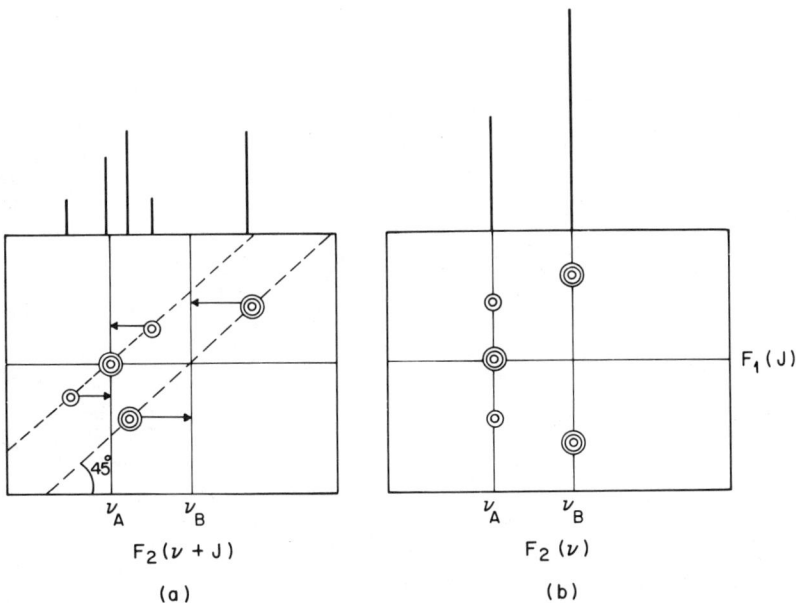

FIGURE 1.51 Schematic homonuclear J-resolved 2D spectra. In (a) the peaks in the t_1 dimension appear at the frequencies $v \pm J$. The spectrum can be rotated (b) such that there is a complete separation of the frequency axes.

information [Fig. 1.51(b)] and a complete separation of J coupling and chemical shift has been achieved.

In many cases the J-axis scale is so much expanded over that of the chemical shift axis—10-fold, for example—that the need for such a rotation is minimal. Vertical slices through such spectra may be generated, and these provide an accurate means for evaluating the spacing and intensities of the spectral components (*vide infra*). In Fig. 1.52 is shown the 500-MHz J-resolved proton spectrum of polycaprolactone.

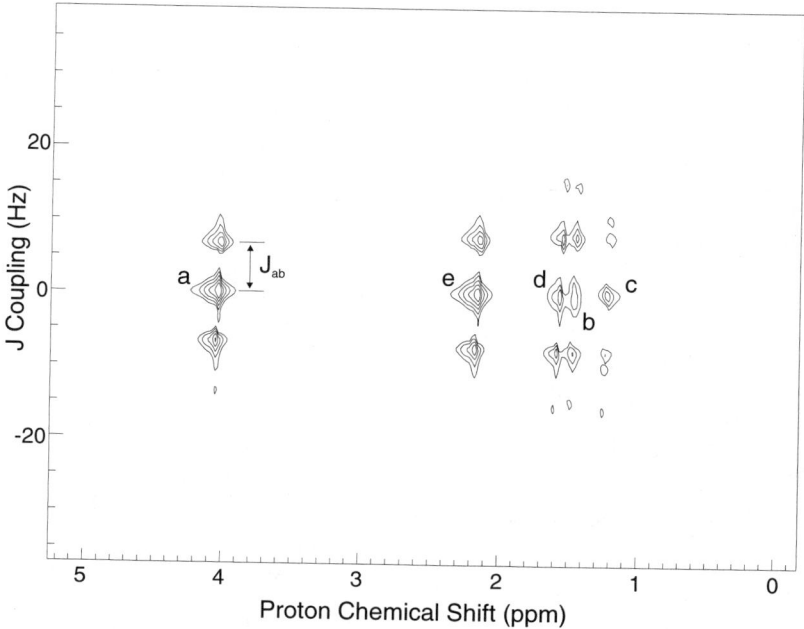

Polycaprolactone

The horizontal axis is the usual chemical shift scale, expressed in ppm from TMS, with assignments indicated. In the vertical direction

FIGURE 1.52 The 2D J-resolved spectrum for polycaprolactone obtained at 500 MHz.

the multiplets appear in contour form, spacings being expressed in hertz on the left-hand axis. Measurement of J_{ab} from this spectrum gives a value of 6.4 Hz for the $H(a)$ protons. It is found that by this procedure one may extract reliable coupling values even from very poorly resolved multiplets to which ordinary 1D measurements cannot be applied.

Heteronuclear J-resolved spectroscopy, commonly involving ^{13}C and ^1H, is extremely useful. By this means one may obtain spectra with carbon chemical shifts displayed along the horizontal axis and $^{13}C-^1H$ multiplets shown vertically. This enables one to simplify complex ^{13}C spectra and to easily identify carbon types, i.e., quarternary C (singlet), CH (doublet), CH_2 (triplet), or CH_3 (quartet).

Two pulse sequences for heteronuclear J-resolved 2D NMR are shown in Fig. 1.53 (86,87). The same sort of data is obtained using both pulse sequences, but the sequence in Fig. 1.53(a) is preferred because it is easier to implement. As in the homonuclear experiment, the chemical shifts are refocused during the t_1 period so the spin system evolves under the influence of the carbon–proton heteronuclear J coupling. The results from such an experiment on poly(propylene oxide) are shown in Fig. 1.54. This spectrum shows how the

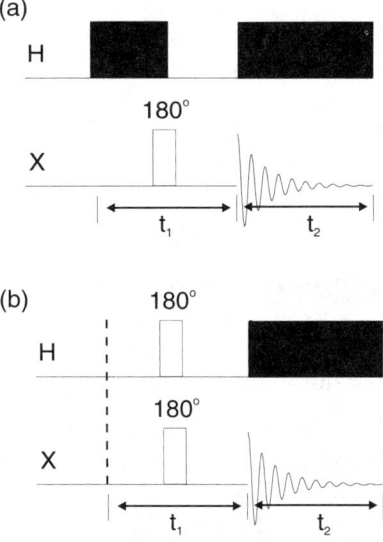

FIGURE 1.53 The pulse sequences for measuring heteronuclear J-resolved spectroscopy using the (a) gated method and (b) the spin-flip method.

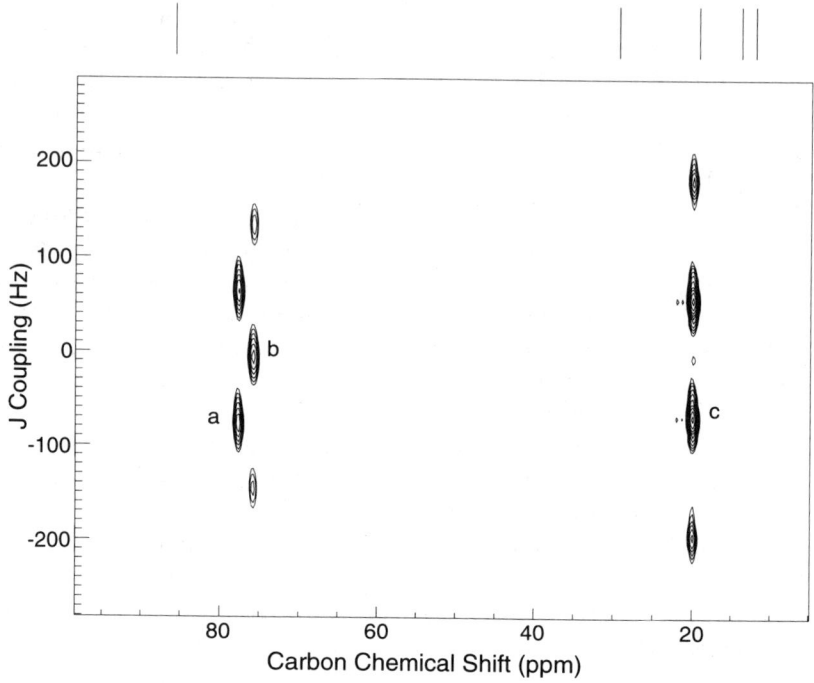

FIGURE 1.54 The 2D ^{13}C–^{1}H J-resolved spectra of poly(propylene oxide).

carbon type (methine, methylene, or methyl) can be directly assigned from the inspection of such a plot.

Poly(propylene oxide)

1.8.7.3 Two-Dimensional Correlated Spectroscopy

The second class of frequently used NMR experiments are *correlated* ones in which the resonances that are frequency labeled in the t_1 period are correlated with their frequencies in the t_2 period. Magnetization exchange may occur either through bonds (J coupling) or through space (dipolar coupling). The magnetization that undergoes a change of frequency between the t_1 and the t_2 periods appears as a cross peak connecting the two frequencies. These experiments

1.8. Experimental Methods

are invaluable for spectral assignments since a single 2D spectrum can provide a complete set of molecular connectivities. They have revolutionized solution structure determination because the crosspeak intensity in experiments where the magnetization arises from through-space dipolar interactions depends on the inverse sixth power of the distance between the two spins. Such correlations can be used both in homonuclear (usually protons) and in heteronuclear (usually protons and carbon) experiments. Heteronuclear correlations can automatically lead to ^{13}C chemical shift assignment if the ^{1}H assignments are known from the homonuclear correlations.

The homonuclear correlation of resonances via through-bond J coupling has been termed COSY (*co*rrelated *s*pectroscop*y*) and is the most frequently used 2D NMR experiment. The pulse sequences for COSY (75) and a commonly used variant called *double-quantum filtered* COSY (88) are shown in Fig. 1.55. Following the initial 90° pulse, the spins are frequency labeled as they evolve during the t_1 period. For those spins that are J coupled to other protons, magnetization transfer is affected by the application of a second 90° pulse prior to data acquisition. The magnetization transfer is not easily understood in terms of the simple vector pictures we have used to illustrate the principles of NMR thus far, but they can be understood using the *product operator formalism* (89). In such an experiment the phases of the pulses are cycled to remove various types of artifacts, and following Fourier transformation with respect to t_1 and t_2 the two-dimensional spectrum is obtained, where the frequency axes F_1 and F_2 show the frequencies of the spins during these periods. COSY spectroscopy greatly simplifies interpretation of complex spectra—particularly of protons by furnishing virtually automatic assignment of the resonances.

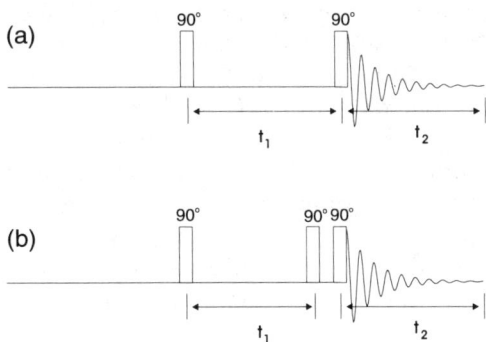

FIGURE 1.55 The pulse sequence for (a) 2D COSY and (b) double-quantum filtered COSY spectroscopy.

The simple COSY pulse sequence shown in Fig. 1.55(a) is extremely useful for the study of small molecules and biomolecules, but has some limitations for the study of polymers. One of the limitations arises because the resolution in polymer spectra is typically poor—primarily due to the complex microstructure—and correlations must be observed between protons that are close in frequency. In the simple two-pulse COSY experiment of Fig. 1.55(a) it is often observed that diagonal peaks are very large and can obscure the correlations between protons that do not have a large chemical shift separation. This problem is partially solved by the *double-quantum filtered* COSY experiment shown in Fig. 1.55(b). The phase of the third pulse in this experiment is cycled in such a way that magnetization that is not transferred by the second 90° mixing pulse is filtered from the spectrum and the size of the diagonal peaks is greatly reduced.

The second problem that is especially important for the study of polymers concerns the structure of the cross peaks connecting the two correlated protons. In COSY spectra the cross peaks have an antiphase structure, which means that parts of the cross peak may have opposite phases. In an AX system, for example, the correlation will consist of four peaks that are separated from each other by the coupling constant and two parts of the multiplet will have a positive phase while two have a negative phase. If the linewidths become large, either from homogeneous or inhomogeneous broadening, the parts of the multiplets can begin to cancel each other and the size of the cross peak will be greatly diminished. This problem can be partially overcome using another COSY variant called TOCSY (*to*tal *c*orrelatation *s*pectrosco*py*) in which all the parts of the multiplet of the cross peak have the same phase (*vide infra*) (90).

Figure 1.56 shows the 500-MHz double-quantum COSY spectrum of polycaprolactone and illustrates how molecular connectivities can be determined from a single 2D spectrum. On the basis of the electronegativity effects on proton chemical shifts discussed earlier, it is expected that the lowest field (least shielded) polycaprolactone resonances will be those next to the oxygen in the main chain. Starting with this resonance at 4.0 ppm, the chemical shift of the second methylene can be identified by the cross peak labeled a,b, since this is the only proton that is *J* coupled to the first proton. Simply by following the connectivity, the spectrum of polycaprolactone can be completely assigned. This is a very simple application of the general principles, but it should be obvious that this technique is very valuable for the identification and characterization of new materials.

There are several variations of the COSY spectrum designed to resolve the assignment ambiguities that might arise from overlap of

1.8. Experimental Methods

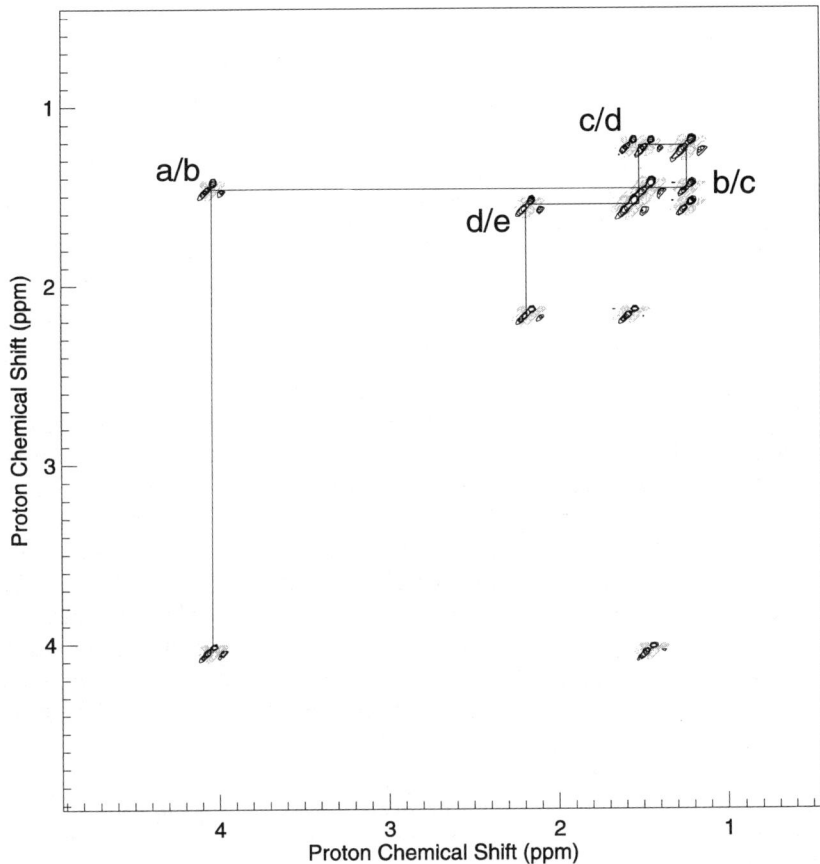

FIGURE 1.56 The 500-MHz proton double-quantum filtered COSY spectrum of polycaprolactone.

the cross peaks in a 2D spectrum. An important extension of the COSY experiment yields the *relayed coherence transfer* spectrum (91). The pulse sequences are more complex than those for COSY but can lead to multiple correlations between the t_1 and the t_2 periods. This results in cross peaks not only to the nearest neighbors, but also to the next nearest ones. Two-dimensional experiments using magnetization transfer in the rotating frame, TOCSY, constitute another class of correlation experiments that are not restricted to a single coherence transfer step. The pulse sequence for TOCSY is shown in Fig. 1.57 (92). This pulse sequence differs from COSY in that the magnetization transfer is affected by the application of a multipulse spin-locking field rather than a 90° pulse. As noted above, the pri-

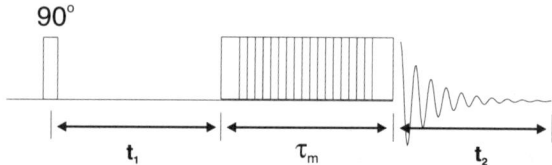

FIGURE 1.57 The pulse sequence for the TOCSY experiment for through-bond J coupling correlations. Magnetization transfer occurs during the MLEV-16 pulse sequence, which is repeated 20–50 times and is sandwiched by trim pulses of about 2 ms in duration.

mary advantage of TOCSY over COSY is that the cross peaks are in the absorption phase and are typically obtained with a higher signal-to-noise ratio than those in COSY spectra. The magnetization transfer during the spin-locking field is not restricted to directly coupled protons, so longer runs of coupled protons can be identified. However, nearest neighbor couplings can be observed by making the spin-locking time too short for extensive magnetization transfer. Figure 1.58 shows the TOCSY spectrum for the same polycarpolactone sample that was used to illustrate COSY spectroscopy. Note the appearance of additional cross peaks that show the relayed coherence transfer and are extremely useful for establishing connectivities and assignments.

In heteronuclear correlated spectroscopy two nuclear species—most commonly ^{1}H and ^{13}C—are correlated through the one-bond heteronuclear J coupling. If for some reason it is easier to assign the ^{13}C spectrum, then the complete assignment of the ^{1}H spectrum can be achieved from the ^{1}H–^{13}C 2D correlation experiment and of course *vice versa*. Figure 1.59 shows the pulse sequence for heteronuclear correlation via the J coupling. In such an experiment the carbon chemical shift appears along one axis and the proton spectrum along the other. Figure 1.60 shows the heteronuclear correlation spectrum for polycaprolactone obtained with this pulse sequence. It is clear that the combination of COSY, TOCSY, and the heteronuclear correlation experiments can quickly lead to the assignment of all the protons and carbons in polymer spectra.

One of the factors limiting the use of heteronuclear correlations is sensitivity, since the sensitivity of ^{13}C is low, and long times are required for the acquisition of a 2D NMR data set. These limitations are partly overcome using *h*eteronuclear *m*ultiple-*q*uantum *c*oherence (HMQC) spectroscopy, in which the proton signals are detected rather than the carbons. The signals in the 2D spectrum still arise from those protons directly bonded to ^{13}C's, but since protons have a

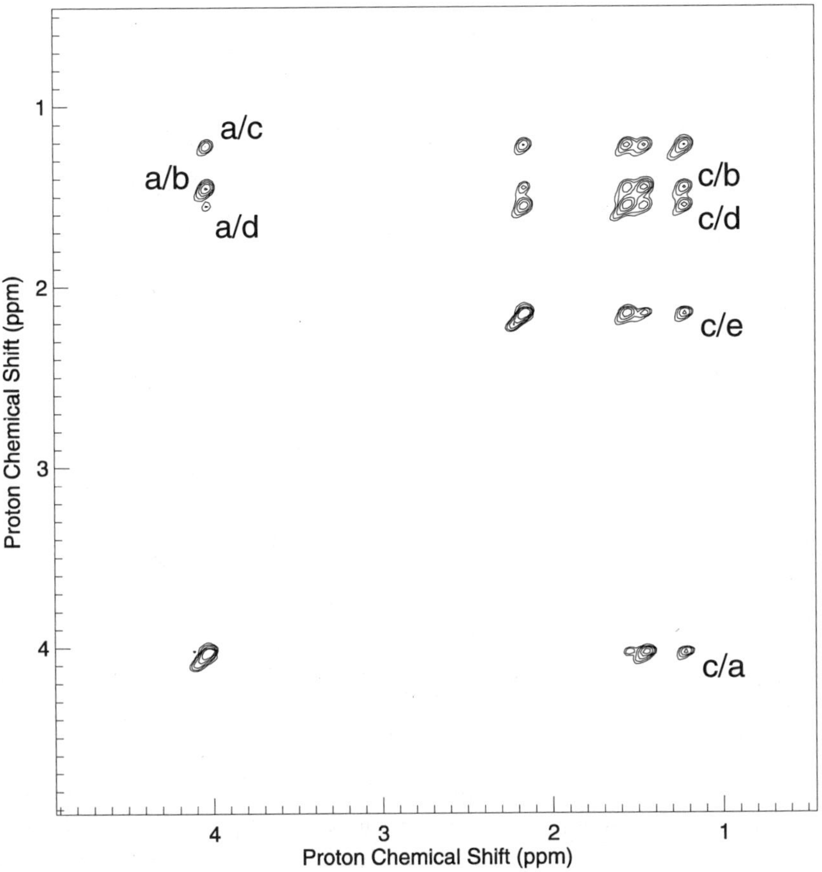

1.8. Experimental Methods

FIGURE 1.58 The 500-MHz proton TOCSY spectrum of polycaprolactone obtained with a spin-lock period of 40 ms using a field strength of 6 kHz.

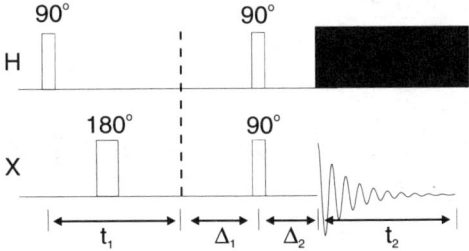

FIGURE 1.59 The pulse sequence for heteronuclear correlation spectroscopy. The delay times Δ_1 and Δ_2 are set equal to $1/2J$ and $1/3J$, respectively.

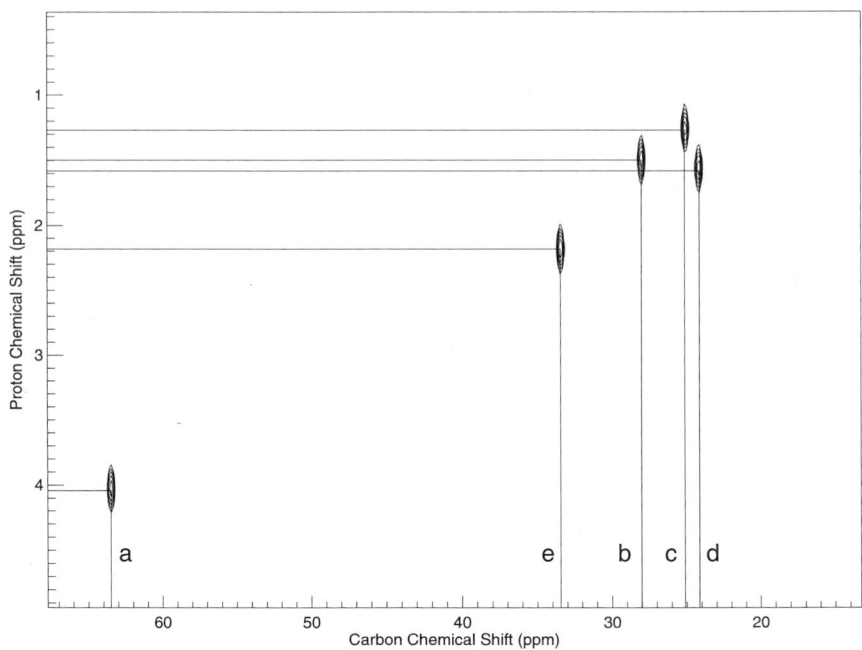

FIGURE 1.60 The carbon–proton heteronuclear correlation spectra of 40 wt% polycaprolactone in benzene-d_6.

larger magnetogyric ratio, the sensitivity is enhanced. In such 2D NMR experiments, an increase in sensitivity greatly decreases the amount of time required to obtain a 2D spectrum with the same signal-to-noise ratio. Figure 1.61 shows the pulse sequence for a simple HMQC experiment (93). Following the initial proton 90° pulse and the delay Δ ($= 1/2J$), the carbon 90° pulse converts the proton magnetization into multiple-quantum coherences that evolve during

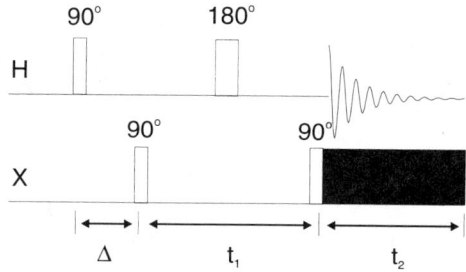

FIGURE 1.61 The pulse sequence diagram for heteronuclear multiple-quantum spectroscopy. Note that the signals of protons directly bonded to ^{13}C's are detected.

1.8. Experimental Methods

the t_1 period. These are converted back into proton longitudinal magnetization by the second 90° carbon pulse and the signals are observed during the t_2 period. Such multiple-quantum coherences are not easily pictured with vector diagrams, but are easy to understand using the product operator formalism (89).

The other important form of correlated spectroscopy is 2D *n*uclear *O*verhauser *e*ffect *s*pectroscopy (NOESY) (94). The format is similar to COSY except that the correlating influence is the direct, through-space dipole–dipole interaction rather than the *J* coupling. This experiment is important for studying the structure of polymers in solution because the intensity of the cross peaks can be related to internuclear distances.

In Fig. 1.62 are shown the three-pulse NOESY sequence and a series of vector diagrams showing how a two-spin system might behave under this sequence. The first 90° pulse, following the preparation period, converts the *z* magnetization of the two chemically shifted spins A and B (which we may think of as protons in the subsequent discussion) into observable magnetization at (a). We suppose for simplicity that spin B coincides in Larmor frequency with the carrier v_0 and so does not precess in the rotating frame during the evolution time t_1, while spin A precesses. At (b) we show the vectors after a particular choice of t_1 equal to $1/2\delta_{AB}$, where δ_{AB} is the chemical shift difference (in hertz) between A and B. During this interval, a 180° phase angle θ will accumulate between the vectors.

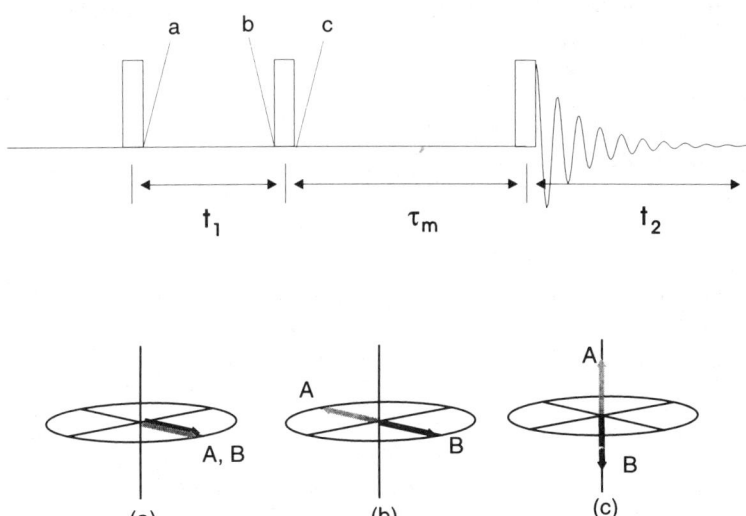

FIGURE 1.62 The pulse sequence and vector diagram for 2D NOESY spectroscopy.

The second 90° pulse at (c) produces the $\pm z$ polarization of the spins, the optimum condition for exchange of magnetization. Note that this situation is similar to that of Fig. 1.38 showing *selective* inversion. As noted above, the zero-quantum "flip-flop" term dominates the relaxation in the slow motion limit expected for most polymers. Other choices of t_1 will result in less-than-optimum z polarization, but since t_1 is systematically incremented over a series of several hundred values (typically 256–512) covering the chemical shifts of all the protons in a multiproton system, each proton will be given the opportunity to experience its optimum t_1 value. The net effect is a frequency labeling of each spin.

Any process that leads to an exchange of magnetization during the mixing time, τ_m, can be measured by this method, including chemical exchange or dipolar relaxation. The mixing time is typically chosen to be on the order of the spin-lattice relaxation time.

The intensities of the diagonal and cross peaks depend on the probability of magnetization exchange and the length of the mixing time τ_m. When τ_m is very short compared to the time constant for magnetization exchange, the off-diagonal peaks will be very weak compared to the diagonal peaks. Another extreme is represented by choosing τ_m to be long enough to allow the z magnetization to come to equilibrium during τ_m. In this case, the frequency-labeled magnetizations A and B have a 50% probability of being converted during τ_m, and the intensities of the diagonal and cross peaks will be equal. If τ_m greatly exceeds the relaxation times of the system, the spin system will come to thermal equilibrium with the lattice (Section 1.4)) and no peaks will be detected, either on or off the diagonal, because the spin system will have lost all memory of the frequency labeling.

NOESY spectroscopy can be used to quantitatively measure the structure and dynamics of polymers in solution. The equation describing the time dependence of the diagonal and cross peaks in the NOESY spectrum as a function of τ_m is

$$A(\tau_m) = A_0 e^{-R\tau_m}, \quad (1.105)$$

where $A(\tau_m)$ is the matrix of peak volumes obtained at mixing time τ_m, A_0 is the matrix of peak volumes at $\tau_m = 0$ (i.e., the diagonal spectrum), and R is the relaxation rate matrix given by

$$R = \begin{bmatrix} \rho_{11} & \sigma_{12} & \sigma_{13} & \cdot \\ \sigma_{21} & \rho_{22} & \sigma_{23} & \cdot \\ \sigma_{31} & \sigma_{32} & \rho_{33} & \cdot \\ \cdot & \cdot & \cdot & \cdot \end{bmatrix}, \quad (1.106)$$

where the ρ_{ii} and σ_{IS} are the terms for selective spin–lattice relaxation and cross relaxation given by Eqs. (1.68) and (1.69). If the

1.8. Experimental Methods

spectrum is well enough resolved that all of the diagonal and cross peaks can be measured or estimated, the peak volumes can be used to solve for relaxation rates, from which distances may be measured and the rates of molecular motion determined. Equation (1.106) can be rewritten as

$$\frac{-\ln[A(\tau_m)/A_0]}{\tau_m} = R \qquad (1.107)$$

and solved as

$$\frac{-T\ln[D]T^{-1}}{\tau_m} = R, \qquad (1.108)$$

where D is the diagonal matrix of eigenvalues of the normalized peak volume matrix and T and T^{-1} are its eigenvector matrix and its inverse. Thus, at a single mixing time it is possible to measure all of the diagonal and cross-relaxation rates, although gathering spectra at several mixing times is usually more desirable (95).

In the event that the peaks are not well enough resolved for this analysis, the cross-relaxation rates can be measured from the initial build-up of cross-peak intensity. The initial rate of cross-peak build-up is approximately given by

$$\sigma_{IS} = \sigma_{IS} \cdot \tau_m + \tfrac{1}{2}\sigma_{IS}^2 \tau_m^2 + \cdots, \qquad (1.109)$$

and in the event that τ_m is short, the cross-relaxation rate can be obtained from the initial build-up.

Once the relaxation rates are obtained from the 2D NOESY analysis, these rates must be interpreted in terms of the molecular dynamics [i.e., the spectral density terms of Eq. (1.75)] and the internuclear distances. The correlation times can be measured from a relaxation time analysis of the system, or, in many cases, they can be estimated from the cross-relaxation rates for pairs of protons separated by a fixed distance, such as geminal methylene protons or neighboring protons on aromatic rings. Errors in this analysis are somewhat minimized by the fact that the inverse sixth power dependence of the distance on the cross-relaxation rate. Figure 1.63 shows the 2D NOESY spectrum for poly(benzyl-L-glutanate) obtained with an 0.5-s mixing time.

NOESY spectroscopy is the most commonly used method for measuring proton–proton distances by 2D NMR. However, there are some circumstances where no cross peaks will be observed. One limitation is that because of the spin–lattice relaxation, the cross-relaxation rates for protons separated by more than 5 Å are too small to give rise to observable cross peaks. A second situation is when the product $\omega_0 \tau_c = 1.18$ and the cross-relaxation term given by Eq. (1.88) is zero.

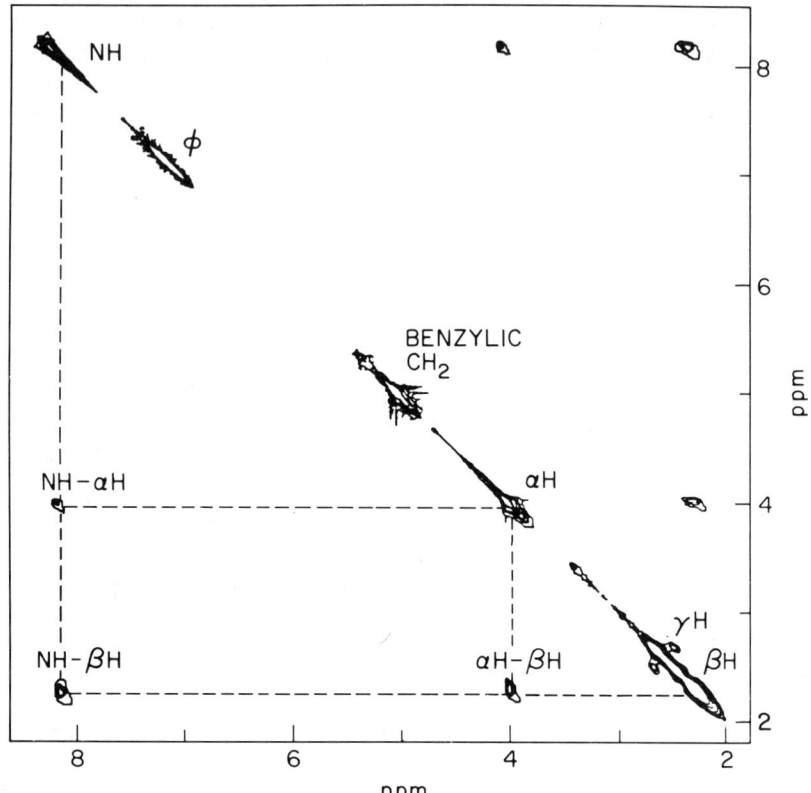

FIGURE 1.63 The 500-MHz NOESY spectrum of poly(benzyl-L-glutamate obtained with a 0.5-s mixing time.

In such a situation the cross-relaxation rates can be measured by *r*otating frame *O*verhauser *e*ffect *s*pectroscop*y* (ROESY) using the pulse sequence shown in Fig. 1.64. This pulse sequence differs from NOESY in that the magnetization transfer occurs under a weak spin-locking field (typically much weaker than required for the mag-

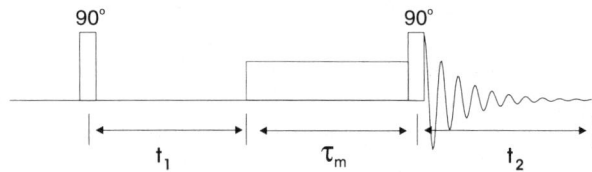

FIGURE 1.64 The pulse sequence diagram for ROESY spectroscopy. The spin-locking field used in this experiment is much weaker than that in TOCSY spectroscopy.

1.8. Experimental Methods

netization transfer via J coupling as in TOCSY spectroscopy). Figure 1.65 shows a plot of the NOE and ROE as a function of correlation time. This plot shows that as the correlation time becomes long, the NOE switches from positive to negative and goes through zero when $\omega_0 \tau_c = 1.18$. By contrast, the ROE is positive at short correlation times and becomes increasingly positive in the slow-motion limit.

1.8.8 Spectral Editing and Simplification

Spectral editing techniques enable one to select and identify specific types of nuclei—most commonly ^{13}C or ^{15}N—on the basis of the local NMR interactions. In solutions the spectral editing is typically based on the differences in *polarization transfer* due to the number of directly bonded protons, while in the solid state the editing and spectral simplifications are due to differences in the strength of the dipolar interactions, which is also due to the number of directly bonded protons. Also included in this section is the method for elimination of spinning sidebands that often complicate the spectra of polymers obtained with magic angle spinning.

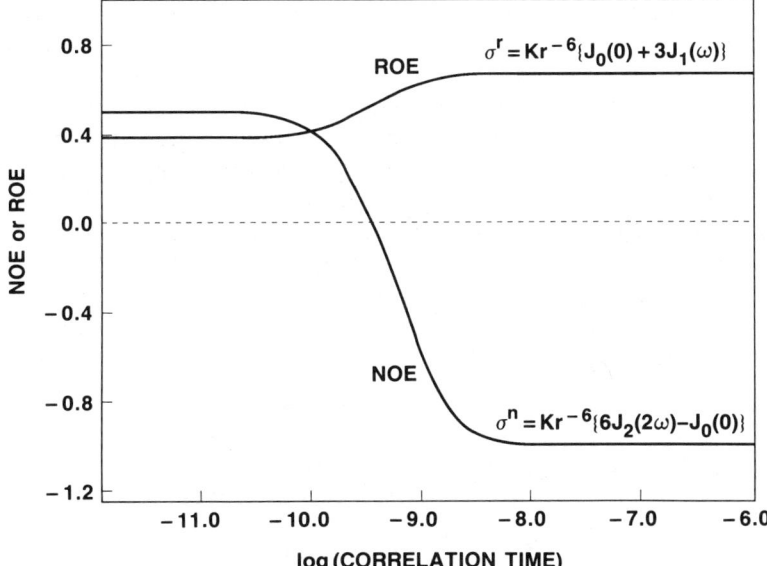

FIGURE 1.65 The proton–proton nuclear Overhauser effect and the rotating frame Overhauser effect as a function of correlation time at a proton frequency of 500 MHz.

108 1. Fundamentals of Nuclear Magnetic Resonance

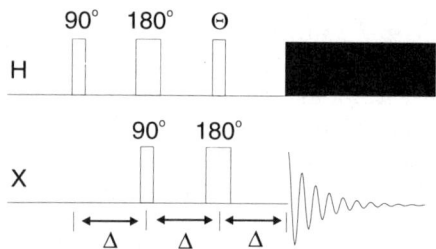

FIGURE 1.66 The pulse sequence diagram for spectral editing using the DEPT pulse sequence.

1.8.8.1 Spectral Editing in Solution

Spectral editing in solution is based on heteronuclear polarization transfer—usually between protons and carbons—to discriminate between atoms with differing numbers of directly attached protons. By suitably choosing the experimental parameters, it is possible using these techniques to obtain subspectra containing methine, methylene, or methyl carbons.

Figure 1.66 shows the pulse sequence diagram for the DEPT (96,97) experiment that is most frequently used for spectral editing. A key feature of this experiment is the delay times Δ, which are set to $1/2J$, during which the multiplets precess until they are out of phase with each other. Their behavior during these periods will depend on the multiplicity, which is determined by the number of directly bonded protons. Through the product operator formalism it is easily demonstrated that the efficiency of the polarization transfer and the phase of the resulting signal depend on the pulse angle θ, given by $\sin \theta \cos^{n-1} \theta$, where n is the number of directly bonded protons. The strategy used to obtain separate subspectra for the methines, methylenes, and methyls is to record spectra with θ equal to 45°, 90°, and 135° and to take linear combinations of these spectra such that the resulting spectra contain only the methine, methylene, or methyl resonances. Figure 1.67 shows the results from a DEPT experiment on 2,2,4-trimethylpentane, in which separate subspectra are obtained for the different types of carbon resonances.

$$\underset{a}{CH_3}\underset{CH_3}{\overset{CH_3}{\diagdown}}\overset{b}{CH}-\overset{}{CH_2}-\underset{CH_3}{\overset{CH_3}{\overset{|}{\underset{|}{C}}}}-\overset{e}{CH_3}$$

2,2,4-Trimethylpentane

1.8. Experimental Methods

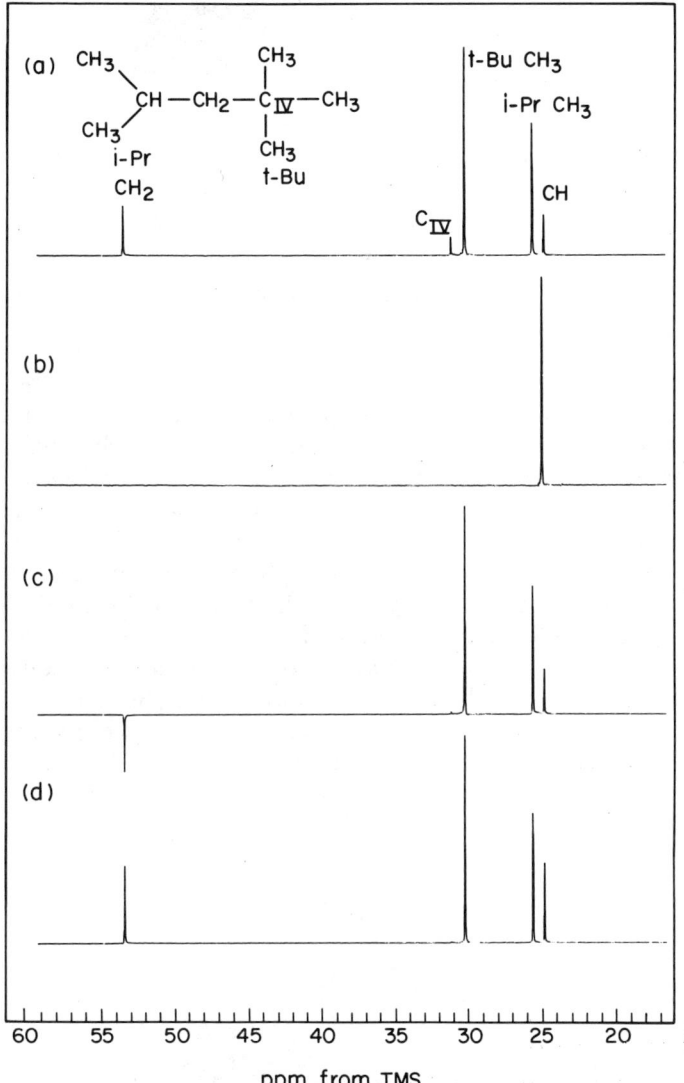

FIGURE 1.67 The DEPT spectrum of 2,2,4-trimethylpentane.

1.8.8.2 Spectral Simplification in the Solid State

Spectral editing and simplification are also important in the solid state, although based on different principles since the scalar couplings are typically not observed in rigid solids. One extremely simple method used for spectral simplification is *dipolar dephasing*. The pulse sequence diagram for this experiment is shown in Fig. 1.68 (98).

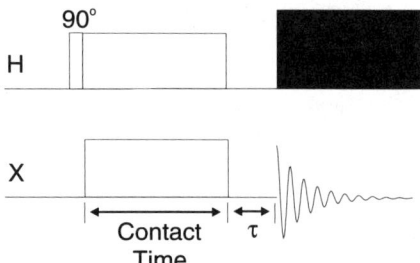

FIGURE 1.68 The pulse sequence diagram for dipolar dephasing.

Dipolar dephasing differs for cross polarization only in the variable delay period τ following the cross polarization and before data acquisition, during which both the proton and the carbon spin-locking fields are turned off. Since there is no proton decoupling during this period, the effective linewidth (and T_2) is that due to the carbon–proton dipolar coupling. Therefore, carbons with directly bonded protons will decay much faster than those without directly bonded protons. The effect is illustrated in Fig. 1.69, which shows the CPMAS spectra of poly(methyl methacrylate) recorded without (a) and with (b) a 200-μs delay between cross polarization and signal acquisition. All of these signals decrease due to relaxation during the delay time τ, but the signals from carbons without directly bonded protons [b and d in Fig. 1.69(a)] are reduced to a much lesser extent. This experiment is typically used to identify nonprotonated carbons and to remove the signals from protonated carbons.

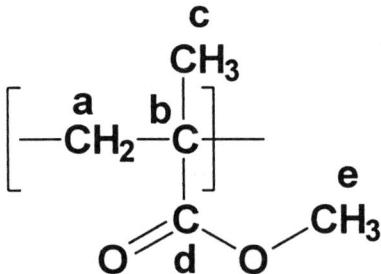

As noted in Section 1.8.5.3, it is frequently not possible to spin polymer samples rapidly enough to completely remove the spinning sidebands for the aromatic and carbonyl signals. Thus, the spectra of polymers with several of these signals may be so overlapped that it is difficult to separate the isotropic signals from the spinning sidebands. While the information contained in the frequencies of the spinning sidebands may contain important information about the chemical shift anisotropy tensor, it is often desirable to simplify the spectrum

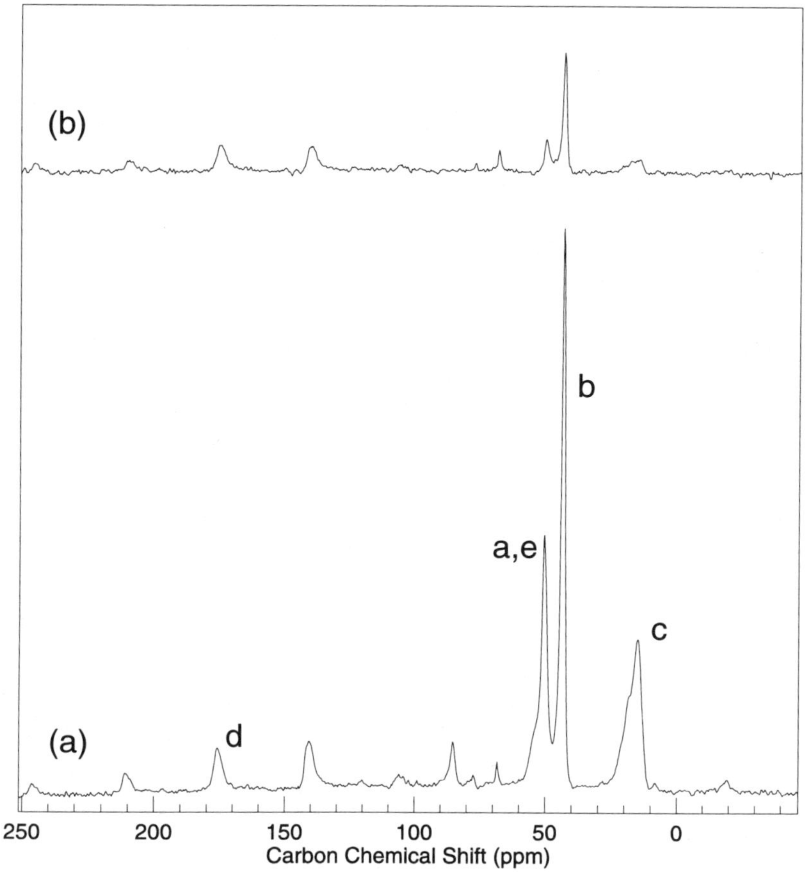

1.8. Experimental Methods

FIGURE 1.69 The CPMAS spectrum of poly(methyl methacrylate) obtained without (a) and with (b) a dipolar dephasing delay of 50 μs.

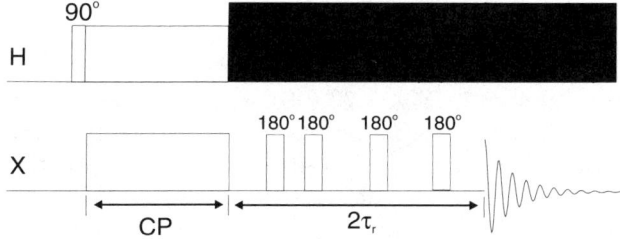

FIGURE 1.70 The pulse sequence diagram for sideband elimination using the TOSS pulse sequence.

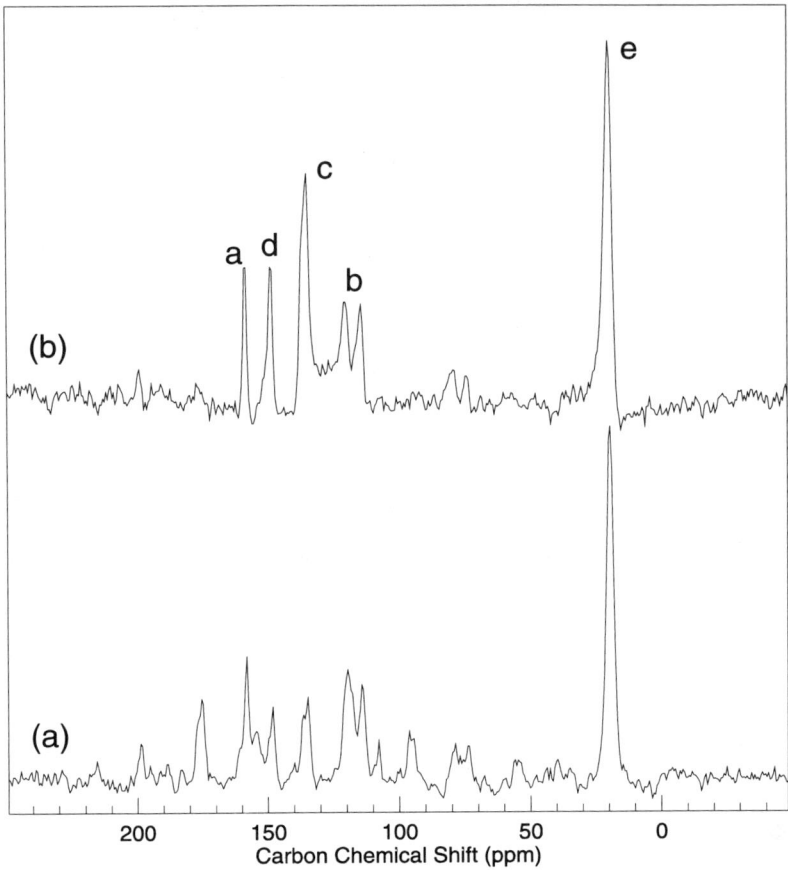

FIGURE 1.71 The 100-MHz CPMAS spectrum of poly(dimethyl phenylene oxide) recorded (a) without and (b) with the TOSS pulse sequence for the elimination of spinning sidebands.

by removing all but the signals at the isotropic chemical shift. This can be accomplished with the *to*tal *s*uppression of *s*idebands (TOSS) pulse sequence shown in Fig. 1.70 (99). This pulse sequence uses four 180° pulses applied at specific points during two rotor cycles to form echoes leading to the cancellation of all signals except those arising from the isotropic chemical shifts. Figure 1.71 shows the CPMAS spectrum of poly(dimethyl phenylene oxide) obtained with a spinning speed of 4065 Hz in which there is extensive overlap between the sidebands of the various aromatic signals. This spectrum is greatly simplified with the application of the TOSS pulse sequence.

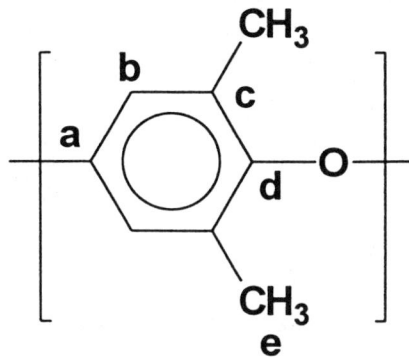

REFERENCES

1. N. L. Alpert, *Phys. Rev.* **72**, 637 (1947).
2. E. M. Purcell, H. C. Torrey, and R. V. Pound, *Phys. Rev.* **69**, 37 (1946).
3. F. Bloch, W. W. Hansen, and M. E. Packard, *Phys. Rev.* **69**, 127 (1946).
4. M. Saunders and A. Wishnia, *Ann. N.Y. Acad. Sci.* **70**, 870 (1958).
5. A. Odajima, *J. Phys. Soc. Jpn.* **14**, 777 (1959).
6. F. A. Bovey, G. V. D. Tiers, and G. Filipovich, *J. Polym. Sci.* **38**, 73 (1959).
7. W. Pauli, *Naturwissenschaften* **12**, 741 (1924).
8. W. P. Slichter, *Adv. Polym. Sci.* **1**, 35 (1958).
9. D. W. McCall, *Acc. Chem. Res.* **4**, 223 (1971).
10. R. R. Ernst and W. A. Anderson, *Rev. Sci. Instrum.* **37**, 93 (1966).
11. N. Bloembergen, E. M. Purcell, and R. V. Pound, *Phys. Rev.* **73**, 679 (1948).
12. F. Bovey, "Nuclear Magnetic Resonance Spectroscopy," 2nd Ed. Academic Press, New York, 1988.
13. G. E. Pake, *J. Chem. Phys.* **16**, 327 (1948).
14. J. H. V. Vleck, *Phys. Rev.* **74**, 1168 (1948).
15. E. R. Andrew and R. Bersohn, *J. Chem. Phys.* **18**, 159 (1950).
16. R. C. Farrar and E. D. Becker, "Pulse and Fourier Transform NMR." Academic Press, New York, 1971.
17. L. P. Lindeman and J. Q. Adams, *Anal. Chem.* **43**, 1245 (1971).
18. J. Jonas and H. S. Gutowsky, *Ann. Rev. Phys. Chem.* **31**, 1 (1950).
19. J. T. Arnold, S. S. Dharmatti, and M. E. Packard, *J. Chem. Phys.* **19**, 507 (1951).
20. G. V. D. Tiers, *J. Phys. Chem.* **62**, 1151 (1958).
21. W. E. Lamb, *Phys. Rev.* **60**, 817 (1941).
22. W. C. Dickinson, *Phys. Rev.* **80**, 563 (1951).
23. H. M. McConnell, *J. Chem. Phys.* **27**, 226 (1957).
24. J. S. Waugh and R. W. Fessenden, *J. Am. Chem. Soc.* **79**, 846 (1957).
25. J. S. Waugh and R. W. Fessenden, *J. Am. Chem. Soc.* **80**, 6697 (1958).
26. C. W. Haigh and R. B. Mallion, *Prog. NMR Spectrosc.* **13**, 303 (1980).
27. H. Spiesecke and W. G. Schneider, *J. Chem. Phys.* **35**, 722 (1961).

28. D. M. Grant and E. G. Paul, *J. Am. Chem. Soc.* **86**, 2984 (1964).
29. D. M. Grant and B. V. Cheney, *J. Am. Chem. Soc.* **39**, 5315 (1967).
30. A. Tonelli, "NMR Spectroscopy and Polymer Microstructure: The Conformational Connection." VCH, New York, 1989.
31. J. A. Pople and M. S. Gordon, *J. Am. Chem. Soc.* **89**, 4253 (1967).
32. R. Ditchfield and P. D. Ellis, "Topics in Carbon-13 NMR Spectroscopy," Vol. 1. Academic Press, New York, 1974.
33. J. B. Stothers, "Carbon-13 NMR Spectroscopy." Academic Press, New York, 1972.
34. G. C. Levy, R. L. Lichter, and G. L. Nelson, "Carbon-13 Nuclear Magnetic Resonance Spectroscopy," 2nd Ed. Wiley (Interscience), New York, 1980.
35. E. Breitmaier and W. Voelter, "Carbon-13 NMR Spectroscopy," 3rd Ed. VCH, Weinheim, 1987.
36. G. Filipovich and G. V. D. Tiers, *J. Phys. Chem.* **63**, 761 (1959).
37. M. Witanowski, L. Stefaniak, and G. A. Webb, *Annu. Rep. NMR Spectrosc.* **11B** (1981).
38. L. Stefaniak, G. A. Webb, and M. Witanowski, *Ann. Rep. NMR Spectrosc.* **18**, 42 (1986).
39. W. von Philipsborn and R. Muller, *Angew. Chem.* **25**, 383 (1983).
40. H. S. Gutowsky, D. W. McCall, and C. P. Slichter, *Phys. Rev.* **4**, 589 (1951).
41. E. L. Hahn and D. W. Maxwell, *Phys. Rev.* **88**, 1070 (1952).
42. M. Karplus, *J. Chem. Phys.* **30**, 11 (1959).
43. M. Karplus, *J. Chem. Phys.* **33**, 1842 (1960).
44. J. N. Shoolery, *J. Chem. Phys.* **31**, 768 (1959).
45. N. Muller, *J. Chem. Phys.* **36**, 359 (1962).
46. M. D. Newton, J. M. Schulman, and M. M. Manus, *J. Am. Chem. Soc.* **96**, 17 (1974).
47. R. K. Harris and N. Sheppard, *Trans. Faraday Soc.* **59**, 606 (1963).
48. D. G. Gorenstein, "P-31 NMR: Principles and Applications." Academic Press, New York, 1981.
49. R. Ernst, *J. Chem. Phys.* **45**, 3845 (1966).
50. H. J. Reich, M. Jautelat, M. T. Meese, F. J. Weigert, and J. D. Robert, *J. Am. Chem. Soc.* **91**, 7445 (1969).
51. J. B. Grutzner and A. E. Santini, *J. Magn. Reson.* **19**, 178 (1975).
52. A. J. Shaka, J. Keeler, and R. Freeman, *J. Magn. Reson.* **53**, 313 (1983).
53. M. H. Levitt and R. Freeman, **43**, 502 (1981).
54. J. S. Waugh, *J. Magn. Reson.* **50**, 30 (1982).
55. J. S. Waugh, *J. Magn. Reson.* **49**, 517 (1982).
56. I. J. Lowe, *Phys. Rev. Lett.* **2**, 285 (1959).
57. E. R. Andrew, A. Bradbury, and R. G. Eades, *Nature (London)* **183**, 1802 (1959).
58. B. C. Gerstein, R. G. Pembleton, R. C. Wilson, and L. M. Ryan, *J. Chem. Phys.* **66**, 361 (1977).
59. T. M. Duncan, "*A Compilation of Chemical Shift Anisotropies.*" The Farragut Press, Chicago, 1990.
60. M. Mehring, "High Resolution NMR in Solids," 2nd Ed. Springer-Verlag, New York, 1983.
61. C. S. Yannoni, *Acc. Chem. Res.* **15**, 201 (1982).
62. L. W. Jelinski, *Macromolecules* **14**, 1341 (1981).
63. S. R. Hartmann and E. L. Hahn, *Phys. Rev.* **128**, 2042 (1962).
64. E. Stejskal, J. Schaefer, M. Sefcik, and R. McKay, *Macromolecules* **14**, 275 (1981).
65. I. Solomon, *Phys. Rev.* **99**, 559 (1959).
66. K. F. Kuhlmann, D. M. Grant, and R. K. Harris, *J. Chem. Phys.* **52**, 3439 (1970).
67. D. A. Torchia, *J. Magn. Res.* **30**, 613 (1978).
68. D. G. Cory and W. M. Ritchey, *Macromolecules* **22**, 1611 (1989).

References

69. J. Schaefer, E. O. Stejskal, and R. Buchdahl, *Macromolecules* **10**, 384 (1977).
70. E. L. Hahn, *Phys. Rev.* **8**, 580 (1950).
71. H. Y. Carr and E. M. Purcell, *Phys. Rev.* **94**, 630 (1954).
72. S. Meiboom and D. Gill, *Rev. Sci. Instrum.* **29**, 688 (1958).
73. M. L. Martin, G. J. Martin, and J. J. Delpuech, "Practical NMR Spectroscopy." Heyden, London.
74. J. Jeener, "*Ampre Int. Summer School.*" Basko Polje, Yugoslavia, 1971.
75. W. Aue, E. Bartnoldi, and R. Ernst, *J. Chem. Phys.* **64**, 2229 (1976).
76. R. R. Ernst, *Chimia* **29**, 179 (1975).
77. R. K. Hester, J. L. Ackerman, B. L. Neff, and J. S. Waugh, *Phys. Rev. Lett.* **34**, 993 (1975).
78. A. Kumar, D. D. Welti, and R. R. Ernst, *Naturwissenschaften* **62**, 34 (1975).
79. R. Freeman and G. A. Morris, *Bull. Magn. Reson.* **1**, 5 (1979).
80. R. Benn and H. Gunther, *Angew. Chem. Int. Ed. Engl.* **22**, 350 (1983).
81. A. Bax and L. Lerner, *Science* **232**, 960 (1986).
82. A. Bax, "Two Dimensional Nuclear Magnetic Resonance in Liquids." Delft University Press and Reidel, Dordrecht, The Netherlands, 1982.
83. H. Kessler, M. Gehrke, and C. Griesinger, *Angew. Chem. Int. Ed. Engl.* **27**, 447 (1988).
84. R. Ernst, G. Bodenhausen, and A. Wokaun, "Principles of Nuclear Magnetic Resonance in One and Two Dimensions." Oxford Univ. Press (Clarendon).
85. R. Freeman and H. D. W. Hill, "Dynamic Nuclear Magnetic Resonance Spectroscopy." Academic Press, New York, 1975.
86. R. Freeman, S. P. Kempsell, and M. H. Levitt, *J. Magn. Reson.* **34**, 663 (1979).
87. G. Bodenhausen, R. Freeman, R. Niedermeyer, and J. Turner, *J. Magn. Reson.* **26**, 133 (1977).
88. M. Rance, O. W. Sorensen, G. Bodenhausen, G. Wagner, R. R. Ernst, and K. Wuthrich, *Biochem. Biophys. Res. Commun.* **117**, 479 (1983).
89. O. W. Sorense, G. W. Eich, M. H. Levitt, G. Bodenhausen, and R. R. Ernst, *Prog. NMR Spectrosc.* **16**, 163 (1983).
90. A. Bax and D. Davis, *J. Magn. Reson.* **63**, 207 (1985).
91. P. H. Bolton and G. Bodenhausen, *Chem. Phys. Lett.* **89**, 139 (1982).
92. L. Braunschweiler and R. Ernst, *J. Magn. Reson.* **53**, 521 (1983).
93. A. Bax, R. H. Griffey, and B. L. Hawkins, *J. Am. Chem. Soc.* **105**, 7188 (1983).
94. J. Jeneer, B. Meier, P. Bachmann, and R. Ernst, *J. Chem. Phys.* **71**, 4546 (1979).
95. P. Mirau, *J. Magn. Reson.* **96**, 480 (1992).
96. M. R. Bendall, D. M. Doddrell, and D. T. Pegg, *J. Am. Chem. Soc.* **108**, 4603 (1981).
97. O. W. Sorensen, S. Donstrup, H. Bildsoe, and H. J. Jakobsen, *J. Magn. Reson.* **55**, 347 (1983).
98. M. H. Frey, S. J. Opella, A. L. Rockwell, and L. M. Gierasch, *J. Am. Chem. Soc.* **107**, 1946 (1985).
99. W. T. Dixon, *J. Chem. Phys.* **77**, 1800 (1982).

2
THE MICROSTRUCTURE OF POLYMER CHAINS

2.1 INTRODUCTION

One of the most significant applications of nuclear magnetic resonance spectroscopy to macromolecules, in which it has proved uniquely powerful, is the observation and quantitative measurement of chain *microstructure*. This term embraces those features of polymer chains which are fixed by their covalent structure, and is generally understood to include the following: head-to-tail versus head-to-head–tail-to-tail isomerism, i.e., regioisomerism; stereochemical configuration; geometrical isomerism; and branching and cross linking. Chain conformation, molecular weight, crystal structure, solid-state morphology, and the gel state are excluded, although of course all these characteristics depend heavily upon microstructure. This topic is of intense practical interest because peaks from the different microstructures can be resolved in the NMR spectrum, providing a detailed and *quantitative* characterization of chain microstructure. It is often observed that the macroscopic properties of polymers are critically dependent on the microstructure.

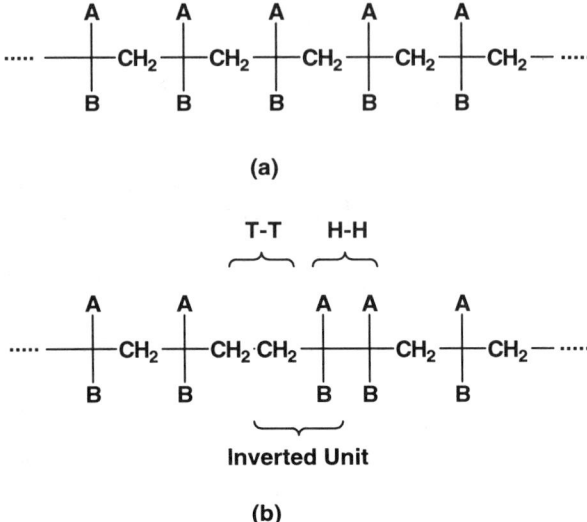

FIGURE 2.1 Chain connectivity in polymers and regioisomerism. The drawings show (a) a head-to-tail orientation and (b) an inverted unit with head-to-tail, head-to-head (H-H), and tail-to-tail (T-T) chain connectivities.

2.2 POLYMER MICROSTRUCTURE

2.2.1 Head-to-Tail versus Head-to-Head–Tail-to-Tail Isomerism: Regioisomerism

Polymers are frequently synthesized from asymmetric monomers, giving rise to the possibility that the incoming monomer can add to the chain in a head-to-tail, head-to-head, or tail-to-tail orientation, as shown in Fig. 2.1.

A head-to-head junction with no accompanying tail-to-tail unit will also arise from the recombination of growing chain radicals, but in this case there can be only one such unit per chain (Fig. 2.2).

The general subject of regioisomerism and its mathematical consequences has been explored by Cais and Sloane (1). They have proposed the terms *isoregic* and *syndioregic* to describe head-to-tail and head-to-head–tail-to-tail sequences, respectively, and this nomenclature seems to have been generally accepted.

2.2.2 Stereochemical Configuration

It is generally understood that stereochemical configuration refers to the *relative handedness* of successive monomer units. In vinyl poly-

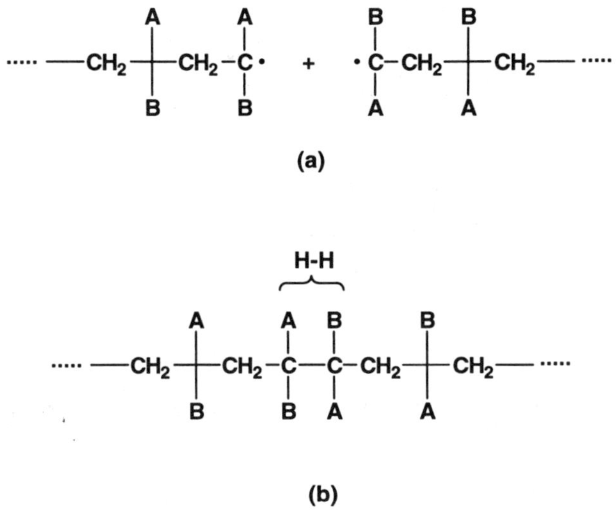

2.2. Polymer Microstructure

FIGURE 2.2 Combination of two chains (a) to form a chain with one head-to-head defect (b).

mers the main-chain substituted carbons, commonly designated as α-carbons, are termed "pseudoasymmetric" since, if the chain ends are disregarded, such carbons do not have the four different substituents necessary to qualify for being truly asymmetric. Nevertheless, they have the possibility of relative handedness. The simplest regular arrangements along a chain are the *isotactic* structure [Fig 2.3(a)], in which all the substituents, here represented by R, are located on the same side of the zigzag plane representing the chain stretched out in an all-*trans* conformation, and the *syndiotactic* arrangement, in which the groups alternate from side to side [Fig. 2.3(b)]. In the *atactic* arrangement, the R groups are placed at random on either side of the zigzag plane [Fig. 2.3(c)]. It should be emphasized that these isomeric forms cannot be interconverted by rotating R groups about the carbon–carbon main chain.

The possibility of such isomerism had been pointed out by Staudinger (2), Huggins (3), and Schildknecht *et al.* (4), but it was not taken very seriously because there seemed to be no effective way to control it, nor was it believed to have any marked effect on polymer properties. Interest became intense, however, when Natta (5) showed that by use of the coordination catalysts developed by Ziegler for the polymerization of ethylene, α-olefins can be polymerized to products having stereoregular structures. A major revolution in polymer science and technology was thus initiated, which continues to this day.

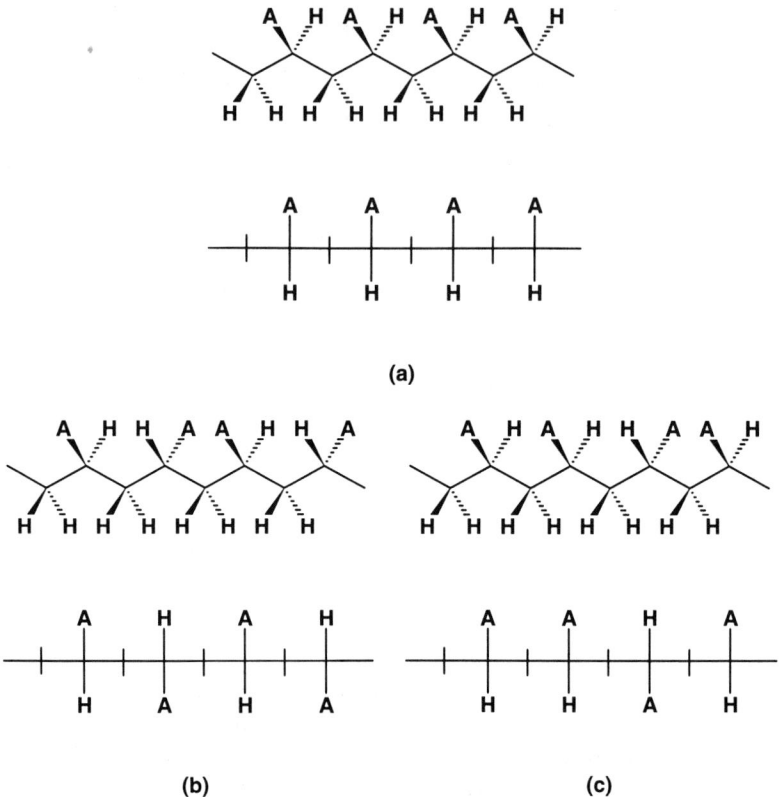

FIGURE 2.3 The structures of (a) isotactic, (b) syndiotactic, and (c) atactic vinyl chains.

1,2-Disubstituted vinyl monomers have additional options in that the polymer chain now has two interleaved systems of nonidentical pseudo-asymmetric carbon atoms (Fig. 2.4). Three limiting configurations can be defined: *disyndiotactic*, *erythrodi-isotactic*, and *threodi-isotactic*, as shown in Fig. 2.5. The prefixes *erytho-* and *threo-* are based on the configurations of the sugars erythrose and threose. A

FIGURE 2.4 The configurations of 1,2-disubstituted monomers.

2.2. Polymer Microstructure

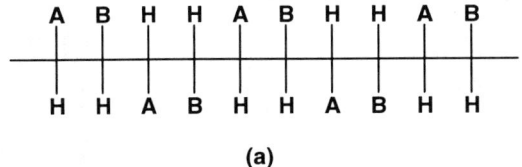

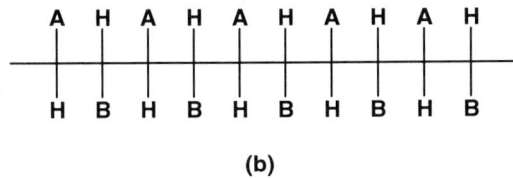

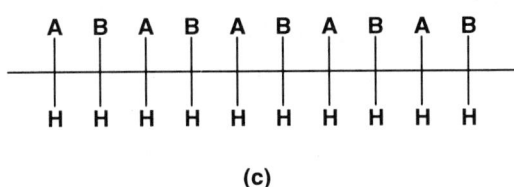

FIGURE 2.5 The structure of (a) disyndiotactic, (b) erythrodi-isotactic, and (c) threodi-isotactic polymer chains.

mnemonic device that may be useful is to recall that if A = B, the *erythro* would be *meso* and the *threo* would be *racemic*, as shown in Fig. 2.6.

There are a very limited number of actual cases in which A and B are equivalent, and we shall consider these later.

FIGURE 2.6 The structure of (a) *meso* and (b) *racemic* chains.

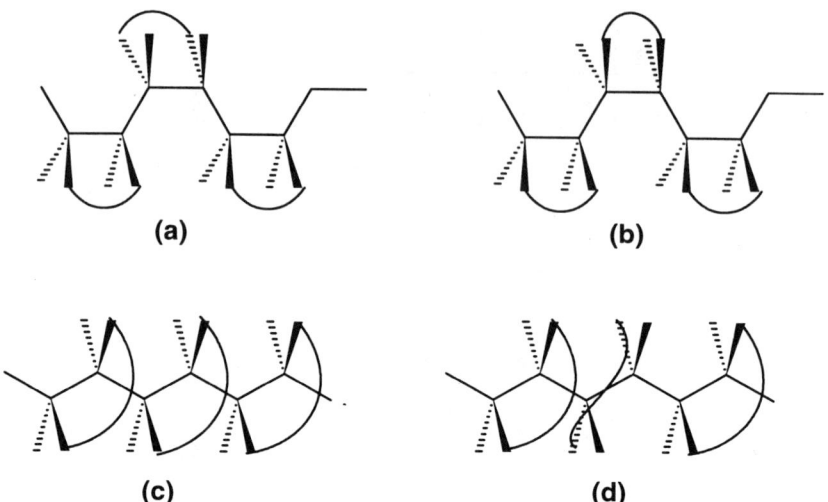

FIGURE 2.7 The structure of (a) erythrodi-isotactic, (b) erythrodisyndiotactic, (c) threodi-isotactic, and (d) threodisyndiotactic chains. The rings are *cis* in (a) and (c) and *trans* in (c) and (d).

Vinyl polymers may acquire true asymmetric centers by ring closure. The ring may be closed between α-carbons, as in poly(vinyl formal):

Poly(vinyl formal)

Ring formation may occur between adjacent carbons, and some of the possibilities are as shown in Fig. 2.7, the rings being represented in a generalized manner. The rings may be symmetrical (A = B) or unsymmetrical (A ≠ B). An example of a symmetrical ring system is polycyclobutene (Fig. 2.8).

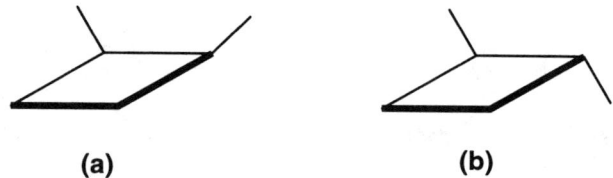

FIGURE 2.8 The structure of (a) *cis*- and (b) *trans*-cyclobutane rings.

2.2. Polymer Microstructure

[Figure showing structures (a) and (b) of cis- and trans-1,4-butadiene]

FIGURE 2.9 The structure of (a) *cis*- and (b) *trans*-1,4-butadiene.

2.2.3 Isomerism in Diene Polymer Chains

The polymerization of diene monomers can produce structures having combinations of geometrical and stereochemical isomerism. Butadiene can give 1,4- or 1,2-enchainment. The 1,4-structures may be *cis* (Z) or *trans* (E) as shown in Fig. 2.9. The 1,2-structures have the same configurational properties as vinyl polymers and may occur in isotactic or syndiotactic sequences (Fig. 2.10).

Polymers of 2-substituted butadienes may show every sort of isomerism we have discussed so far. 1,4-Units may exhibit the head-to-tail and head-to-head–tail-to-tail structures shown in Fig. 2.11. The proportion of inverted units may be substantial, for example, when A = Cl (polychloroprene or neoprene).

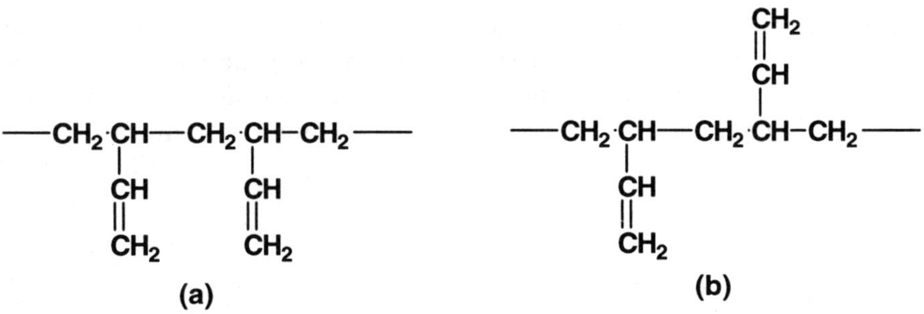

[Figure showing structures (a) and (b) of isotactic and syndiotactic 1,2-polybutadiene]

FIGURE 2.10 The structure of (a) isotactic and (b) syndiotactic 1,2-polybutadiene.

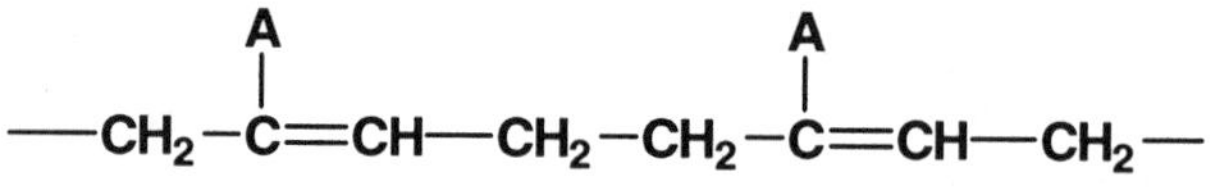

$$-CH_2-\underset{\underset{A}{|}}{C}=CH-CH_2-CH_2-\underset{\underset{A}{|}}{C}=CH-CH_2-$$

(a)

$$-CH_2-CH=\underset{\underset{A}{|}}{C}-CH_2-CH_2-\underset{\underset{A}{|}}{C}=CH-CH_2-CH_2-CH=\underset{\underset{A}{|}}{C}-CH_2-$$

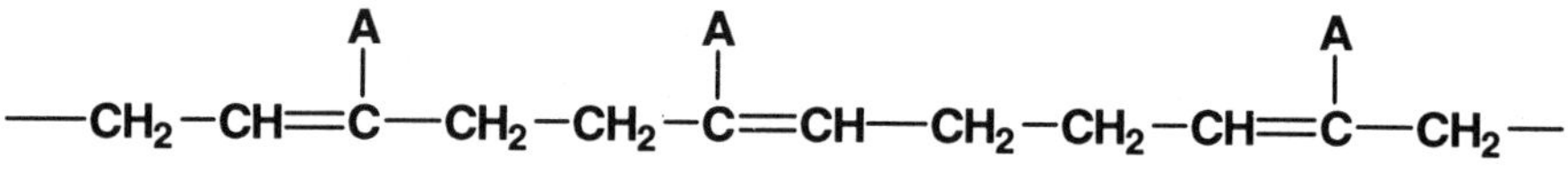

(b)

FIGURE 2.11 The structure of (a) head-to-tail and (b) head-to-head–tail-to-tail isomers of 2-substituted butadienes.

In addition to 1,4-addition, in either a *cis* or a *trans* fashion (Fig. 2.12), such monomers have a choice between 1,2- and 3,4-addition (Fig. 2.13), both of which may occur in isotactic or syndiotactic sequences (and head-to-tail and head-to-head–tail-to-tail sequences).

When $A = CH_3$ we have, of course, polyisoprene. In nature, this occurs as the *cis* isomer—natural rubber—and as the *trans* isomer, called *balata* or *gutta percha*. Each is very pure, without a trace of the other isomer. (They are formed by complex, enzyme-catalyzed processes not involving isoprene as such.) Balata is a semicrystalline plastic rather than a rubber, but is no longer produced commercially. Both isomeric chains can also be prepared synthetically, but are then not entirely stereochemically pure.

The 1-substituted 1,3-butadienes such as 1,3-pentadiene can yield all of the types of isomers we have just discussed and still others in addition. Thus, Natta (5) prepared an optically active *cis*-1,4-poly-1,3-pentadiene; this is possible with a suitably chosen optically active catalyst because the chain now has true asymmetric centers.

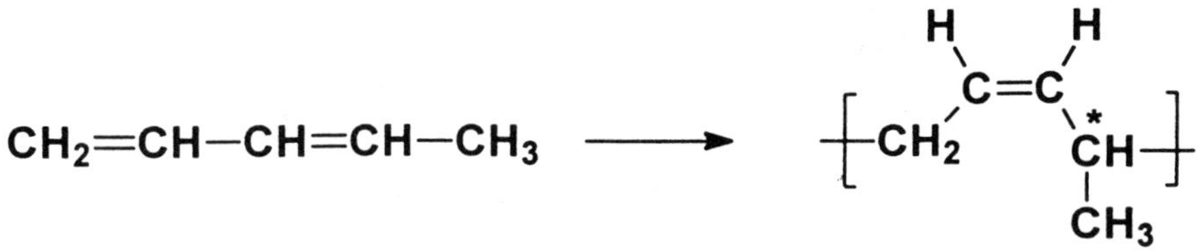

$$CH_2=CH-CH=CH-CH_3 \longrightarrow \left[CH_2 \underset{}{\overset{HH}{\underset{}{\overset{\diagdown\diagup}{C=C}}}} \underset{\underset{CH_3}{|}}{\overset{*}{CH}} \right]$$

2.2. Polymer Microstructure

```
    H       A              H      CH₂—
     \     /                \    /
      C=C                    C=C
     /     \                /    \
  —CH₂    CH₂—           —CH₂    A

    (a)                     (b)
```

FIGURE 2.12 The structure of (a) *cis* and (b) *trans* 2-substituted butadienes.

Such monomers can yield *cis*-isotactic, *trans*-isotactic, *cis*-syndiotactic, and *trans*-syndiotactic structure, but as yet these have not been prepared and established.

Dienes of the ACH = CHCH = CHA type can give rise to the isomeric chains shown in Fig. 2.14 by 1,4-addition. If the substituents are unlike, ACH = CHCH = CHB, the structures shown in Fig. 2.15 can be generated.

These structures may be termed di-isotactic; one can also imagine syndiotactic structures with A or A and B alternating from side to side of the plane of the main-chain carbons. Natta and co-workers (6,7) have prepared optically active crystalline polymers from esters of *trans–trans* sorbic acid:

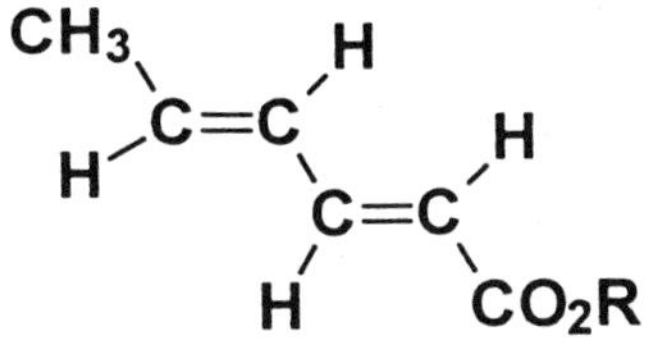

```
    CH₃      H
      \    /
       C=C         H
      /    \      /
     H      C=C
           /    \
          H     CO₂R
```

trans-trans Sorbic acid

```
        A                        
        |                       
  —CH₂—C—              —CH₂—CH—
        |                     |
        CH                    C—A
        ‖                     ‖
        CH₂                   CH₂

       (a)                    (b)
```

FIGURE 2.13 The (a) 1,2- and (b) 3,4-additions of 2-substituted butadienes.

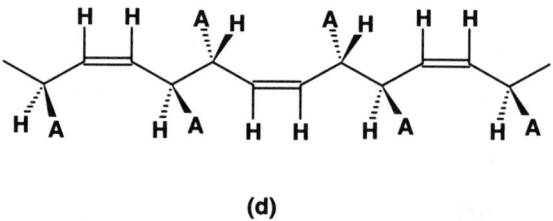

FIGURE 2.14 The structure of (a) *trans*-DL, (b) *trans-meso*, (c) *cis*-DL, and (d) *cis-meso* diene chains that arise from 1,4-addition.

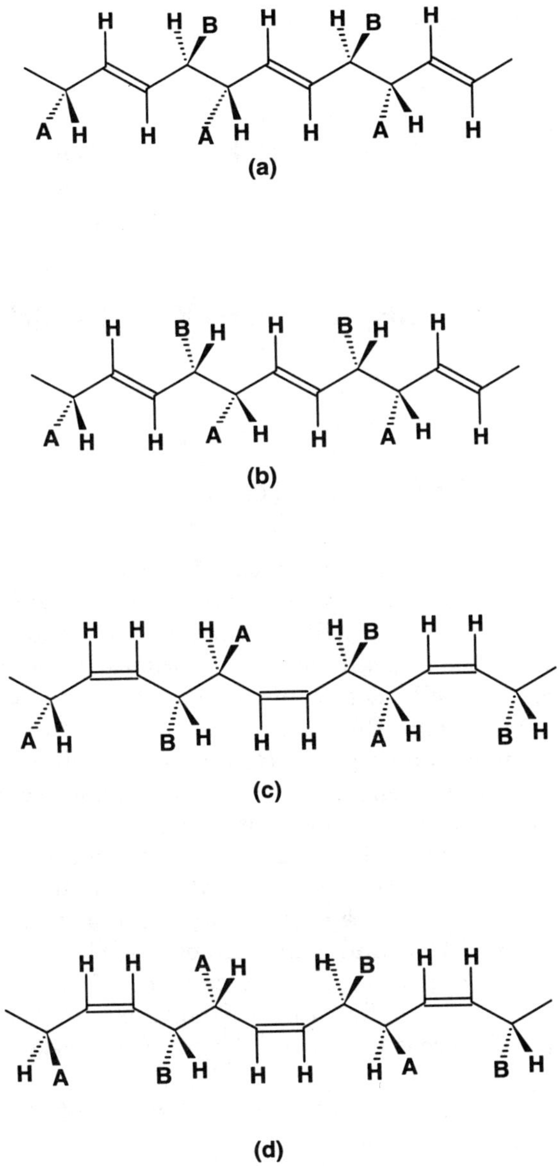

2.2. Polymer Microstructure

FIGURE 2.15 The structures of (a) *trans*-DL-erythro, (b) *trans*-DL-threo, (c) *cis*-DL-erythro, and (d) *cis*-DL-threo chains arising from polymerization of unsymmetric diene monomers.

2.2.4 Vinyl Polymers with Optically Active Sidechains

Vinyl and related monomers may have true asymmetric centers in the sidechains, for example,

$$R-CH_2-\underset{\underset{R}{|*}}{CH}-(CH_2)_n CH=CH_2$$

$$R-CH_2-\underset{\underset{R}{|*}}{CH}-(CH_2)_n O-CH=CH_2$$

$$R-CH_2-\underset{\underset{R}{|*}}{CH}-(CH_2)_n \underset{\underset{}{}}{\overset{\overset{O}{\|}}{C}}H$$

(It may be noted that aldehydes can be polymerized through the carbonyl double bond and thus in effect behave like vinyl monomers.) These asymmetric centers will be carried over into the polymers. Polymerization of the optically active enantiomer (D or L; or R and S) will lead to an optically active polymer that may also be isotactic, syndiotactic, or atactic in the usual sense.

Polymerization of the racemic (RS) monomer will generally result in a copolymer of the R and S monomer units, both coexisting in the same chain. However, it has been observed that some RS monomers, for example, RS-4-methyl-1-hexene, can be polymerized by nonchiral catalysts to yield isotactic polymers which can be separated into poly(R) and poly(S) chains, a very surprising and intriguing observation (8). It is evident that under the influence of this type of catalyst a growing chain ending in an R unit can only add another R unit, while a growing chain ending in an S unit can only add another S unit. This type of process has been termed *stereoselective polymerization*.

There is another type of selectivity which is exerted by a *chiral* catalyst, such as

$$\left[\begin{array}{c} CH_3 \\ \diagdown CH-CH_2 \\ | \\ CH_2 \\ | \\ CH_3 \end{array}\right]_3 Al/TiCl_4$$

2.2. Polymer Microstructure

If we use the chiral catalyst with the same monomer, (R, S)-4-methyl-1-hexene, we obtain a result similar to that observed with the nonchiral catalyst. But if the asymmetric center is one carbon closer to the double bond, as in methyl-1-pentene, the catalyst will now select one enantiomer (not necessarily the one with the same handedness as the catalyst) and will ignore the other.

NMR ordinarily cannot distinguish a poly(R) chain from a poly(S) chain, not being sensitive to chirality as such unless the chiral pair is dissolved in a solvent which is itself chiral (9). In addition, junctions between blocks of units of opposite configuration can in principle be detected.

2.2.5 Polymers with Asymmetric Centers in the Main Chain

It is not possible, as we have seen, for vinyl monomers to have truly asymmetric main-chain carbons, but there are other types of monomers, polymerizing in an analogous manner, which do have such carbons. Much studied examples are polypropylene oxide (and polyalkylene oxides generally), polysulfones, and in general polymers with a three-atom main-chain repeat.

$$\left[\text{NH} - \underset{\underset{\text{R}}{|}}{\text{CH}} - \underset{\underset{\text{O}}{\|}}{\text{C}} \right]$$

Poly-α-amino acids

If optically active monomers are polymerized, the resulting polymer will be optically active, and may also be isotactic, syndiotactic, atactic, or heterotactic, as shown in Fig. 2.16. There are many important polymers in this class, including poly-α-amino acids and polylactic acids:

$$\left[\text{O} - \underset{\underset{\text{R}}{|}}{\text{CH}} - \underset{\underset{\text{O}}{\|}}{\text{C}} \right]$$

Poly(lactic acids)

2.2.6 Branching and Cross Linking

Branching is a highly important structural variable that has received substantial theoretical and experimental study but only limited observation and measurement. Only polyethylene and poly(vinyl chlo-

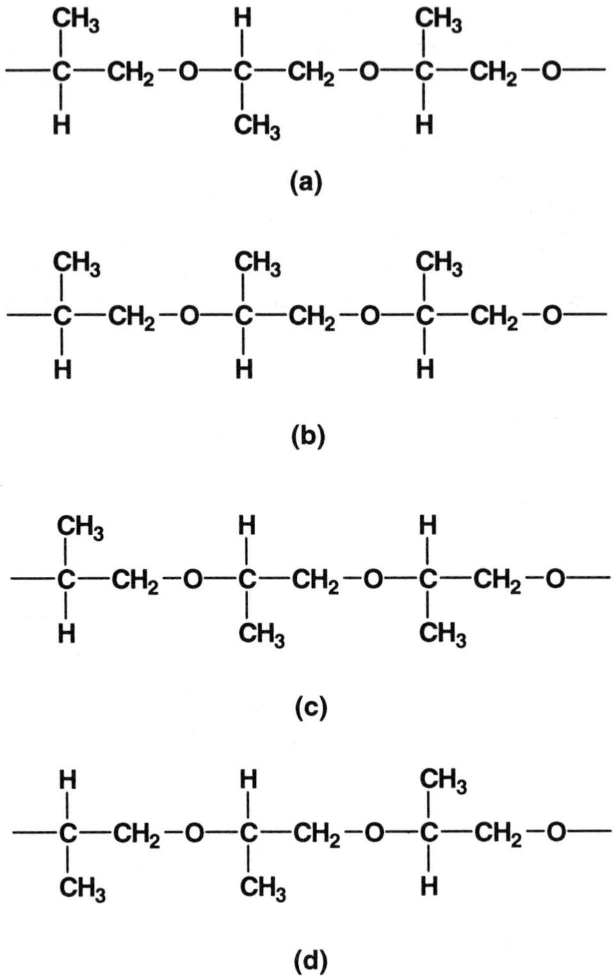

(a)

—C(CH₃)(H)—CH₂—O—C(CH₃)(CH₃...

FIGURE 2.16 The structure of (a) isotactic (*RRR* or *SSS*), (b) syndiotactic (*RSR* or *SRS*), (c) heterotactic-1 (*RRS* or *SSR*), and (c) heterotactic-2 (*SRR* or *RSS*) sequences in poly(propylene oxide).

ride) have been investigated in great detail. Branching may be produced deliberately by the introduction of diene and divinyl and polyvinyl monomers as comonomers. These comonomers yield double bonds in the copolymer chains, which can polymerize to yield branches and cross links as shown in Fig. 2.17. Branching may also be introduced by processes that are under less specific control and involve chain transfer reactions of various types. Such reactions are particularly to be expected for highly reactive polymer radicals that are not

2.2. Polymer Microstructure

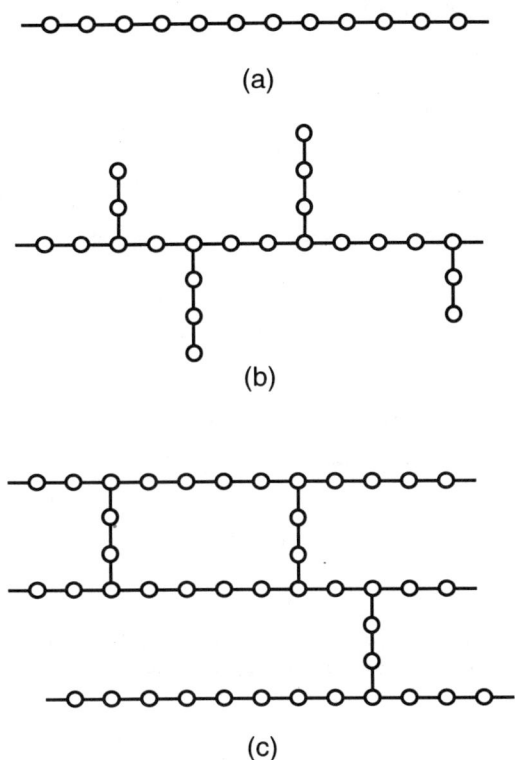

FIGURE 2.17 Chain architecture in polymers for (a) linear, (b) branched, and (c) cross-linked chains.

stabilized by resonance, such as those from ethylene, vinyl chloride, and vinyl acetate.

In condensation, polymer reactions and related systems such as poly-epoxides and polyurethanes cross linking is of great importance and may be followed by solid-state NMR techniques on materials that have passed well beyond the soluable state. Such studies are described in Chapter 4.

2.2.7 COPOLYMER SEQUENCES

Copolymer sequence is another feature that can have an important impact on the macroscopic behavior of a polymer synthesized from more than a single monomer. Some of the possible copolymer architectures are shown in Fig. 2.18, and include random, alternating, block, and graft copolymerizations. Chain architecture can have a

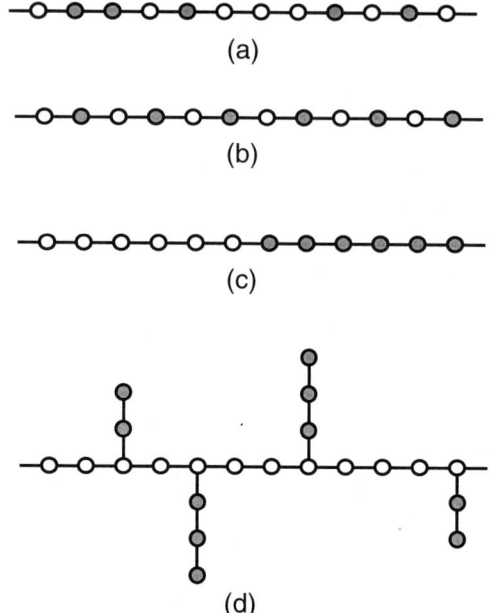

FIGURE 2.18 Chain architecture in copolymers. The drawings show (a) random, (b) alternating, (c) block, and (d) graft copolymers.

profound effect on the appearance of the NMR spectra. In addition to such chain architectures, copolymers exhibit the microstructural variations found in other polymers, including stereochemical isomerism and regioisomerism. This is illustrated in Fig. 2.19, which shows the six types of dyads that may be found in copolymers.

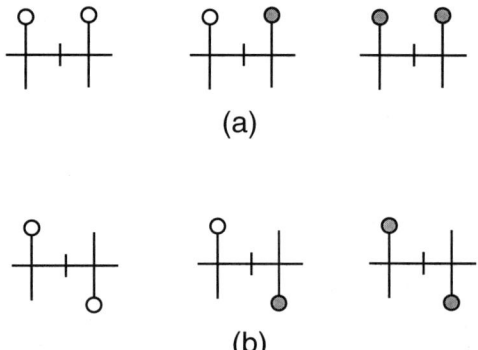

FIGURE 2.19 The structure of (a) *meso* and (b) racemic copolymer sequences.

2.2.8 OTHER TYPES OF ISOMERISM

It perhaps goes without saying that in polymers other than those already mentioned, there are many types of structural isomerism which can usually be readily determined by NMR, as for example, the presence of *ortho-* or *meta*-linked units in aromatic polyesters, isomerism in phenol–formaldehyde resins, polyamides, polysulfones, etc. Studies of this kind are often very useful, but for the most part involve problems in isomer identification which are no different from those of the corresponding small molecules. They do not take advantage of the unique ability of NMR to provide sequence information.

2.3 CONFIGURATIONAL STATISTICS AND PROPAGATION MECHANISM

NMR spectroscopy is extremely useful for microstructural characterization and for determination of the propagation mechanism. In proton, and especially carbon, NMR, peaks can be assigned for long configurational sequences, often to the pentad level or higher. The intensities of the peaks for the different sequences are sensitive to the propagation mechanism and can be used to distinguish between mechanisms in favorable cases.

2.3.1 BERNOULLI AND FIRST-ORDER PROPAGATION MODELS

In Fig. 2.20 is shown the building up of a chain by Bernoulli trial and first-order Markov steps. It is perhaps generally understood what each of these types of sequences implies concerning the mechanism producing it, but we shall nevertheless consider these distributions in some detail here, because one sometimes finds statements in the literature concerning their interpretation that are inconsistent or misleading. In Bernoulli trial propagation, the chain end is not represented as having any particular stereochemistry; i.e., not only is it not D and L (a terminology generally to be avoided), but it is also unimportant whether it is m or r. This process is thus like reaching at random into a large jar containing balls marked m and r, the fraction of m balls being P_m. The outcome of the choice does not depend on the previous choice, the jar being assumed to be large enough so that neither species is appreciably depleted. It is sometimes said of this mechanism that the addition "is influenced only by the end unit of the growing chain." Such statements should not be understood to imply that it is influenced by the *stereochemistry* of the end of the growing

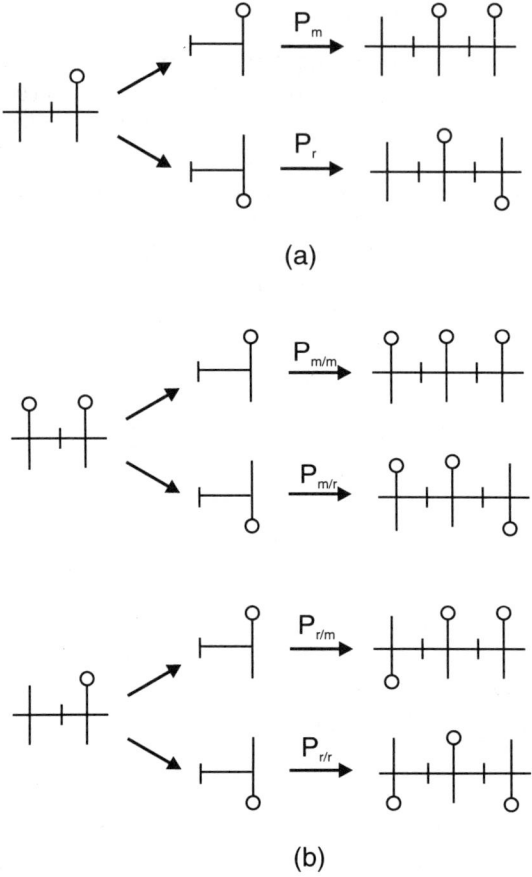

FIGURE 2.20 Polymerization diagrams for (a) Bernoullian and (b) first-order Markov polymerization.

chain, for *one* monomer chain unit considered alone as in Fig. 2.20(a) has no stereochemistry.[1]

The first-order Markov sequence is generated by propagation steps in which the adding monomer *is* influenced by the stereochemistry of the growing chain end, which may be m or r. We now have four probabilities characterizing the addition process: $P_{m/m}$, $P_{r/m}$, $P_{m/r}$, and $P_{r/r}$. (The designation $P_{r/m}$ means here the probability that the

---

[1] It is probable that in free radical propagation (and possibly anion and cationic as well) the growing radical chain end is actually trigonal, i.e., "flat," and that its stereochemistry is decided as the next monomer adds. These considerations do not affect the formal statistical arguments.

2.3. Configurational Statistics and Propagation Mechanism

monomer adds in m fashion to an r chain end, etc.) For the triad frequencies we have

$$(mm) = \frac{P_{m/m}P_{r/m}}{P_{m/r} + P_{r/m}} \tag{2.1}$$

$$(mr) = \frac{2P_{m/r}P_{r/m}}{P_{m/r} + P_{r/m}} \tag{2.2}$$

$$(rr) = \frac{P_{r/r}P_{m/r}}{P_{m/r} + P_{r/m}}. \tag{2.3}$$

We also have the relationships

$$P_{m/m} + P_{m/r} = 1 \tag{2.4}$$

and

$$P_{r/r} + P_{r/m} = 1. \tag{2.5}$$

There are thus only two independent probabilities, which we shall take as $P_{r/m}$ and $P_{m/r}$. The propagation may be characterized if we can determine values of these conditional probabilities. Of course, one can also easily imagine second-order Markov processes, characterized by four conditional probabilities and implying an influence of an additional chain unit, and still higher-order Markov propagation steps. One can also imagine non-Markov processes, one of which we shall discuss later.

Concerning the fitting of data to postulated mechanisms, we can say the following (10):

- (i) From dyad information alone, i.e., (m) and (r), any mechanism can be fitted but none can be tested. By "tested," we mean that a polymer can be shown to be consistent or inconsistent with a given model at a given level of sequence discrimination. But there is always the possibility that examination with higher discrimination may reveal inconsistencies with the proposed model.
- (ii) From triad information (which implies dyad as well), i.e., (mm), (mr), and (rr), a Bernoulli model can be tested and Markov models of any order can be fitted. In addition, certain non-Markov models can be tested, as we shall see.
- (iii) From tetrad information, a first-order Markov model can be tested, and higher orders fitted.
- (iv) From *pentad* information, a second-order Markov model can be *tested*, and higher orders fitted.

The extension of such statements to higher sequences is obvious.
From Fig. 2.20 we find that

$$(mm) = P_m^2 = (m)^2 \tag{2.6}$$

$$(rm) = 2P_m(1 - P_m) = 2(r)(m) \tag{2.7}$$

$$(rr) = (1 - P_m)^2 = (r)^2. \tag{2.8}$$

These relationships may be used for testing peak intensities for conformity to Bernoullian statistics. Dyad data provide useful support through the dyad–triad necessary relationships, but are not necessary.

One point should be emphasized in connection with such fitting to supposed Bernoullian schemes. Triad intensities will fit Bernoullian frequencies so long as the placement of the adding monomer unit is independent of whether the end is m or r; it can be influenced to an arbitrarily high degree by the stereochemistry of the next unit from the end and still appear Bernoullian by this test. Thus, for example, we may imagine that by a second-order Markov process, we may generate chains with repeating $mmrr$ units:

$$\cdots mmrrmmrrmmrr \cdots .$$

By observation of only diad and triad intensities, such a polymer would appear to be random or "atactic," since $(m) = (r)$ and $(mm) = (rr) = 0.5(mr)$. From the observation of tetrad frequencies, however, it would become obvious that this is not the case, since from Fig. 2.20 we find that for a random polymer,

$$(mmm) = (rrr) = (rmr) = (mrm) = \tfrac{1}{8}$$

and

$$(mmr) = (rrm) = \tfrac{1}{4},$$

while for the above polymer we have only

$$(mmr) = (rrm) = \tfrac{1}{2}$$

since the other tetrad sequences are absent.

To illustrate how the NMR peak intensities can be used to investigate the polymerization mechanism, we compare the intensities reported for two synthesis of poly(methyl methacrylate) (11). The intensities for the free radical poly(methyl methacrylate), polymer I, are listed in Table 2.1, and the intensities for the anionically polymerized material, polymer II, are listed in Table 2.2. The conformity to Bernoullian statistics is tested by application of Eqs. (2.6)–(2.8), or in

2.3. Configurational Statistics and Propagation Mechanism

TABLE 2.1
The Peak Intensities for Stereosequences in Poly(methyl methacrylate) (I) from a Free Radical Polymerization (14)

	Stereochemistry	Observed	Bernoulli trial, $P_m = 0.24$
Dyad			
	(m)	0.24	
	(r)	0.76	
Triad			
	(mm)	0.04	0.06
	(mr)	0.36	0.36
	(rr)	0.60	0.58
Tetrad			
	(mmm)	~0.00	0.01
	(mmr)	0.07	0.09
	(rmr)	0.19	0.20
	(mrm)	0.04	0.04
	(mrr)	0.23	0.23
	(rrr)	0.43	0.44
Pentad			
	$(mmmm)$	~0.00	0.003
Peak 1	$(rmmr)$	0.05	0.03
Peak 2	$(mmmr)$	0.02	0.02
	$(rmrr)$		
Peak 3	$(mmrr)$	0.25	0.27
Peak 4	$(rmrm)$	0.06	0.07
Peak 5	$(mmrm)$	0.02	0.02
Peak 6	(rrr)	0.40	0.39
Peak 7	$(mrrr)$	0.14	0.16
Peak 8	$(mrrm)$	0.05	0.03

a slightly different way, by using Eqs. (2.9) and (2.10), obtained from Eqs. (2.1)–(2.5):

$$P_{m/r} = \frac{(mr)}{2(mm) + (mr)} [= (1 - P_m)] = P_r \tag{2.9}$$

$$P_{r/m} = \frac{(mr)}{2(rr) + (mr)} [= P_m]. \tag{2.10}$$

The relationships in brackets are true if the statistics are indeed Bernoullian, so that in this case $P_{m/r} + P_{r/m} = 1$. (We shall desig-

TABLE 2.2
The Peak Intensities for Stereosequences in Poly(methyl methacrylate) (II) from an Anionic Polymerization (14)

	Stereochemistry	Observed	First-order Markov	Second-order Markov	Coleman–Fox
Dyads					
	(m)	0.82			
	(r)	0.18			
Triads					
	(mm)	0.75			
	(mr)	0.14			
	(rr)	0.11			
Tetrad					
	(mmm)	0.70	0.69	0.70	0.70
	(mmr)	0.09	0.13	0.09	0.09
	(rmr)	0.03	0.006	0.03	0.02
	(mrm)	0.04	0.03	0.04	0.04
	(mrr)	0.06	0.09	0.06	0.06
	(rrr)	0.07	0.07	0.07	0.08
Pentad					
Peak 1	(mmmm)	0.75	0.75	0.77	0.74
	(mmmr)				
	(rmmr)				
Peak 2	(rmrr)	0.07	0.09	0.06	0.07
	(mmrr)				
Peak 3	(rmrm)	0.07	0.05	0.08	0.08
	(mmrm)				
Peak 4	(rrr)	0.07	0.04	0.05	0.07
Peak 5	(mrrr)	0.04	0.07	0.05	0.05
	(mrrm)				

nate $P_{m/r} + P_{r/m}$ as P_Σ in some contexts). From measurements of α-methyl resonance intensities we find (Table 2.1)

$$(mm) = 0.04 \qquad P_{m/r} = \frac{0.36}{0.08 + 0.36} = 0.82$$

$$(mr) = 0.36 \qquad P_{r/m} = \frac{0.36}{1.2 + 0.36} = 0.23 = P_m$$

$$(rr) = 0.60 \qquad P_\Sigma = 0.05.$$

The polymer is thus Bernoullian within the probable experimental error.

2.3. Configurational Statistics and Propagation Mechanism

When this test is applied to the spectrum of polymer II (see triad data in Table 2.2) we find

$$P_{m/r} = \frac{0.14}{1.50 + 0.14} = 0.085$$

$$P_{r/m} = \frac{0.14}{0.22 + 0.14} = 0.39$$

$$P_\Sigma = 0.475.$$

Thus, as already indicated, this polymer clearly does not conform to Bernoullian statistics.

A more searching test of the configurational statistics uses higher-order configurational sequences, and in Table 2.1 the tetrad and pentad intensities for polymer I are compared to the predicted intensities based on a Bernoullian model. They agree within experimental error, assuming a value of P_m of 0.24 (This value is slightly higher than than obtained from the preceding triad analysis and should be regarded as more nearly correct, since the frequencies of higher sequences are of course more sensitive measures of P_m.)

In order to explore further the statistics of polymer II, we must examine the fitting of tetrad and pentad sequence frequencies to the first-order Markov model. We shall find that we must also consider more complex propagation models: (i) second-order Markov and (ii) possible non-Markovian models, of which we select the Coleman–Fox model.

2.3.2 Fitting of Longer Sequences of the First-Order Markov Model

The peak intensities from longer configurational sequences can be used to test more complex models. To do so we first recast the triad relations of Eqs. (2.6) and (2.8) in terms of $P_{m/r}$ and $P_{r/m}$:

$$(mm) = \frac{(1 - P_{m/r})P_{r/m}}{P_{m/r} + P_{r/m}} \qquad (2.11)$$

$$(mr) = \frac{2P_{m/r}P_{r/m}}{P_{m/r} + P_{r/m}} \qquad (2.12)$$

$$(rr) = \frac{(1 - P_{r/m})P_{r/m}}{P_{m/r} + P_{r/m}}. \qquad (2.13)$$

For tetrad sequences,

$$(mmm) = \frac{(1 - P_{m/r})^2 P_{r/m}}{P_{m/r} + P_{r/m}} \left[= \frac{(mm)^2}{(m)} \right] \qquad (2.14)$$

$$(mmr) = \frac{2 P_{m/r}(1 - P_{m/r}) P_{r/m}}{P_{m/r} + P_{r/m}} \left[= \frac{(mm)(mr)}{(m)} \right] \qquad (2.15)$$

$$(mrm) = \frac{P_{m/r} P_{r/m}^2}{P_{m/r} + P_{r/m}} \left[= \frac{(mr)^2}{4(r)} \right] \qquad (2.16)$$

$$(mrr) = \frac{2 P_{m/r} P_{r/m}(1 - P_{r/m})}{P_{m/r} + P_{r/m}} \left[= \frac{(mr)(rr)}{(m)} \right] \qquad (2.17)$$

$$(rmr) = \frac{P_{m/r}^2 P_{r/m}}{P_{m/r} P_{r/m}} \left[= \frac{(mr)^2}{4(m)} \right] \qquad (2.18)$$

$$(rrr) = \frac{P_{m/r}(1 - P_{r/m})^2}{P_{m/r} + P_{r/m}} \left[= \frac{(rr)^2}{(r)} \right]. \qquad (2.19)$$

It is desirable and sufficient to show that four (not five) of these six relationships hold, but this is not always experimentally feasible because of overlapping and masking of resonances. The relationships in brackets hold for Bernoullian statistics and for first-order Markov statistics. They can therefore be useful for testing deviations from first-order Markov statistics. Thus for polymer II we find (see Table 2.2) that $(mm) = 0.75$ and $(m) = 0.82$, and so $(mm)^2/(m) = 0.69$; the observed value of (mmm) is 0.70, in agreement with Eq. (2.14) within experimental error. As we shall see shortly, more stringent tests can be applied.

For pentad sequences,

$(mmmm)$

$$= \frac{P_{r/m}(1 - P_{m/r})^3}{P_{m/r}(1 - P_{r/m}) + 2 P_{m/r} P_{r/m} + P_{r/m}(1 - P_{m/r})} \left[= \frac{(mmm)^2}{(mm)} \right] \qquad (2.20)$$

2.3. Configurational Statistics and Propagation Mechanism

$(mmmr)$
$$= \frac{2P_{m/r}P_{r/m}(1-P_{m/r})^2}{P_{m/r}(1-P_{r/m}) + 2P_{m/r}P_{rm} + P_{r/m}(1-P_{m/r})}$$
$$\left[= \frac{(mmm)(mmr)}{(mm)} \right] \quad (2.21)$$

$(mmrm)$
$$= \frac{2P_{m/r}P_{r/m}^2(1-P_{m/r})}{P_{m/r}(1-P_{r/m}) + 2P_{m/r}P_{r/m} + P_{r/m}(1-P_{m/r})}$$
$$\left[= \frac{2(mmr)(mrm)}{(mr)} \right] \quad (2.22)$$

$(mmrr)$
$$= \frac{2P_{m/r}P_{r/m}(1-P_{r/m})(1-P_{m/r})}{P_{m/r}(1-P_{r/m}) + 2P_{m/r}P_{r/m} + P_{r/m}(1-P_{m/r})}$$
$$\left[= \frac{(mmr)(mrr)}{(mr)} \right] \quad (2.23)$$

$(mrmr)$
$$= \frac{2P_{m/r}^2 P_{r/m}^2}{P_{m/r}(1-P_{r/m}) + 2P_{m/r}P_{r/m} + P_{r/m}(1-P_{m/r})}$$
$$\left[= \frac{4(mrm)(rmr)}{(mr)} \right] \quad (2.24)$$

$(mrrm)$
$$= \frac{P_{m/r}P_{r/m}^2(1-P_{r/m})}{P_{m/r}(1-P_{r/m}) + 2P_{m/r}P_{r/m} + P_{r/m}(1-P_{m/r})} \left[= \frac{(mrr)^2}{4(rr)} \right]$$
$$(2.25)$$

$(mrrr)$
$$= \frac{2P_{m/r}P_{r/m}(1-P_{r/m})^2}{P_{m/r}(1-P_{r/m}) + 2P_{m/r}P_{r/m} + P_{r/m}(1-P_{m/r})}$$
$$\left[= \frac{(mrr)(rrr)}{(rr)} \right] \quad (2.26)$$

$(rrrr)$

$$= \frac{P_{m/r}(1 - P_{r/m})^3}{P_{m/r}(1 - P_{r/m}) + 2P_{m/r}P_{r/m} + P_{r/m}(1 - P_{m/r})} \left[= \frac{(rrr)^2}{(rr)} \right]. \quad (2.27)$$

Again, the bracketed relationships hold for Bernoullian and first-order Markov but not for higher Markovian (or non-Markovian) statistics. Because of necessary relations among sequence frequencies, it is necessary to show that six of the unbracketed relationships hold in order to establish consistency with first-order Markov statistics. This is at present even less likely to be experimentally possible than for the tetrad sequences, usually because of multiple overlaps.

It might be appropriate to summarize the four possible types of chains that can be described in first-order Markov terms (polymers resembling the last two can of course be produced by Bernoullian propagation, the limits being the corresponding pure polymer types):

1. "Block": $P_{m/r} < 0.5$, $P_{r/m} < 0.5$, $P_\Sigma < 1$
2. Heterotactic-like ("alternating"): $P_{m/r} > 0.5$, $P_{r/m} > 0.5$, $P_\Sigma > 1$
3. Isotactic-like: $P_{m/r} < 0.5$, $P_{r/m} > 0.5$
4. Syndiotactic-like: $P_{m/r} > 0.5$, $P_{r/m} < 0.5$

Note that for a "pure" heterotactic polymer, $P_{m/r} = P_{r/m} = 1.0$.

Coleman and Fox (12,13) have defined a useful quantity, the *persistence ratio*, as

$$\rho = \frac{2(m)(r)}{(mr)}. \quad (2.28)$$

For first-order Markov statistics, ρ is equal to $(P_{m/r} + P_{r/m})^{-1}$. For polymer II, we find that ρ is equal to 2.1, the deviation in excess of unity expressing the tendency of the r and m species to propagate themselves preferentially, i.e., in a nonrandom manner.

Another criterion for Bernoullian statistics, which employs only triad data, is

$$4\frac{(mm)(rr)}{(mr)}. \quad (2.29)$$

This quantity is unity for a Bernoullian chain. This is a very sensitive test, and should be applied only to very accurate data, especially if (mm) or (rr) is small.

2.3.3 More Complex Propagation Models

In order to explore further the statistics of polymer II, we must examine complex propagation models. We shall discuss two: (i) second-order Markov and (ii) Coleman–Fox (one of many possible non-Markovian) models.

2.3.3.1 Second-Order Markov Model

This model requires the specification of eight conditional probabilities, for we must consider the influence of the relative configurations of the last three pseudoasymmetric centers of the growing chain end. For convenience, we designate these as

$$P(mm/m) = \alpha, \quad P(mm/r) = \bar{\alpha}$$
$$P(mr/m) = \beta, \quad P(mr/r) = \bar{\beta}$$
$$P(rm/m) = \gamma, \quad P(rm/r) = \bar{\gamma}$$
$$P(rr/m) = \delta, \quad P(rr/r) = \bar{\delta},$$

where $P(mm/m)$ is the probability of a monomer adding in m fashion to a chain ending in mm, and so on. There are actually four independent probabilities since

$$\alpha + \bar{\alpha} = 1, \quad \gamma + \bar{\gamma} = 1$$
$$\beta + \bar{\beta} = 1, \quad \delta + \bar{\delta} = 1.$$

This model reduces to the first-order Markov model when $\alpha = \gamma$ and $\beta = \delta$. Putting

$$s = \bar{\alpha}\bar{\beta} + 2\bar{\alpha}\delta + \gamma\delta, \tag{2.30}$$

we find that the observable dyad through pentad frequencies are

$$(m) = (\bar{\alpha} + \gamma)\delta/s \tag{2.31}$$

$$(r) = (\bar{\beta} + \delta)\alpha/s \tag{2.32}$$

$$(mm) = \gamma\delta/s \tag{2.33}$$

$$(mr) = 2\bar{\alpha}\delta/s \tag{2.34}$$

$$(rr) = \bar{\alpha}\bar{\beta}/s \tag{2.35}$$

$$(mmm) = \alpha\gamma\delta/s \tag{2.36}$$

$$(mmr) = 2\bar{\alpha}\gamma\delta/s \tag{2.37}$$

$$(mrm) = \bar{\alpha}\beta\delta/s \tag{2.38}$$

$$(mrr) = 2\bar{\alpha}\bar{\beta}\delta/s \tag{2.39}$$

$$(rmr) = \bar{\alpha}\bar{\gamma}\delta/s \tag{2.40}$$

$$(rrr) = \bar{\alpha}\bar{\beta}\bar{\delta}/s \tag{2.41}$$

$$(mmmm) = \alpha^2\gamma\delta/s \left[= \frac{(mmm)^2}{(mm)} \right] \tag{2.42}$$

$$(mmmr) = 2\alpha\bar{\alpha}\gamma\delta/s \left[= \frac{(mmm)(mmr)}{(mm)} \right] \tag{2.43}$$

$$(mmrm) = 2\bar{\alpha}\beta\gamma\delta/s \left[= \frac{2(mmr)(mrm)}{(mr)} \right] \tag{2.44}$$

$$(mmrr) = 2\bar{\alpha}\bar{\beta}\gamma\delta/s \left[= \frac{(mmr)(mrr)}{(mr)} \right] \tag{2.45}$$

$$(mrmr) = 2\bar{\alpha}\beta\bar{\gamma}\delta/s \left[= \frac{4(mrm)(rmr)}{(mr)} \right] \tag{2.46}$$

$$(mrrm) = \bar{\alpha}\bar{\beta}\delta^2/s \left[= \frac{(mrr)^2}{4(rr)} \right] \tag{2.47}$$

$$(mrrr) = 2\bar{\alpha}\bar{\beta}\delta\bar{\delta}/s \left[= \frac{(mrr)(rrr)}{(rr)} \right] \tag{2.48}$$

$$(rmmr) = \bar{\alpha}^2\gamma\delta/s \left[= \frac{(mmr)^2}{4(mm)} \right] \tag{2.49}$$

$$(rmrr) = 2\bar{\alpha}\bar{\beta}\bar{\gamma}\delta/s \left[= \frac{2(mrr)(rmr)}{(mr)} \right] \tag{2.50}$$

$$(rrrr) = \bar{\alpha}\bar{\beta}\bar{\delta}^2/s \left[= \frac{(rrr)^2}{rr} \right]. \tag{2.51}$$

The bracketed relationships hold for Bernoullian, first-order Markov, and second-order Markov statistics, but not for higher-order

2.3. Configurational Statistics and Propagation Mechanism

Markov models. The following expressions are also useful:

$$\alpha = \frac{(mmm)}{(mm)}, \qquad \bar{\alpha} = \frac{(mmr)}{2(mm)}$$

$$\beta = \frac{2(mrm)}{(mr)}, \qquad \bar{\beta} = \frac{(mrr)}{(mr)}$$

$$\gamma = \frac{(mmr)}{(mr)}, \qquad \bar{\gamma} = \frac{2(rmr)}{(mr)}$$

$$\delta = \frac{(mrr)}{2(rr)}, \qquad \bar{\delta} = \frac{(rrr)}{(rr)}.$$

2.3.3.2 Coleman–Fox Model

Another significant mechanistic scheme, which does not contemplate any influence of the chain-end stereochemistry on the mode of addition of the next monomer unit, is that put forward by Coleman and Fox (12–14), primarily to explain the "block-like" configuration exhibited in varying degrees by most if not all propagating species which deviate from Bernoullian statistics. (We do not consider here Ziegler–Natta catalysis.) Mechanistically, one can plausibly account for the tendency of an m sequence to propagate itself, for one may easily imagine, as many authors have suggested, that in Grignard- or metal–alkyl-initiated polymerizations the counterion chelates with the incoming monomer and the chain end in such a way as to favor m placements. It is not quite so easy to see why r placements should tend to propagate nonrandomly, although "penultimate" effects or complexing can again be imagined to account for this. Instead, Coleman and Fox have proposed that "block" configurations are generated because the propagating chain end may exist in two (or possibly more) states, corresponding to chelation by the counterion and the interruption of this chelation by solvation. The intervals between the arrival and departure of the solvating species (usually an added ether such as tetrahydrofuran) are imagined to be longer than that corresponding to an average propagation step, but not necessarily very much longer since block lengths are often quite short. It is assumed that polymerization takes place in a homogeneous phase, which is probably reasonable for the metal–alkyl-initiated polymerizations to which this treatment has been applied. In each state, the propagation is assumed to be Bernoullian, but the probability of m placement is taken to be different in each state.

To express this scheme quantitatively, a notation scheme is employed. The frequency of m placement is

$$(m) = \frac{(\lambda_a k_{1i} + \lambda_b k_{2i})}{(\lambda_a k_1 + \lambda_b k_2)},$$

where

λ_a = rate constant for State 2 → State 1 of the chain end

λ_b = rate constant for State 1 → State 2 of the chain end

k_{1i} = rate constant for m placement in State 1

k_{2i} = rate constant for m placement in State 2

$k_1 = k_{1i} + k_{1s}$, where k_{1s} is the rate constant for r placement in State 1

$k_2 = k_{2i} + k_{2s}$, where k_{2s} is the rate constant for r placement in State 2

$$a = \frac{\lambda_a \lambda_b k_1 k_2}{(\lambda_a k_1 + \lambda_b k_2)^2} \left[\frac{k_{1i}}{k_1} - \frac{k_{2i}}{k_2} \right]^2$$

$$b = \frac{k_{1i} k_{2i}}{k_1 k_2}$$

$$c = \frac{k_{1s} k_{2s}}{k_1 k_2}$$

$$d = \frac{\lambda_a k_1 + \lambda_b k_2}{k_1 k_2}$$

$$x = \frac{[M]}{[M] + d},$$

where [M] is the monomer concentration in the polymerizing system.

Coleman and Fox (12) have described how the rate constants could be determined individually, assuming that λ_a and λ_b, but not the k's, may depend on the concentration of the initiator.

Such a polymerizing system will *in general* generate chains which cannot be described by first-order Markov statistic or by Markov statistics of any finite order. Sequence frequencies are expressed by a series of relationships (12).

2.3. Configurational Statistics and Propagation Mechanism

For triads,

$$(mm) = (m)^2 + ax \tag{2.52}$$

$$(mr) = 2(m)(r) - 2ax \tag{2.53}$$

$$(rr) = (r)^2 + ax. \tag{2.54}$$

These equations should be compared to Eqs. (2.6)–(2.8).

For tetrads,

$$(mmm) = \frac{(mm)^2 + abx^2}{(m)} \tag{2.55}$$

$$(mmr) = \frac{(mm)(mr) - 2abx^2}{(m)} \tag{2.56}$$

$$(mrm) = \frac{(mr)^2 + 4acx^2}{4(r)} \tag{2.57}$$

$$(mrr) = \frac{(mr)(rr) - 2acx^2}{(r)} \tag{2.58}$$

$$(rmr) = \frac{(mr)^2 + 4abx^2}{4(m)} \tag{2.59}$$

$$(rrr) = \frac{(r)^2 + acx^2}{(r)}. \tag{2.60}$$

Comparison with Eqs. (2.14)–(2.19) (bracketed relationships) show clearly the deviation from expectation for Bernoullian and first-order Markov propagation.

For pentads,

$$(mmmm) = \frac{(mmm)^2 + AM^2}{(mm)} \tag{2.61}$$

$$(mmmr) = \frac{(mmm)(mmr) - 2AM^2}{(mm)} \tag{2.62}$$

$$(mmrm) = \frac{2(mmr)(mrm) - 4AMR}{(mm)} \tag{2.63}$$

$$(mmrr) = \frac{(mmr)(mrr) - 4AMR}{(mr)} \tag{2.64}$$

$$(mrmr) = \frac{4(rmr)(mrm) - 4AMR}{(mr)} \qquad (2.65)$$

$$(mrrm) = \frac{(mrr)^2 + 4AR^2}{4(rr)} \qquad (2.66)$$

$$(mrrr) = \frac{(rrr)(mrr) - 2AR^2}{(rr)} \qquad (2.67)$$

$$(rmmr) = \frac{(mmr)^2 + 4AM^2}{4(mm)} \qquad (2.68)$$

$$(rmrr) = \frac{2(mrr)(rmr) + 4AMR}{(mr)} \qquad (2.69)$$

$$(rrrr) = \frac{(rrr)^2 + AR^2}{(rr)}. \qquad (2.70)$$

Here, $A = ax$, $M = bx$, and $R = cx$, as defined previously. These relationships should be compared to Eqs. (2.42)–(2.51) to exhibit deviations from second-order Markov (as well as Bernoullian and first-order Markov) statistics.

In Table 2.2, the observed dyad, triad, tetrad, and pentad intensities (in the last case, largely summations of superimposed peaks, unfortunately) of polymer II are compared to the predictions of first-order and second-order Markov mechanisms, and of a Coleman–Fox two-state mechanism. For the first-order Markov mechanism, the sequence frequencies are calculated from the values of the conditional probabilities obtained from the triad analysis. To test for conformity to second-order Markov statistics, the eight conditional probabilities are calculated from the observed triad and tetrad intensities:

$$\alpha = 0.94, \quad \bar{\alpha} = 0.06$$
$$\beta = 0.56, \quad \bar{\beta} = 0.44$$
$$\gamma = 0.63, \quad \bar{\gamma} = 0.37$$
$$\delta = 0.28, \quad \bar{\delta} = 0.64.$$

These values constitute also a test of first-order Markov statistics since, as we have seen, $\alpha = \gamma$ and $\beta = \delta$ if these apply. It is clear that this is not the case. From the method of calculation, it is necessary that $\beta + \bar{\beta} = 1$, and so this agreement does not constitute a test of second-order Markov statistics. The fact that $\alpha + \bar{\alpha} = 1$ and $\delta + \bar{\delta} = 1$ does constitute a valid test, however. A further test is indicated in the fourth column of Table 2.2, in which pentad intensi-

2.3. Configurational Statistics and Propagation Mechanism

ties, calculated from Eqs. (2.42)–(2.51), are listed. The agreement with experimental values is quite good except that the predicted ratio of *rrr* (peak 4) to *mrr-mrrm* (peak 5) is somewhat lower than that observed.

The calculation of a consistent set of constants for the Coleman–Fox mechanism is discussed in Frisch et al. (14), to which the reader is referred for details. The quantities a/b, b/c, and (m) are obtained from experiment by an appropriate choice among Eqs. (2.55) through (2.60) by comparison with the corresponding Markov relations. If we let $P_{m_1} = k_{1i}/k_1$ be the probability of isotactic placement in State 1, $P_{m_2} = k_{2i}/k_2$ be the probability of isotactic placement in State 2, and $W_1 = \lambda_a k_1 (\lambda_a k_1 + \lambda_b k_2)$ be the weight fraction of polymer produced in State 1, we find for polymer II that $P_{m_1} = 0.96$, $P_{m_2} = 0.21$, and $W_1 = 0.81$, State 1 being arbitrarily taken as that in which isotactic placement predominates. In the last column of Table 2.2, the tetrad and pentad frequencies calculated from these parameters are shown. The agreement with experimental observation is at least as good as that of the second-order Markov model, and is in fact rather better for peaks 4 and 5, which are not correctly predicted by the latter model. Although the close similarity in the predictions of the two models means that this result does not provide conclusive evidence for the two-state model, the parameters obtained are very reasonable. This implies that about 80% of the polymer is produced under conditions such that isotactic placement is nearly exclusively preferred. The rest of the time, isotactic placement has the probability characteristic of a free radical or uncomplexed anionic chain end. Syndiotactic sequences contain $n_r - 1$ *rr* triads and n_r *r* dyads, assuming (as always) semi-infinite chains. If $N(n_m)$ is the number of isotactic sequences of length n_m, then there are $\Sigma N(n_m)(n_m - 1)$ *mm* triads. Similarly, there are $\Sigma N(n_r)(n_r - 1)$ *rr* triads in the syndiotactic sequence of length n_r. (The summations run from 1 to infinity over n_m and n_r.) Since the numbers of isotactic and syndiotactic sequences must be equal, $N(n_m) = \Sigma N(n_r)$. Therefore

$$\frac{(mm)}{(rr)} = \frac{\Sigma N(n_m)(n_m - 1)}{\Sigma N(n_r)(n_r - 1)} = \frac{\Sigma N(n_m) n_m - \Sigma N(n_m)}{\Sigma N(n_r) n_r - \Sigma N(n_r)}$$

$$= \frac{(\bar{n}_m - 1)}{(\bar{n}_r - 1)} \tag{2.71}$$

since

$$\bar{n}_m = \frac{\Sigma n_m N(n_m)}{\Sigma N(n_m)} \tag{2.72}$$

and

$$\bar{n}_r = \frac{\Sigma n_r N(n_r)}{\Sigma N(n_r)}, \qquad (2.73)$$

where $\bar{n}_m$ and $\bar{n}_r$ are the number of average lengths of the isotactic and syndiotactic sequences, respectively. It further follows, as is obvious, that

$$\frac{(m)}{(r)} = \frac{(\bar{n}_m)}{(\bar{n}_r)}. \qquad (2.74)$$

From these relations it is found that

$$\bar{n}_m = \frac{[1-(mm)/(rr)]}{[1-(mm)(r)/(rr)(m)]} \qquad (2.75)$$

$$\bar{n}_r = \frac{[1-(mm)/(rr)]}{[(m)/(r)-(mm)/(rr)]}. \qquad (2.76)$$

The intensity of the heterotactic triad resonance is an inverse measure of block lengths, since there is one heterotactic triad between isotactic and syndiotactic blocks and the number of such units is thus equal to the number of blocks,

$$\Sigma N(n_m) + \Sigma N(n_r) = 2\Sigma N(n_m) = 2\Sigma N(n_r) \qquad (2.77)$$

$$(mr) = \frac{2\Sigma N(n_m)}{\Sigma N(n_m)(n_m-1) + \Sigma N(n_r)(n_r-1) + 2\Sigma N(n_m)}$$

$$= \frac{2\Sigma N(n_m)}{\Sigma N(n_m)n_m + \Sigma N(n_r)n_r}, \qquad (2.78)$$

so that

$$(mr)^{-1} = \tfrac{1}{2}(\bar{n}_m + \bar{n}_r) = \bar{n}, \qquad (2.79)$$

where $\bar{n}$ is the average length of all blocks. Thus a prominent mr resonance implies short blocks. Since

$$\frac{\bar{n}_m}{(\bar{n}_m - \bar{n}_r)} = (m) \qquad (2.80)$$

$$\frac{\bar{n}_r}{(\bar{n}_m + \bar{n}_r)} = (r) \qquad (2.82)$$

it follows from Eqs. (2.6)–(2.8) that

$$\bar{n}_m = \frac{2(m)}{(mr)} = 1 + \frac{2(mm)}{(mr)} \qquad (2.83)$$

$$\bar{n}_r = \frac{2(r)}{(mr)} = 1 + \frac{2(rr)}{(mr)}. \qquad (2.84)$$

For Bernoullian chains, it follows that

$$\bar{n}_m = \frac{1}{(1 - P_m)} \qquad (2.85)$$

$$\bar{n}_r = \frac{1}{P_m}. \qquad (2.86)$$

For any statistics, the persistence length is given by

$$\rho = \bar{n}_m(r) = \bar{n}_r(m). \qquad (2.87)$$

For first-order Markov statistics

$$\bar{n}_m = \left[(r)(P_{m/r} + P_{r/m})\right]^{-1} \qquad (2.88)$$

$$\bar{n}_r = \left[(m)(P_{m/r} + P_{r/m})\right]^{-1}. \qquad (2.89)$$

Extension to higher-order Markov statistics is obvious, but such relations are not particularly useful since block lengths can be obtained directly from triad (or dyad–triad) frequencies from Eqs. (2.82) and (2.83).

2.4 PROPAGATION ERRORS

A problem of some significance in relation to the propagation mechanism is the nature of the configurational defects introduced into highly stereoregular chains. Even the most efficient catalyst systems can be expected to be something less than perfect in their stereoregulating abilities, and the nature of the errors introduced, if it can be determined, has a bearing on the interpretation of catalyst action.

One may imagine two ways in which a minor configurational "error" may be introduced into an isotactic chain. The first is a "template" propagation error [Fig. 2.21(a)], for which the (*mmr*):(*mrr*) ratio is 1:1. A "steric" propagation error [Fig. 2.21(b)] is also possible, for which one will observe a 2:1 (*mmr*):(*mrm*) ratio and no *mrr*. Also, no *rr* resonance will be observed.

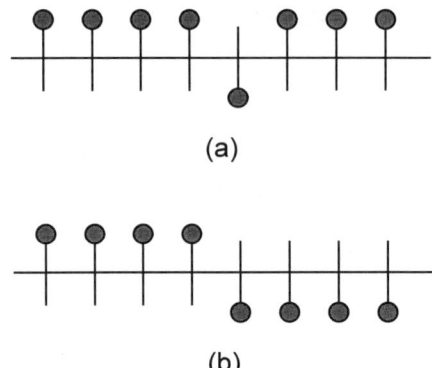

FIGURE 2.21 Schematic illustration of (a) "template" and (b) "steric" propagation errors.

The first of these implies, but does not rigorously prove, that the propagation is under the control of an asymmetric catalyst site which is sensitive to the symmetry of the growing chain end and corrects an r propagation error by restoring the absolute relationship of chain to site. The second structure implies that the principal correcting force is the steric nature of the chain end itself, possibly under the influence of a coordinating counterion; once a new m unit is added beyond the r error, propagation proceeds as before.

Zambelli (15) demonstrated in an ingenious way that template control probably prevails in some isotactic propylene polymerizations. This was accomplished by copolymerizing ^{13}C-enriched ethylene in an amount small enough to ensure that virtually all of the ethylene units are inserted between propylene units. If the structure produced is exclusively of the type shown in Fig. 2.21(a) or is exclusively of the type shown in Fig. 2.21(b), the ethylene resonance will consist of two equal peaks corresponding to the two equally probably positions of the carbon-13.

If, on the other hand, there is no requirement that the propylene unit added beyond the ethylene unit is added as in Fig. 2.21(a), i.e., no template control, both will be produced, probably with nearly equal frequency since one can scarcely imagine that chain-end stereochemical influences will be able to bridge the intervening ethylidene unit. In this case, four ethylene peaks should be observed. In fact, two equal peaks were observed. This, in the absence of model studies, does not rigorously prove the generation of case (a) [Fig. 2.21(a)] rather than (b) [Fig. 2.21(b)] but clearly makes it highly probable that "template" control in fact occurs in this case. Thus, the nature of the propagation error evidently depends on the particular catalyst system employed.

2.4.1 Copolymer Polymerization

The NMR analysis of copolymer structure, like that of stereochemical configuration, can be carried out to varying degrees of complexity, depending on the length of the sequences that can be observed. If the copolymerization is not strictly alternating or block-like, the dyad, triad, etc., sequence can be represented as $m_1 m_1$, $m_2 m_2$, etc., if we ignore the stereochemistry. The dyad probabilities are given by

$$[m_1 m_1] = F_1 P_{11} \tag{2.90}$$

$$[m_1 m_2] = [m_2 m_1] = 2 F_1 (1 - P_{11}) \tag{2.91}$$

$$[m_2 m_2] = F_2 P_{22}, \tag{2.92}$$

where F_1 and F_2 are the overall mole fractions of monomer 1 and monomer 2 in the chain and P_{11}, P_{12}, and P_{22} are the conditional probabilities of adding monomer 1 to a chain ending in monomer 1, etc. The mole fractions are given by

$$F_1 = \frac{r_1 f_1^2 + f_1 f_2}{r_1 f_1^2 + 2 f_1 f_2 + r_2 f_2^2}, \tag{2.93}$$

and the reactivity ratios r_1 and r_2 are given by

$$r_1 = \frac{k_{11}}{k_{12}} \tag{2.94}$$

and

$$r_2 = \frac{k_{22}}{k_{21}}, \tag{2.95}$$

where k_{11}, k_{12}, and k_{22} are the rate constants for adding a monomer of one type to a polymer chain. As noted before, the combination of copolymer sequence with other types of microstructure can make the spectra of these materials extremely complex.

REFERENCES

1. R. E. Cais and N. J. A. Sloane, *Polymer* **24**, 179 (1983).
2. H. Staudinger, "Die Hochmolecularen Organischen Verbindungen," Springer-Verlag, Berlin and New York, 1932.
3. M. L. Huggins, *J. Am. Chem. Soc.* **66**, 1991 (1944).

4. C. E. Schildknecht, S. J. Gross, H. R. Davidson, J. M. Lambert, and A. O. Zoss, *Ind. Eng. Chem.* **40**, 2104 (1948).
5. G. Natta, *J. Am. Chem. Soc.* **77**, 1708 (1955).
6. G. Natta, M. Farina, M. Donati, and M. Peraldo, *Chim. Ind.* **42**, 1363 (1960).
7. G. Dall'Asta, G. Mazzanti, G. Natta, and L. Porri, *Macromol. Chem.* **56**, 224 (1962).
8. P. Pino, F. Ciardlli, and G. P. Lorenzi, *J. Polym. Sci., Part C* **4**, 21 (1963).
9. W. H. Pirkle and D. J. Hoover, *Top. Stereochem.* **13**, 263 (1982).
10. H. L. Frisch, C. L. Mallows, and F. A. Bovey, *J. Chem. Phys.* **45**, 1565 (1966).
11. F. A. Bovey, "High Resolution NMR of Macromolecules." Academic Press, New York, 1972.
12. B. D. Coleman and T. G. Fox, *J. Chem. Phys.* **38**, 1065 (1963).
13. B. D. Coleman and T. G. Fox, *J. Polym. Sci., Part A* **1**, 3183 (1963).
14. H. L. Frisch, C. L. Mallows, F. Heatley, and F. A. Bovey, *Macromolecules* **1**, 533 (1968).
15. A. Zambelli, "NMR Basic Principles and Progress," Vol. 4. Springer-Verlag, New York, 1971.

3

THE SOLUTION CHARACTERIZATION OF POLYMERS

3.1 INTRODUCTION

One of the main uses of NMR in polymer science is materials characterization, which provides the link between the synthesis of new materials and the structure–property relationships. In many polymers the thermal, mechanical, optical, and electronic properties depend on the chain microstructure, which can be characterized in great detail by ^{1}H, ^{13}C, ^{29}Si, ^{19}F, ^{31}P, and ^{15}N NMR in solution. We will see how the microstructures detailed in Chapter 2 can be observed, assigned, and characterized.

The level to which the microstructure can be elucidated depends on the resolution and the methods available for establishing the resonance assignments. The resolution depends on a number of factors, including the nuclei under observation, chain dynamics, concentration, temperature, and solvent. The resolution is best when observing a nuclei with a wide range in chemical shifts, such as ^{13}C, ^{19}F, ^{29}Si, ^{31}P, or ^{15}N. However, the sensitivity for such nuclei is not as good as that for protons (Table 1.1) so high-solution concentrations, often on the order of 10–30 wt%, are required. Fortunately, many polymers are soluable to this degree and this does not present a serious experimental limitation.

The resolution also depends on the linewidth, and a greater microstructural characterization is possible with sharper lines. As shown in Section 1.4.2 [Eq. (1.26)] the linewidth depends inversely on the spin–spin relaxation time T_2, which depends on the correlation time τ_c. For rigid molecules T_2 scales with the third power of the molar volume, so large molecules are expected to have extremely broad lines. This expectation is one of the reasons why polymer NMR developed more slowly than some of the other applications. Fortunately, most polymers are not rigid, but experience segmental flexibility on length scales ranging from one bond to several monomer units. This decreases the effective correlation time to a degree that the observed linewidths are often in the range of 1 to 10 Hz and high-resolution structures can be observed. The side chains are less constrained than the main chain, and these resonances are often sharper.

The segmental flexibility of polymers makes it possible to observe high-resolution spectra under conditions that might not be expected to yield such good results. In Section 1.4.1 (Eq. 1.17) we noted that the correlation times scale with the solution viscosity, and so broad lines are expected for viscous solutions. We also noted that high concentrations are required for solution characterization by relatively insensitive nuclei such as ^{13}C, and, because of chain entanglements, the concentrated polymer solutions often have a high viscosity. However, in many cases the distances between entanglements are much larger than the length scale of segmental flexibility, so high-resolution spectra may still be observed. In such cases, the linewidths depend on the microscopic viscosity rather than the macroscopic viscosity. The most extreme demonstration of this effect is in swollen cross-linked materials, where high-resolution spectra can be obtained as long as the distance between cross links is greater than the length scale of segmental flexibility.

The observation of chain microstructure most frequently utilizes proton and carbon NMR spectroscopy, although we will see that other nuclei can also yield important structural information. This is primarily due to the fact that the acquisition of proton and carbon NMR spectra is routine on most spectrometers and that carbons and protons make up a majority of the atoms in the chain. In addition, the NMR sensitivity of protons is higher than any other nuclei, so small amounts of materials are required. However, because proton chemical shifts typically span only 10 ppm, the resolution is not as good as that for carbons. In many cases, the lines are broad in the proton spectra because of inhomogeneous broadening, so while the correlation times may be short and the T_2's long, few resonances are resolved. Such broadening occurs when the chemical shift separations for various

3.1. Introduction

microstructures are smaller than the linewidths. Conditions can sometimes be improved by increasing the temperature or changing the solvent, but often these changes do not lead to great improvements.

Inhomogeneous broadening is less of a problem in carbon NMR because the resonances are spread out over 200 ppm, so the resonances for the various structures are more likely to be resolved. However, higher concentrations are required to observe the carbon spectrum, and the spin–lattice relaxation times are often much longer, and may be as long as several minutes. In addition, care must be taken to obtain quantitative measurements of the structures due to nuclear Overhauser effects (Section 1.8.6.4).

One of the most fundamental problems involved in using NMR methods for characterization of polymer microstructure is assignment of the resonances. A variety of methods have been used to establish resonance assignments, including the use of model compounds, the *γ-gauche* effect, the calculation of expected chemical shifts from empirical rules, and two-dimensional NMR. The comparison of the polymer spectra with model compounds or polymers was one of the first methods used to establish resonance assignments. Assignments of the m and r resonances, for example, can be made by comparison to the isotactic and syndiotactic polymers, if these materials are available. Similarly, stereosequence assignments have also been made by comparison to the NMR spectra of small molecules in which the various diastereomers can be separated and compared. The synthesis of new materials is of course an extremely difficult and time-consuming way to establish the resonance assignments.

For those structures in which large differences in chemical shift are expected, it is possible to compare the observed chemical shifts with those calculated on the basis of empirical rules established in small molecules. Examples of where these types of methods are used include establishing the assignments for regioisomers, branches, and cross-linking sites. It is now possible to buy computer programs for personal computers which quickly perform such calculations.

An approach that has also been used with great success in vinyl polymers is the *γ-gauche* effect with the *rotational isomeric state* (RIS) model, where comparison of the calculated with observed spectra can lead to the assignment of many of the resonances. Although such calculations are not perfect, they are often close enough to be convincing. In addition, if the polymerization mechanism is known, then the intensities can be used to assist in the assignments, as shown in Section 2.3.

Two-dimensional NMR is a more recent and extremely useful method for establishing the assignments in polymers. It has the advantage that the synthesis of new materials is not required, but such experiments are more demanding in terms of the spectrometer and the amount of time required to obtain the data. The general strategy in such studies is to correlate nearest neighbors, next nearest neighbors, and even long sequences. Also, if the assignments are known for the carbons, then the proton spectra can be directly assigned by heteronuclear correlations.

In addition to using solution NMR methods to characterize the microstructure of polymers, these methods can be used to study the time-averaged solution structure. Generally such studies use 2D NMR to measure the averaged J coupling constants which depend on the torsional angle between two protons separated by three bonds (Section 1.6.2) or the distances between pairs of protons. Such studies are an important measure of the conformation of polymers and the structure of intermolecular polymer complexes, and have contributed greatly to our understanding of the forces controlling miscibility in polymer blends.

3.2 RESONANCE ASSIGNMENTS

The assignment of resonances in the spectra of polymers is often the first step in materials characterization. This task is difficult because of the complex microstructure that may be observed for even the simplest polymers. Unfortunately some of the resonances are only shifted slightly from the main peaks and are difficult to observe and assign.

3.2.1 Resonance Assignments from Model Compounds

The comparison of the NMR spectra of polymers with those of small model compounds was one of the first methods used to establish resonance assignments. While this method can be very effective, it has the drawback that many molecules often have to be synthesized and separated from each other before their spectra can be required. To establish the stereosequence assignments in vinyl polymers, for example, small molecules like the 2,4,6-trisubstituted heptanes have been used (1). The isomers exist as

2,4,6-trimethyl heptane

3.2. Resonance Assignments

diastereomers and can be chromatographically separated and purified. In a very early example, many of the assignments for polypropylene were obtained by Zambelli et al. (2) using heptamethhyl heptadecane model compounds labeled at the 9 position, such as

$$3(S),5(R),7(RS),9(RS),11(RS),13(R),15(S)\text{-}$$

heptamethyl heptadecane.

~~~
    3    5    7    9   11   13   15
    |    |    |    |    |    |    |
   CH3  CH3  CH3 ¹³CH3 CH3  CH3  CH3
~~~

3,5,7,9,11,13,15-heptamethyl heptadecane

Another important source of resonance assignments is the comparison of the spectra with stereochemically "pure" materials. This is illustrated in Fig. 3.1, which compares the carbon spectra for isotactic and syndiotactic polypropylene with those of the atactic material (3). It is clear that some of the resonance assignments can be established directly from this spectral comparison. These spectra also illustrate that some of the resonances (the methyls) are much more sensitive to stereosequence effects than are others.

The comparison with the isotactic and syndiotactic polymers leads directly to the assignments of the *mmmm* and *rrrr* stereosequences, but not many of the others that appear in Fig. 3.1. One approach that has been effectively used to obtain additional assignments is the epimerization of stereochemically pure materials (4,5). Epimerization is the inversion of the stereochemistry at methine carbons that has been observed in the presence of certain catalysts. Epimerization appears to be a random process, so the appearance of stereosequences from the isotactic or syndiotactic material can be calculated statistically.

This method has been used with great success with polypropylene, where epimerization of the isotactic and syndiotactic materials gives rise to new resonances in the methyl region (4). Syndiotactic polypropylene gives rise to three resonances with an intensity ratio of 2:2:1 that can be assigned to *rrmm*, *rrrm*, and *rmmr*, while in isotactic polypropylene these resonances are assigned to *mmmr*, *mmrr*, and *mrrm*. By consideration of the necessary pentad–pentad relationships

$$2rmmr + rrrm = mmrm + mmrr$$

$$2mrrm + mrrr = mmrr + rmrr,$$

160   3. *The Solution Characterization of Polymers*

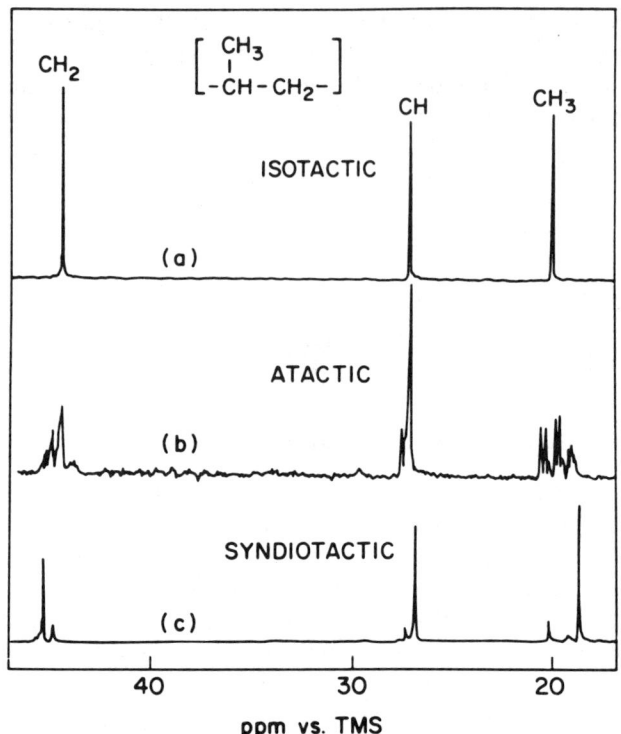

FIGURE 3.1 The 25-MHz carbon NMR spectra of (a) isotactic, (b) atactic, and (c) syndiotactic polypropylene. Reprinted with permission (3).

the three remaining stereosequences *mmrm*, *rmrr*, and *rmrr* can be assigned.

3.2.2 Assignment by Intensity and Chemical Shift Calculations

The assignments for some polymers can be made by comparison of the peak intensities with those calculated from propagation models or by comparison of the chemical shifts calculated by the methods described below. If the polymerization mechanism is known, then a partial assignment of the resonances may be obtained by comparison of the intensities with those calculated from the propagation models discussed in Section 2.3. One of the limitations of this approach is that these models often predict approximately equal amounts of two or more stereosequences and they cannot be distinguished on the basis of intensity alone. Also, as shown by Tables 2.1 and 2.2, the intensi-

3.2. Resonance Assignments

ties of some peaks are very small and they are not easy to observe in the spectra.

In some situations it is possible to establish the chemical shift assignments by comparing the chemical shifts with those calculated with empirical rules or the γ-*gauche* effect. Among the situations where the empirical rules are important are the assignments for end groups, branches, cross links, and regioisomers, where the chemical structures and resonance positions of the defects may be quite different from those of the main chain. The γ-*gauche* effect has been effectively used to establish stereosequence assignments.

Over the past several decades the chemical shifts for a wide variety of compounds have been reported, and empirical rules have been developed to relate the chemical structure to the chemical shifts, particularly for carbon NMR (6–10). In Section 1.5.2 we discussed some of these, i.e., the α, β, and γ effects on the chemical shifts of hydrocarbons. The chemical shifts for hydrocarbons are given by (10)

$$\delta_C = B + \Sigma A_l n_l + \Sigma S_l, \qquad (3.1)$$

where δ_C is the calculated chemical shift, B is a constant given by the chemical shift of methane (-2.3 ppm), n_l is the number of carbons at position l away from the carbon of interest, A_l is the additive shift due to carbon l, and S_l is a term included to account for branching. The shift parameters A_l are given in Table 3.1 for the α to ε carbons, and Table 3.2 contains the correction terms for branching. The correction terms depend on the degree of substitution of a given carbon and its neighbor. The nomenclature is such that for a tertiary carbon next to a secondary carbon the correction term 3°(2°)

TABLE 3.1
The Parameters for Calculating the Chemical Shift of Alkanes Using Emperical Additivity Relationships (10)

Carbon position	A_l (ppm)
α	9.1 ± 0.10
β	9.4 ± 0.10
γ	−2.5 ± 0.10
δ	0.3 ± 0.10
ε	0.1 ± 0.10

TABLE 3.2
The Corrective Term S_{lk} for Calculating the Chemical Shift of Branched Alkanes Using Emperical Chemical Shift Relationships (10)

	S_{lk} (ppm)
1°(3°)	-1.10 ± 0.20
1°(4°)	-3.35 ± 0.35
2°(3°)	-2.50 ± 0.25
2°(4°)	-7.5
3°(2°)	-3.65 ± 0.15
3°(3°)	-9.45
4°(1°)	-1.50 ± 0.10
4°(2°)	-8.35

would be used. This method can be illustrated by calculating the carbon chemical shift for the third carbon in 2-methyl hexane (10).

$$\overset{\beta}{CH_3}-\overset{\alpha}{C}-\overset{\alpha}{\overset{*}{C}H_2}-\overset{\beta}{CH_2}-\overset{\gamma}{CH_2}-CH_3$$
$$\underset{\beta}{\overset{|}{CH_3}}$$

2-methyl hexane

This carbon has two α, three β, and one γ neighbor, and is a 2° carbon next to a 3° carbon, so the chemical shift calculated from Tables 3.1 and 3.2 is

$$\delta_C = B + 2A_\alpha + 3A_\beta + A_\gamma + S[2°(3°)]$$
$$= -2.3 + 18.2 + 28.2 - 2.5 - 2.5$$
$$= 39.1,$$

which may be compared with the observed value of 39.45 ppm. The uncertainties in the predicted chemical shifts can be estimated from the uncertainties given in Tables 3.1 and 3.2. This methodology can be extremely useful for calculating the predicted chemical shifts for branches or cross links as well as the expected chemical shifts for head-to-head and tail-to-tail defects in polyolefins.

These empirical chemical shift rules exist not only for hydrocarbons, but also for substituted hydrocarbons, dienes, aromatic rings, and many of the other chemical constituents of polymers, and interested readers are referred to the original literature (6–10).

3.2. Resonance Assignments

Another chemical shift calculation, the γ-*gauche* effect, has been extremely useful for establishing the stereosequence resonance assignments in a wide variety of polymers. As noted in Section 1.5.2, the γ-*gauche* effect is conformational in origin and is believed to arise from the proximity of a carbon atom with substituents separated by three intervening bonds (the γ neighbor). While there is no universal agreement on the origin of the γ-*gauche* effect, it has been suggested that it arises from van der Waals interactions between the attached hydrogens leading to polarization of the CH bonds (10). These atoms are closest in the *gauche* (3 Å) conformation and more distant (4 Å) in the *trans* conformation. In most polymers there is relatively free rotation about the carbon–carbon bonds in the main chain, and many *gauche* and *trans* conformations are sampled on the time scale of the carbon chemical shifts, so the measured shift depends on the average number of *gauche* and *trans* conformers. The measured chemical shift has contributions from all of the γ-*gauche* interactions.

As might be expected for van der Waals or other types of through-space interactions, the magnitude of the induced shift depends on the group at the γ position. Table 3.3 compares the γ-*gauche* effects that have been deduced from small-molecule studies (11) for methyl groups, hydroxyl groups, and chlorine atoms. Note that the magnitude of the γ-*gauche* effect is large enough that such induced shifts can be easily observed.

Calculation of the γ-*gauche* effect requires several steps, including calculating the relative energies of the *gauche* and *trans* conformations, constructing an RIS model, calculating the conformational probabilities, and summing up all of the γ-*gauche* interactions. Such calculations give a relative value of the chemical shifts, not the absolute values as in the calculations involving the substituent effects discussed above.

The power of this approach is illustrated with the small molecule 2,4,6-trichloropeptane, which is a model compound for poly(vinyl chloride). The carbon spectrum obtained at 33°C is shown in Fig. 3.2

TABLE 3.3
The Chemical Shifts Induced by the γ-*gauche*
Effect for Several Groups (11)

Group	$\Delta \delta_c$ (ppm)
$-CH_3$	-5.2
$-OH$	-7.2
$-Cl$	-6.8

164   3. The Solution Characterization of Polymers

(12). The bond probabilities can be calculated from the three-state RIS model for poly(vinyl chloride) for the *mm* (I), *mr* (H), and *rr* (S), stereosequences, and Fig. 3.2 compares the calculated and observed chemical shifts for the methine, methylene, and methyl resonances. The close correlation between the calculated and the observed spectra

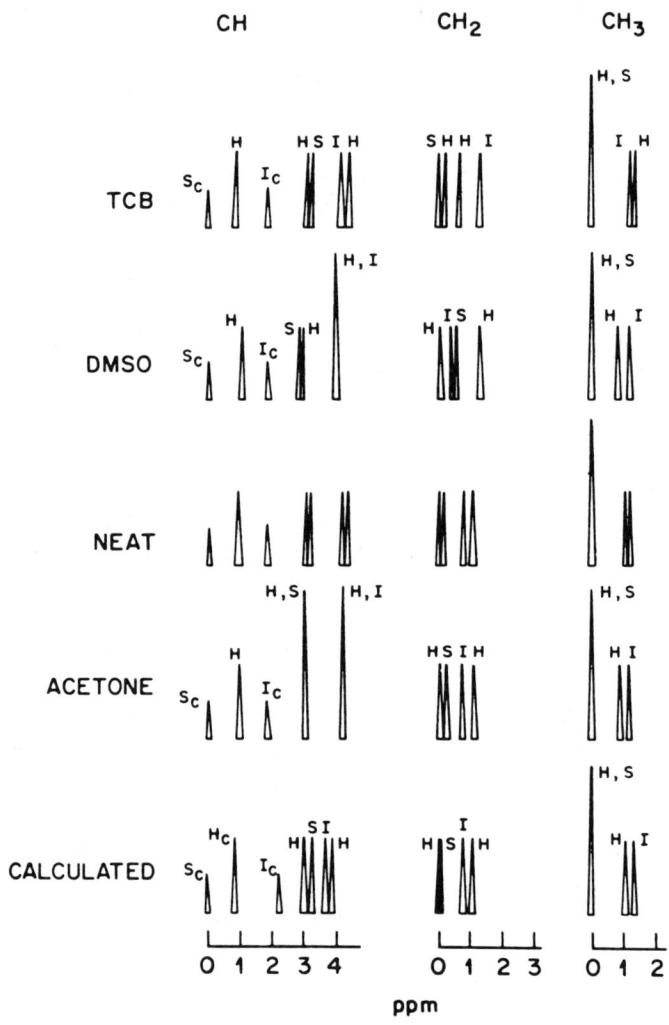

FIGURE 3.2 Comparison of the calculated and observed chemical shifts of 2,4,6-trichloroheptane neat and in various solvents at 33°C. The chemical shifts are referenced to the furthest downfield at 0 ppm. Reprinted with permission (12).

illustrates how this method can be used to assign the spectra of extremely complex stereopolymers.

2,4,6-trichloroheptane

3.2.3 Two-Dimensional NMR

In the past several years two-dimensional NMR methods (Section 1.8.7) have emerged as a powerful way to establish assignments in polymers without the synthesis of new materials or the purification of model compounds. A variety of methods have been employed which make use of both through-bond and through-space connectivities to establish correlations between nearest neighbors. More sophisticated methods using *relayed coherence transfer*, which have the power to correlate atoms separated by a larger number of intervening bonds, have been used to establish connectivities between atoms that are more distant in the polymer chain. Two-dimensional J-resolved NMR spectroscopy (Section 1.8.7.2) has also been used to aid in assignments through the identification of the carbon type (methine, methylene, methyl).

One of the simplest and most powerful methods is the correlation of nearby resonances via the three-bond J coupling between protons attached to neighboring carbon atoms, using the COSY (13) or double-quantum filtered COSY (14) pulse sequence, as illustrated in Fig. 1.56 for polycaprolactone in dilute solution. For polycaprolactone the proton signals are well resolved and the connectivities can be traced through the five methylene groups. Once the protons have been assigned, the carbon assignments can be obtained using heteronuclear correlation via the one-bond carbon–proton coupling. Thus, with two routine 2D NMR experiments, the assignments for all of the atoms can often be established. In a similar way these methods can be used to trace the connectivities for long side-chain groups in polymers.

Two-dimensional NMR is also an important method for the assignment of defect structures in polymers, since defects often have a pattern of cross peaks in the 2D spectra that differs from that of material with normal chain connectivities. This can be illustrated by considering the cross peaks expected for head-to-head (H-H) and tail-to-tail (T-T) defects in a vinyl polymer chain.

$$\overset{\overset{\displaystyle\text{H-H}}{\overbrace{}}}{\underset{\underset{\displaystyle\text{T-T}}{\underbrace{}}}{-CH_2-\underset{|}{\overset{X}{C}H}-CH_2-\underset{|}{\overset{X}{C}H}-\underset{|}{\overset{X}{C}H}-CH_2-CH_2-\underset{|}{\overset{X}{C}H}-CH_2-\underset{|}{\overset{X}{C}H}-}}$$

If we measure nearest neighbor proton–proton connectivities using COSY or NOESY spectroscopy, only methine–methylene correlations are observed for the normal head-to-tail connected monomers, while direct methine–methine and methylene–methylene correlations can be observed for the head-to-head and tail-to-tail units. Thus, the defect has an identifiable signature in the 2D spectra. One of the limitations of 2D NMR is that the spectra take much longer to acquire than 1D spectra, and high polymer concentrations are often required. If the defects are present at concentrations below 1%, it may be difficult to identify them by 2D NMR.

Two-dimensional NMR is also important for stereosequence assignments in polymers, through the use of both short- and long-range through-bond and through-space correlations. The general principles for the use of 2D NMR for stereosequence assignments can be illustrated using the sequence $mmrr$, which contains the mm, mr, and rr triads. If three peaks are observed, then they can be assigned using 2D NMR without reference to model compounds. Since the mm and rr sequences are both neighbors to the mr sequence, both will show correlations to the mr resonance, and so this peak is assigned. For vinyl polymers, the methylene protons in m and r dyads can be distinguished on the basis of symmetry, since the methylene protons in an r dyad are in a symmetric environment and therefore are equivalent, while those in m dyads are in a nonsymmetric environment and are nonequivalent. The two-bond geminal J coupling between the nonequivalent methylene protons is much larger (~ 15 Hz) than the three-bond vicinyl couplings (2–11 Hz) and gives rise to strong correlations in the COSY spectra. The assignments to m or r dyads can be established by the correlation between the methine protons and either two or one peak in the methylene region. The m and r dyads can also be distinguished in heteronuclear correlation spectra, since the methylene carbons will be correlated with one proton chemical shift in r dyads and two proton chemical shifts in m dyads.

This simple example illustrates how 2D NMR can be used to establish the stereosequence assignments. Most practical examples are more complex, but can be analyzed using the same general approach. In addition, nuclei other than carbons or protons, most notably fluorines, can also be used in a similar manner. In cases where the resolution is very good, correlations between pentads, or even higher sequences, can be observed.

In addition to characterization of polymer sequence, 2D NMR can also be used to study chain conformation and structure through the measurement of conformationally sensitive J coupling constants or intermolecular distances. As noted in Section 1.6.2, the three-bond J coupling constants depend on the torsional angle between the two protons attached to neighboring carbons, and vary between 11 Hz for the *trans* and 2 Hz for the *gauche* conformations. Since most chains undergo rapid interconversion between conformational states, the coupling constant is a measure of the relative populations of the *gauche* and *trans* conformations. This conformation information is difficult to extract from the 1D spectra because of spectral overlap, but can be directly measured by 2D J-resolved spectroscopy (Section 1.8.7.2). In some cases the magnitude of the coupling, and the conformation, can be inferred from the intensity of the correlations in the COSY spectra (15).

In a similar way, NOESY spectra provide information about the average chain conformation from the distances measured between pairs of protons. The NOESY correlations depend on the inverse sixth power of the internuclear distances and are extremely sensitive to distances in the range of 2–5 Å (16). Again the average chain conformation is extracted from an average of the distances in the *gauche* and *trans* states. The NOESY experiment has also been used to study the structure of associating polymers in solution, primarily through intermolecular NOEs, making it possible to identify the interacting functional groups leading to miscibility in polymer blends (17).

3.3 MICROSTRUCTURAL CHARACTERIZATION OF POLYMERS IN SOLUTION

NMR spectroscopy has been extensively used for microstructural characterization of polymers in solution. These studies are often motivated by a desire to understand structure–property relationships at the molecular level and to understand how changes in the synthetic methodology affect the structure of materials. It is often observed that polymers with similar monomer compositions have different mechanical, thermal, optical, or electrical properties as a consequence of differences in polymer microstructure. NMR is unique in its ability to characterize polymers at the molecular level so that such structure–property relationships can often be understood.

3.3.1 REGIOISOMERISM

Regio isomers (regio denoting directional) are sometimes incorporated into polymer chains as a consequence of the polymerization mechanism. As might be expected, the changes in chemical structure for reversed monomer units and their neighbors can have a large effect on the NMR spectra as well as on the material's properties. The changes in chemical shifts near regiodefects are often large enough that these resonances are shifted well away from the head-to-tail resonances, even in the proton NMR spectra where the spread in chemical shifts is smaller than that observed by carbon, fluorine, silicon, or phosphorus NMR.

Figure 3.3 shows the high-resolution proton NMR spectra of two poly(3-hexyl thiophenes) obtained from two synthetic methodologies optimized to produce regiorandom and regioregular polymers (18). The regioregular material contains 98.5% head-to-tail–head-to-tail monomers and gives rise to a single sharp peak for the protons at the four position at 6.977 pmm. The regiorandom material contains approximately equal amounts of the head-to-tail–head-to-head, head-to-tail–head-to-tail, tail-to-tail–head-to-tail, and tail-to-tail–head-to-head monomers, giving rise to four resonances in the aromatic region.

Poly(3-hexyl thiophene)

Regioisomerism is more difficult to detect by proton NMR in vinyl polymers because signals from the methine and methylene protons in the main chain and side chain often occur over a broad frequency range and the defect resonances are likely to overlap with the main-chain signals. However, defect structures often have a signature that can be detected as an unusual pattern of cross peaks in the two-dimensional NMR spectrum. This behavior is illustrated in the 2D NOESY spectrum of a copolymer of styrene and 2-phenyl-1,1-di-

3.3. Microstructural Characterization of Polymers in Solution

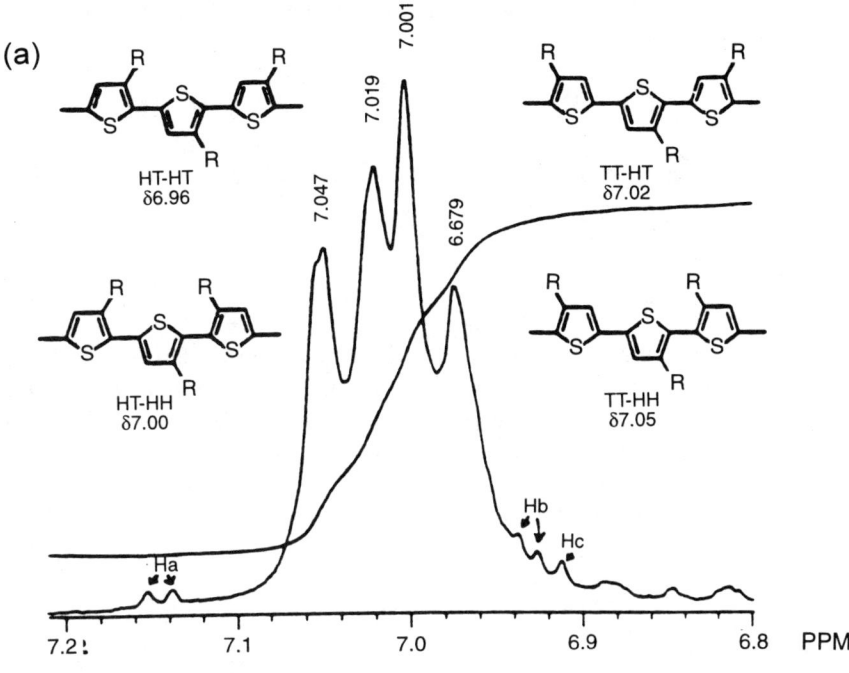

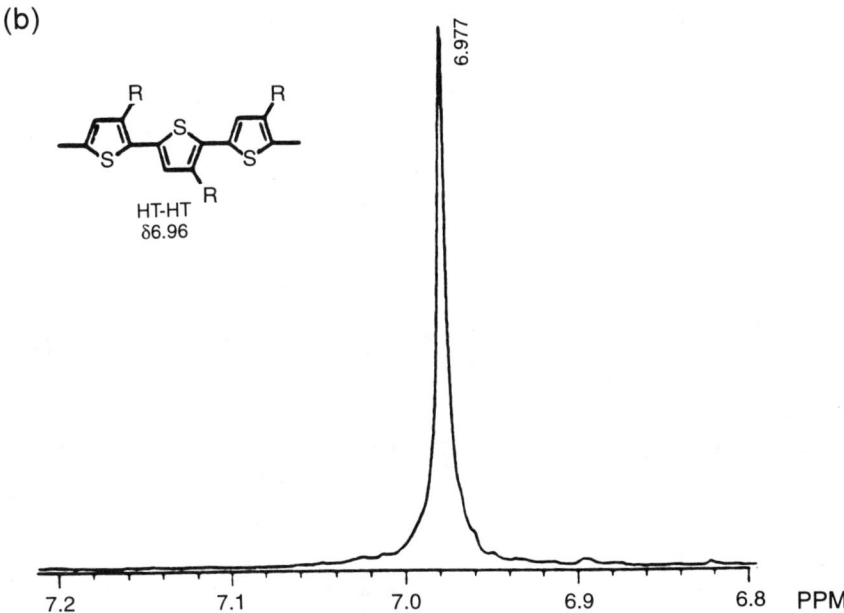

FIGURE 3.3 The high-resolution proton NMR spectra of (a) regioirregular and (b) regioregular poly(3-hexyl thiophene). Reprinted with permission (18).

cyanoethene, which can have head-to-tail (A) or head-to-head (B) monomer additions.

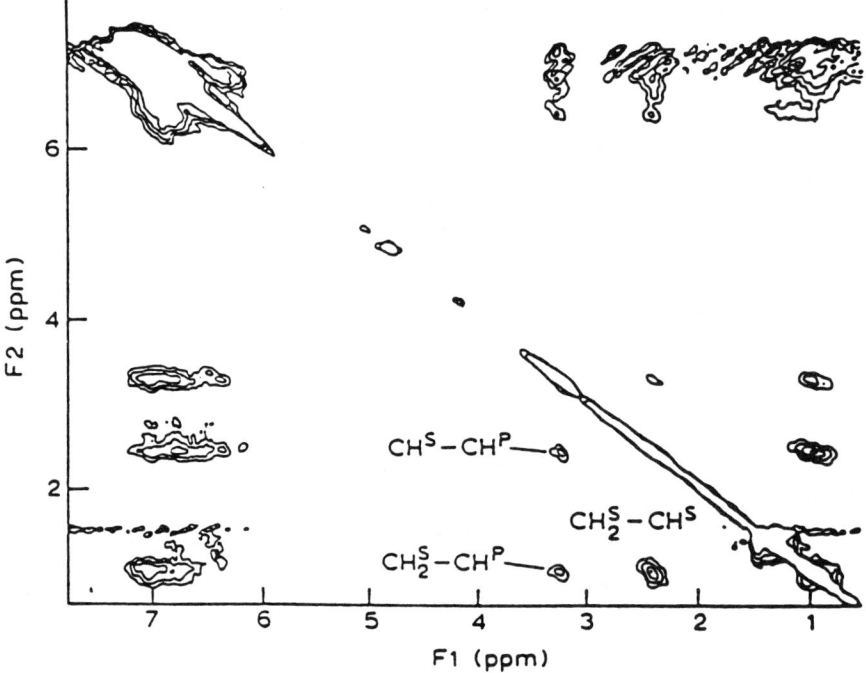

These two structures can be distinguished by the methine–methine cross peaks in a COSY or NOESY spectrum which can be observed only in the defect structure. Figure 3.4 shows the 2D NOESY spectrum of the poly(styrene-co-2-phenyl-1,1-dicyanoethene) polymer containing 38 mol% dicyanoethene units (19). In addition to the expected phenyl–main-chain and methine–methylene cross peaks, correlation

FIGURE 3.4 The 500-MHz 2D NOESY spectrum of the poly(styrene-co-2-phenyl-1,1-dicyanoethene) copolymer obtained with a 0.4-s mixing time. Note the correlations between the methine protons (CH^S–CH^P) that are indicative of directional isomers. Reprinted with permission (19).

between the methine peaks from the inverted units can also be observed. Once these peaks have been identified, the carbon spectrum can be directly assigned by heteronuclear correlation spectroscopy.

^{13}C NMR has been extensively used to investigate regioisomerism in polypropylene and other vinyl polymers. Polypropylene is a commercially important material, and its synthesis by some synthetic routes is known to produce polymers containing regiodefects. The assignments for polypropylene are hampered by the overlap from the many signals in the methine and methylene region.

A combination of methods has been used to establish the regioisomer assignments in polypropylene, including chemical shift calculations, rotational isomeric state and γ-gauche effect calculations, and 1D and 2D NMR methods. A nomenclature has been developed to describe the specific atoms in the main chain and side chains that is also used to describe copolymer structure. The designations S, T, and P refer to secondary (CH_2), tertiary (CH), and primary (CH_3) carbons while two Greek subscripts refer to the distance along the chain to the nearest methyl groups in each direction. When four subscripts are given, they refer to the distances to two methyl groups along the chain. Thus, the designation $S_{\gamma\alpha\alpha\gamma}$ refers to a secondary carbon with methyl groups at the α and δ position in both directions along the chain, and the designation $T_{\beta\beta}$ refers to a methine carbon with methyl groups at the β position in both directions along the chain. Using this terminology, the methyl, methylene, and methine carbons of polypropylene are designated $P_{\beta\beta}$, $S_{\delta\alpha\alpha\delta}$, and $T_{\beta\beta}$.

The spectrum of polypropylene is likely to be complex due to the possibility of both stereochemical isomerism and regioisomerism. The effects of regioisomerism are much larger than those due to stereochemical isomerism, as shown by the chemical shifts for the head-to-head and defect resonances listed in Table 3.4. Such large shifts can lead to the overlap of the methine and methylene signals for the regioregular and defect structures, making the spectra difficult to assign.

TABLE 3.4
The Chemical Shifts Observed for Regioregular and Regioisomers of Polypropylene (11)

Carbon	Carbon chemical shift (ppm)		
	H-T	H-H	T-T
CH	28.5	37.0	—
CH_2	46.0	—	31.3
CH_3	20.5	15.0	—

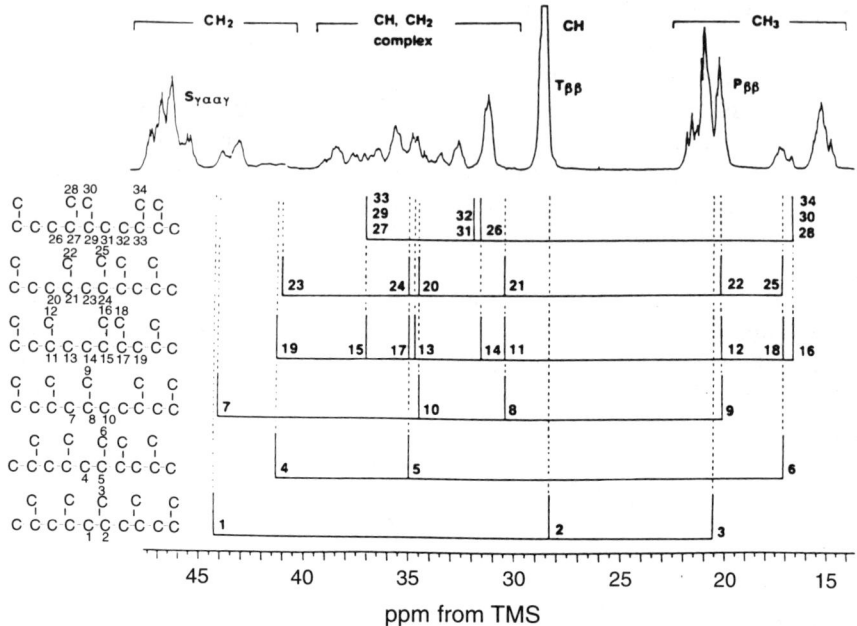

FIGURE 3.5 Comparison of the 100-MHz carbon spectrum of regioirregular polypropylene with the chemical shifts calculated for a variety of microstructures. Reprinted with permission (20).

Figure 3.5 compares the 100-MHz carbon NMR spectrum for regioirregular polypropylene with the chemical shifts calculated for a variety of different types of monomer insertions into the main chain (20). In this instance the chemical shifts were calculated using both the empirical chemical shift rules and the γ-gauche effect for the various structures, which used an RIS model for polypropylene and incorporated the effects of tacticity. Two features are obvious from this comparison. First, it may be observed that the spectrum is extremely complex in these types of polymers, making the signals difficult to assign. Second, it may be observed that there is extensive overlap between the methylene and the methine carbon signals, and the assignments of these signals to methine or methylene carbons in the first step in the assignment procedure.

The assignment of the carbon signals to a particular type of carbon can be established using the DEPT or INEPT methods described in Section 1.8.8.1. It may be recalled that by using these methods, the spectra may be edited by properly choosing the phase angle for one of the pulses such that only one type of carbon signal appears in the spectra. This experiment, for regioirregular polypropylene, is illus-

3.3. Microstructural Characterization of Polymers in Solution

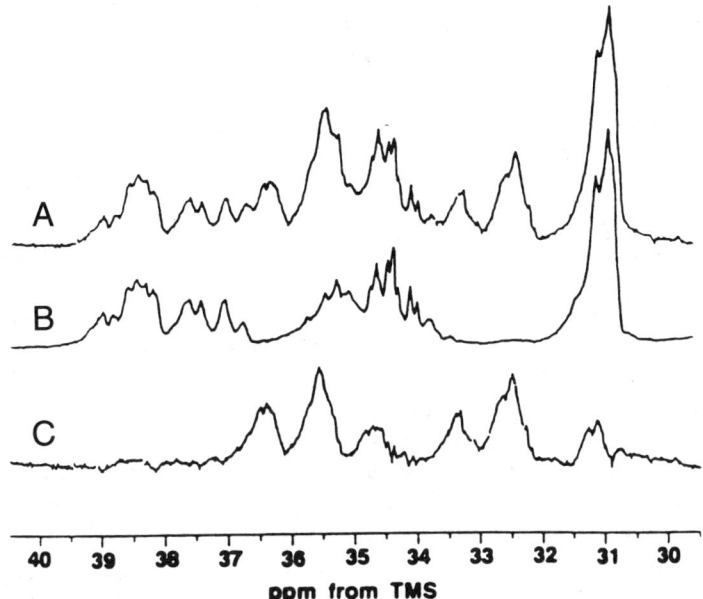

FIGURE 3.6 The INEPT spectra of regioirregular polypropylene (12% inverted units) showing (a) methylene and methine carbon resonances, (b) the methine resonances only, and (c) the methylene resonances only. Reprinted with permission (20).

trated in Fig. 3.6, which compares with normal ^{13}C spectrum (a) with that obtained using the INEPT pulse sequence (b), in which only the methine carbon signals appear. Since this region is known not to contain any methyl signals, the spectrum of only the methylene carbons (c) can be obtained from the difference of spectra (a) and (b).

If the lines are well resolved, it may be possible to use the calculated chemical shifts to establish the assignments. However, this is not possible in polypropylene containing both regioisomerism and stereochemical isomerism, and while the assignments can be proposed from the chemical shift calculations, they cannot be definitively established.

It is possible in principle to use the proton–proton correlations (COSY or NOESY) to identify regioisomers in polymers, as was demonstrated for the styrene copolymer shown above (Fig. 3.4). However, in many polymers, such as polypropylene, the range in proton chemical shifts is not large enough that the cross peaks which are a signature of inverted units can be identified and assigned. These limitations can be overcome using the so-called 2D INADEQUATE experiment, in which correlations are observed between neighboring carbon atoms *via* the one-bond carbon–carbon coupling (21). The coupling constants are large (30–70 Hz) so that nearby signals can be effectively correlated, but the sensitivity is extremely poor because correlations are observed only for neighboring ^{13}C atoms. Since the natural abundance of ^{13}C is 1.1%, the chances of having nearest neighboring carbon atoms is 1000 times lower than for protons. Thus, it is possible to use INADEQUATE to trace out the chain connectivity, but high sample concentrations are required and the signal-to-noise ratio of the spectra is often much lower than that obtained in other 2D NMR experiments.

Figure 3.7 shows the pulse sequence for the 2D INADEQUATE experiment (21). Double-quantum coherences are created during the $90°-\tau-180°-\tau-90°$ part of the pulse sequence, where τ is $1/2J_{CC}$, and evolves during the t_l period before it is converted to observable magnetization with the final pulse. The format of the final spectrum differs from that of other 2D correlation spectra in that there are no strong diagonal peaks. Instead, horizontal lines connect the resonances of carbon atoms that are directly bonded to each other.

The 100-MHz 2D INADEQUATE spectrum of regioirregular polypropylene is shown in Fig. 3.8 (20). Note that many peaks are observed and that the signal-to-noise ratio is poor, even for a 24% w/v sample run for many hours. However, even with these limitations it is possible to trace out the connectivity and establish the peak assignments. The structural formula in Fig. 2.8 shows a defect, and the horizontal and vertical lines show how it is possible to trace out the chain connectivity even when the spectra are extremely complex. A similar strategy has been used to make assignments in regioisomers of poly(1-butene) (22).

Poly(propylene oxide) is another polymer that has been extensively analyzed by ^{13}C NMR, in terms of both its stereochemistry (3) and its regioisomerism (23). The stereochemistry for poly(propylene oxide) is more complex than that for vinyl polymers because its has three atoms in the main chain, and can have heterotactic isomers in addition to the isotactic and syndiotactic materials (Section 2.2.5). As in the case of polypropylene, the proton spectra of poly(propylene

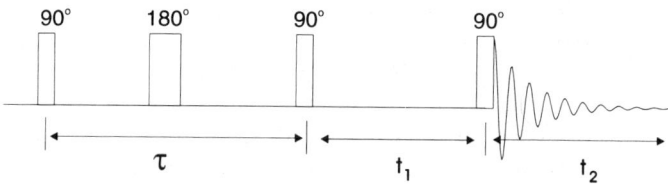

FIGURE 3.7 The pulse sequence for the correlation of directly bonded ^{13}C resonances using carbon–carbon double-quantum coherences (INADEQUATE) (21).

3.3. Microstructural Characterization of Polymers in Solution

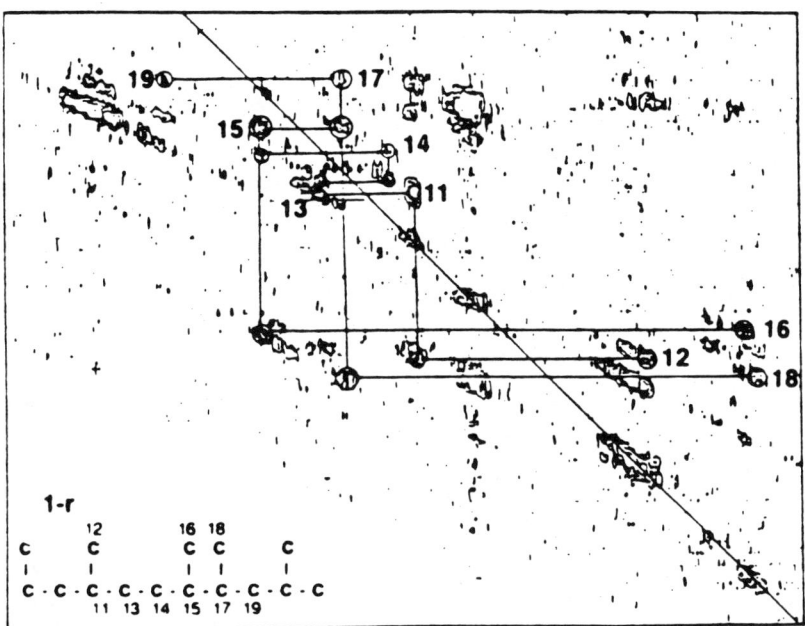

FIGURE 3.8 The 2D INADEQUATE spectrum of regioregular polypropylene showing the connectivities for a head-to-head defect. The horizontal lines show the correlations between directly bonded carbon atoms. Reprinted with permission (20).

oxide) are extremely complex with the overlap of the methine and methylene protons, and are of limited use for establishing the chemical shift assignments for the regioisomers.

The assignments for poly(propylene oxide) have been made primarily on the basis of chemical shift calculations and DEPT NMR spectra. As in polypropylene, assignments can be made only when the carbon types (methine vs methylene) have been established. Figure 3.9 shows 50.3-MHz DEPT spectra of poly(propylene oxide) (M_n = 4000) observed at 23°C (23) which have been edited to show the (a) methine and methylene, (b) methine, and (c) methylene resonances only. In such low-molecular-weight materials, there is the possibility that some of the observed peaks are due to end groups rather than regioisomers. The end groups were identified by comparing the spectra for higher and lower molecular weight polymers. Table 3.5 lists the chemical shifts calculated from the combination of the empirical rules and the γ-*gauche* effect for the head-to-tail and regiodefect resonances. Note that the range in chemical shifts can be quite large, as much as 4.7 ppm for some of the methine and methylene carbons.

TABLE 3.5
The Calculated Carbon Chemical Shifts for Poly(propylene oxide)
Using the γ-gauche Effect (23)

$$\begin{array}{c}
12345\\
CH_3CH_3CH_3CH_3CH_3\\
-CH_2-CH-O-CH_2-CH-O-CH-CH_2-O-CH_2-CH-O-CH_2-CH-O-\\
1122334455\\
abcd
\end{array}$$

Carbon		Dyad[a]				$\Delta\delta^b$ (ppm)
		a	b	c	d	
CH_3,	1	m	—	—	—	0.00
	2	r	r	—	—	0.45
	2	m	r	—	—	0.49
	2	r	m	—	—	0.75
	2	m	m	—	—	0.79
	3	—	r	—	—	0.53
	3	—	m	—	—	0.82
	4	—	—	—	m	0.02
	4	—	—	—	r	0.04
	5	—	—	—	m	0.00
CH_2,	1	m	—	—	—	0.00
	2	m	m	—	—	−0.15
	2	r	m	—	—	−0.19
	2	r	r	—	—	0.20
	2	m	r	—	—	0.25
	3	—	m	—	—	4.40
	3	—	r	—	—	4.73
	4	—	—	—	r	4.75
	5	—	—	—	m	0.00
CH,	1	m	—	—	—	0.00
	2	m	m	—	—	−4.49
	2	r	m	—	—	−4.45
	2	m	r	—	—	−4.49
	2	r	r	—	—	−4.45
	3	—	m	—	—	−4.48
	3	—	r	—	—	−4.48
	4	—	—	—	r	−0.25
	4	—	—	—	m	−0.27
	5	—	—	—	m	0.00

[a] The dash (—) indicates either m or r dyad placement.
[b] The chemical shift relative to the chemical shift in the head-to-tail monomer.

3.3. Microstructural Characterization of Polymers in Solution

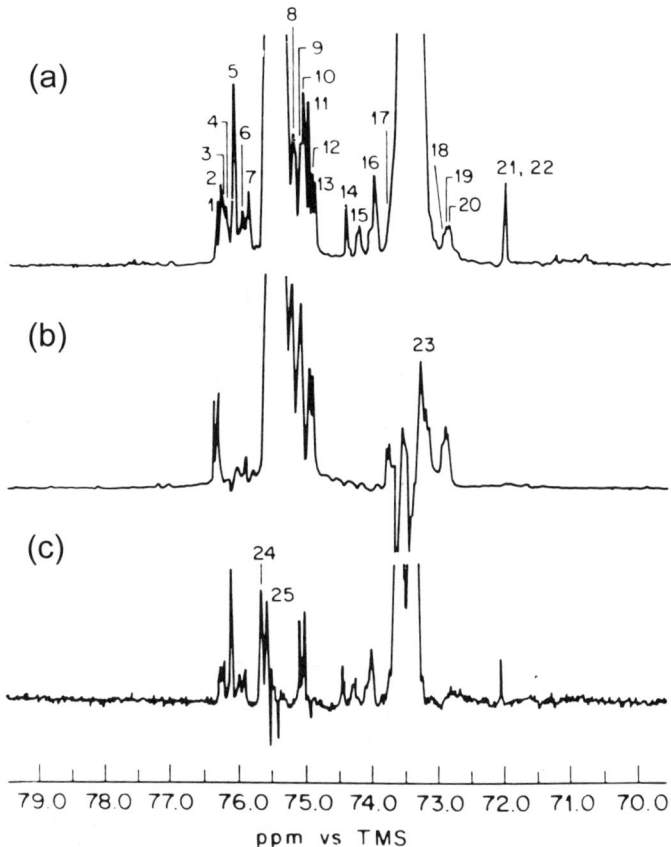

FIGURE 3.9 The 50-MHz carbon DEPT spectra of poly(propylene oxide) (MW = 4000) showing (a) all resonances, (b) the methine resonances only, and (c) the methylene resonances only. The assignments for the peaks labeled 1–22 are compiled in Table 3.6. Reprinted with permission (23).

The assignments for the peaks labeled in Fig. 3.9 are listed in Table 3.6, along with the assignments for numerous end groups.

Poly(vinylidine fluoride) synthesized by free radical polymerization is another material where regiodefects are often observed. In commercial materials the defect level is about 3–6%, but material can be synthesized containing no defects or a very high level (23%) (24,25). In addition to the chemical shift calculations, which may also be applied to fluorine-containing polymers, it is also possible to characterize these materials by ^{19}F NMR, which has a higher sensitivity and a larger chemical shift range. Figure 3.10 compares the ^{13}C NMR

TABLE 3.6
The Carbon NMR Assignments for the Methine and Methylene
Carbons of Poly(propylene oxide) at 23°C (23)

Resonance	δ_c (ppm)	Assignment[a]	
1	76.36	–CH–	E
2	76.29	–CH–	E
3	76.26	–CH$_2$–	3,4
4	76.21	–CH$_2$–	3,4
5	76.10	–CH$_2$–	E
6	75.98	–CH$_2$–	3,4
7	75.88	–CH$_2$–	3,4
8	75.24	–CH–	E
9	75.13	–CH–	E
10	75.08	–CH–, –CH$_2$–	E
11	75.02	–CH$_2$–	E
12	74.96	–CH–	E
13	74.91	–CH–	E
14	74.46	–CH$_2$–	2
15	74.26	–CH$_2$–	2
16	74.02	–CH$_2$–	2
17	73.82	–CH–	2,3
18	72.97	–CH–	2,3
19	72.93	–CH–	2,3
20	72.87	–CH–	2,3
21	72.06	–CH$_2$–	E
22	72.03	–CH$_2$–	E
23	73.30	–CH–	2,3
24	75.65	–CH$_2$–	E
25	75.57	–CH$_2$–	E

[a] E indicates a chain end; 2,3,4 indicates a head-to-tail–tail-to-tail defect structure with the nomenclature shown in Table 3.5.

spectra of poly(vinylidine fluoride) with the chemical shifts calculated for the combination of RIS models with the γ-*gauche* effect and the empirical chemical shift rules (26). Note that a slightly more complex procedure is required to observe these spectra, since the carbon resonances can be split from both the carbon–proton and the carbon–fluorine one-bond couplings. The high-resolution spectra in Fig. 3.10 were obtained with simultaneous carbon–proton and carbon–fluorine spin decoupling.

$$\text{—CF}_2\text{–CH}_2\overset{1}{\text{–CF}}_2\text{–CH}_2\overset{A}{\text{–CF}}_2\overset{B}{\text{–CF}}_2\text{–CH}_2\overset{2}{\text{–CH}}_2\overset{3}{\text{–CF}}_2\overset{C}{\text{–CH}}_2\overset{4}{\text{–CF}}_2\overset{D}{\text{–CH}}_2\text{—}$$

The regiodefects in poly(vinylidine fluoride) are also amenable to analysis by ^{19}F NMR, and Fig. 3.11 compares the spectrum obtained

3.3. Microstructural Characterization of Polymers in Solution

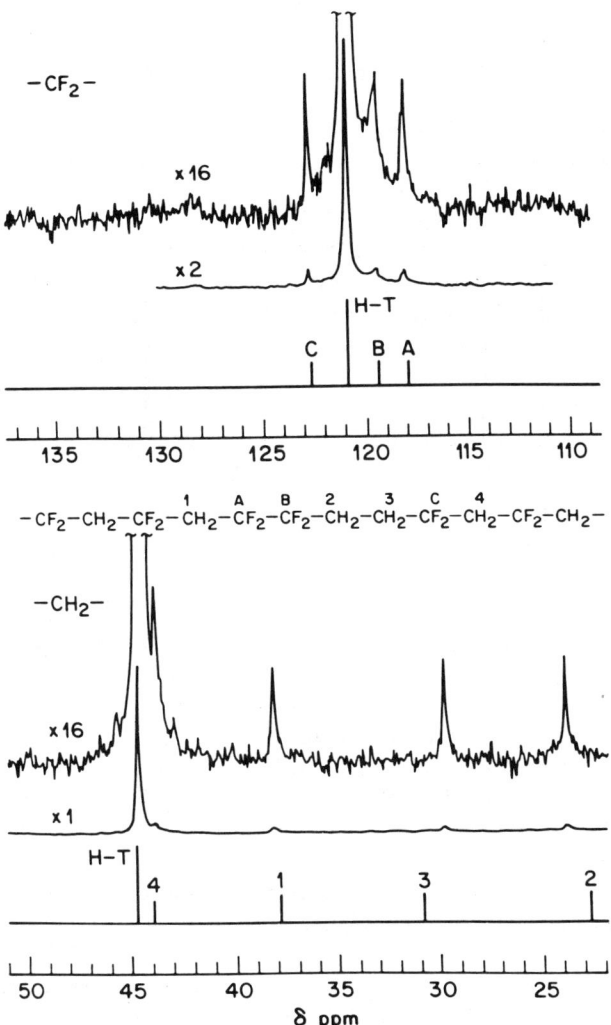

FIGURE 3.10 Comparison of the ^{19}F and ^{1}H decoupled carbon NMR spectra of poly(vinylidine fluoride) with the calculated chemical shifts for defect structures. Reprinted with permission (26).

at 188.2 MHz with the calculated chemical shifts (27). The head-to-tail signal is assigned as the largest peak in the spectrum, and is the only signal observed in the isoregic material. These materials can also be analyzed by 2D correlated NMR *via* the three- and four-bond ^{19}F–^{19}F J couplings, which are on the order of 3–10 Hz (28). Figure 3.12 shows the 2D COSY spectrum for a poly(vinylidine fluoride) sample

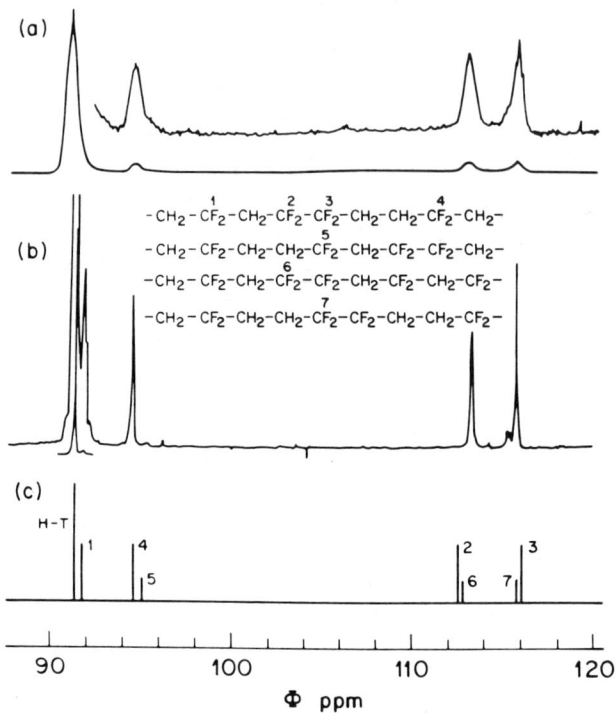

FIGURE 3.11 Comparison of the (a) 84.6-MHz and (b) 188-MHz ^{19}F NMR spectrum with the (c) calculated chemical shifts for the head-to-tail (H-T) and other defect structures. Reprinted with permission (27).

containing 18% regio-defects, where the cross peaks lead to an unambigious assignment for all of the major peaks in the ^{19}F NMR spectrum (25).

Poly(vinyl fluoride) containing regiodefects has also been studied by ^{19}F NMR, and many of the regiosequence defect resonances can be assigned from the 188-MHz 2D COSY spectrum shown in Fig. 3.13 (28). By comparison with the regioregular material, the spectral region between 178 and 183 ppm can be assigned to the stereosequence peaks for the head-to-tail monomers A and D.

As we will see later (Section 3.3.2), the four-bond ^{19}F–^{19}F J couplings are large enough that cross peaks are observed between the stereosequences in the regioregular polymer, and it is these cross peaks that are observed in this region of the COSY spectrum. In the

3.3. *Microstructural Characterization of Polymers in Solution* 181

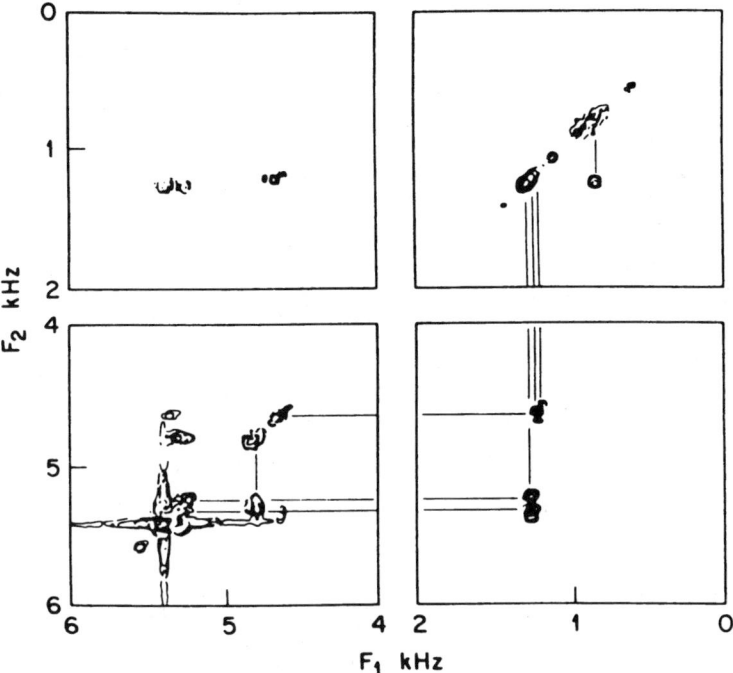

FIGURE 3.12 The 188-MHz fluorine 2D COSY spectrum of poly(vinylidine fluoride) containing 18% inverted monomers. Reprinted with permission (25).

higher field portion of the spectrum cross peaks are observed *via* the three-bond ^{19}F–^{19}F couplings in head-to-head units (B and C). Note, however, that some of the expected cross peaks (A–D and A–B) are not observed. It was suggested that this may arise from the cancellation of cross peaks with complex multiplet structures in the COSY spectrum.

3.3.2 STEREOCHEMICAL ISOMERISM

NMR spectroscopy has been extensively used to investigate stereochemical isomerism in polymers in an attempt to understand polymer structure–property relationships on the molecular level. Many of the same methods used to identify and characterize the directional defects in polymers (Section 3.3.1) are used to establish the stereochemical resonance assignments in polymers.

Stereochemical isomerism in polypropylene has been extensively studied by NMR, and the understanding of stereochemical isomerism in polypropylene has been aided both by the development of 2D NMR

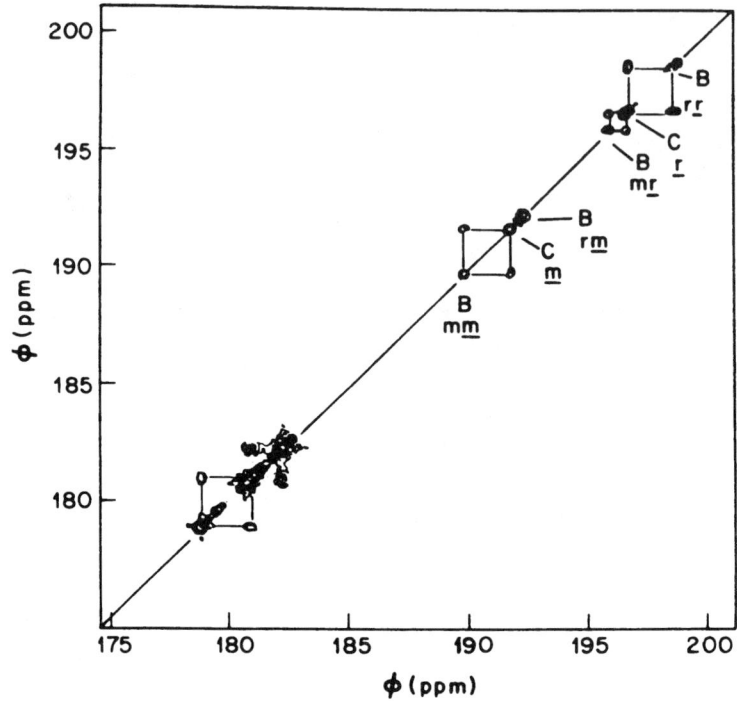

FIGURE 3.13 The 188-MHz fluorine 2D COSY spectrum of poly(vinyl fluorine) in DMF-d_7 containing regiosequence defects. Reprinted with permission (28).

methods and by the synthesis of many model polymers. The initial assignments of the isotactic and syndiotactic resonances were established by comparison of isotactic and syndiotactic polypropylene with the atactic material, as shown in Fig. 3.1 (3). A closer examination of the methyl region of atactic polypropylene (Fig. 3.14) shows that more than 20 resonances are observed (3), indicating that the methyl carbons in polypropylene are sensitive to stereosequences to the heptad level. By comparison of the spectra for the isotactic, syndiotactic, and atactic material, the peaks labeled I and S in Fig. 3.14 can be assigned to the *mmmmmm* and *rrrrrr* stereosequences.

Many of the assignments of stereosequences in polypropylene have been established with the aid of the *γ-gauche* effect. Figure 3.15 shows the 90-MHz carbon spectrum of atactic polypropylene in 1,2,4-trichloro-benzene obtained at 100°C and the heptad chemical shifts calculated with the Suter–Flory RIS model for polypropylene (29) and the *γ-gauche* effect (30). By fitting the intensities of the predicted peaks to those observed in the spectrum, it is possible to deduce the polymerization mechanism, (Section 2.3). Figure 3.16 compares the

3.3. Microstructural Characterization of Polymers in Solution

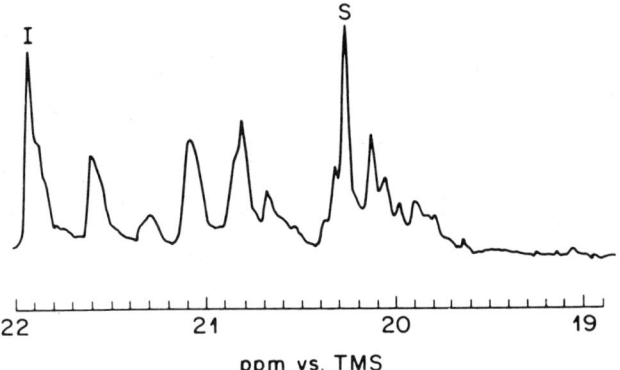

FIGURE 3.14 The 90-MHz carbon NMR spectrum of atactic polypropylene in n-heptane at 67°C. Reprinted with permission (3).

carbon spectrum with a spectrum calculated from the chemical shifts of the 36 heptad stereosequences where each peak was assumed to have the same linewidth (0.1 ppm) and the intensities were adjusted to give the best fit between the calculated and the observed spectra (3). These results show that the polymerization statistics do not

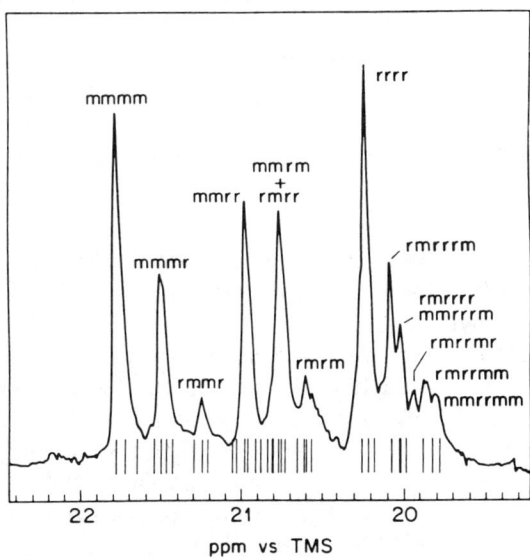

FIGURE 3.15 Comparison of the 90-MHz carbon spectrum of atactic polypropylene with the chemical shifts calculated from the RIS model and the γ-gauche effect. Reprinted with permission (3).

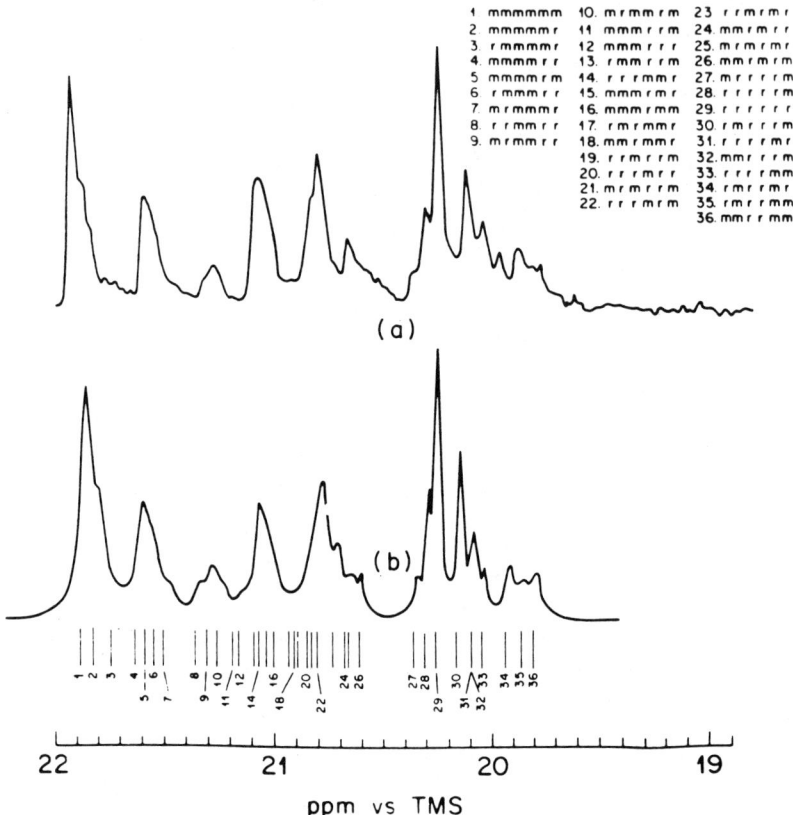

FIGURE 3.16 Comparison of the (a) observed 90-MHz carbon spectrum of atactic polypropylene with (b) the spectrum simulated by adjusting the intensities and using the chemical shifts calculated from the γ-gauche effect. Reprinted with permission (3).

conform to a simple Bernoullian or first-order Markov process (30), but can be described by a two-site model for Ziegler–Natta polymerization of polypropylene (31).

The ^{13}C NMR spectra are more highly resolved in so-called hemiisotactic material, in which assignments to the undecad level have been reported (32). Hemi-isotactic polymers are those produced by a synthetic method in which the stereochemistry at every other pseudoasymmetric center is random. The spectra of these materials are considerably simplified since the number of possible stereoisomers is greatly reduced by the fixed stereochemistry at every other pseudoasymmetric center. Figure 3.17 compares the observed and simulated 50-MHz carbon spectra of hemi-isotactic polypropylene from which these assignments were deduced (32). Once these assignments

3.3. Microstructural Characterization of Polymers in Solution

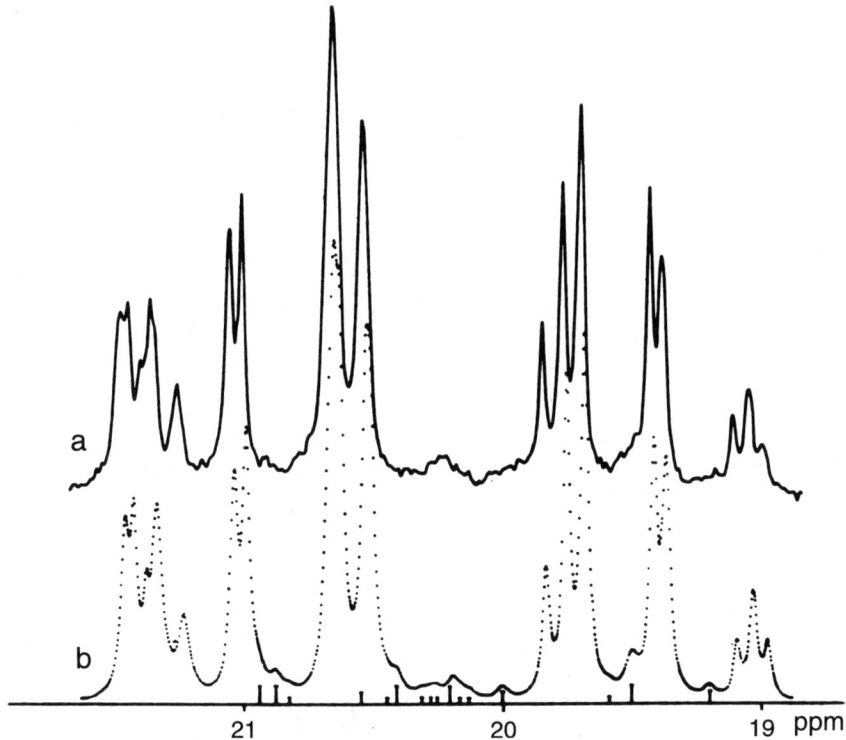

FIGURE 3.17 The observed (a) and simulated (b) 50-MHz carbon NMR spectra of hemi-isotactic polypropylene. Reprinted with permission (32).

have been established, the assignments for the proton signals can be directly deduced by proton–carbon heteronuclear correlations, as shown in Fig. 3.18 (33).

Poly(methyl methacrylate) is another example of a polymer that has been extensively studied by solution NMR. Unlike polypropylene, the proton NMR spectrum of poly(methyl methacrylate) is sufficiently well resolved that it can be used for materials characterization. Figure 3.19 shows the 500-MHz proton NMR spectra of poly(methyl methacrylate) acquired at 100°C at three different magnifications (34). At the lowest setting, only the signals from the isotactic stereosequences are observed. Note that there is a very large chemical shift separation between the nonequivalent methylene protons labeled *e* and *t*. The *e* and *t* notations refer to *erythro* and *threo*, which are *syn* and *anti* to the ester group in the *trans–trans* conformation. As the magnification is increased in Fig. 3.19, an increasing number of peaks are visible in the proton spectra, demonstrating that proton NMR can be an effective tool for the characterization of low levels of

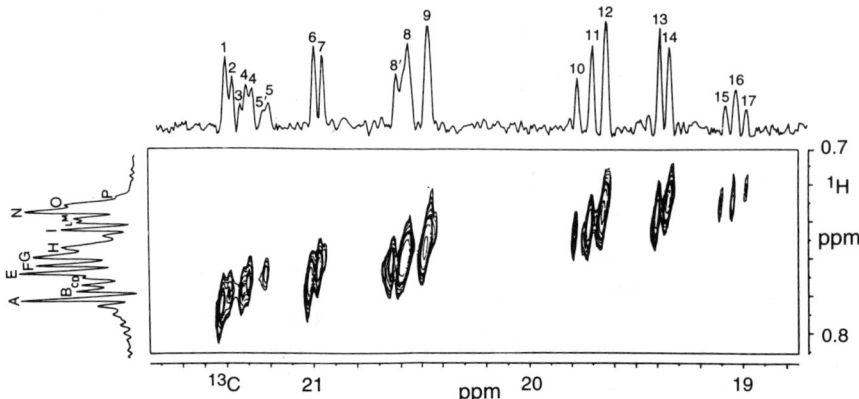

FIGURE 3.18 The 2D heteronuclear correlated spectrum of hemi-isotactic polypropylene. The carbon chemical shifts are shown along the bottom axis and the proton chemical shifts are shown along the side axis for the methyl groups. Reprinted with permission (33).

stereosequences, or other defect structures. In such well-resolved spectra, it may be possible to identify and characterize defect structures well below the 1% level.

The chemical shift difference between the *erythro* and the *threo* methylene protons is large in poly(methyl methacrylate), and is a signature of *m* centered stereosequences (Section 3.2.3). Such resonances can be easily identified by strong cross peaks in the COSY spectrum (Section 1.8.7.3) because the two-bond geminal coupling constant (~15 Hz) is much larger than the three-bond vicinyl proton–proton coupling constants (2–11 Hz). This behavior is illustrated in the COSY spectrum of atactic poly(methyl methacrylate) shown in Fig. 3.20, where strong e–t cross peaks are observed for *m* centered stereosequences (34). These peaks can also be identified as the largest peaks in the NOESY spectrum since the geminal proton distances (1.77 Å) are much shorter than the vicinyl proton–proton separations (2.2–3.0 Å).

The resolution in the carbon spectrum of poly(methyl methacrylate) is also good, as shown by the 25.2-MHz spectrum shown in Fig.

3.3. *Microstructural Characterization of Polymers in Solution* 187

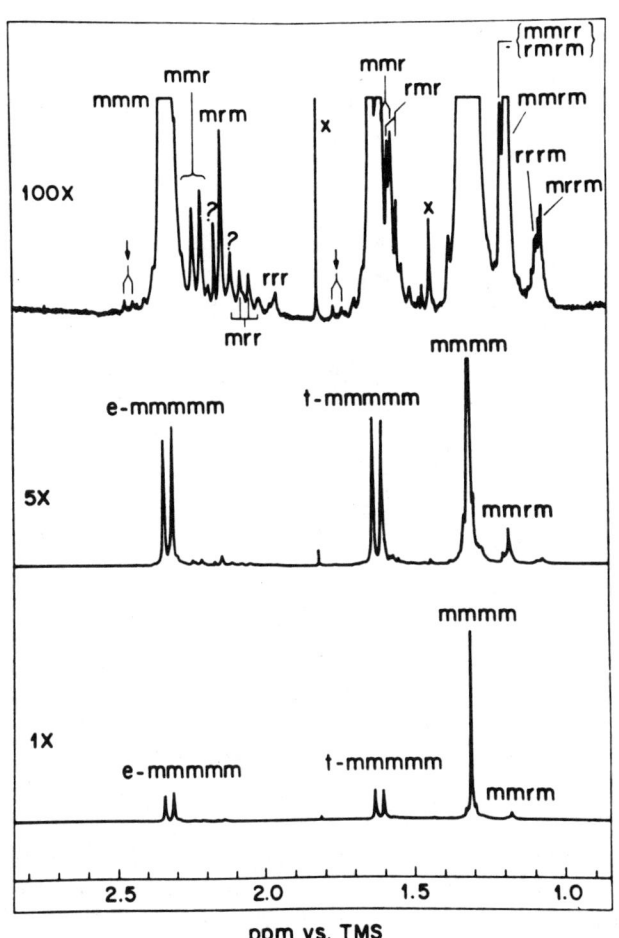

FIGURE 3.19 The 500-MHz proton NMR spectra of isotactic poly(methyl methacrylate) in chlorobenzene-d_5 acquired at 100°C. The spectra are shown at higher and higher magnification. Reprinted with permission (34).

3.21 (35). Poly(methyl methacrylate) was among the first polymers studied by NMR, and the initial assignments were established by comparing the peak intensities with the intensities calculated with a Bernoullian distribution. Polymers prepared by another synthetic route have been shown to be consistent with second-order Markov or Coleman–Fox behavior (36).

More recently poly(methyl methacrylate) has been studied by 2D NMR methods and site-specific ^{13}C NMR labeling of the carbonyl group. Figure 3.22 shows the heteronuclear chemical shift correlation spectrum for poly(methyl methacrylate) that is 90% enriched at the

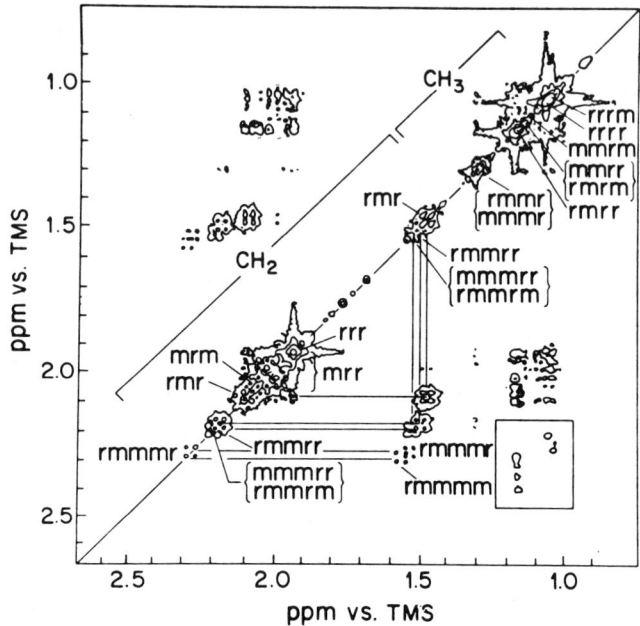

FIGURE 3.20 The 500-MHz 2D COSY spectrum of atactic poly(methyl methacrylate). Note the correlation between the methylene protons in *m* centered stereosequences. The peaks enclosed in the box are due to four-bond CH_2–CH_3 couplings that are observed for some conformations (Section 3.4.1). Reprinted with permission (34).

carbonyl carbon (37). Note that only the carbonyl region of the carbon spectrum is shown and that the correlations arise from the three-bond proton–carbon couplings that are in the range of 5–10 Hz (10). The same pulse sequence is used for this experiment as for heteronuclear correlations via the one-bond carbon–proton couplings, but the delay times (Fig. 1.59) required for the long-range correlations are much longer (on the order of $1/2J$) (38). This decreases the sensitivity due to T_2 relaxation during this time period, so such experiments are feasible only for isotopically labeled samples.

The three-bond carbon–proton correlations are observed in both directions along the chain for the methylene and methyl protons, as well as for the side-chain methoxy protons. The assignment strategy is based on the pentad-level resolution observed in the carbonyl region. Some of these pentads consist of coupled tetrads, so two sets of correlations are expected. For example, the *mrmm* pentad contains *mrm* and *rmm* tetrads, so two sets of correlations are expected. As in the COSY spectra, the carbon signals for *m* centered stereosequences are correlated with two proton signals. Figure 3.23 shows the carbonyl region of the carbon spectrum along with cross sections through

3.3. Microstructural Characterization of Polymers in Solution

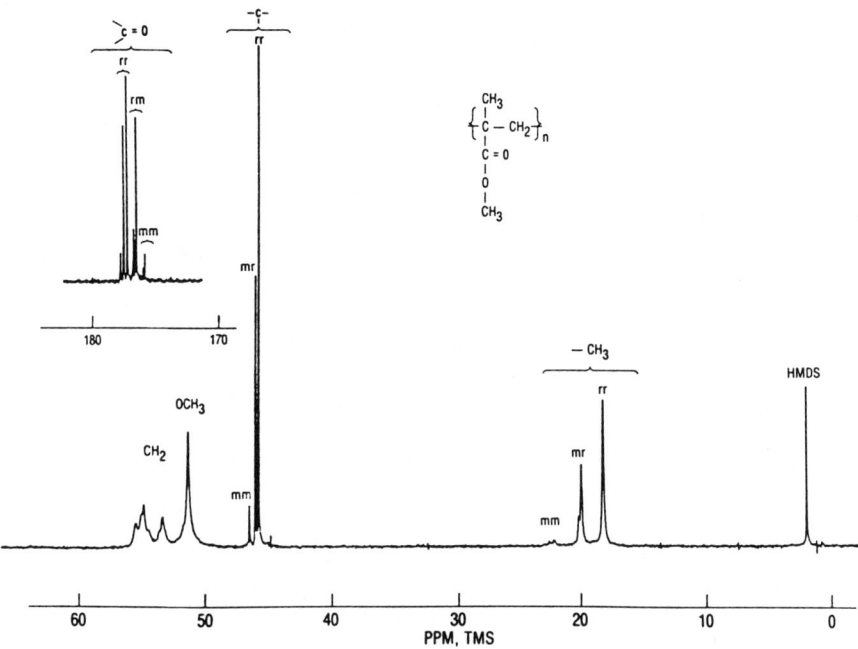

FIGURE 3.21 The 25.2-MHz carbon NMR spectrum of poly(methyl methacrylate) in 1,2,4-trichlorobenzene at 120°C. Reprinted with permission (35).

the 2D correlation spectrum shown in Fig. 3.22. It is possible to confirm the assignments by tracing through the connectivities in such a spectrum. A subsequent study showed that it is possible to improve the sensitivity of this experiment with detection of the proton rather than the carbon signals (39), because of the inherent sensitivity of the protons relative to the carbons.

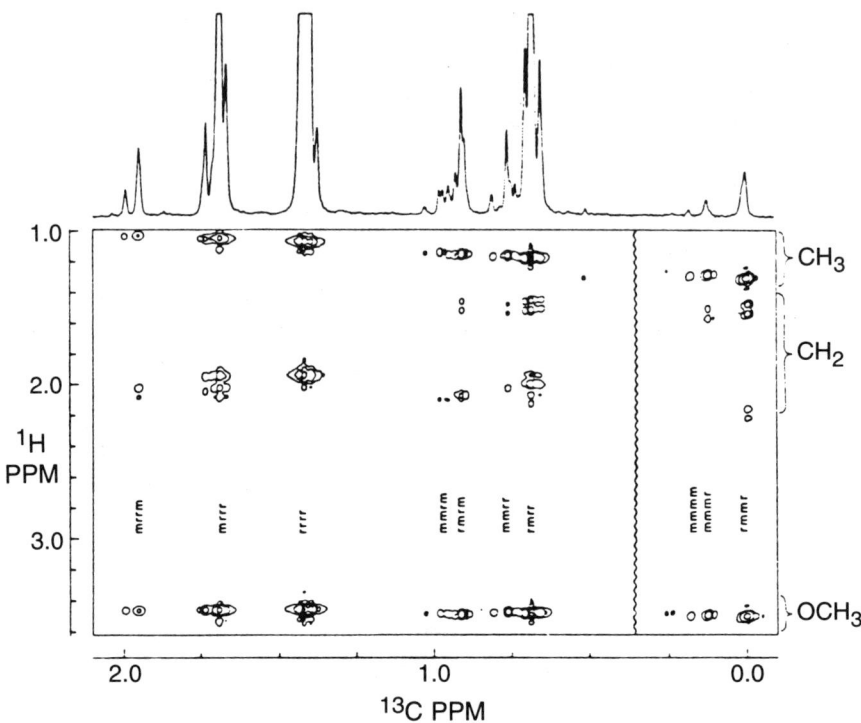

FIGURE 3.22 The 1D spectrum and the carbon–proton heteronuclear correlation spectrum of the carbonyl region of poly(methyl-[carbonyl-^{13}C]methacrylate). Reprinted with permission (37).

Poly(vinyl chloride) is a commercially important polymer that has been extensively studied by carbon and proton NMR. Stereosequence assignments were initially made by comparing the peak intensities with those expected from Bernoullian statistics and model compounds (40). Figure 3.24 shows the carbon spectrum of atactic poly(vinyl chloride) and the chemical shifts calculated from the γ-gauche effect (12) using the RIS models for poly(vinyl chloride) (41,42) to calculate the bond probabilities. Good agreement is observed between the calculated and the observed spectra in the methine carbon region (54–58 ppm), while much poorer agreement is observed in the methylene region (45–48 ppm). Such behavior might be expected since it has been observed that the methylene chemical shifts of model compounds are extremely sensitive to solvent effects (12), which are of course not included in the RIS calculations. It is possible to choose a solvent system consisting of tricholorobenzene and dimethylsulfoxide that shows a closer agreement between the calculated and the experimental spectra (11).

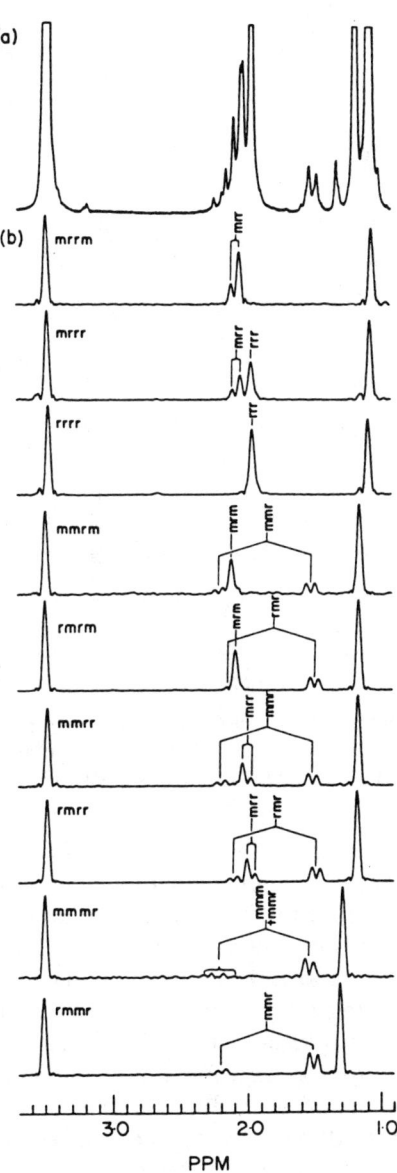

FIGURE 3.23 The (a) 1D spectrum and (b) cross sections through the carbon–proton heteronuclear correlation spectrum of poly(methyl-[*carbonyl*-^{13}C]methacrylate) showing the carbonyl correlations to the methoxy, methylene, and methyl protons. Reprinted with permission (37).

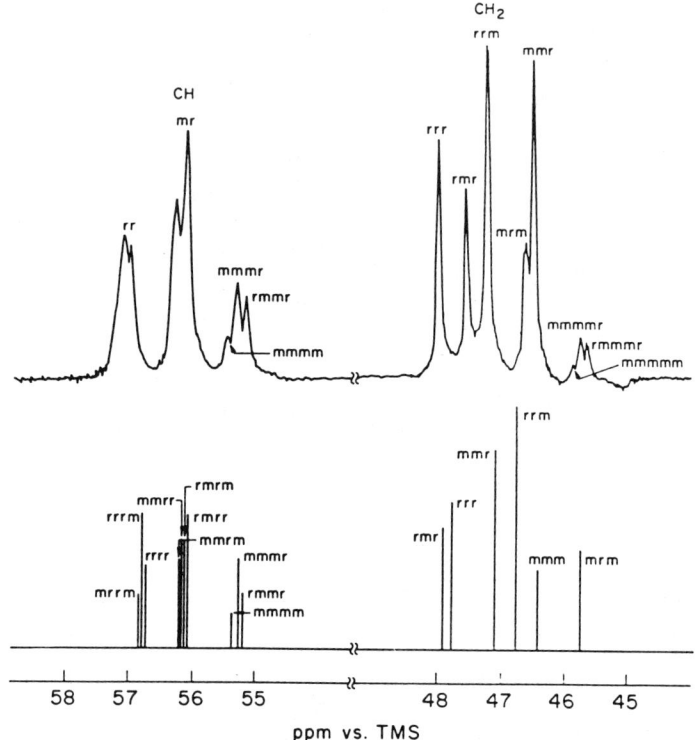

FIGURE 3.24 Comparison of the experimental spectrum of atactic poly(vinyl chloride) in 1,2,4-trichlorobenzene at 120°C with the chemical shifts calculated from the RIS models and the γ-gauche effect. Reprinted with permission (12).

More recently 2D NMR methods have been used to confirm the chemical shift assignments for poly(vinyl chloride) and to correct some of the carbon assignments proposed in the early studies. The strategy for the assignments is to use relayed coherence methods to correlate triad and tetrad resonances. For example, the *rr* triad methine peak of poly(vinyl chloride) should show correlations to the *rrr* and *rrm* tetrad methylene peaks. These tetrads can be distinguished since the *rrm* tetrad will also show a correlation to the *mr* methine triad peak while the *rrr* stereo-sequence will not.

Figure 3.25 shows the pulse sequence for the heteronuclear correlation experiment with relayed coherence transfer (43) that differs from the heteronuclear correlation pulse sequence shown in Fig. 1.59 in two important aspects. Note that after the t_1 period, during which the magnetization evolves under the influence of the proton chemical shifts, that proton magnetization is transferred between coupled

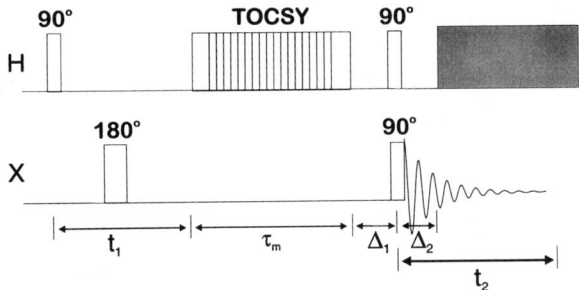

FIGURE 3.25 The pulse sequence diagram for the phase-sensitive heteronuclear spin-lock RELAY experiment. The proton magnetization evolves during the t_1 period and undergoes TOCSY-type spin exchange via the three-bond proton J couplings during the spin-lock period. Note that acquisition begins before the decoupler is turned on (43).

protons using a TOCSY-type (Section 1.8.7.3) magnetization transfer. Note also that magnetization is transferred from protons to carbons with an INEPT-like pulse sequence. The result of the TOCSY magnetization transfer is that in the final spectrum the carbon resonances are not only correlated to their directly bound protons, but also to any protons that are scalar coupled to them (relayed coherence). Using the INEPT carbon–proton magnetization transfer it is possible to record phase-sensitive spectra that have much higher resolution than the absolute value 2D spectra.

Figure 3.26 shows the 125-MHz heteronuclear correlation spectrum for poly(vinyl chloride) in 1,2,4-trichlorobenzene obtained at 90°C with a spin-lock time of 18 ms (44). Peaks are observed in four spectral regions that are marked A–D. The peaks in areas A and C arise from one-bond magnetization transfer of the methine and methylene carbons and their directly bonded protons, and are similar to the peaks observed in a normal heteronuclear correlation spectrum. The relayed peaks in areas B and D arise from correlations between methylene carbons and methine protons and between methine carbons and methylene protons.

Figure 3.27 shows an expanded plot of regions A and B and may be used to illustrate the assignment strategy for vinyl polymers (44). The

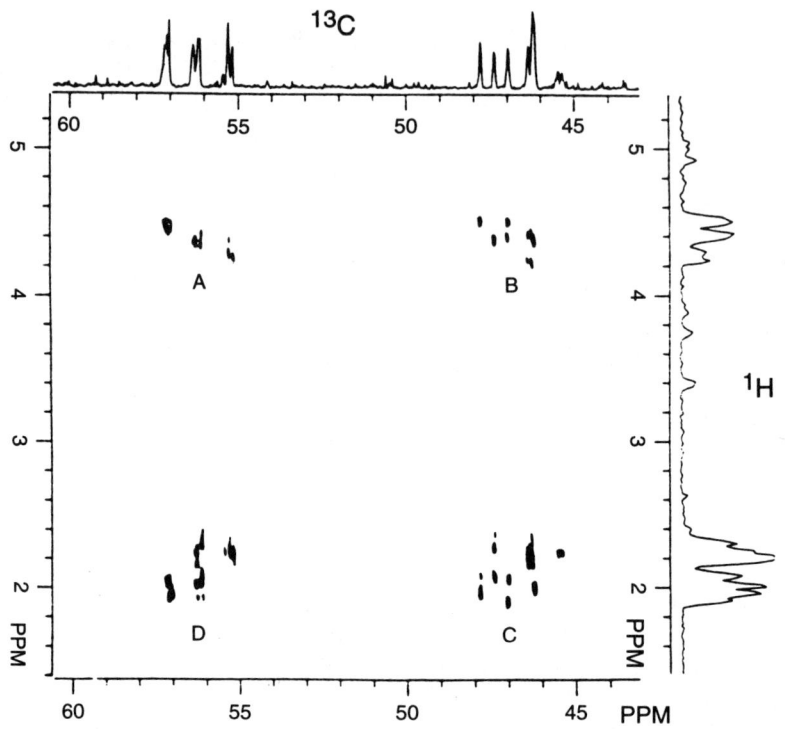

FIGURE 3.26 The heteronuclear spin-lock RELAY spectrum of poly(vinyl chloride) in 1,2,4-trichlorobenzene at 90°C obtained with an 18-ms spin-lock mixing time. Reprinted with permission (44).

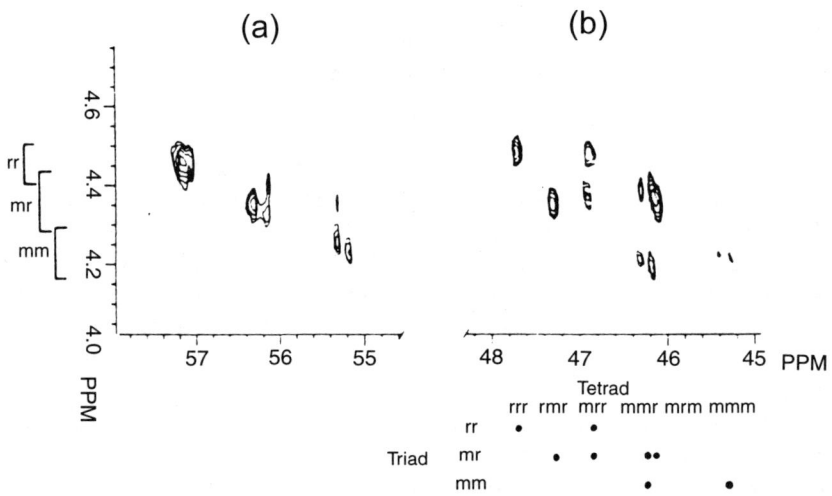

FIGURE 3.27 An expanded section (regions A and B) of the heteronuclear spin-lock RELAY spectrum shown in Fig. 3.26. Reprinted with permission (44).

tetrads mmm and rrr can be distinguished from the others in that they are not correlated with the mr triad, and can be distinguished from each other since the methylene protons in the rrr stereosequence are equivalent, while those in the mmm stereosequence are not. By this means the highest field methylene peak is assigned to rrr and the lowest field to mmm. Four of the six tetrad sequences contain the mr triad, and two of these, mmr and rrm, also show correlations with the mm and rr triad peaks. The remaining sequence, mrm and rmr, can be distinguished since the methylene protons in the mrm stereosequence are equivalent, while those in the rmr stereosequence are not. The nonequivalence of the m centered stereosequences is not obvious from this plot, but can be observed in an expanded plot.

Two-dimensional NMR has been used to establish stereosequence assignments in poly(vinyl fluoride) using the four-bond $^{19}F-^{19}F$ coupling constants (28). The fluorine spectrum of isoregic poly(vinyl fluoride) shows three groups of signals spread over 10 ppm that can be assigned to the mm, mr, and rr triads. In the mr and rr regions there are four and three resolved peaks, respectively, while higher level splitting is not resolved in the mm region. Figure 3.28 shows the 188-MHz fluorine 2D COSY spectrum of poly(vinyl fluoride) in DMF-d_7, obtained at 130°C, from which the stereosequence assignments

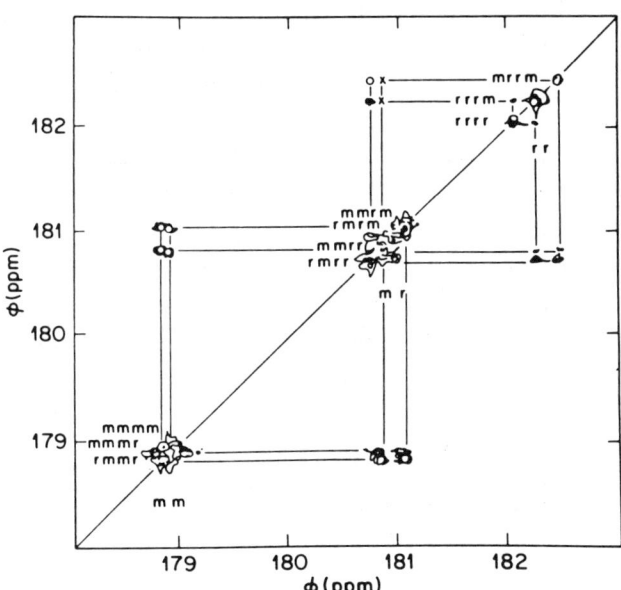

FIGURE 3.28 The 188-MHz 2D COSY spectrum of isoregic poly(vinyl fluoride) in dimethylformamide-d_7 at 130°C. X denotes peaks that are visible at lower contour levels. Reprinted with permission (28).

can be established (28). Cross peaks within the *rr* region show the connectivities between the *rrrr* and *rrrm* sequences, which can be distinguished since the *rrrm* sequence also shows a cross peak to the *mr* region. Thus, the remaining peak in the *rr* region must be *mrrm*. In the *mr* region the *mmrr* sequence can be identified as the only peak showing correlations to both the *mm* and the *rr* region. The sequence *mmrm* can be identified by the correlations to the *mm* region, and the *rmrm* peak shows correlations only within the *mr* region.

Although only one resolved resonance is observed in the *mm* region, the increased resolution in 2D NMR makes it possible to establish the assignments within this overlapping envelope of peaks. Figure 3.29 shows an expanded region of the 2D COSY spectrum for the *mm* region in which three peaks are resolved along the diagonal (28). The stereosequence *mmmm* can be identified as the peak that does not show correlations to the *mr* region, and the *mmmr* and *rmmr* assignments are made on the basis of the peak intensities.

3.3.3 GEOMETRIC ISOMERISM

In addition to regioisomerism and stereoisomerism, chain microstructure in diene polymers can have contributions from geometric isomerism, as noted in Section 2.2.3. In fact, the microstructure of these materials can be extremely complex due to the simultaneous presence of the three types of isomerism. The assignments can be established using the methods used to identify and assign the regio- and

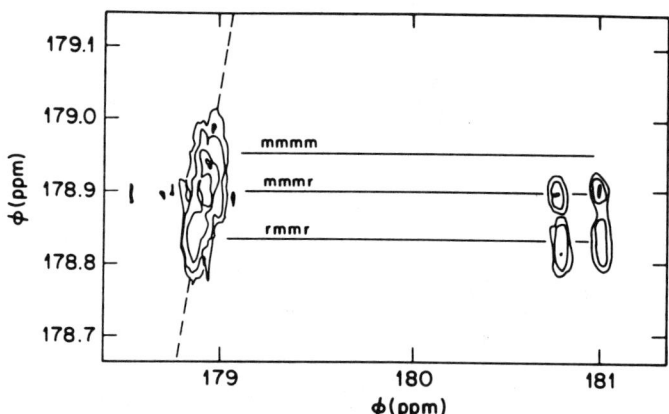

FIGURE 3.29 An expanded section of the 2D COSY spectrum of isoregic poly(vinyl fluoride) shown in Fig. 3.28 showing the correlations to *m* centered stereosequences. Reprinted with permission (28).

3.3. Microstructural Characterization of Polymers in Solution

stereoisomers, namely comparison with model compounds and polymers, chemical shift calculations, and 2D NMR.

Among the diene polymers that have been extensively investigated are the polybutadienes, polyisoprenes, and neoprenes. Figure 3.30 shows an early carbon spectral comparison between the *cis* and the *trans* forms of 1,4-polybutadiene (45). This spectrum illustrates that the β carbon chemical shifts are extremely sensitive to geometric isomerism, and the chemical shift difference between the *cis* and the *trans* form is about 8 ppm.

The carbon NMR spectrum of polybutadiene synthesized by a free radical mechanism is known to be extremely complex because of the statistical incorporation of *cis*, *trans*- and 1,2-butadiene units. This is illustrated in Fig. 3.31, which shows the 50-MHz carbon spectrum for free radical polybutadiene (46). At the lowest gain setting, several major and minor peaks are visible, including the methylene signals for *cis* (b) and *trans* (d) 1,4-butadiene units. At a higher gain setting, 14 signals in the aliphatic region can be observed and assigned using the calculated chemical shifts. The inset to Fig. 3.31 is a computer simulation of the aliphatic part of the spectrum for a polymer that contains *cis*, *trans*, and 1,2-units in a ratio of 23:58:19 using Bernoul-

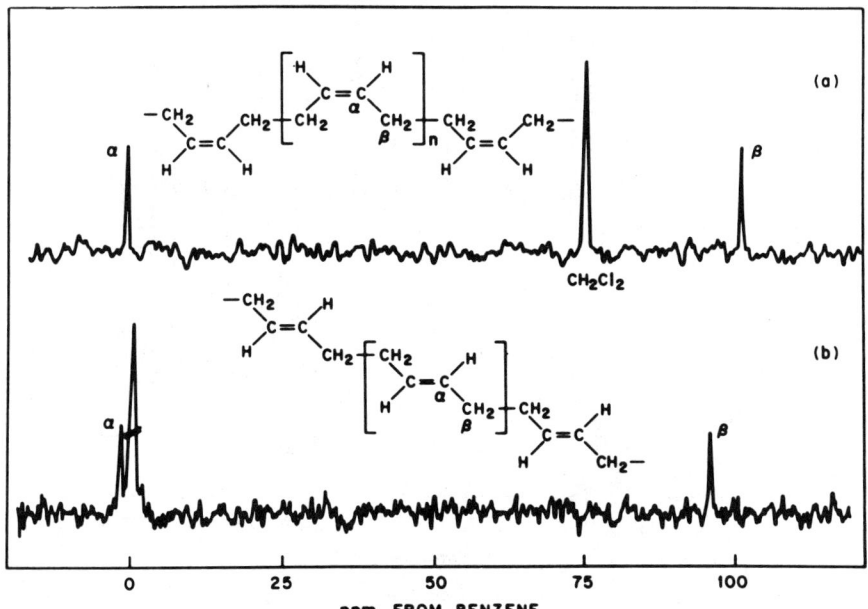

FIGURE 3.30 The 15-MHz carbon NMR spectra of (a) *cis* in polybutadiene dichloromethane and (b) *trans* polybutadiene in benzene. Reprinted with permission (45).

lian statistics. The good agreement between the observed and the calculated spectra shows that the incorporation of geometric isomers into the growing chain can be described as a Bernoullian process similar to the incorporation of pseudoasymmetric centers.

Polyisoprene is another important diene polymer that has been extensively studied by NMR methods. Figure 3.32 compares the 50-MHz carbon NMR spectra of *cis*- and *trans*-1,4-polyisoprene from natural sources (46). Both polymers are produced enzymatically and are extremely pure. Note that the chemical shifts for the methylene and methyl resonances are extremely sensitive to geometric isomerism. Synthetic *cis*- and *trans*-1,4-polybutadiene typically contain a high fraction of geometric defects (2–6%) that can be easily visualized by ^{13}C NMR.

Poly(2-chlorobutadiene), more commonly known as neoprene, was one of the first commercially successful rubber polymers. Again the micro-structure is extremely complex because of the possible inclu-

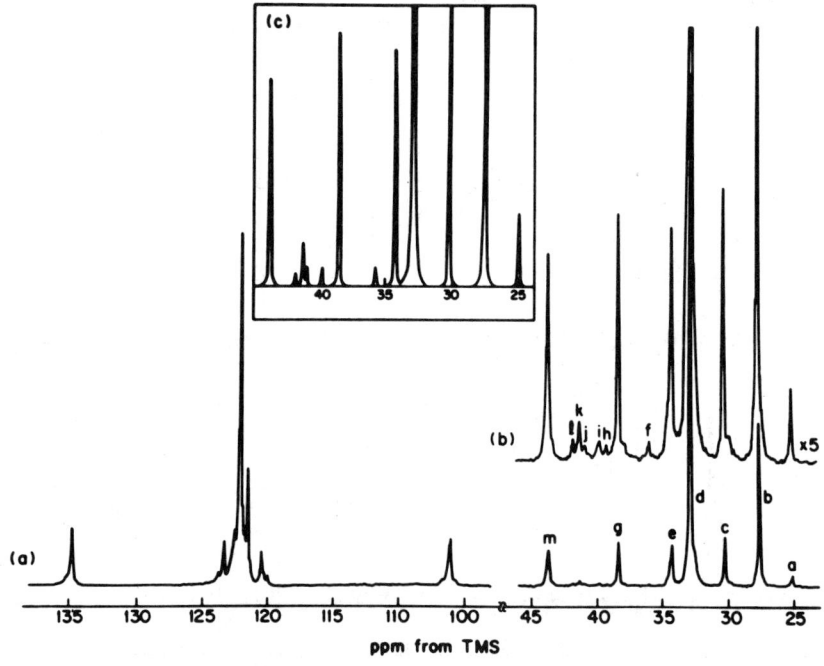

FIGURE 3.31 The experimental and simulated 50-MHz carbon NMR spectra of free radical polymerized polybutadiene. Spectrum (a) shows the experimental spectrum, (b) shows the aliphatic region at a higher gain, and (c) shows the simulated spectrum of the aliphatic region based on a random distribution of *cis*, *trans*, and 1,2-units in a proportion of 23:58:19. Reprinted with permission (46).

3.3. Microstructural Characterization of Polymers in Solution

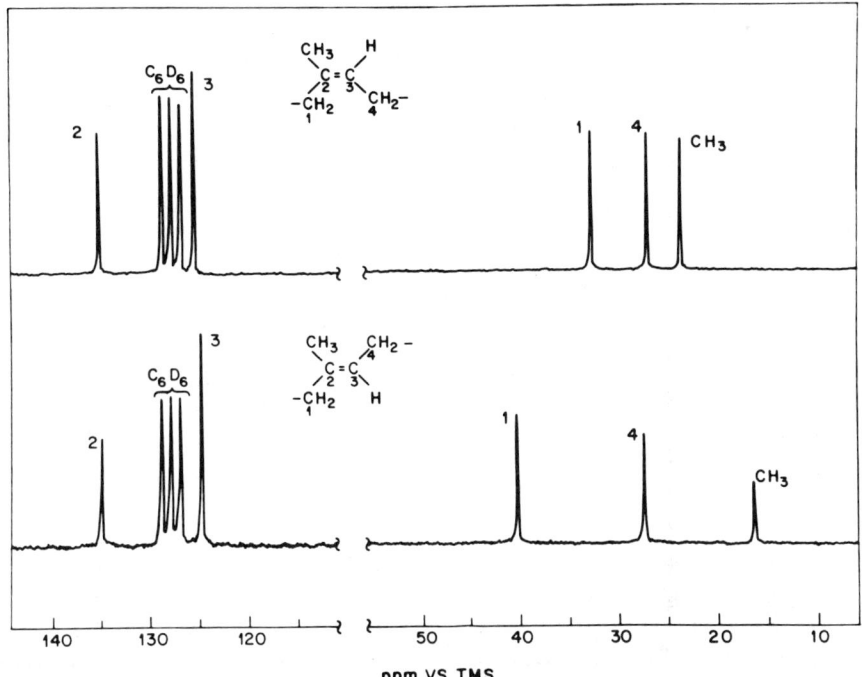

FIGURE 3.32 The 50-MHz carbon NMR spectra of (a) cis and (b) trans-1,4-polyisoprene obtained from natural sources. Reprinted with permission (46).

sion of cis-1,4-, trans-1,4-, 1,2-, and 3,4-butadiene units, in addition to isomerized chlorobutadiene units. The spectrum is further complicated by regioiso-merism. The 22-MHz carbon spectrum of neoprene is shown in Fig. 3.33 along with some of the carbon assignments (46). The polymer contains predominantly 1,4-units, and about 12% of them are regioisomers. The occurrence of 1,2- and 3,4-units in this sample is below the noise level of the NMR experiment.

3.3.4 BRANCHES AND END GROUPS

NMR spectroscopy has been an important tool for the identification and characterization of chain branches and end groups. This is a topic of intense interest because low levels of these structures can have a large effect on the crystallinity in some polymers and can be potentially reactive sites in otherwise inert materials. The methods used to study branches and chain ends by NMR are similar to those used for the characterization of regioisomers and stereoisomers. The 2D and other sophisticated methods are often not useful because the defect

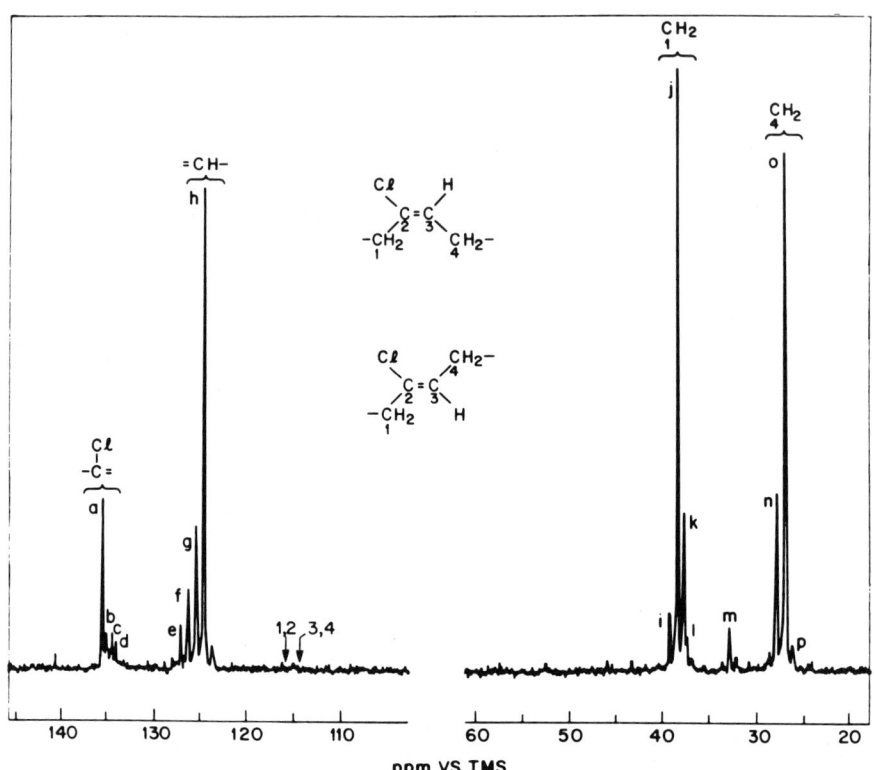

FIGURE 3.33 The 22-MHz carbon NMR spectrum of neoprene in CDCl$_3$. Reprinted with permission (46).

structures occur at such low concentrations that the cross peaks would not be visible above the noise. In such situations the chemical shifts can be assigned using chemical shift calculations. In some cases the assignments can be established using 2D methods on model compounds.

Branch formation has been extensively investigated in polyethylene, because such structures are known to reduce both the melting temperature and the crystallinity. The number and types of branches are also sensitive to the synthetic method. Figure 3.34 shows the 50-MHz carbon NMR spectrum of high-pressure polyethylene (46). The largest peak, which is off scale at 30.0 ppm, is due to main-chain methylene carbons that are four or more bonds removed from a branch site. Numerous other small resonances have been assigned, as noted in Fig. 3.34. The size and distribution of branches have been analyzed for many polyethylenes, and the types of branches and

3.3. Microstructural Characterization of Polymers in Solution

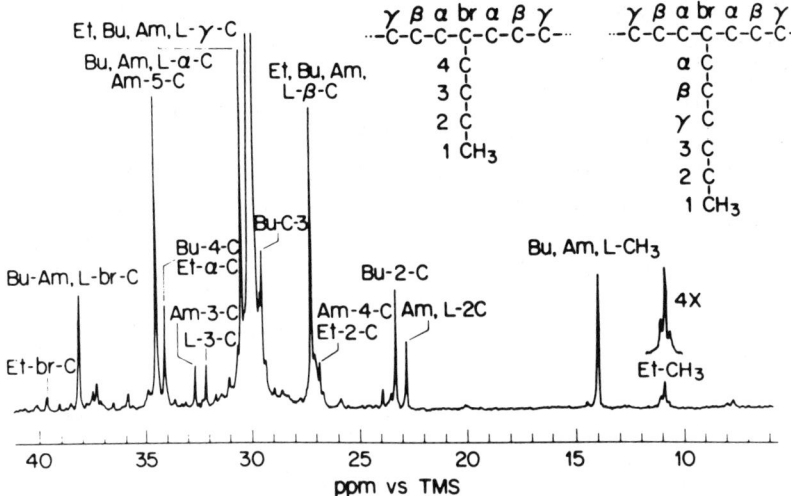

FIGURE 3.34 The 50-MHz carbon NMR spectrum of high-pressure polyethylene in 1,2,4-trichlorobenzene at 110°C. The assignments for some of the branch structures are shown. Reprinted with permission (46).

number per 1000 main-chain carbons are listed in Table 3.7 for high-pressure polyethylene (M_n = 18,400, M_w = 129,000).

Branch formation is also important for poly(vinyl chloride). Poly(vinyl chloride) is amorphous, so branching is not expected to greatly affect the melting temperature, but the branch points are important sites for the initiation of chemical decomposition. Because of its commercial importance, the branching and defect structures in poly(vinyl chloride) have been extensively investigated.

TABLE 3.7
The Frequency of Branch Formation in Hi-Pressure Polyethylene[a]

Branch type	Branches per 1000 backbone carbons
$-CH_3$	0.00
$-CH_2CH_3$	1.0
$-CH_2CH_2CH_3$	0.0
$-CH_2CH_2CH_2CH_3$	9.6
$-CH_2CH_2CH_2CH_2CH_3$	3.6
Hexyl or longer	5.6
Total	19.8

[a] F. Bovey and L. Jelinski, *J. Phys. Chem.* **89**, 571 (1985).

202   3. The Solution Characterization of Polymers

The carbon spectrum of poly(vinyl chloride) is too complex due to stereochemical isomerization to directly yield information about branches. The spectrum can be greatly simplified by reductive dechlorination with lithium aluminum hydride (47) or the free radical reagent tri-n-butyltin hydride (48). The resulting hydrocarbon polymer can then be analyzed in the same manner as polyethylene. The departing chlorines can be replaced by deuterons instead of protons so that the carbon atoms that were previously bonded to chlorines will now appear as triplets due to the carbon–deuteron coupling [since deuterium is a spin-1 nuclei (Section 1.6.1)]. Figure 3.35 com-

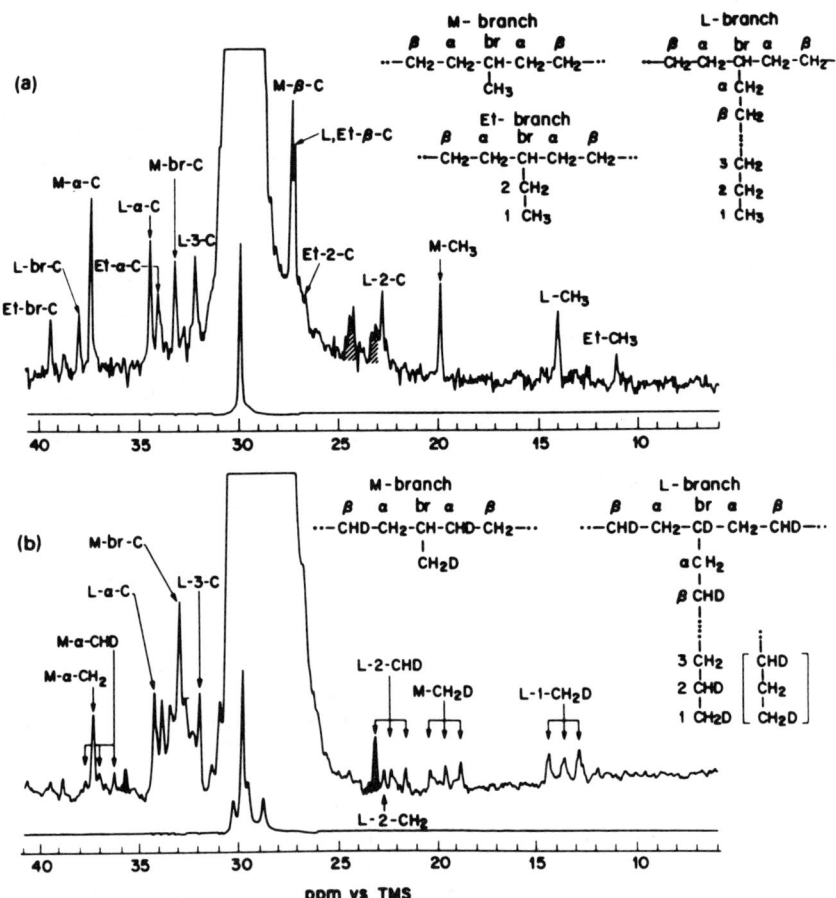

FIGURE 3.35 The 25-MHz carbon NMR spectrum of poly(vinyl chloride) reduced with (a) tri-n-butyl tin hydride or (b) tri-n-butyl tin deuteride. Reprinted with permission (46).

3.3. Microstructural Characterization of Polymers in Solution

pares the 125-MHz carbon spectra of poly(vinyl chloride) reduced with tri-n-butyltin hydride in which the chlorines are replaced with either (a) protons or (b) deuterons (46). The main defects are the methyl branches (two to three per 1000 main-chain carbons), with a smaller proportion of the ethyl and butyl branches.

In highly branched materials, the branch sites are present in high enough concentrations that they are amenable to analysis by 2D NMR methods. One such example is polysuccinimide, which is obtained by thermal polymerization of aspartic acid, and can be obtained in the linear or branched form, depending on the presence and concentration of the catalyst phosphoric acid (49). The branched sites can be identified by proton, carbon, and nitrogen NMR spectroscopy. In the proton spectrum, the branches are characterized by amide resonances (7–12 ppm) that are J-coupled to methine protons, a cross-peak pattern that can be easily identified with 2D COSY experiments, such as that shown in Fig. 3.36 (49). The box marked X shows the amide–methine cross peaks, and the box marked Y shows the cross peak between these methine protons and the methylenes in the branched unit. Further evidence for the branching is obtained from the ^{15}N spectrum for the ^{15}N-labeled polymer shown in Fig. 3.37. In the uncatalyzed material a peak is observed around 100 ppm that can

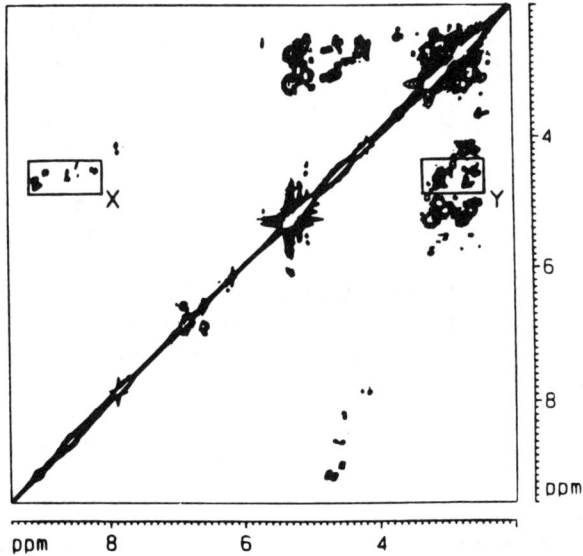

FIGURE 3.36 The 500-MHz 2D COSY spectrum of polysuccinimide containing a high concentration of branched sites in DMSO-d_6. Reprinted with permission (49).

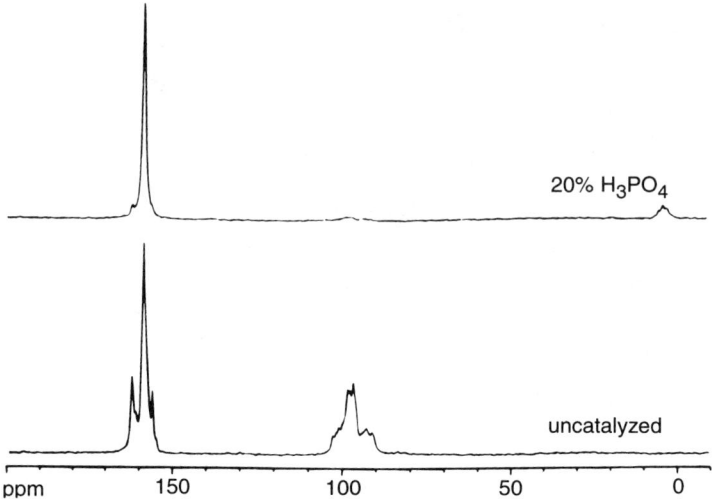

FIGURE 3.37 The 50-MHz nitrogen NMR spectra of (a) linear and (b) branched polysuccinimide. Reprinted with permission (49).

be assigned to the amide nitrogens in the branch sites by ^{15}N/^{1}H HMQC (49).

The study of branch and cross-linked sites is frequently limited by their low concentration and the overlap of the branch resonances with the main-chain signals of the polymer. One approach to overcoming this limitation is to use branched materials in which the branches are specifically labeled with an NMR-active nuclei, as in the ^{15}N labeling in the above experiment. Among the possible NMR-active nuclei, ^{2}H, ^{13}C, and ^{15}N are preferred because they introduce the smallest perturbations. Solution-state deuterium NMR has been employed using the deuterated cross-linking reagent allyl methacrylate, as shown in Fig. 3.38 (50). The chemical shifts follow a pattern similar to that of the proton spectra, but the signals are much broader because deu-

3.3. Microstructural Characterization of Polymers in Solution

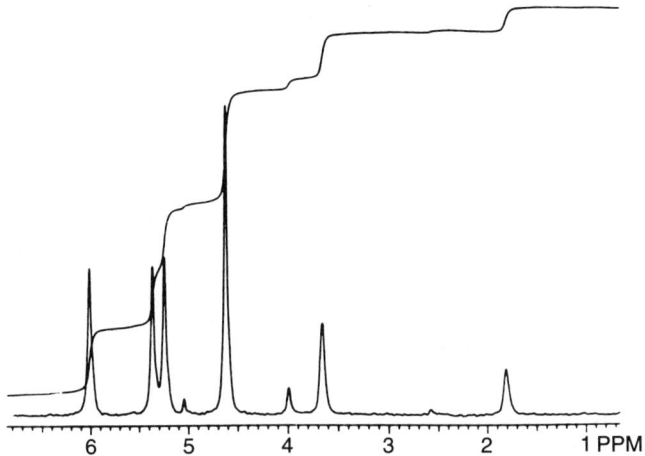

FIGURE 3.38 The 61-MHz solution-state deuterium NMR spectrum of allyl methacrylate-d_5 in THF. The small peaks at 1.8 and 3.65 ppm are due to the natural abundance signals from the solvent. Reprinted with permission (50).

terium is a spin-1 nucleus. Figure 3.39 compares the deuterium spectra for acrylate polymers cross linked with (a) deuterated and (b) protonated allyl methacrylate at the 0.5–1% level. The signals observed for the sample cross linked with the protonated divinyl cross-linking agent are the natural abundance deuterium signals (0.015%) for the solvent and the polymer. The chemistry of the cross-linking reaction can be studied by observing changes in the deuterium spectra, and the results show that about 30% of the monomer is cross linked (50). Such spectra require long acquisition times, but the signal of the cross-linking reagent is clearly visible above the background deuterium signals. This experiment would be difficult to perform by ^{13}C labeling because the signals from the cross-link reagent would be approximately as large as the natural abundance signals from the polymers.

$$CH_2=C\begin{matrix}CH_3\\ \\C-O-CD_2-CD=CD_2\\\parallel\\O\end{matrix}\qquad 4.6\quad 6.0\quad 5.4, 5.$$

allyl methacrylate

The degree and site of branching in novolac polymers used in microlithography can have important consequences for the behavior

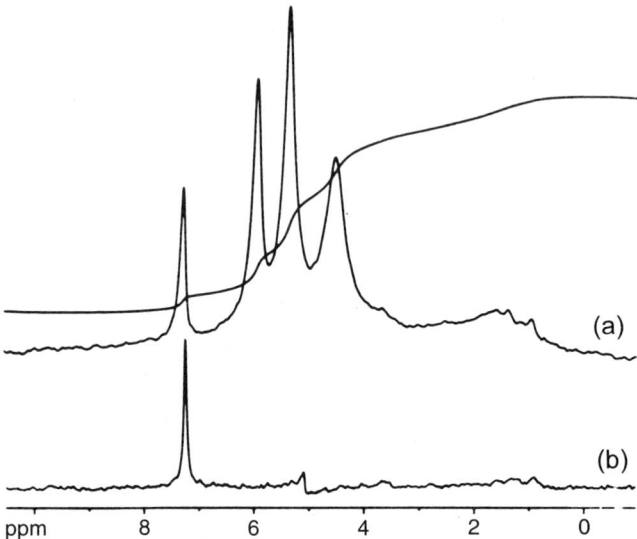

FIGURE 3.39 The 61-MHz deuterium NMR spectrum of polyacrylates reacted with (a) deuterated or (b) protonated allyl methacrylate. The broad resonances between 1 and 3 ppm are from the natural abundance signals in the polymer, and the signal at 7.2 ppm is the natural abundance signal from the chloroform solvent. Reprinted with permission (50).

of these materials. These polymers are synthesized via acid-catalyzed condensation of phenols and formaldehyde and can have complex structures with a variety of possible branching sites. Although it is difficult to study branching in the high-molecular-weight materials, 2D NMR methods have been used to study branching in model compounds. Figure 3.40 shows the 2D INADEQUATE (21) spectrum of two model compounds from which it is possible to trace out the carbon connectivities from the $^{13}C-^{13}C$ J couplings (51).

II III

Recently there has been an interest in highly branched materials, known as dendritic macromolecules, which are branched in three dimensions (52). Figure 3.41 shows an example of dendrimers based on 1,3,5-benzenetricarboxylic acid (53). These molecules grow by at-

3.3. Microstructural Characterization of Polymers in Solution 207

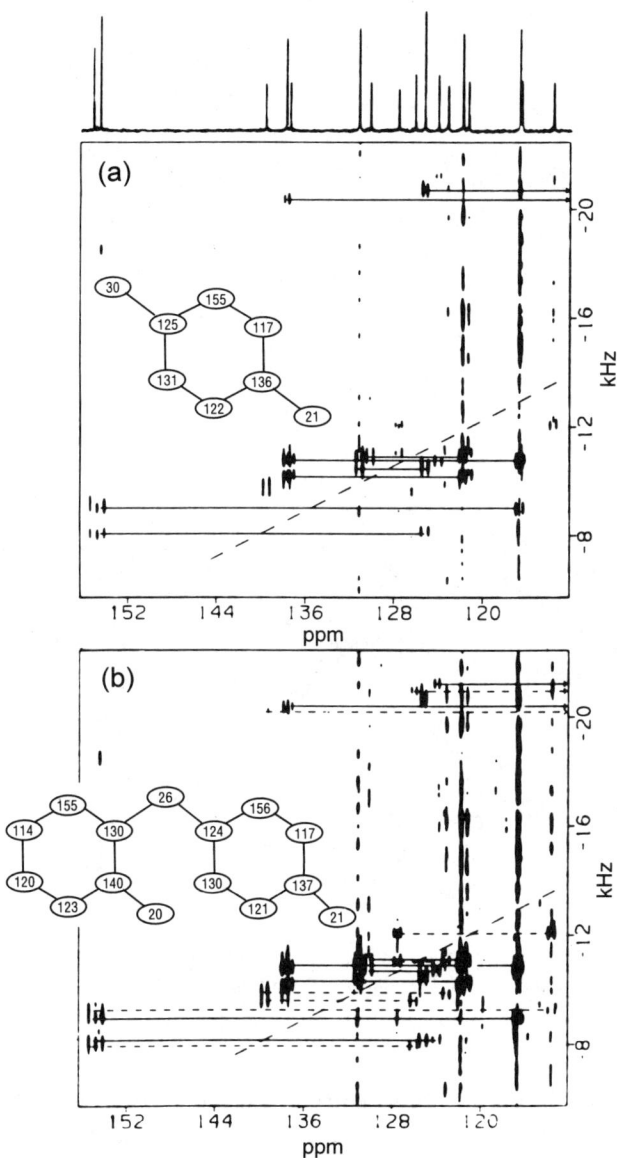

FIGURE 3.40 The 100-MHz carbon 2D INADEQUATE spectrum of mixtures of novolak model compounds. Spectrum (a) shows the correlations at a high contour level in which the connectivities for the symmetric model compound **II** are connected by horizontal lines. Spectrum (b) shows the same data at a lower contour level, and the solid and dashed horizontal lines show the connectivities for the two rings of compound III. Reprinted with permission (51).

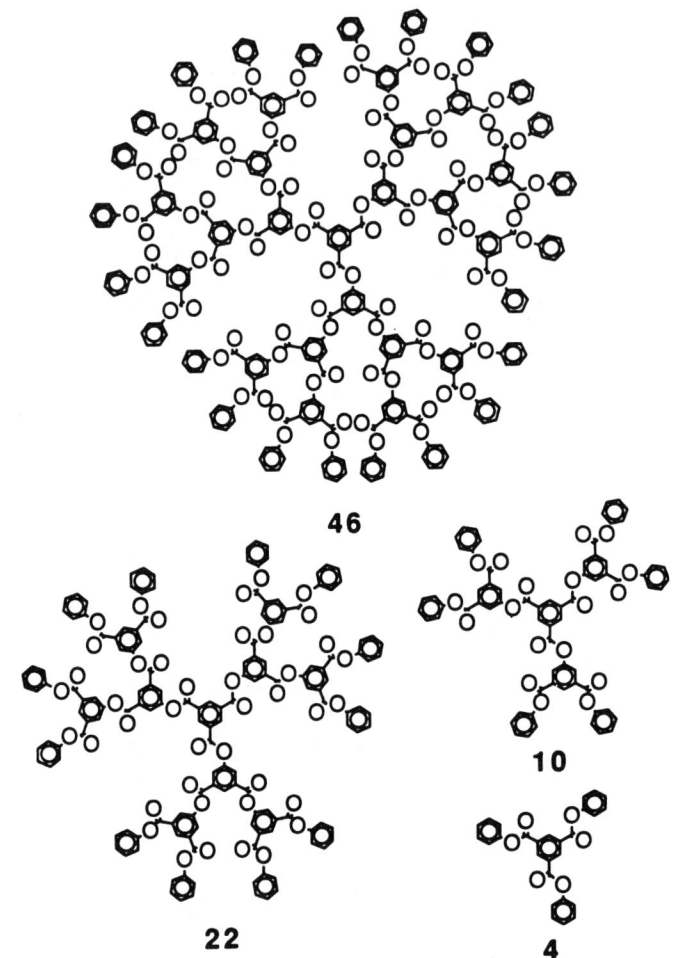

FIGURE 3.41 A schematic diagram of sequential generations of dendritic macromolecules containing 4, 10, 22, and 46 monomer units. Reprinted with permission (53).

taching three monomers to a central monomer to give a first-generation polymer with four monomer units. These macromolecules grow exponentially in size, and the second, third, and fourth generations contain 10, 22, and 46 monomer units. Figure 3.42 shows the 360-MHz proton spectra of dendrimers for several generations (53).

The NMR characterization of end groups in polymers requires a methodology similar to that used to study branching because, like the branches and some other types of defects, the end-group resonances may be very small and may overlap the main-chain resonances. A

3.3. Microstructural Characterization of Polymers in Solution

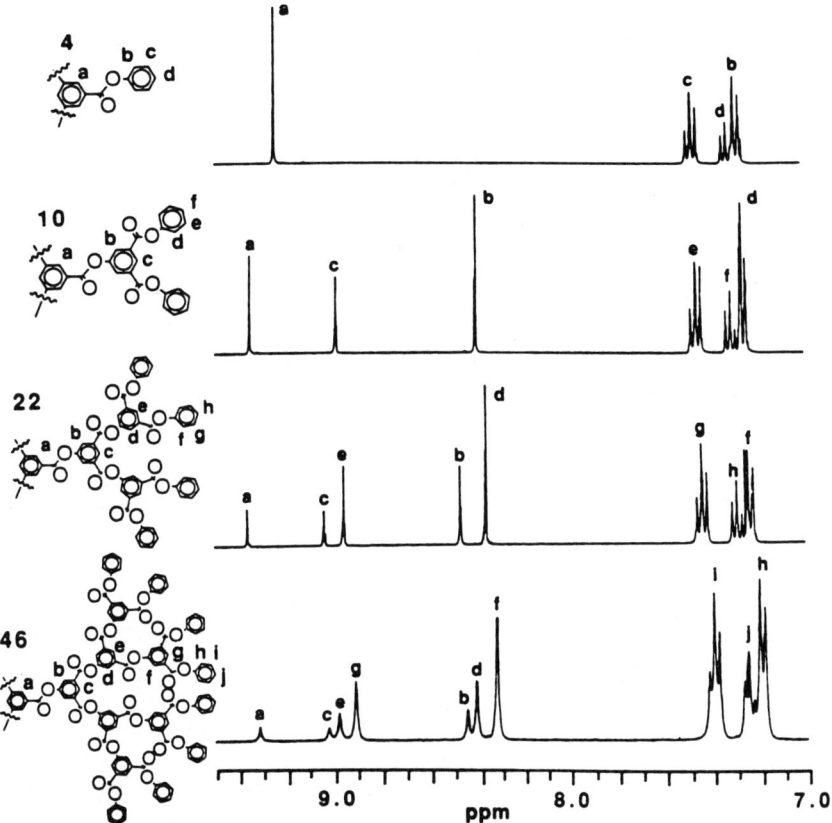

FIGURE 3.42 The 360-MHz proton NMR spectra of dendritic macromolecules containing 4, 10, 22, and 46 monomer units. Reprinted with permission (53).

critical difference, however, is that the relative fraction of end-group resonances will scale with molecular weight. This is illustrated in Fig. 3.43, which compares with 50-MHz carbon NMR spectra for poly(propylene oxide) of molecular weights (a) 4000 and (b) 1000 (23). In the spectrum of the 4000-molecular-weight material seven resonances are observed that could arise from either end groups or other types of defects, such as regio-isomers. Comparison of this spectrum with that obtained for the 1000-molecular-weight material shows that several of the peaks are much larger compared to those of the main chain, while the ratio for several others remains unchanged. From this type of a comparison we can assign the peaks that change in intensity relative to the main-chain signals as end groups. These and other experiments show that the resonances labeled four, five, and six arise from methyl groups in head-to-head–tail-to-tail defects.

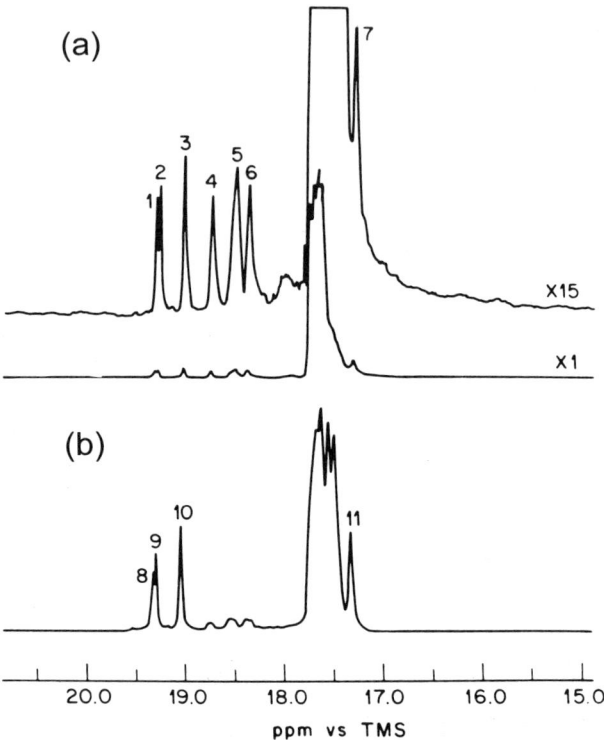

FIGURE 3.43 The methyl region of the 50-MHz carbon NMR spectra of atactic poly(propylene oxide) of molecular weight (a) 4000 and (b) 1000. Reprinted with permission (23).

Once these end groups have been assigned, they can be used to calculate the molecular weight, provided that care is taken to acquire spectra that can be quantitatively compared. As noted in Section 1.8.6.4, it is important to wait five times longer than the longest T_1 and to acquire the spectra without NOE enhancement for quantitative comparisons. This is an important consideration because the T_1 values for the end groups may be considerably longer than those for the main-chain resonances, especially in higher molecular weight polymers. In the 4000-molecular-weight poly(propylene oxide), for example, the T_1's for some of the end-group resonances are twice as long as those for the main-chain resonances (23). The molecular weight calculated from the end groups from the spectra in Fig. 3.43 is 5400, compared with the value of M_n = 4000 calculated by the manufacturer.

The NMR spectra as a function of molecular weight can be used to identify end-group resonances, but cannot be directly used to deter-

3.3. Microstructural Characterization of Polymers in Solution

mine the structure of the end groups. For this, the chemical shift calculations, model compounds, INEPT or DEPT editing, or 2D NMR methods must be employed. Model compounds in combination with 2D NMR can be used to establish the chemical shift assignments. Figure 3.44 shows the 2D COSY spectrum for a model compound for a polyamide from a study in which the proton and carbon assignments were established with a combination of the 2D COSY and HETCOR experiments (54). If all the peaks from the model compound are observed in the high-molecular-weight material, then it can be reasonably concluded that such end groups are present.

An alternative strategy for the identification of end groups is to label or derivatize the end groups with an NMR-active nuclei, preferably one in which the chemical shift is extremely sensitive to the structure. One reaction that has been used in the analysis of end groups in poly(2,6-dimethyl-1,4-phenylene oxide) is (55)

This reaction is quantitative and can be carried out in an NMR tube, and the ^{31}P chemical shifts of the final products are extremely sensitive to the end-group structure. The phosphorus spectra were recorded for a wide variety of possible end groups, and Fig. 3.45 shows the ^{31}P spectra of two derivatized poly(2,6-dimethyl and phenylene oxide) resins in which the end groups **2**, **4**, and **8** were observed (55). By comparing the spectra with those obtained from polymers with a known amount of end groups, the molecular weights can also be calculated by this method.

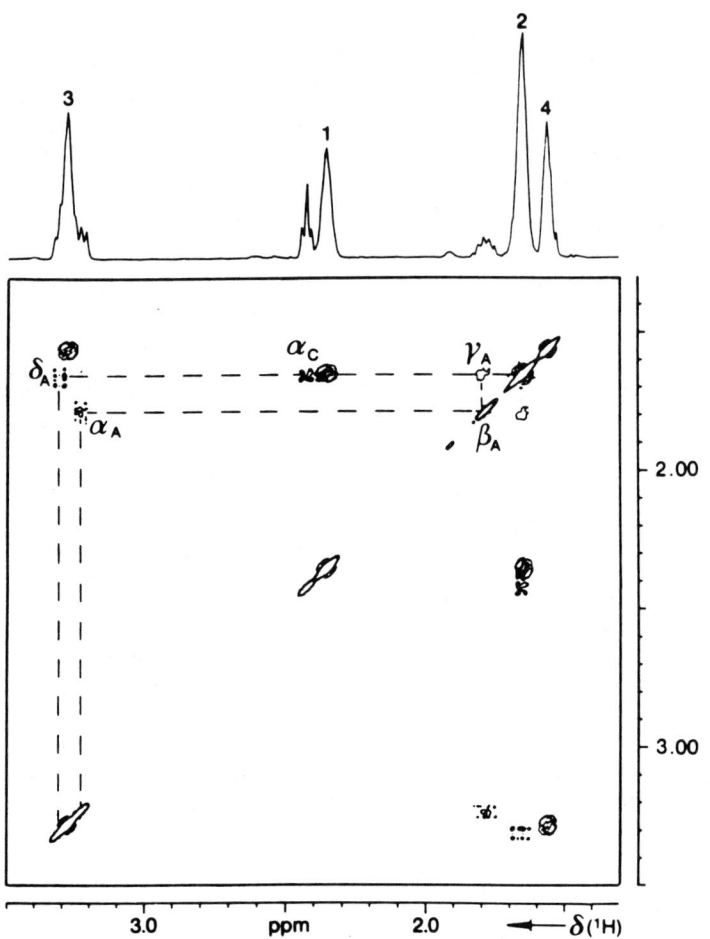

FIGURE 3.44 The 400-MHz 2D COSY spectrum of polyamide-4,6 at 50°C in DCOOD. Reprinted with permission (54).

3.3.5 COPOLYMER CHARACTERIZATION

The sequence and microstructural characterization of copolymers have been extensively investigated in an effort to understand structure–property relationships. The microstructure of copolymer chains can be extremely complex due to the combination of the normal chain microstructure, including regioisomerism, stereoisomerism, branching, geometric isomerization, etc.; monomer sequence; and chain architecture. Among the most important chain architectures are the random, alternating, block, and graft copolymers as shown in Fig. 2.18. Many copolymers are commercially important materials, such as

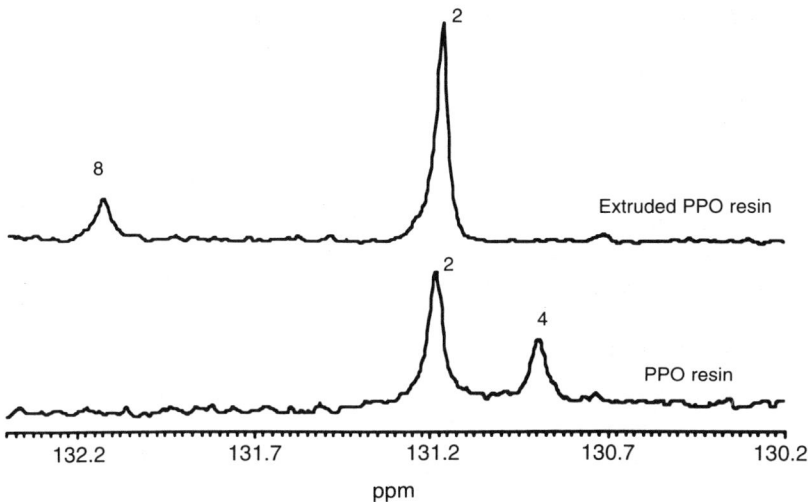

FIGURE 3.45 The 121-MHz ^{31}P spectra of two poly(2,6-dimethyl-1,4-phenylene oxide) resins derivatized with 1,3,2-dioxaphospholanyl chloride. Reprinted with permission (55).

ethylene–propylene copolymers, and have been investigated in great detail.

Chain architecture can have a large effect on the NMR spectra of copolymers. The NMR spectra of copolymers in regular, repeating sequence architectures, such as alternating copolymers, can be greatly simplified relative to those with a random sequence. This is illustrated in the 500-MHz proton NMR spectra of random sequence and alternating 50:50 copolymers of styrene and methyl methacrylate shown in Fig. 3.46 (56). In such materials, the spectrum from the alternating polymers is much simpler because the strict monomer alternation creates a uniform magnetic environment compared to the variety of local magnetic environments in the random copolymer. Although the sequence is well defined in the alternating copolymer, the resonances are split by stereochemical isomerism.

The stereoisomeric assignments for the 50:50 alternating styrene–methyl methacrylate copolymer have been established with a combination of carbon and proton NMR spectroscopy. Figure 3.47 shows the 125-MHz carbon spectra for the alternating copolymer for two spectral regions (56). Note the apparent 1:2:1 splitting for the carbonyl, phenyl C_1, the methine carbon resonances for the copolymer. Such a ratio is expected to arise from the co-isotactic, co-heterotactic, and co-syndiotactic resonances in a copolymer with random stereochemistry. These and the proton signals can be assigned by 2D

NMR since the distances between the phenyl protons in the styrene units and the methyl or methoxy protons depend on the stereochemistry. In the co-isotactic sequences the phenyl protons are closest to the methoxy protons, while they are most distant in the co-syndiotactic stereosequences.

Co-syndiotactic

Co-isotactic

Co-heterotactic

Figure 3.48 shows the 500-MHz 2D NOESY spectrum for the 50:50 alternating styrene–methyl methacrylate copolymer from which many of the spectra assignments can be determined (56). Note the intense cross peak between the phenyl protons and one of the methoxy protons that must be assigned to the co-isotactic sequence. A weaker cross peak is observed to the center methoxy peak (co-heterotactic), and cross peaks are absent to the most downfield methoxy signal, which must be assigned to the co-syndiotactic stereosequences. In a similar way the stereosequence resonance assignments for the methyl and methylene protons can be established.

The structure of block copolymers can also be well defined by NMR methods. The spectra of block copolymers appear mostly as a composite of the NMR spectra of the two block components, and may be simple or complex, depending on the microstructure of the block polymers. The only differences that are typically visible are the resonances shifted from the main-chain signals that arise from the junction between blocks. This behavior is illustrated in Fig. 3.49, which shows the 59-MHz ^{29}Si spectrum of a block copolymer of tetramethyl-*p*-silphenylenesiloxane and dimethyl siloxane (57). The resonances a and d can be assigned to the tetramethyl-*p*-silphenylenesiloxane and dimethyl siloxane blocks by comparison to the spectra of the homopolymers, and the resonances b and c to the tetramethyl-*p*-silphenylenesiloxane and dimethyl siloxane units at the block junction. The small resonance e has been assigned to tetramethyl-*p*-

3.3. *Microstructural Characterization of Polymers in Solution* 215

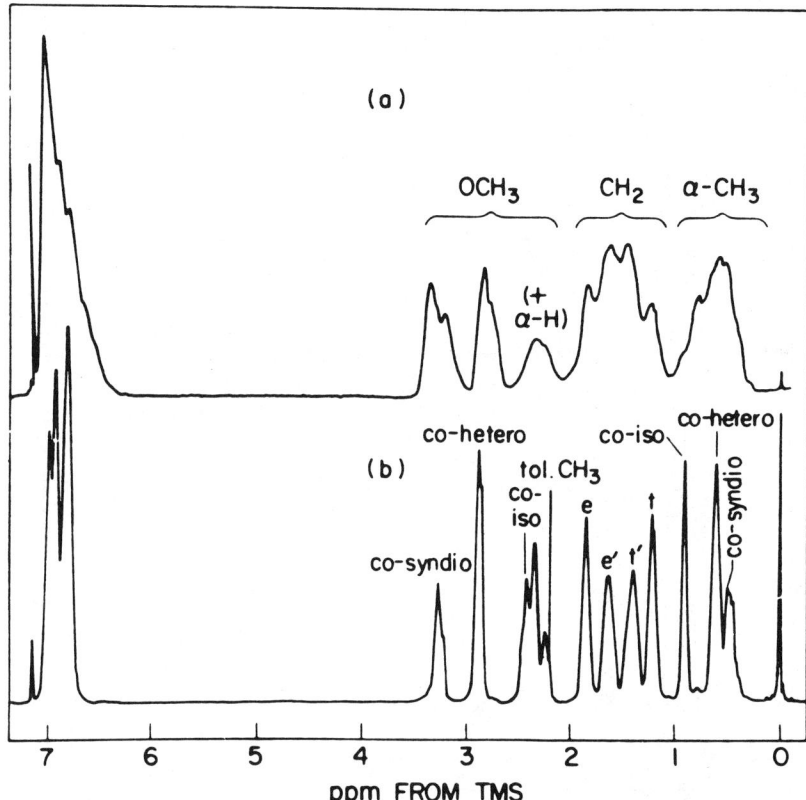

FIGURE 3.46 The 500-MHz proton NMR spectra of the (a) random and (b) alternating 50:50 poly(styrene-co-methyl methacrylate) copolymers. Reprinted with permission (56).

silphenylenesiloxane units alternating with dimethyl siloxane units by the synthesis of the strictly alternating copolymer.

If the spectra such as those shown in Fig. 3.49 are collected in a manner such that the peak intensities can be quantitatively compared, the average block length ($\bar{n}_{TMPS}$ and $\bar{n}_{DMS}$) can be calculated directly from the peak intensities, taking into account that each tetramethyl-*p*-silphenylene-siloxane monomer has two silicons. The

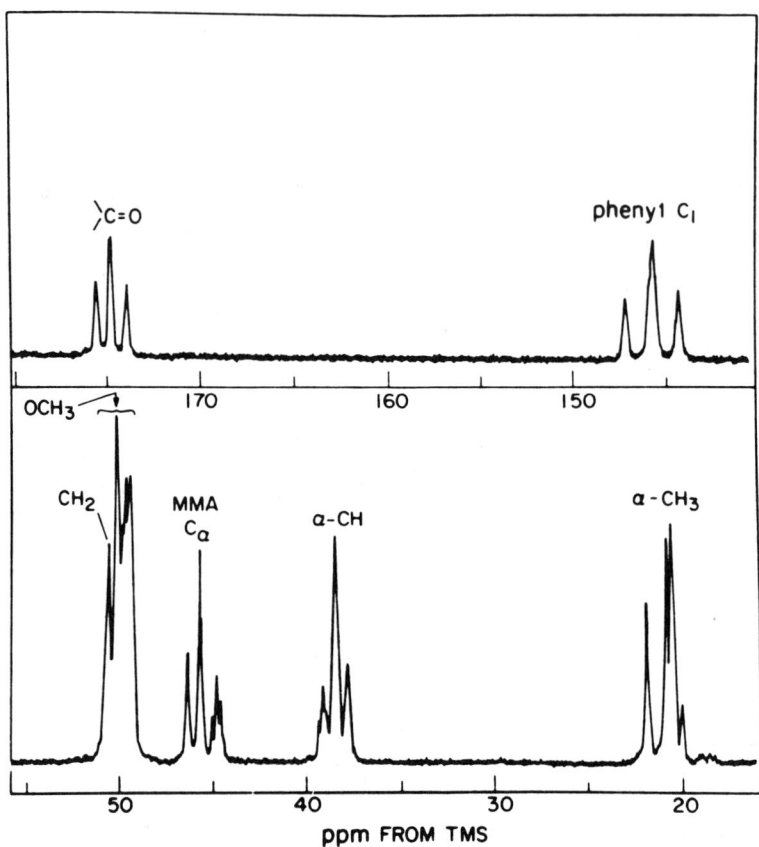

FIGURE 3.47 The 125-MHz carbon NMR spectrum of the alternating styrene–methyl methacrylate copolymer. Reprinted with permission (56).

results calculated from Fig. 3.49 show that the average block lengths for tetramethyl-p-silphenylenesiloxane and the dimethyl siloxane units are 4 and 7 units. Of course as the blocks become long the signals from the junctions may be difficult to detect as they become very small relative to the main block resonances. This behavior is illustrated in Fig. 3.50, which shows the 75-MHz carbon spectrum of the triblock polymer t-butyl ethylene oxide–ethylene oxide–t-butyl ethylene oxide, where the block length is 48 units for the t-butyl ethylene oxide units and 40 for the ethylene oxide units (58). By comparison of the spectra with shorter block lengths to identify the end groups, the resonances between 86.4 and 86.7 ppm can be assigned to the t-butyl ethylene oxide units at the junction between the

3.3. Microstructural Characterization of Polymers in Solution

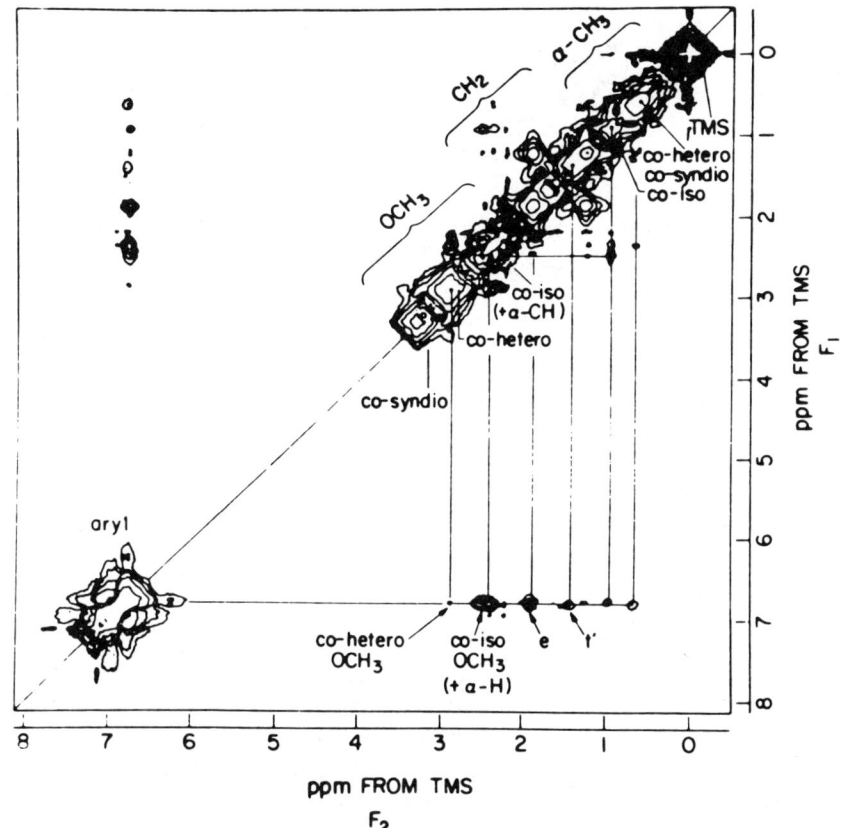

FIGURE 3.48 The 500-MHz proton 2D NOESY spectrum of the alternating styrene–methyl methacrylate copolymer obtained with a 0.3-s mixing time. Reprinted with permission (56).

blocks. As the block lengths get larger, the resonances at the junction will be increasingly more difficult to detect.

The NMR spectra of statistical copolymers can be extremely complex due to the combination of monomer sequence, regioisomerism, stereochemical isomerism, and the presence of other defect structures, and the assignment of such spectra remains a fundamental challenge. The methods used to establish the resonance assignments are similar to those used to study other types of isomerism, and include comparison with model compounds and polymers, chemical shift calculations, and 2D NMR methods.

Ethylene–propylene copolymers have been extensively characterized by NMR methods. Figure 3.51 shows the 25-MHz carbon NMR

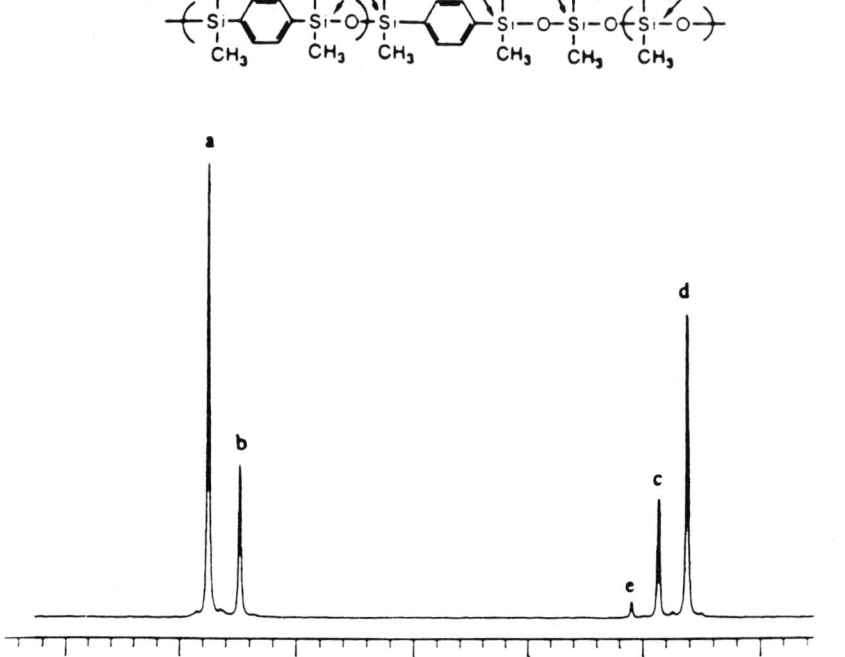

FIGURE 3.49 The 59-MHz silicon NMR spectrum of the tetramethyl-p-silphenylenesiloxane–dimethyl siloxane block copolymer. Reprinted with permission (57).

spectrum of a 50:50 ethylene–propylene copolymer along with some of the assignments established primarily through the use of chemical shift calculations (35). The nomenclature is similar to that introduced in Section 3.3.1 for the description of regioisomerism, and the designations (such as $\alpha\alpha$ or $\gamma\delta$) refer to the distance in each direction along the chain to the nearest branch site. The analysis of the ethylene–propylene copolymer is complicated by regioisomerism and stereochemical isomerism for neighboring propylene units. Carbon-13 NMR is often the method of choice for the analysis of these spectra since much higher resolution is typically observed than for the proton spectra (59). Once the carbon chemical shifts have been established, the proton resonance assignments can be directly determined by heteronuclear 2D correlation experiments (60).

The analysis of copolymer microstructure is similar in many ways to the analysis of stereochemical isomerism, and among the features

3.3. Microstructural Characterization of Polymers in Solution

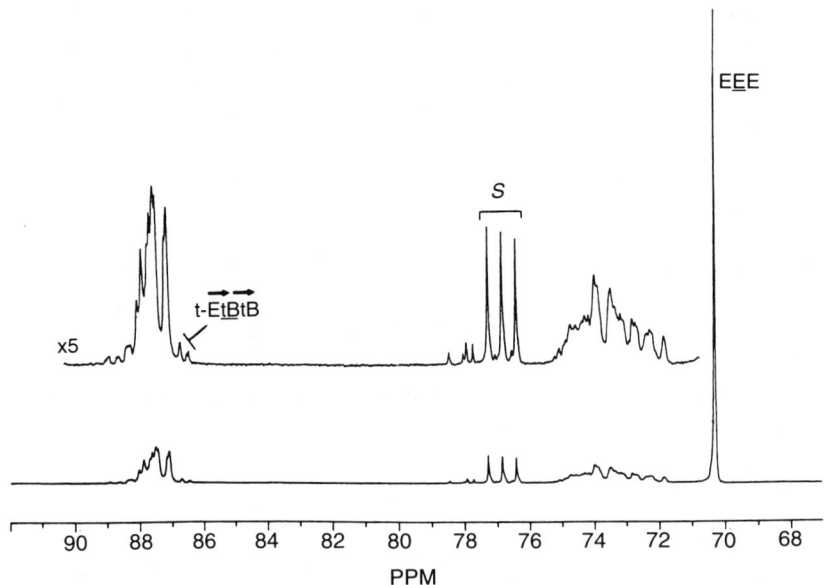

FIGURE 3.50 The 75-MHz carbon NMR spectrum of the t-butyl ethylene oxide and ethylene oxide block copolymer. The peaks labeled S arise from the solvent. Reprinted with permission (58).

that can be determined by NMR are the mole fractions of the monomers, the average sequence length, and the correspondence of the polymerization to a type of statistics. Figure 3.52 shows the 25-MHz carbon NMR spectrum of the 60:40 ethylene–vinyl acetate copolymer that has been extensively studied by NMR (35,61). The concentrations of the VVV, VVE, and EVE triads were obtained from the methine and carbonyl resonance intensities as a function of monomer feed ratio, and a comparison of these values with those calculated using Bernoullian statistics is compiled in Table 3.8. These results show a close correlation to the values calculated from Bernoullian statistics, as do the average sequence lengths as a function of feed ratio.

In some favorable cases the proton spectra of statistical polymer are well enough resolved that they are amenable to analysis by proton NMR. One example is the vinylidine chloride–isobutylene copolymer, the 200-MHz proton NMR spectrum of which is shown in Fig. 3.53 (62). Several methylene resonances are observed that can be assigned by comparison to the homopolymers or by using NOEs between neighboring monomer units. For example, the sequence VVVIV can be identified by methylene–methylene cross peaks in the 2D NOESY spectra between the VVVI and the VVIV tetrad sequences. The assignments can be made directly from the NOESY spectra, such as that shown in Fig. 3.54. By following an assignment strategy similar to that used for stereosequence assignments, the assignments for all of the copolymer peaks can be established, and the results are listed in Table 3.9.

$$\begin{array}{ccccc} \text{Cl} & \text{Cl} & \text{Cl} & \text{CH}_3 & \text{Cl} \\ | & | & | & | & | \\ -\text{C}-\text{CH}_2-\text{C}-\text{CH}_2-\text{C}-\text{CH}_2-\text{C}-\text{CH}_2-\text{C}- \\ | & | & | & | & | \\ \text{Cl} & \text{Cl} & \text{Cl} & \text{CH}_3 & \text{Cl} \\ \text{V} & \text{V} & \text{V} & \text{I} & \text{V} \end{array}$$

3.4 THE SOLUTION STRUCTURE OF POLYMERS

High-resolution NMR spectroscopy in solution can be used to study chain structure, conformation, and the structure of associations of polymers. Information about the conformation is primarily obtained for the proton–proton coupling constants or the internuclear distances that are typically measured by 2D NMR. Most polymer chains undergo a rapid interconversion between conformers, so the conformations measured by these methods yields a time-averaged picture of the chain conformation.

3.4.1 CHAIN CONFORMATION

Information about the average chain conformation of polymers in solution can be obtained by measurement of either the conformationally sensitive proton–proton coupling constants or the intra- or intermolecular distances. As noted in Section 1.6.2, the three-bond proton–proton coupling constants depend on the torsional angle between the protons. From the Karplus relationship (63,64) [Eq. (1.58)] it is estimated that the values for the *trans* and *gauche* orientations

3.4. The Solution Structure of Polymers

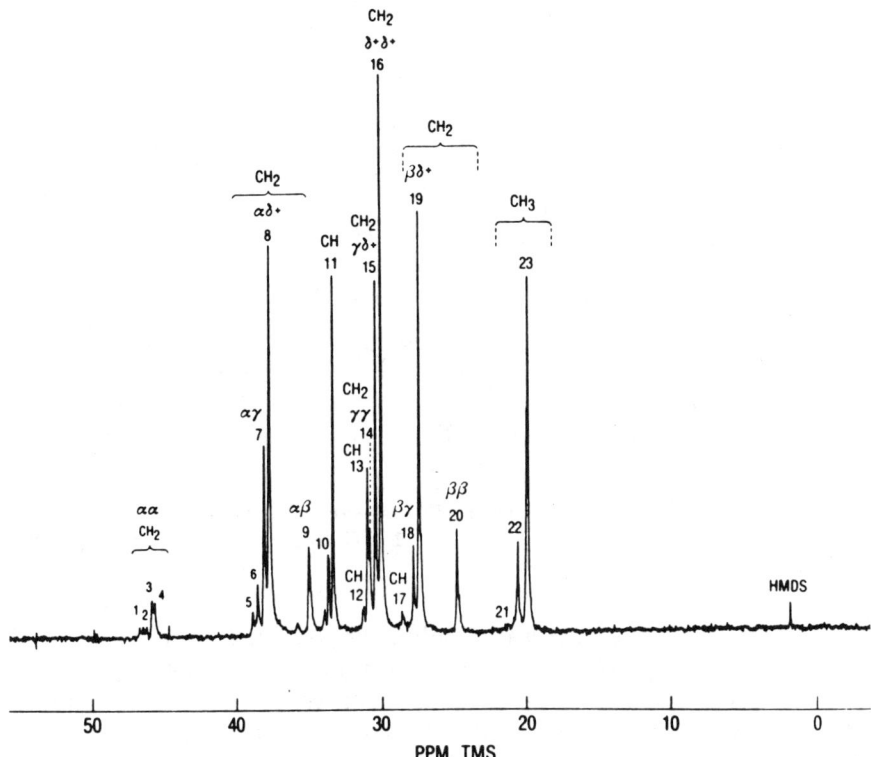

FIGURE 3.51 The 25-MHz carbon NMR spectrum of the 50:50 ethylene–propylene copolymer in 1,2,4-trichlorobenzene at 120°C. Reprinted with permission (35).

of the protons are 11 and 2 Hz, and for flexible polymers the observed value is a population weighted average of the conformational states.

In most instances the lines in the proton spectra of most polymers are too broad for the direct observation of the proton–proton coupling constants, and 2D NMR experiments are used. As shown in Section 1.8.7.2, the chemical shifts and coupling constants can be separated into two dimensions with 2D J-resolved NMR spectroscopy (65). Figure 3.55 shows the 500-MHz proton 2D J-resolved spectra of poly(propylene oxide) and illustrates how the coupling constants can be directly measured from the spectra (66). The results show that the methine–methylene coupling constants are 5 Hz for the atactic polymer. Assuming that the coupling constants for the *trans* and *gauche* conformations are 11 and 2 Hz, it can be calculated that populations of *gauche*, *gauche*$^+$, and *trans* conformers are approximately equal.

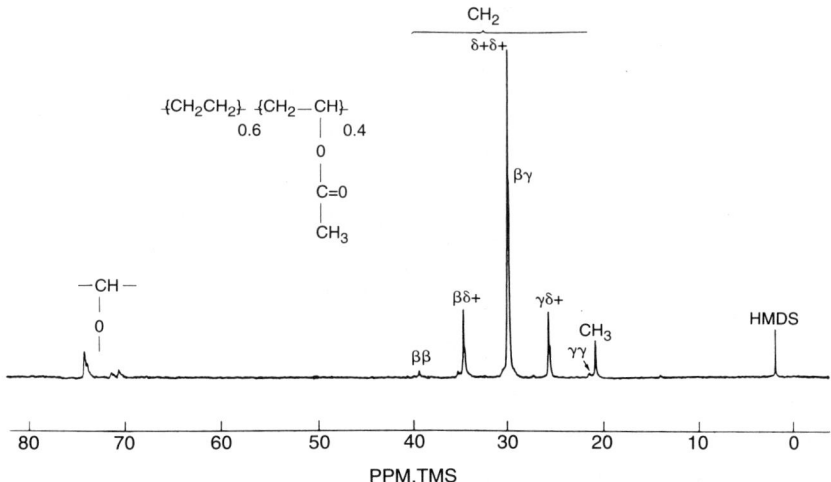

FIGURE 3.52 The 25-MHz carbon NMR spectrum of the 40:60 ethylene–vinyl acetate copolymer in 1,2,4-trichlorobenzene at 120°C. Reprinted with permission (35).

In many cases the spectra of polymers are well enough resolved that the coupling constants can be easily measured by the 2D NMR methods. In other cases the overlap is so severe that the coupling constants cannot be directly measured using the simple J-resolved spectra, and more complex experiments must be employed. Figure 3.56 shows a 3D NMR experiment using a combination of NOESY spectroscopy and J-resolved spectroscopy to measure the proton–proton coupling constants (67). A 3D NMR experiment is similar to a 2D experiment in that it consists of a series of pulses and delays during

TABLE 3.8
Comonomer Triad Distributions for Ethylene–Vinyl Acetate Copolymers (35)

| | | Comonomer triads | | |
Mole fraction vinyl acetate	dos vs calc.	EVE	VVE	VVV
0.73	Observed	0.06–0.07	0.30–0.32	0.36–0.37
	Calculated	0.06	0.32	0.36
0.48	Observed	0.09–0.12	0.24–0.26	0.11
	Calculated	0.13	0.24	0.14
0.34	Observed	0.12–0.15	0.15–0.17	0.04
	Calculated	0.15	0.17	0.05
0.10	Observed	0.07–0.09	0.01–0.03	0
	Calculated	0.08	0.02	0

3.4. The Solution Structure of Polymers

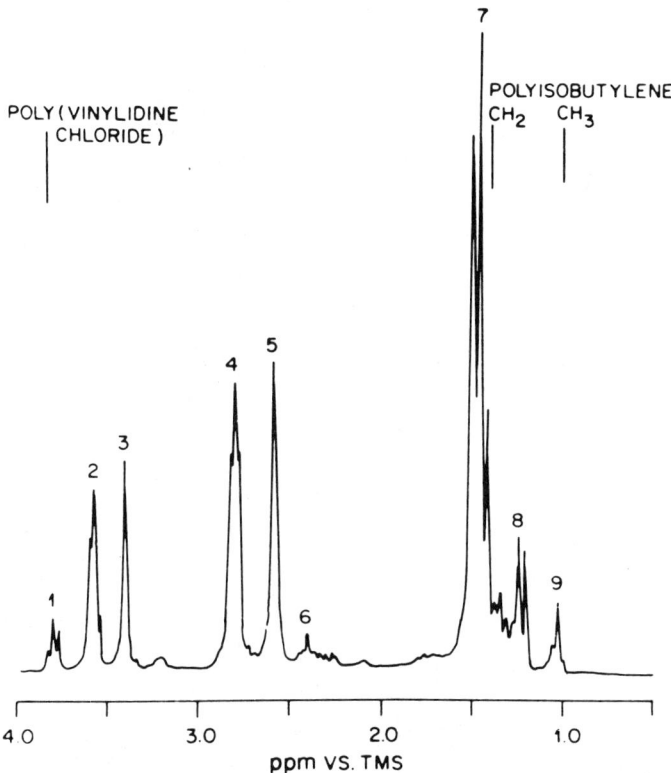

FIGURE 3.53 The 200-MHz proton NMR spectrum of the 65:35 vinylidine chloride–isobutylene copolymer at 40°C in CDCl$_3$. Reprinted with permission (62).

which the magnetization evolves under the influence of the chemical shifts and coupling constants or is transferred via nuclear Overhauser effects. In practice a 3D NMR experiment is acquired and processed as a series of 2D experiments. After the normal data processing in the t_2 and t_3 dimensions, the data consist of a series of 2D spectra in which the peaks are modulated as a function of the t_1 delay. Fourier transformation of the modulations of the peak intensities yields the full three-dimensional NMR spectrum, which consists of the intensities as a function of the three NMR frequencies. It is typically easier to view the data as 2D cross sections through the 3D data.

The 3D experiment shown in Fig. 3.56 combines the NOESY spectra with the J-resolved spectra. After Fourier transformation in the t_2 and t_3 dimensions a spectrum is obtained in which the NOESY cross peaks are modulated by the J coupling constants. The great

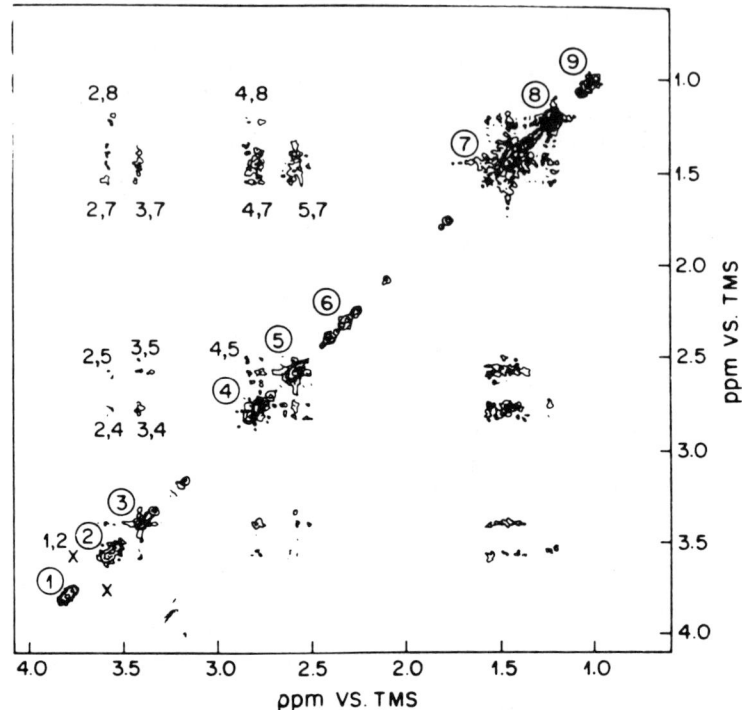

FIGURE 3.54 The 2D NOESY spectrum of the 65:35 vinylidene chloride–isobutylene copolymer at 40°C in $CDCl_3$ obtained with a mixing time of 0.5 s. Reprinted with permission (62).

TABLE 3.9
The Proton Chemical Shift Assignments for the Vinylidine Chloride–Isobutylene Copolymer (62)

Peak	Carbon type	Assignment	δ_H (ppm)
1	CH_2	VVVV	3.80
2	CH_2	VVVI	3.58
3	CH_2	IVVI	3.41
4	CH_2	VVIV	2.80
5	CH_2	VVII + IVIV	2.59
6	CH_2	IVII	2.33
7	CH_3	VIV	1.46
8	CH_3	VII	1.24
9	CH_3	III	1.03

3.4. The Solution Structure of Polymers

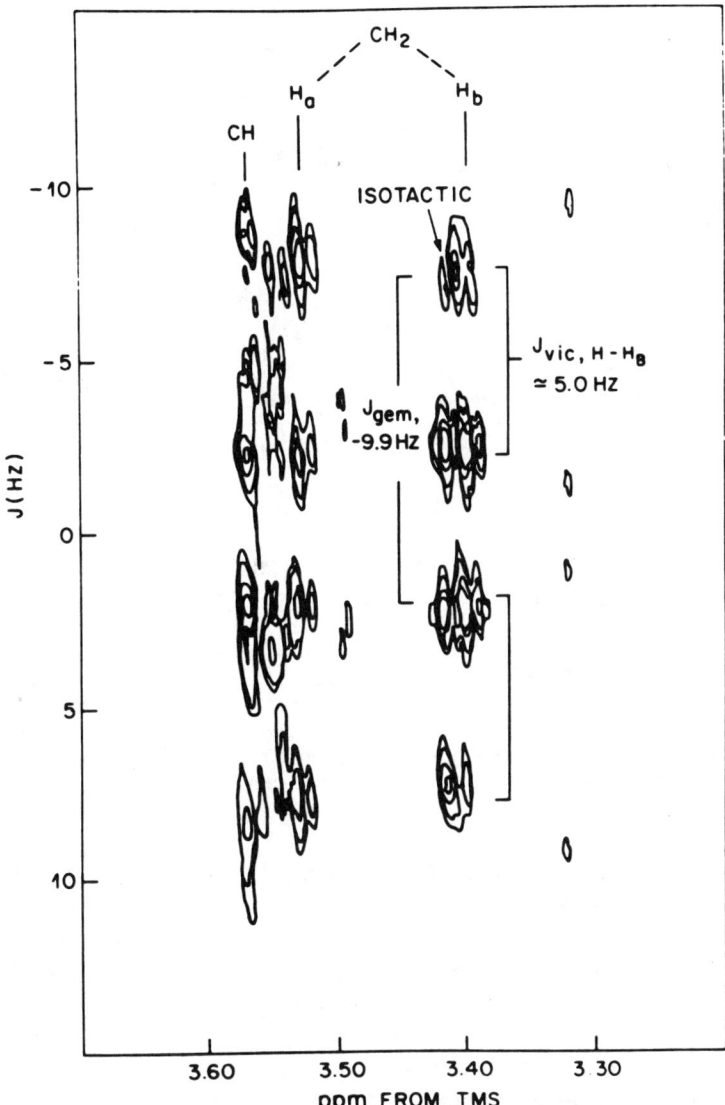

FIGURE 3.55 The 500-MHz 2D J-resolved NMR spectrum of the downfield region of atactic poly(propylene oxide) showing the couplings for the methine and methylene protons. Reprinted with permission (66).

increase in resolution in this experiment comes from the fact that the coupling constants can be obtained from the well-resolved cross peaks rather than from the diagonal peaks.

Three-dimensional NOESY/J-resolved NMR spectroscopy has been used to investigate chain conformation in the poly(vinylidine cyanide-

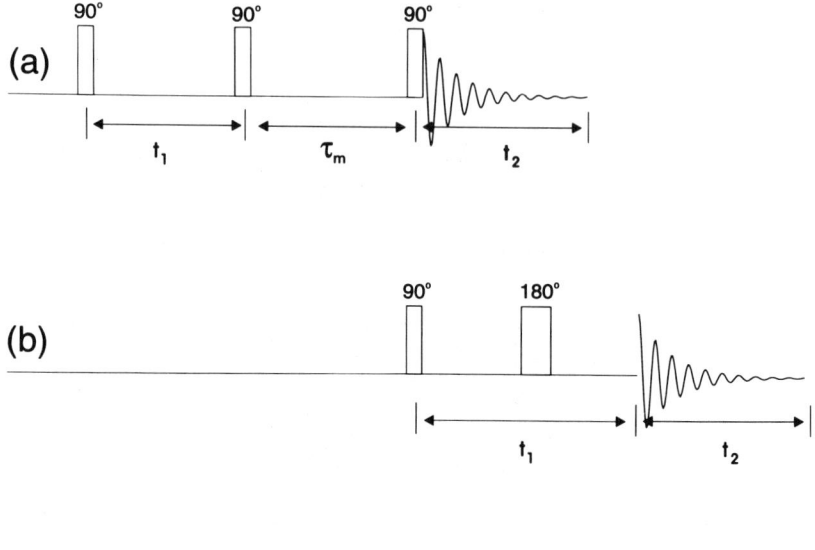

FIGURE 3.56 The pulse sequence diagram for (a) 2D NOESY, (b) 2D J-resolved, and (c) 3D NOESY/J-resolved NMR spectroscopy.

alt-vinyl acetate) copolymer (67). Figure 3.57 shows the 2D NOESY spectrum obtained from the first t_1 point prior to Fourier transformation along the t_1 dimension. There are four groups of resonances that can be assigned to the methine (5.7 ppm), two groups of methylene protons (2.4 and 2.6 ppm), and the acetate peak (1.8 ppm). In principle the J coupling constants could be obtained from the 2D J-resolved spectra of the methine protons, but in practice the coupling pattern is too complex to easily extract the coupling constants. The coupling constants can be obtained for the methylene protons from the cross peaks marked A and B in Fig. 3.57. Figure 3.58 shows a cross section through the 3D spectra from which the methylene couplings can be measured. The methylene protons are nonequivalent and show both geminal and vicinyl couplings. Both peaks are split by a 14-Hz geminal coupling, and cross peak A is split by a 10-Hz vicinyl coupling, while the vicinyl coupling of resonance B is too small to measure (> 3 Hz) in such an experiment. These results are different than those observed for poly(propylene oxide) shown above, and

3.4. The Solution Structure of Polymers

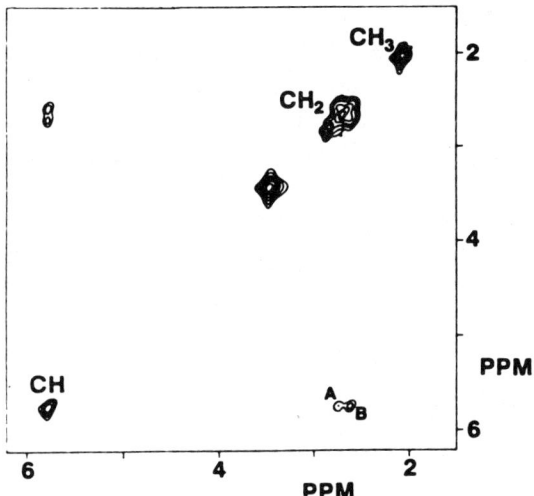

FIGURE 3.57 The 500-MHz 2D NOESY spectrum of the alternating vinylidine cyanide–vinyl acetate copolymer obtained with a 0.5-s mixing time. Reprinted with permission (67).

demonstrate that the orientation of the protons about the methine–methylene bond is approximately 90% *trans* (67).

$$\left[-CH_2-CH-CH_2-C-\right] \atop {\underset{\underset{CH_3}{O=C}}{O}} \quad {\underset{CN}{\overset{CN}{|}}}$$

In certain favorable circumstances the chain conformation of polymers can be inferred from the observation of cross peaks in the 2D COSY spectrum. As noted in Section 1.8.7.3, COSY cross peaks are only observed when the coupling constants are larger than the linewidths, and in most polymers it is difficult to observe correlations in the COSY spectra for coupling constants smaller than 1 Hz (68). As with the three-bond couplings, the four-bond couplings are also extremely sensitive to conformation, and only in certain (all *trans*) conformations will such coupling constants be large enough to give rise of 2D COSY cross peaks.

Figure 3.59 shows the 250-MHz 2D COSY spectrum of a poly(methyl methacrylate) oligomer in which such four-bond J coupling cross peaks are observed between methylene protons and the methyl protons (15). As noted above, the four-bond couplings are also sensitive to conformation, and the coupling constants are only in the range that can be measured by 2D COSY for the "W" or *trans* orientation of the protons. Thus, the observation of such cross methylene–methyl cross peaks can be observed only for poly(methyl methacrylate) with a *trans* orientation of these groups. These cross peaks are also observed in high-molecular-weight poly(methyl methacrylate), as shown in Fig. 3.20 (34).

The chain conformation of polymers can also be determined from the proton–proton separations measured by 2D NOESY spectroscopy (16,68). In polymers with a well-defined conformation, like proteins or helical polymers, these distances can be used to uniquely determine the conformation. In flexible polymers, such measurements can be used to determine the average conformation that is specified by the fraction of *trans* and *gauche* conformers.

Figure 3.60 shows the 500-MHz 1D and 2D NOESY spectra of a poly(benzyl-L-glutamate) 20-mer in chloroform with 5% trifluoroacetic acid (69), conditions under which this polymer adopts an α-helical conformation (70). In the study of rigid helical polymers by high-resolution NMR it is necessary to study oligomers since the high-molecular-weight polymers may have long correlation times and very broad lines. The peak intensities from the 2D NOESY spectra can be used to measure distances since they depend on the inverse sixth power of the internuclear separation. In simple cases the distances can be measured from the build-up of cross-peak intensities as a function of the NOESY mixing time. Such data are shown in Fig. 3.61 for several of the cross peaks, and the distances measured from the build-up of cross-peak intensity are listed in Table 3.10 along with the distances expected for a typical α-helix (69). These data show that the distances can be accurately measured by 2D NOESY NMR and confirm the α-helical structure of this polymer in solution.

3.4. The Solution Structure of Polymers

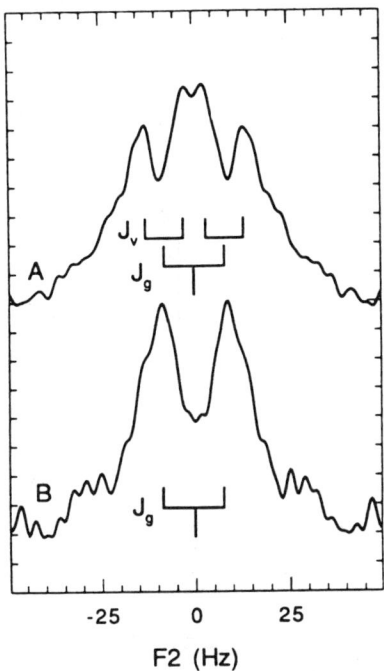

FIGURE 3.58 Cross sections through the 500-MHz 3D NOESY spectrum of the alternating vinylidine cyanide–vinyl acetate copolymer obtained with a 0.5-s mixing time for the cross peaks marked A and B in Fig. 3.57. Reprinted with permission (67).

The same NOESY approach can be used to study the chain conformation of flexible polymers. Figure 3.62 shows an expanded plot of the methylene region of the 2D NOESY spectrum of the alternating styrene–methyl methacrylate copolymer shown in Fig. 3.46 (71). The largest cross peaks arise from the geminal methylene protons that are separated by 1.78 Å, while the smaller cross peaks arise from the conformationally sensitive methylene–methylene interactions. Figure 3.63 shows the cross-peak intensities as a function of mixing time along with the best fit of the cross-relaxation rates (solid lines) that were obtained by solving the relaxation rate matrix at each mixing time [Eq. (1.108)]. The distances can be directly calculated from the relaxation rates and are compiled in Table 3.11. These data are best fit with a polymer conformation that has a fraction of *trans* conformers of 0.53 and in which there is a 20° deviation from the *trans* and *gauche* conformations to accommodate the bulky phenyl groups (71).

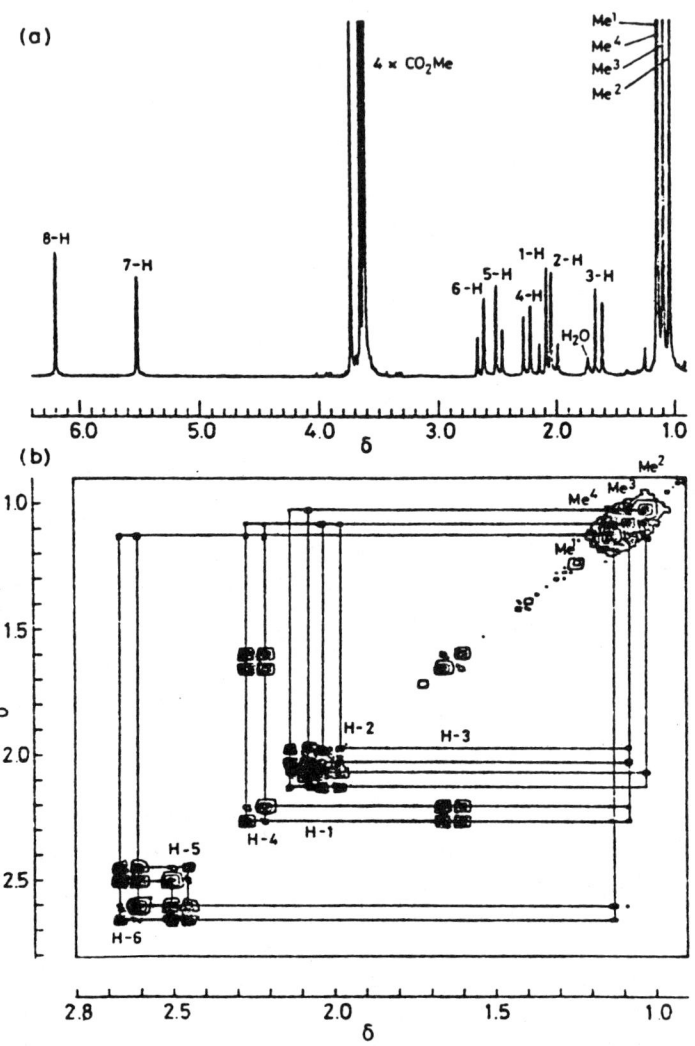

FIGURE 3.59 The 250-MHz (a) 1D and (b) 2D COSY spectra of the poly(methyl methacrylate) oligomer showing the methylene–methyl cross peaks arising from four-bond couplings. Reprinted with permission (15).

3.4. *The Solution Structure of Polymers* 231

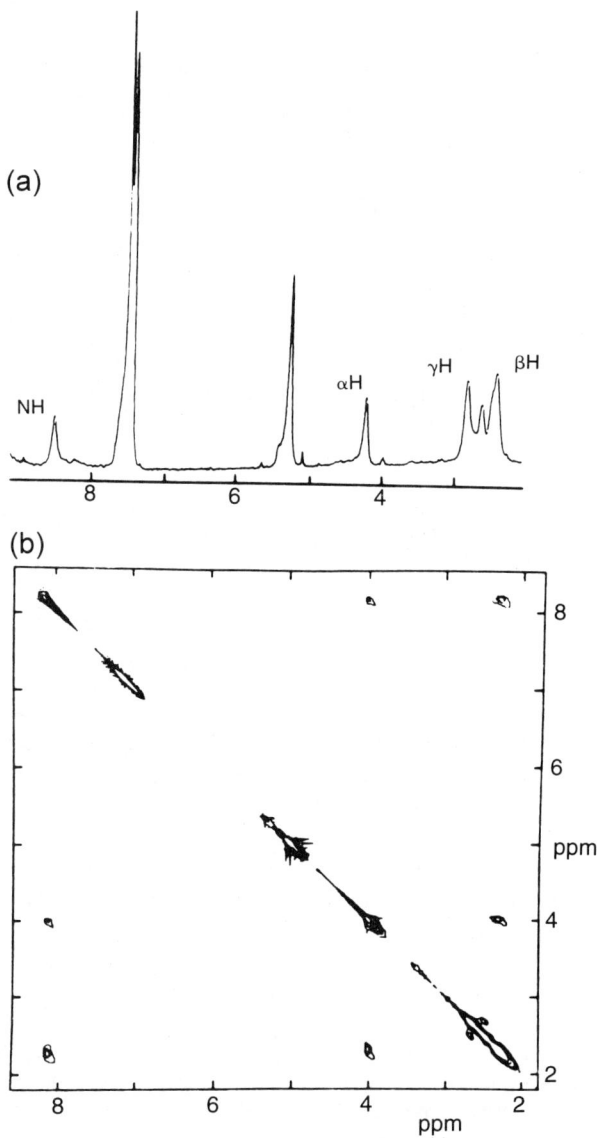

FIGURE 3.60 The 500-MHz 1D (a) and 2D (b) NOESY spectra of the poly(benzyl-L-glutamate) 20-mer in 95:5 chloroform:trifluoroacetic acid obtained with a 0.076-s mixing time. Reprinted with permission (69).

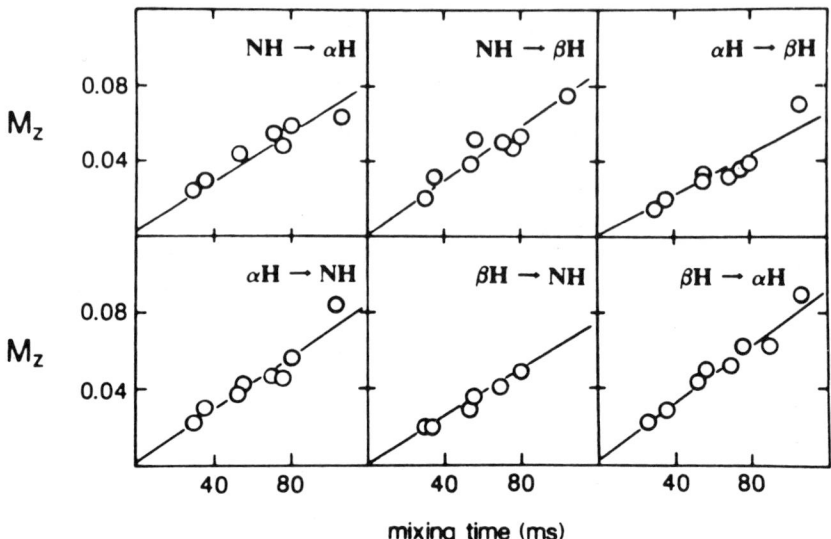

FIGURE 3.61 The rise in cross-peak volume as a function of mixing time for several of the cross peaks in the 2D NOESY spectrum of the poly(benzyl-L-glutamate) 20-mer in 95:5 chloroform:trifluoroacetic acid. Reprinted with permission (69).

3.4.2 Intermolecular Association of Polymers

The 2D NOESY methods have been used not only to investigate chain conformation in polymers, but also to identify the interacting groups and to measure the interaction strength in associating polymers that form miscible polymer blends. Most polymers are not miscible in the solid state because the entropy of mixing is not favorable, and molecular-level mixing of the polymer chains is usually observed only when there are favorable intermolecular interactions (72). However, it is typically not easy to identify the favorable interactions from the

TABLE 3.10
Comparison of the Expected and Observed Intermolecular Distances for the Poly(benzyl-L-glutamate) 20-mer in the α-Helical Conformation (69)

Interaction	Distance (Å)	
	Observed	Expected
NH–αH	2.20	2.2–2.8
NH–βH	2.20	2.0–4.1
αH–βH	2.1	2.0–3.5

3.4. The Solution Structure of Polymers

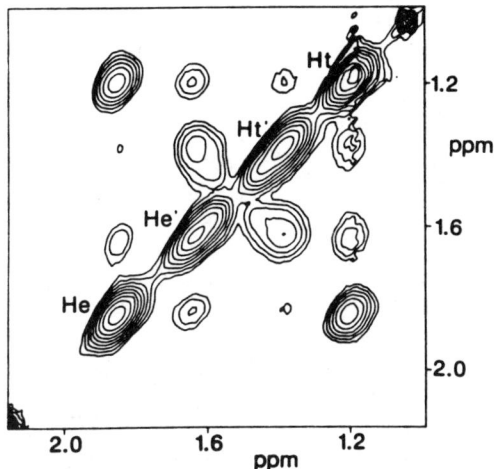

FIGURE 3.62 The 500-MHz 2D NOESY spectrum of the methylene region of the alternating styrene–methyl methacrylate copolymer obtained at 65°C with a 0.2-s mixing time. Reprinted with permission (71).

chemical structure of the polymers. It has been observed that 2D NOESY can be used to study the interacting groups by observing which groups show intermolecular NOEs (17,73–75), and the strength of the intermolecular interaction can be estimated from the polymer concentration required for the observation of the intermolecular NOEs (76).

Two-dimensional NOESY methods were first found to be useful for the study of intermolecular interactions in a study of poly(methyl methacrylate-co-4-vinyl pyridine) and poly(styrene-co-styrenesulfonic acid) in dilute dimethyl sulfoxide solution (74). It was proposed that the normally immiscible polystyrene and poly(methyl methacrylate) could be made miscible by introducing monomers that are able to form strong intermolecular associations. Evidence for the intermolecular association was obtained from the 2D NOESY spectrum shown in Fig. 3.64, that shows cross peaks between the aromatic signals of polystyrene at 7 ppm and the methoxy signals of poly(methyl methacrylate) at 3.2 ppm. Since magnetization transfer between these two groups can arise only from through-space dipolar interactions that result from two groups being in close proximity (<5 Å), the observation of these cross peaks can be considered proof of intermolecular association.

Since these initial observations, 2D NOESY methods have been extensively used to investigate intermolecular interactions in polymers capable of forming miscible polymer blends. In many blends, it

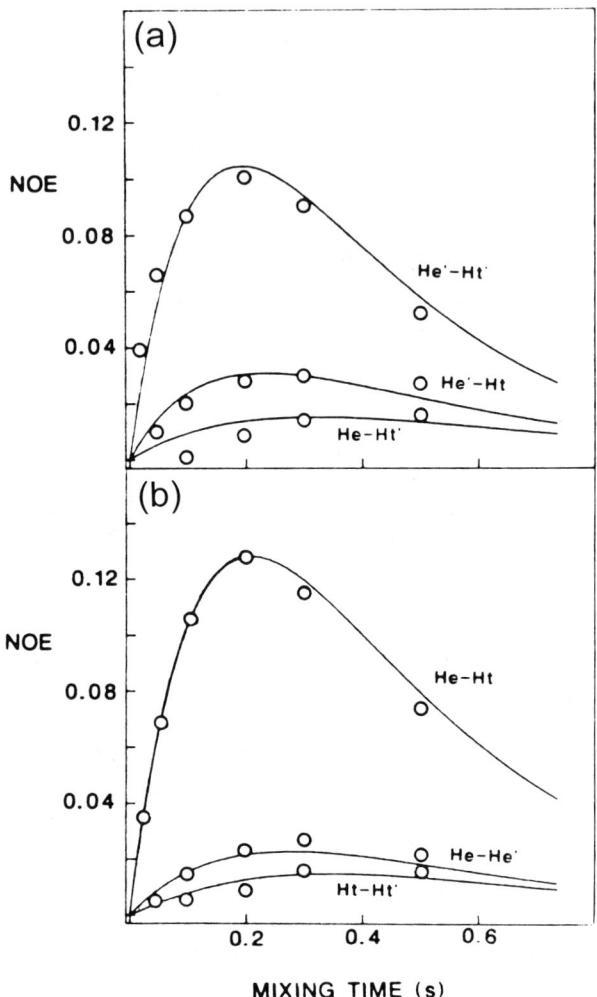

FIGURE 3.63 A plot of the cross-peak volume as a function of mixing time for the alternating styrene–methyl methacrylate copolymer at 65°C for the (a) $H_{e'} - H_{t'}$, $H_{e'} - H_t$, and $H_e H_{t'}$ and (b) $H_e - H_t$, $H_e - H_{e'}$ and $H_t - H_{t'}$ cross peaks. The solid lines are calculated from the relaxation rates at several mixing times. Reprinted with permission (71).

has been observed that high polymer concentrations are required for the observation of the intermolecular NOEs because the interactions that drive miscibility are often extremely weak. This is illustrated in the 2D NOESY spectra of mixtures of polystyrene and poly(methyl vinyl ether) shown in Fig. 3.65, where cross peaks between the aromatic protons of polystyrene (7 ppm) and the methoxy and me-

3.4. The Solution Structure of Polymers

TABLE 3.11
The Methylene–Methylene Distances Measured from the 2D NOESY Spectra of the Alternating Styrene–Methyl Methacrylate Copolymer (71)

Interaction	Distance (Å)
H_e–$H_{e'}$	2.41–2.65
H_e–$H_{t'}$	2.59–2.97
$H_{e'}$–H_t	2.32–2.40
$H_{t'}$–H_t	2.82–3.04

thine protons of poly(methyl vinyl ether) (3.2 and 3.4 ppm) are only observed at concentrations above 25 wt% (17,25). An analysis of the lineshapes of the cross peaks shows that miscibility is driven by interactions of the side groups, not the main-chain atoms (73).

The same NOESY methods have been used to investigate the structure of the complex formed by poly(vinyl chloride) and

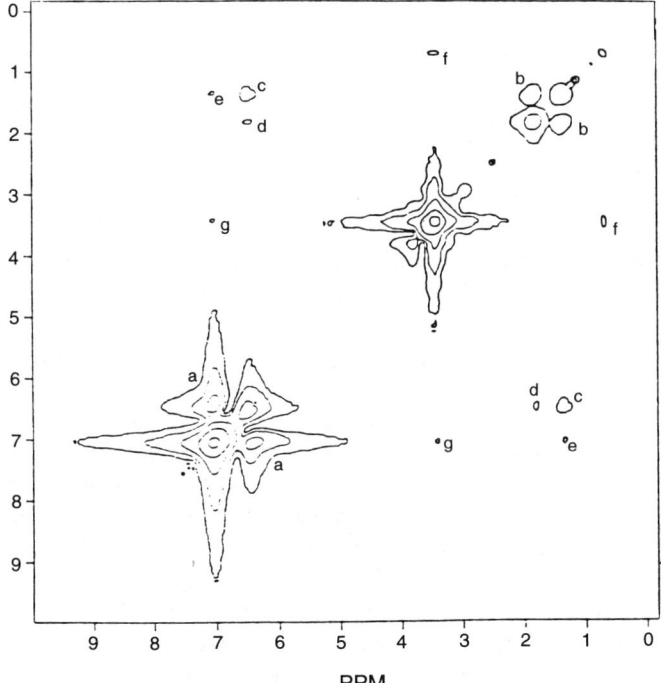

FIGURE 3.64 The 300-MHz 2D NOESY spectrum of the mixture of poly(methyl methacrylate-co-4-vinyl pyridine) and poly(styrene-co-styrenesulfonic acid) in deuterated dimethyl sulfoxide at 85°C. Reprinted with permission (74).

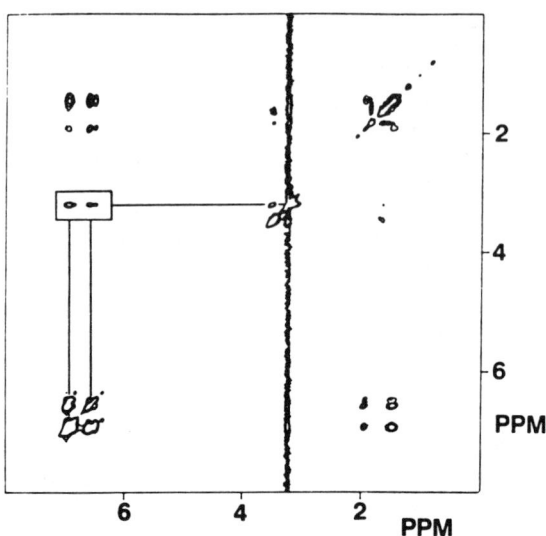

FIGURE 3.65 The 500-MHz 2D NOESY spectrum of the 40 wt% mixture of polystyrene and poly(methyl vinyl ether) in deuterated toluene at 76°C obtained with a mixing time of 0.75 s. Reprinted with permission (17).

poly(methyl methacrylate), as shown by the 2D spectrum shown in Fig. 3.66 (77). Again, the intermolecular interactions must be extremely weak since intermolecular cross peaks are observed only at concentrations above 25 wt%. In this case there are several intermolecular cross peaks so it is possible to more precisely define the structure of the intermolecular complex. The intra- and interchain distances were measured by solving the relaxation rate matrix as a function of mixing time, and Table 3.12 lists the results. These data show that the strongest intermolecular contact is between the poly(methyl methacrylate) methoxy protons and the poly(vinyl chloride) methylene protons. Further insight into the forces driving complex formation can be obtained from NOESY studies as a function of solvent. It has been proposed that these polymers are miscible due to an interaction between the poly(methyl methacrylate) carbonyl and the electron-deficient methine protons of poly(vinyl chloride). Such a weak sort of hydrogen bond should be sensitive to competition by better hydrogen bonding species, such as water, and Fig. 3.67 compares the 2D NOESY spectra for the poly(methyl methacrylate)/poly(vinyl chloride) mixture if deuterated tetrahydrofuran in the absence and presence of water (77). The presence of a small amount of water competes with the intermolecular interactions, and no intermolecular cross peaks are observed.

3.4. The Solution Structure of Polymers

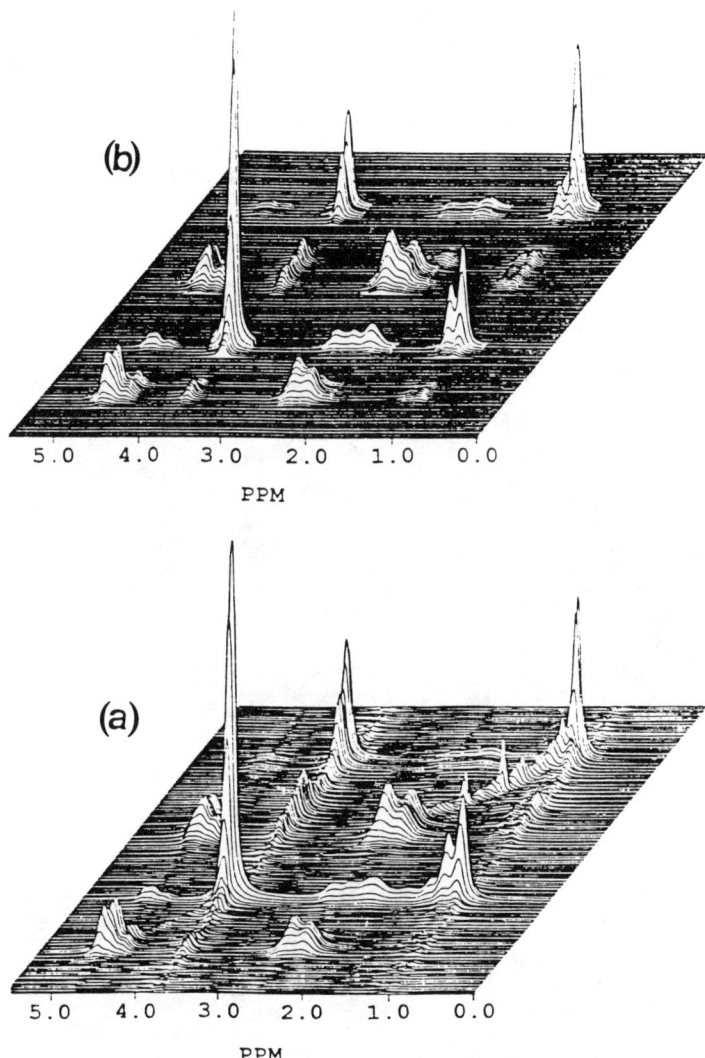

FIGURE 3.66 The (a) experimental and (b) simulated 500-MHz 2D NOESY spectra for the 38 wt% mixture of poly(methyl methacrylate) and poly(vinyl chloride) at 30°C for a 0.5-s mixing time. Reprinted with permission (77).

NOESY studies of polymers able to form intermolecular hydrogen bonds show that intermolecular cross peaks are observed at much lower concentrations. In the complex of poly(acrylic acid) and poly(ethylene oxide) at low pH, for example, the cross peaks are observed below 1 wt% concentration (76). However, the concentration

TABLE 3.12
Interchain Separations (Å) Measured from the 2D NOESY
Spectra for the Solution Mixture of Poly(methyl methacrylate)
and Poly(vinyl chloride) (77)

Poly(vinyl chloride)	Poly(methyl methacrylate)		
	CH_2	CH_3	OCH_3
CH	4.0–4.3	4.5–5.0	3.6–4.0
CH_2	3.5–3.7	3.9–4.0	3.1

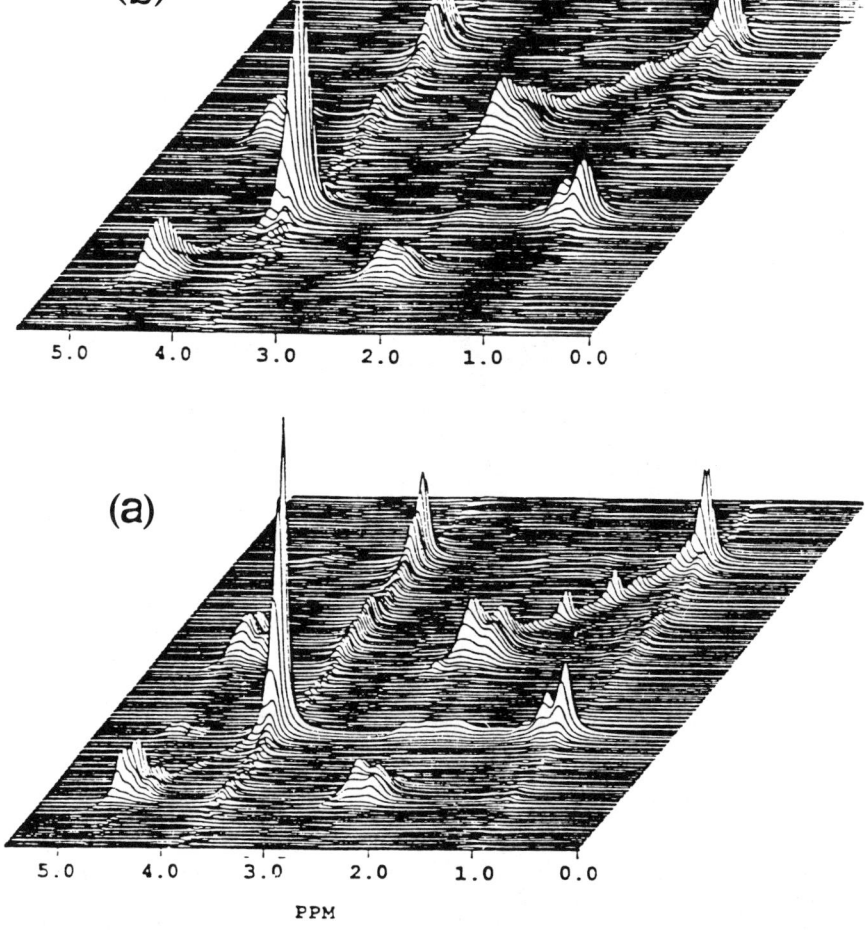

FIGURE 3.67 The 500-MHz 2D NOESY spectra of the 35 wt% mixture of poly(vinyl chloride) and poly(methyl methacrylate) in the (a) absence and (b) presence of a small amount of water. Reprinted with permission (77).

at which the hydrogen bonding cross peaks are observed can depend on a number of factors, including the structure of the hydrogen bonding groups, self-association, and the competition between intra- and intermolecular interactions. This is illustrated in a study of hydrogen bonding in a series of polymers containing p-hydroxystyrene and p-acetoxystyrene in homopolymers, copolymers, and terpolymers (78). Intermolecular hydrogen bonding cross peaks are observed for mixtures of poly(p-hydroxystyrene) and poly(p-acetoxystyrene) at concentrations above 2 wt%. Such hydrogen bonds can arise only from intermolecular interactions, and this can be compared with the poly(p-hydroxystyrene-co-acetoxystyrene) copolymer where the hydrogen bonding cross peaks do not appear to depend on concentration and are observed at concentrations below 0.2 wt%. The hydrogen bonding cross peaks observed at such low concentrations appear to rise from intramolecular interactions. The introduction of sulfone into the poly(p-hydroxystyrene-co-acetoxystyrene) copolymer eliminates the intramolecular hydrogen bond formation and greatly inhibits intermolecular association. Such NOESY studies provide an important insight into hydrogen bond formation in these systems that do not behave as expected from the chemical structure of the polymers.

REFERENCES

1. A. Zambelli, P. Locatelli, and G. Bajo, *Macromolecules* **12**, 154 (1979).
2. A. Zambelli, P. Locatelli, G. Bajo, and F. A. Bovey, *Macromolecules* **8**, 687 (1975).
3. A. Tonelli and F. Schilling, *Acc. Chem. Res.* **14**, 233 (1981).
4. F. C. Stehling and J. R. Knox, *Macromolecules* **8**, 595 (1975).
5. A. Dworak, W. J. Freeman, and H. J. Harwood, *Polym. J.* **17**, 351 (1985).
6. D. M. Grant and E. G. Paul, *J. Am. Chem. Soc.* **86**, 2984 (1964).
7. D. M. Grant and B. V. Cheney, *J. Am. Chem. Soc.* **39**, 5315 (1967).
8. H. N. Cheng and M. A. Bennett, *Makromol. Chem.* **188**, 135 (1987).
9. H. N. Cheng and M. A. Bennett, *Makromol. Chem.* **188**, 2665 (1987).
10. E. Breitmaier and W. Voelter, "Carbon-13 NMR Spectroscopy," 3rd Ed. VCH, Weinheim, 1987.
11. A. Tonelli, "NMR Spectroscopy and Polymer Microstructure: The Conformational Connection." VCH, New York, 1989.
12. A. E. Tonelli, F. C. Schilling, W. H. Starnes, L. Shepherd, and I. M. Plitz, *Macromolecules* **12**, 78 (1979).
13. W. Aue, E. Bartnoldi, and R. Ernst, *J. Chem. Phys.* **64**, 2229 (1976).
14. M. Rance, O. W. Sorensen, G. Bodenhausen, G. Wagner, R. R. Ernst, and K. Wuthrich, *Biochem. Biophys. Res. Commun.* **117**, 479 (1983).
15. P. Cacioli, D. Hawthorne, S. Johns, D. Solomon, E. Rizzardo, and R. Willing, *J. Chem. Soc., Chem. Commun.* 1355 (1985).
16. J. Jeneer, B. Meier, P. Bachmann, and R. Ernst, *J. Chem. Phys.* **71**, 4546 (1979).
17. P. Mirau and F. Bovey, *Macromolecules* **23**, 4548 (1990).

18. T.-A. Chen and R. D. Rieke, *J. Am. Chem. Soc.* **114**, 10087 (1992).
19. G. B. Kharas, P. A. Mirau, K. Watson, and H. J. Harwood, *Polym. Int.* **28**, 67 (1992).
20. T. Asakura, N. Nakayama, M. Demura, and A. Asano, *Macromolecules* **25** 4876 (1992).
21. A. Bax, R. Freeman, and T. Frenkiel, *J. Am. Chem. Soc.* **103** 2102 (1981).
22. T. Asakura and N. Nakayama, *Polym. Commun.* **33**, 650 (1992).
23. F. C. Schilling and A. E. Tonelli, *Macromolecules* **19**, 1337 (1986).
24. R. E. Cais and J. M. Kometani, *Org. Coat. Appl. Polym. Sci. Proc. Am. Chem. Soc.* **48**, 216 (1983).
25. R. E. Cais and J. M. Kometani, *Macromolecules* **18** 1354 (1985).
26. A. E. Tonelli, F. C. Schilling, and R. E. Cais, *Macromolecules* **14**, 560 (1981).
27. A. E. Tonelli, F. C. Schilling, and R. E. Cais, *Macromolecules* **15**, 849 (1982).
28. M. Bruch, F. Bovey, and R. Cais, *Macromolecules* **17**, 2547 (1984).
29. U. W. Suter and P. J. Flory, *Macromolecules* **8**, 765 (1975).
30. F. C. Schilling and A. E. Tonelli, *Macromolecules* **13**, 270 (1980).
31. T. Inoue, Y. Itabashi, R. Chuju, and Y. Doi, *Polymer* **25**, 1640 (1984).
32. G. Di Silvestro, P. Sozzani, B. Savare, and M. Farina, *Macromolecules* **18**, 928 (1985).
33. A. De Marco, P. Sozzani, G. Di Silvestro, and M. Farina, *Macromolecules* **22**, 2154 (1989).
34. F. Schilling, F. Bovey, M. Bruch, and S. Kozlowski, *Macromolecules* **18**, 1418 (1985).
35. J. C. Randall, "Polymer Sequence Distribution Carbon-13 NMR Method." Academic Press, New York, 1977.
36. B. D. Coleman and T. G. Fox, *J. Polym. Sci., Part A* **1**, 3183 (1963).
37. G. Moad, E. Rizzardo, D. H. Solomon, S. R. Johns, and R. I. Willing, *Macromolecules* **19**, 2494 (1986).
38. G. E. Martin and A. S. Zektzer, "Two-Dimensional NMR Methods for Establishing Molecular Connectivity." VCH, New York, 1988.
39. J. Kotyk, P. Berger, and E. Remsen, *Macromolecules* **23**, 5167 (1990).
40. C. J. Carmen, *Macromolecules* **6**, 725 (1973).
41. A. D. Williams and P. J. Flory, *J. Am. Chem. Soc.* **91**, 3118 (1969).
42. P. J. Flory and C. J. Pickles, *J. Chem. Soc. Faraday. Trans. 2* **69**, 632 (1973).
43. A. Bax, D. Davis, and S. K. Sarkar, *J. Magn. Reson.* **63**, 230 (1985).
44. M. Crowther, N. Szeverenyi, and G. Levy, *Macromolecules* **19** 1333 (1986).
45. W. M. Dutch and D. M. Grant, *Macromolecules* **3**, 165 (1970).
46. F. Bovey, "Chain Structure and Conformation of Macromolecules." Academic Press, New York, 1982.
47. M. H. George, R. J. Grisenthwaite, and R. F. Hunter, *Chem. Ind. (London)*, 1114 (1958).
48. W. H. Starns, F. C. Schilling, K. B. Abbas, I. M. Plitz, R. L. Hartless, and F. A. Bovey, *Macromolecules* **12**, 13 (1979).
49. S. K. Wolk, G. Swift, Y. H. Paik, K. M. Yocom, R. L. Smith, and E. S. Simon, *Macromolecules* **27**, 7613 (1994).
50. S. K. Wolk and E. Eisenhart, *Macromolecules* **26**, 1086 (1993).
51. L. E. J. Bogan and S. K. Wolk, *Macromolecules* **25**, 161 (1992).
52. D. A. Tomalia, A. M. Naylor, and W. A. I. Goddard, *Angew. Chem., Int. Ed. Engl.* **29**, 138 (1990).
53. T. M. Miller, E. W. Kwock, and T. X. Neenan, *Macromolecules* **25**, 3143 (1992).
54. K. de Vries, H. Linssen, and G. V. D. Velden, *Macromolecules* **22**, 1607 (1989).
55. K. P. Chan, D. S. Argyropoulos, D. M. White, G. W. Yeager, and A. S. Hay, *Macromolecules* **27**, 6371 (1994).

56. S. Heffner, F. Bovey, L. Verge, P. Mirau, and A. Tonneli, *Macromolecules* **19**, 1628 (1986).
57. E. Williams, J. Wengrovius, V. Van Valkenburgh, and J. Smith, *Macromolecules* **24** 1445 (1991).
58. F. Heatley, G. E. Yu, J. Lawrance, and C. Booth, *Eur. Polym. J.* **11**, 1249 (1994).
59. H. N. Cheng and M. A. Bennett, *Anal. Chem.* **56**, 2320 (1984).
60. H. N. Cheng and G. H. Lee, *Polym. Bull.* **12**, 463 (1984).
61. T. K. Wu, D. W. Ovenall, and G. S. Reddy, *J. Polym. Sci. Polym. Phys. Ed.* **12**, 901 (1974).
62. M. D. Bruch and F. A. Bovey, *Macromolecules* **17**, 978 (1984).
63. M. Karplus, *J. Chem. Phys.* **30**, 11 (1959).
64. M. Karplus, *J. Am. Chem. Soc.* **85**, 2870 (1959).
65. G. Bodenhausen, R. Freeman, R. Niedermeyer, and J. Turner, *J. Magn. Reson.* **26**, 133 (1977).
66. M. Bruch, F. Bovey, R. Cais, and J. Noggle, *Macromolecules* **18**, 1253 (1985).
67. P. Mirau, S. Heffner, and F. Bovey, *Macromolecules* **23**, 4482 (1990).
68. R. Ernst, G. Bodenhausen, and A. Wokaun, "Principles of Nuclear Magnetic Resonance in One and Two Dimensions." Oxford Univ. Press (Clarendon), Oxford, 1987.
69. P. Mirau and F. Bovey, *J. Am. Chem. Soc.* **108**, 5130 (1986).
70. F. A. Bovey and F. C. Schilling, *Macromol. Rev.* **9**, 1 (1975).
71. P. Mirau, F. Bovey, A. Tonelli, and S. Heffner, *Macromolecules* **20**, 1701 (1987).
72. D. Walsh and S. Rostami, *Adv. Polym. Sci.* **70**, 119 (1985).
73. P. Mirau, H. Tanaka, and F. Bovey, *Macromolecules* **21**, 2929 (1988).
74. A. Natansohn and A. Eisenberg, *Macromolecules* **20**, 323 (1987).
75. M. Crowther and G. Levy, *Macromolecules* **21**, 2924 (1988).
76. P. Mirau, S. Heffner, G. Koegler, and F. Bovey, *Poly. Int.* **26**, 29 (1991).
77. G. Kogler and P. Mirau, *Macromolecules* **25**, 598 (1992).
78. S. A. Heffner, M. E. Galvin-Donoghue, E. Reichmanis, L. Gerena, and P. A. Mirau, in press (1994).

4

THE SOLID-STATE NMR OF POLYMERS

4.1 INTRODUCTION

Most uses for polymers are in the solid state, and understanding the properties of solid polymers has been an important factor driving the development of solid-state NMR methods. These methods have been developed over the past two decades and are now an important part of polymer science. The focus of many of these studies is a molecular-level understanding of polymers in their functional state. The resolution of solid-state NMR is less than that for solutions, but it has been shown that much useful information can be measured by this technique. As with solution NMR, there have been dramatic improvements in our ability to characterize materials by solid-state NMR with the introduction of 2D NMR and improvements in the NMR spectrometers. Solid-state NMR has become such an important tool for polymer characterization that it is now considered a routine characterization tool.

Solid-state NMR has been used for polymer characterization because it can provide information about polymers over a wide range of length scales. On the most detailed level is the information about chain conformation that can be extracted from the chemical shifts. Unlike solutions, where there is a large amount of molecular motion, the chains are more rigid in crystalline polymers and amorphous polymers below their glass transition temperatures, so the chemical shifts (through the *γ-gauche* effect) reflect the actual chain conformation rather than the average chain structure. The organization of polymer chains on a longer length scale (20–200 Å) can be measured using proton *spin diffusion*. In such an experiment the proton signals in one section of the sample (such as the crystalline domains) are excited and the transfer of magnetization to other domains can be measured.

NMR has also proved useful for the study of multiphase solid polymers because polymers in different environments have different molecular dynamics and different NMR relaxation times. These differences can be exploited such that the NMR signals from a particular phase of the material can be observed. In this way it is possible to selectively observe the crystalline, amorphous, interfacial, and rubbery materials. As we will see later (Chapter 5), the measurement of such relaxation rates has also greatly contributed to our understanding of the molecular dynamics of polymers.

In Section 1.8.5 we noted that the acquisition of solid-state NMR spectra was more complex than that for solutions, primarily because in solution the chemical shift anisotropy and the dipolar interactions are averaged nearly to zero by rapid chain motion for the polymers in solution. In solids, the chain motion is restricted by the polymer matrix, and high-resolution spectra can be observed only by using rapid magic angle spinning (Section 1.8.5.3) to average the chemical shift anisotropy and high-power irradiation during signal acquisition to average the carbon–proton dipolar interactions. In some cases more sophisticated NMR methods are used, usually consisting of applying a series of pulses in which the phases are alternated to remove a particular type of interaction. For example, in experiments where spin diffusion would complicate the results, it is possible to apply a series of pulses to the protons that quench spin diffusion. In other experiments it is possible to obtain information about oriented polymers by synchronizing the pulses and signal acquisition to the sample rotation.

As noted earlier, the resolution in solid-state NMR is typically not as good as that for solutions. In favorable cases the linewidths can be as small as 50 Hz, but they are often larger. This arises in part because the averaging of the chemical shift anisotropy and the carbon–proton dipolar interactions by magic angle spinning and high-powered decoupling is not as efficient as the averaging by chain motion in solution. Also, the linewidths have contributions from variations in the magnetic susceptibility due to grain boundaries, etc. The lines are sharper for crystalline polymers than for amorphous materials because they exist in a more uniform environment. The polymer matrix restricts chain motion so the γ-*gauche* effects are not averaged as in solution, so the lines of amorphous polymers are inhomogeneously broadened by γ-*gauche* effects from a distribution of conformations.

The wide-line deuterium NMR spectrum is most often used to study the molecular dynamics of polymers because the lines are so broad due to quadrupolar interactions that the different sites on the polymer chain cannot be resolved from each other. However, deuterium NMR can be used to measure the fraction of rigid or crystalline material since the T_1's are sensitive to the molecular dynamics. As noted in Section 1.8.5.4, the cross-polarization magic angle spinning (CPMAS) methods most often used to study solid polymers cannot be used for quantitative purposes since the cross-polarization intensity depends not only on the number of nuclei, but also in their cross-polarization dynamics. It is usually difficult to acquire quantitative spectra in the solid state because the spin–lattice relaxation times are often extremely long.

4.2 CHAIN CONFORMATION IN THE SOLID STATE

4.2.1 SEMICRYSTALLINE POLYMERS

A wide variety of NMR methods have been used to study semicrystalline polymers, including the acquisition of carbon, silicon, nitrogen, phosphorus, and proton spectra; NMR relaxation studies; and multipulse and other NMR methods. Chain conformation is most commonly studied by carbon NMR with cross polarization and magic angle spinning because carbon is abundant in most polymers and the spectra can be routinely acquired. Also, since the γ-*gauche* effect has been extensively studied in polymers, the chemical shifts can provide important information about the chain conformation.

Polyethylene is perhaps the polymer most extensively studied by solid-state NMR. Polyethylene exists in all all-*trans* conformation in the crystalline phase and the spectra have been recorded as a function of crystallinity and sample preparation (1). Figure 4.1 shows the solid-state spectrum of polyethylene acquired using magic angle spinning with and without cross polarization, and shows the behavior that is typically expected (2). The largest peak in the spectrum acquired with cross polarization is a sharp resonance at 33.6 ppm that corresponds to the crystalline polyethylene in the all-*trans* conformation, and the broader resonance located at 2.5 ppm to higher field is assigned to the amorphous material. Note that in the spectrum acquired without cross polarization that the largest peak appears to be the amorphous material, since the relaxation times for the crystalline material are very long and the spectra do not reflect the

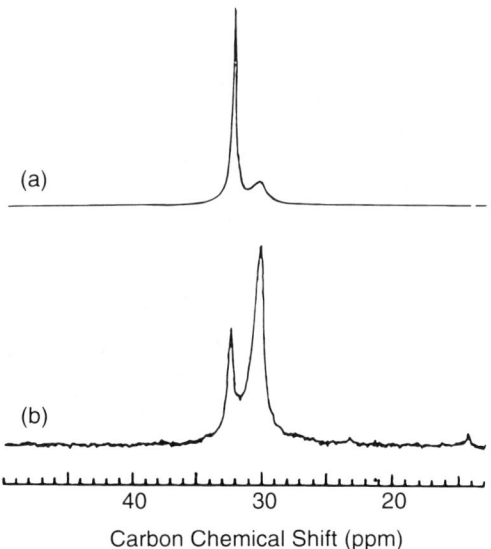

FIGURE 4.1 The carbon spectra of polyethylene acquired (a) with and (b) without cross polarization. Both spectra were acquired with magic angle spinning. Adapted with permission (2).

number of nuclei in the crystalline and amorphous phases. These spectra illustrate how the spectra of a particular phase in a complex material can be emphasized by the proper choice of NMR parameters and experiments, and that extreme care must be taken in the quantitative analysis of these materials. The sample for which the spectra are shown in Fig. 4.1 was shown to be 68% crystalline by X-ray diffraction.

The chemical shifts for the crystalline and amorphous fractions of polyethylene are in accord with those calculated from the *γ-gauche* effect (3). In polyethylene in the melt or in solution, about 40% of the bonds are expected to be in the *gauche* conformation, so an upfield shift of 5 ppm is expected. While this prediction is in the right direction, the magnitude of the shift is considerably less than expected. Among the possible explanations for this observation are that chains in the amorphous fractions have a different fraction of *gauche* bonds than expected on the basis of energy calculations, or that other factors, such as intermolecular interactions, also affect the chemical shifts in the solid state. This proposal is consistent with the observation that the chemical shifts for methylene carbons in *n*-alkanes can vary as much as 1.3 ppm even though they have an all-*trans* conformation (4).

4.2. Chain Conformation in the Solid State

The effect of chain conformation on NMR spectra is particular striking for isotactic and syndiotactic polypropylene. Both polymers are crystalline, but the isotactic polymer adopts a ...*gtgtgt*... 3_1 helical conformation, while the syndiotactic polymer forms a 2_1 helix with a ...*ggttggtt*... conformation. In the syndiotactic polymer half of the methylene groups lie along the interior of the helix and are in a *gauche* arrangement with their γ neighbor, and half the methylene groups lie on the exterior of the helix and are *trans* to their γ neighbor (3). In isotactic polypropylene, the methylene groups are *trans* to one γ neighbor and *gauche* to another. The effects of such conformations are evident in Fig. 4.2, which compares the CPMAS spectra for isotactic and syndiotactic polypropylene (5). A single resonance is observed for the methylene groups in the isotactic polymer, while two resonances separated by 8.7 ppm are observed for the syndiotactic material. This difference in chemical shift for the methylene carbons in the syndiotactic polymer is approximately as large as two γ-*gauche* effects (3). The methylene resonance for isotactic polypropylene appears midway between the two peaks in syndiotactic polypropylene, as expected for a methylene group that has one γ-*gauche* effect.

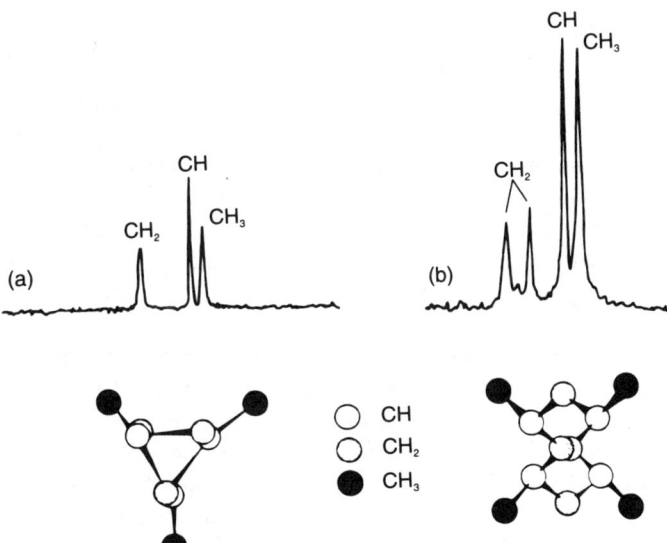

FIGURE 4.2 The solid-state NMR spectra of (a) isotactic and (b) syndiotactic polypropylene acquired with cross polarization and magic angle spinning. Adapted with permission (5).

In some polymers the crystal structure depends on the conditions for crystal growth. Figure 4.3 shows the 50-MHz carbon CPMAS spectra of poly(diethyl oxetane) for samples crystallized at 60, 35, and 0°C (6). Form I [Fig. 4.3(a)] is obtained at high temperature and is characterized by an all-*trans* chain conformation and a melting temperature of 73°C. Form II [Fig. 4.3(c)] is obtained from the low-temperature crystallization and has a ...*ttggttgg*... conformation and a melting temperature of 57°C. The spectra of samples crystallized at high or low temperature are characterized by a single component, while the samples crystallized at intermediate temperatures [Fig 4.3(b)] contain a mixture of the two forms.

$$\left[-CH_2-\underset{\underset{CH_2-CH_3}{|}}{\overset{\overset{CH_2-CH_3}{|}}{C}}-CH_2-\right]$$

Poly(diethyl oxetane)

The chemical shifts observed for forms I and II of poly(diethyl oxetane) are not all in good agreement with those calculated from the γ-*gauche* effect (5). While the chemical shifts expected for the quaternary and methyl carbons are close to the calculated values, the shifts for the methylene carbons are not. The main-chain methylene is predicted to be more upfield in form II, but resonantes 3.9 ppm downfield, and the side-chain methylene is shifted upfield by 5.9 ppm instead of downfield. This shows that factors other than the γ-*gauche* effect can effect the chemical shifts in the solid state. Among the possible explanations for this observation are that interchain packing or changes in the valence angles in the different conformations can affect the chemical shift. It is unlikely that the chemical shift differences can be attributed solely to chain packing effects since much smaller chain packing effects are observed in model hydrocarbon polymers.

Chain packing is known to affect the chemical shifts of some polymers in the crystalline state. This was first demonstrated by the observation of the chemical shifts for the interior methylene carbons in n-alkanes which exist in an all-*trans* conformation but show some chemical shift differences (4). Figure 4.4 shows the effect of crystal packing on the CPMAS carbon spectra of isotactic polypropylene (7). When polypropylene is annealed above 150°C, the stable α polymorph is obtained, while unidirectional crystallization under a strong temperature gradient yields the β form. In both forms the chains adopt a 3_1 helix with a ...*gtgtgt*... conformation, so the differences

4.2. Chain Conformation in the Solid State

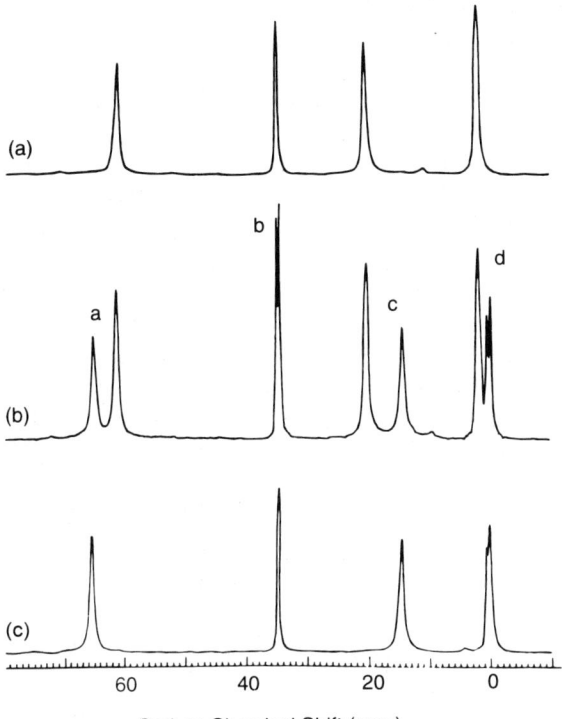

FIGURE 4.3 The 50-MHz solid-state carbon NMR spectra of poly(diethyl oxetane) crystallized at (a) 60°C, (b) 35°C, and (c) 0°C. All spectra were obtained with cross polarization and magic angle spinning. Adapted with permission (6).

in the NMR spectra must result from chain packing effects. Such packing effects are a consequence of the packing of helixes with different handedness in alternate rows in the α form, while helixes of the same handedness are clustered in the β form. The 2:1 peak intensity ratios for the methylene and methyl carbons in the α form correspond to the ratio of nonequivalent packing sites. The two sites have interchain packing distances of 5.28 and 6.14 Å in the α form and 6.36 Å in the β form.

Nuclei other than carbon have also been used to study the structure of crystalline polymers (8). The NMR characterization of silicon- and phosphorus-containing polymers is now quite routine because ^{31}P and ^{29}Si are spin-1/2 nuclei with a relatively high natural abundance and a large chemical shift range. Nitrogen also has a large chemical shift range, but the low natural abundance has limited the use of ^{15}N NMR for the study of polymers.

The ^{29}Si chemical shifts are extremely sensitive to chain conformation and have been used to study both silane and siloxane polymers. Silane polymers have recently been of interest as potential materials for microlithography and because of unusually long wavelength UV absorption due to "sigma conjugation" (9). Two crystalline forms of polysilanes have been reported, and the chain conformation is extremely sensitive to the side-chain groups. Figure 4.5 compares the 39-MHz ^{29}Si NMR spectra of polysilanes with hexyl, pentyl, and butyl side chains (10). In all cases two helical forms are observed, a well-ordered form I and a disordered form II. In form I for poly(di-n-hexyl silane) the chain adopts an all-*trans* conformation and shows a resonance around -22 ppm, while form I of poly(di-n-pentyl silane) and poly(di-n-butyl silane) both adopt a 7_3 helical conformation and show resonances around -25 ppm. The chemical shifts for the disordered form II helix appear in the range of $-23-24$ ppm.

$$\begin{array}{cccc} R & R & R & R \\ | & | & | & | \\ -Si- & Si- & Si- & Si- \\ | & | & | & | \\ R & R & R & R \end{array}$$

Poly(alkyl silane)

^{29}Si NMR has added considerably to our understanding of the crystal structure of poly(dimethyl siloxane). From the volume of the unit cell and the observation of an 8.3-Å fiber repeat, a model was proposed containing a two-fold helix with a regular repeat of the $tts^+s^-g^+g^+$ conformation (11). When the number of γ-*gauche* interactions is counted for the carbons and silicons in such a model, three distinct ^{29}Si and four ^{13}C resonances are expected. This proposal is inconsistent with the low-temperature silicon spectra shown in Fig. 4.6, in which two peaks are observed that can be assigned to the crystalline [Fig. 4.6(b)] and amorphous [Fig. 4.6(a)] fractions (12). The assignment of the amorphous signals at 2.16–2.26 ppm is based on the observation of these signals in the spectra without cross polarization. The fiber repeat and the observation of a single crystalline resonance in both the silicon and the carbon spectra are consistent with an extended helical structure with 35°–40° rotations away from the all-*trans* conformation of the backbone.

$$\left[-O-\underset{\underset{CH_3}{|}}{\overset{\overset{CH_3}{|}}{Si}}- \right]$$

Poly(dimethyl siloxane)

4.2. Chain Conformation in the Solid State

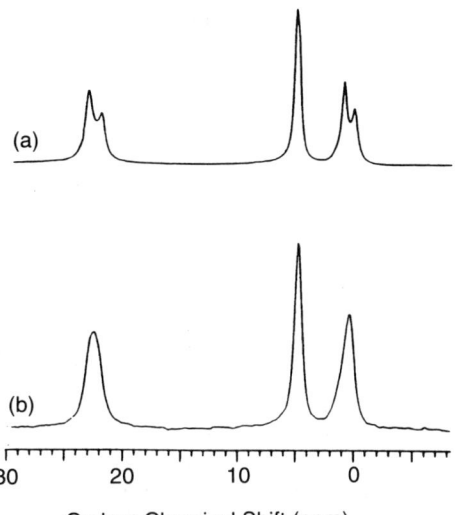

FIGURE 4.4 The solid-state carbon NMR spectra of the (a) α and (b) β forms of isotactic polypropylene acquired with cross polarization and magic angle spinning. Adapted with permission (7).

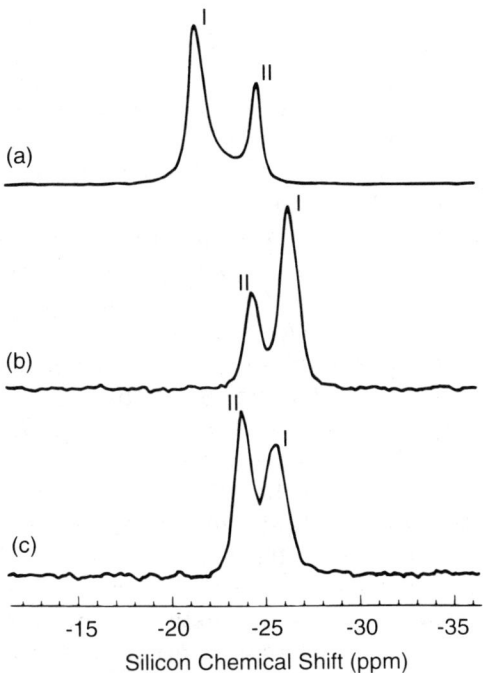

FIGURE 4.5 The 39.75-MHz ^{29}Si magic angle spinning spectra of (a) poly(di-n-hexylsilane), (b) poly(di-n-pentylsilane), and (c) poly(di-n-butylsilane) taken near the midpoint of the form I to form II transition. Adapted with permission (10).

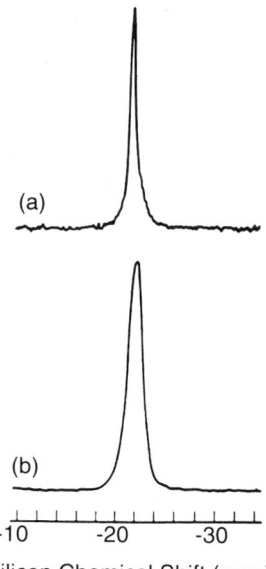

FIGURE 4.6 The solid-state ^{29}Si NMR spectrum of poly(dimethyl siloxane) acquired without (a) and with (b) cross polarization and magic angle spinning: Adapted with permission (12).

As noted above, ^{15}N NMR can be used for polymer characterization, but is limited by low natural abundance and signal-to-noise ratios. ^{15}N has the advantage that the nitrogen chemical shifts are extremely sensitive to the conformation, as illustrated by the 20-MHz natural abundance nitrogen NMR spectra for nylon-6 shown in Fig. 4.7 (13). It is well known that nylon-6 can be prepared in several crystal forms, and Fig. 4.7 shows that significant chemical shift differences are observed between the α and the γ forms, and that these forms can be identified by their chemical shifts in samples containing a mixture of the two forms. In this case, ^{15}N NMR is more sensitive to the crystal structure than is ^{13}C NMR, since it was not possible to distinguish between the two forms in the carbon spectra.

In the above examples we have seen how the NMR spectra are extremely sensitive to the chain conformation. Relatively large differences in the chemical shifts are observed if there is a large difference in conformation between the crystalline and the amorphous material. However, in some instances the resonances from the phases are not shifted relative to each other, and other NMR methods must be used to observe the spectra of the crystalline and the amorphous phases. It is often easy to distinguish between the phases

4.2. Chain Conformation in the Solid State

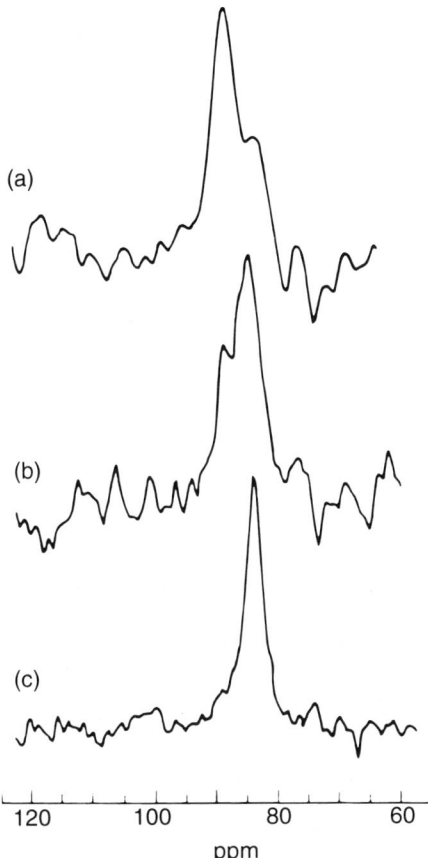

FIGURE 4.7 The solid-state ^{15}N spectra of nylon-6 (a) predominantly the γ form, (b) predominantly the α form, and (c) the α form. Adapted with permission (13).

because there are large differences in the molecular dynamics between the crystalline and the amorphous phases, so they have different relaxation times.

Several different methods have been developed to differentiate between the crystalline and the amorphous material based on the differences in the cross-polarization dynamics, the T_1's the $T_{1\rho}$, or the rates of dipolar dephasing (14). One of the simplest methods is to use the differences in the spin–lattice relaxation times to increase the relative intensity of the crystalline or amorphous phases. This can be accomplished either by choosing the delay time between acquisitions to be short, so that the crystalline component will be saturated, or by using the inversion-recovery pulse sequence and choosing the delay

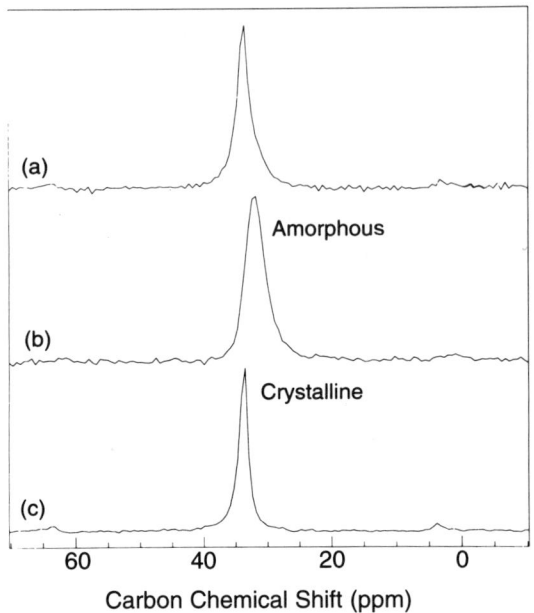

FIGURE 4.8 The spectra of polyethylene gathered under different conditions to emphasize the crystalline or amorphous fraction. The spectra show (a) the cross-polarization magic angle spinning spectra, (b) the single-pulse spectrum with a recycle delay time (0.25 s) that is much shorter than the T_1 for the crystalline fraction, and (c) a cross-polarization T_1 experiment in which the delay time is long (0.25 s) compared to the relaxation time of the amorphous fraction.

time between the excitation and the observe pulse such that one of the signals is nulled. The difference in T_1 can be used to discriminate between the crystalline and the amorphous fractions of polyethylene. Spectra showing predominantly the amorphous form can be obtained by using a short delay time, since the T_1's for the crystalline and amorphous phases are approximately 100 and 0.17 s (15). The proper choice of the delay time in the cross-polarization T_1 experiment can also be used to obtain a spectrum of predominantly one phase. Figure 4.8 compares the cross-polarization spectrum of a polyethylene sample with the spectrum obtained with a long relaxation delay, during which the signal from the amorphous material has decayed away. The discrimination between phases with different dynamics is not restricted only to carbon NMR, but may be used with any nuclei for which there is a large difference in the spin–lattice relaxation times.

Dipolar dephasing (Section 1.8.8.2) is another method used to distinguish between the crystalline and the amorphous phases, since one consequence of the more rigid crystalline environment is that the dipolar interactions are much stronger. The important part of the

4.2. Chain Conformation in the Solid State

dipolar dephasing pulse sequence is the period after cross polarization during which the signals decay due to the proton–carbon dipolar interactions (16). Figure 4.9 compares the CPMAS spectrum of poly(oxymethylene) with that obtained using dipolar dephasing (17). In this case, the chemical shift difference between the crystalline and the amorphous phases is too small to be observed without such methods. Very similar results have also been reported for poly(ethylene oxide), where the crystalline and amorphous phases can be distinguished on the basis of dipolar dephasing, the T_1's, or the $T_{1\rho}$ relaxation times (18).

4.2.2 Amorphous Polymers

NMR has been less extensively used to study chain conformation in amorphous polymers because by their nature they have less well-defined conformation. However, some observations have provided a fundamental insight into the behavior of these materials. The molecular dynamics of amorphous polymers have been extensively studied, as will be discussed in Section 5.3.

The spectral features of amorphous polymers are usually more poorly resolved than are those of crystalline materials. This is illustrated in the comparison of the crystalline and amorphous phases of polyethylene shown in Fig. 4.1 and Fig. 4.8. Polystyrene is a typical amorphous material and the spectrum is shown in Fig. 4.10. In such a spectrum the only well-resolved peak is the nonprotonated aromatic

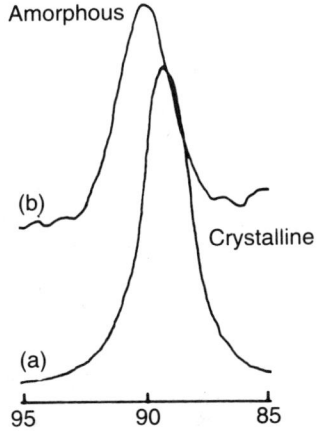

FIGURE 4.9 The solid-state carbon NMR spectra of poly(oxymethylene) obtained with magic angle spinning and (a) cross polarization and (b) dipolar dephasing. Adapted with permission (17).

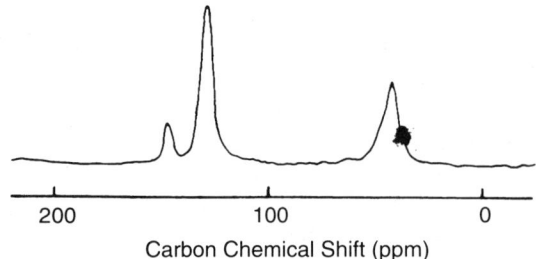

FIGURE 4.10 The solid-state carbon spectrum of polystyrene.

signal at 146 ppm. All of the other protonated aromatic signals resonate at 126 ppm, and the higher field peak contains the overlapping methine and methylene carbon signals. Note that the signals are considerably broader than those for the crystalline polymers. This is due primarily to conformational effects, since there is a distribution of conformations that are slowly interchanging on the NMR time scale, so the lines are inhomogeneously broadened by the γ-*gauche* effect. Because the linewidths are due to inhomogeneous line broadening, the spectra do not improve significantly by going to high magnetic field as expected for small molecules in solution. Thus, the highest magnetic fields are not always required for the study of amorphous polymers by cross polarization and magic angle spinning.

Figure 4.11 shows the proton NMR spectrum of polystyrene acquired with combined rotation and multipulse NMR (CRAMPS) (19),

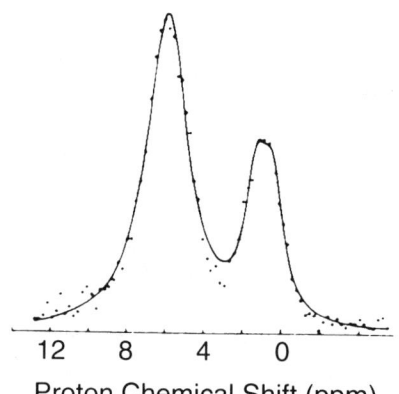

FIGURE 4.11 The solid-state proton NMR spectrum of polystyrene obtained with combined rotation and multipulse decoupling (CRAMPS). Adapted with permission (19).

4.2. Chain Conformation in the Solid State

a method for observing the proton spectrum in the solid state (20). The resolution is typically worse in the proton spectrum because it is difficult to observe protons while eliminating the proton–proton dipolar interactions, and because the range in chemical shifts is typically only 10 ppm for protons.

Although in general the lines from amorphous polymers are broad, information related to chain conformation can be extracted from certain features of the magic angle spinning spectrum. This is illustrated in Fig. 4.12, which shows the CPMAS spectrum of poly(2,6-dimethyl phenylene oxide) (21). This spectrum differs from that observed in solution in that the signal for the protonated carbon signal appears as a doublet, demonstrating that there is conformation asymmetry about the C–O–C bond for the polymer in the solid state. The intramolecular barrier to bond rotation is low, so that in solution there is conformational averaging that is fast on the NMR time scale and the signal appears as a singlet. In the solid state the glassy

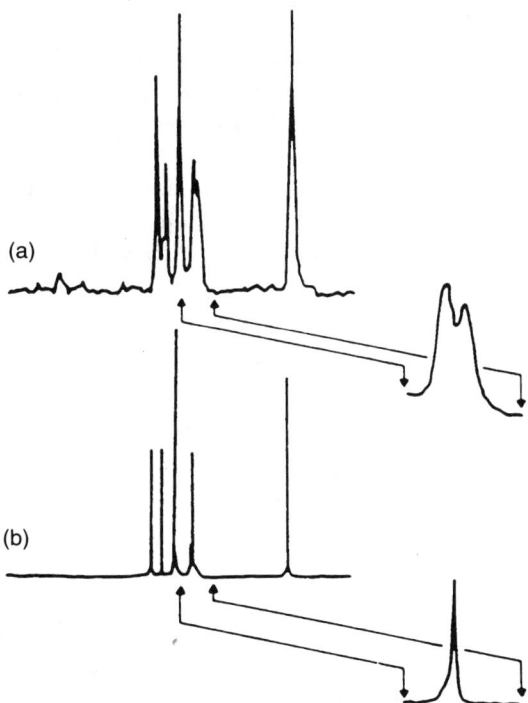

FIGURE 4.12 The 100-MHz solid-state cross-polarization and magic angle spinning spectra of poly(2,6-dimethyl phenylene oxide) in (a) the solid state and (b) solution [adapted with permission (21)].

polymer matrix restricts the conformation averaging, and the asymmetry in the chain conformation becomes apparent (21).

Poly(2,6-dimethyl phenylene oxide)

Although conformation information is difficult to extract from the CPMAS spectra of amorphous polymers, there are many possible applications for solid-state NMR in the study of amorphous polymers. Solid-state NMR is an important tool for the study of curing and reactivity in polymers, as illustrated in Fig. 4.13, which shows how CPMAS spectra can be used to monitor the degree of cure in polyether epoxy resins (22). The top spectrum shows a large peak at 51 ppm that is due to the unreacted epoxy after 2 weeks of curing. This peak disappears after 9 months, or after 2 weeks if the curing is done in the presence of poly(propylene glycol), which changes the reaction mechanism.

Similar methods have been used to study the curing of the resole type of polymers generated by the reaction of phenols with formaldehyde. The properties of these materials are dependent on the extent of reaction, which can be measured by the intensities of the CH_2 and OCH_2 peaks as shown in Fig. 4.14 (23).

While carbon NMR is often the method of choice for monitoring the cure in such polymers because natural abundance carbon spectra are easily acquired, other nuclei often provide a valuable insight into the chemistry of curing. Figure 4.15 shows the effect of cure time and temperature on the ^{15}N NMR spectra of poly(amic acid) oligomers

4.2. Chain Conformation in the Solid State

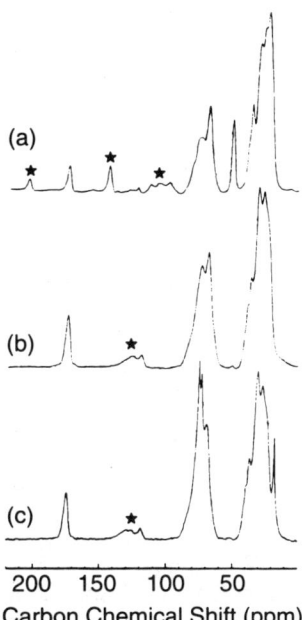

FIGURE 4.13 The solid-state carbon cross-polarization and magic angle spinning spectra of polyether epoxy resins. Spectra (a) and (b) are from the polyether epoxy resin after curing for 2 weeks and 9 months; spectrum (c) is after curing for 2 weeks in the presence of poly(propylene glycol). ★, spinning sideband. Adapted with permission (22).

that have been isotopically labeled (24). Using standard solid-state NMR methods (such as dipolar dephasing) and a comparison with model compounds, the four major signals in the ^{15}N spectra at 150, 110, 28, and 22 ppm can be assigned to the imide and amide nitrogens in the ring compounds, and the terminal amide and amine nitrogens. Once such peaks are assigned, these NMR methods can be used to relate the curing conditions to the structure–property relationships in these materials.

Poly(amic acid)

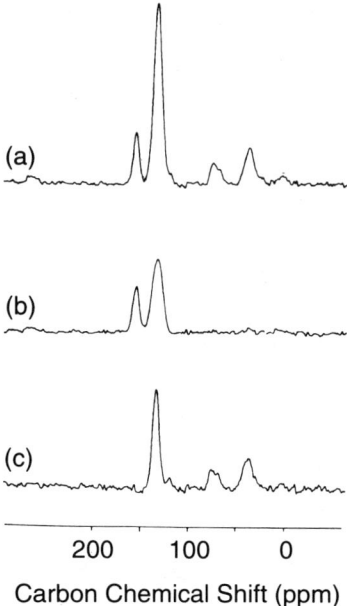

FIGURE 4.14 The solid-state carbon NMR spectra of cured resole-type resins. Spectrum (a) shows the CPMAS spectrum, (b) shows the subspectra of the nonprotonated carbons obtained with dipolar dephasing, and (c) is the difference spectrum showing the protonated carbon resonances. Adapted with permission (23).

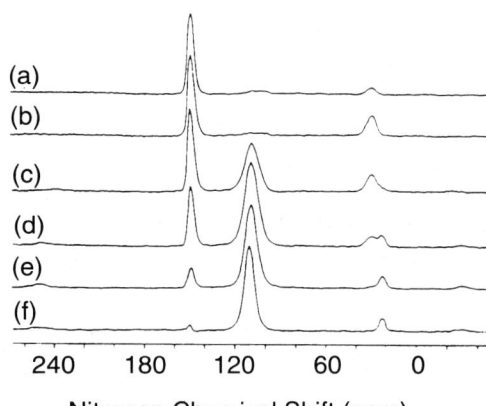

FIGURE 4.15 The 30-MHz ^{15}N NMR spectra of isotopically labeled poly(amic acid) polymers as a function of cure time and temperature. The conditions are (a) 400°C for 60 min, (b) 350°C for 60 min, (c) 150°C for 60 min, (d) 150°C for 30 min, (e) 85°C for 90 min, and (f) 85°C for 30 min. Adapted with permission (24).

4.2.3 SOLID–SOLID PHASE TRANSITIONS

Some polymers are stable in more than a single crystalline conformation, and solid-state NMR is an effective means to monitor solid–solid phase transitions because the changes in conformation are often accompanied by changes in chemical shifts due to the γ-gauche effect or crystal packing effects. 1,4-*trans*-Polybutadiene has long been known to adopt two crystalline forms that can be interconverted by changing the temperature (25). The form I conformation is observed at ambient temperature and is well defined by X-ray diffraction. The form I conformation is $s^\pm$–t–$s^\pm$ for the bonds between the double bonds, where $s^\pm$ represents the skew ($\pm 109°$) conformation. Form II is observed above 75°C, where the chains are packed in a hexagonal array but with a lower density and the equatorial reflections are blurred. Figure 4.16 shows the effect of temperature on the CPMAS spectrum of 1,4-*trans*-polybutadiene, where the conversion from form I to form II can be easily followed by NMR (26). The form II resonances appear upfield from the form I peaks, and the mixtures of the two forms can be easily observed at intermediate temperatures. Several models for the form II conformation have been proposed, including one in which the skew angle is decreased from $\pm 109°$ to $\pm 80°$ to account for the contraction in chain-axis repeat distance observed in the X-ray diffraction (27). It has also been proposed that the chains are conformationally disordered or that they are rapidly rotating about their long axis (28). The NMR spectra and relaxation time measurements are most consistent with a model in which the chains are disordered in the form II conformation (26).

$$\left[\begin{array}{c} H \diagdown \diagup CH_2 \\ C=C \\ \diagup CH_2 \quad H \diagdown \end{array} \right]$$

1,4-*trans*-Polybutadiene

Poly(butylene terephthalate) fibers are known to undergo a reversable crystal–crystal transition from the α to the β phase under unaxial extension. Although poly(butylene terephthalate) is a commercially important material that has been studied in both phases by a number of methods, there is still some controversy over the chain conformation in the different crystalline forms. It has been suggested that the extension forces the butylene segment into an all-*trans* conformation, and such a conformational change should be detected by the carbon chemical shifts through the γ-gauche effect. The magni-

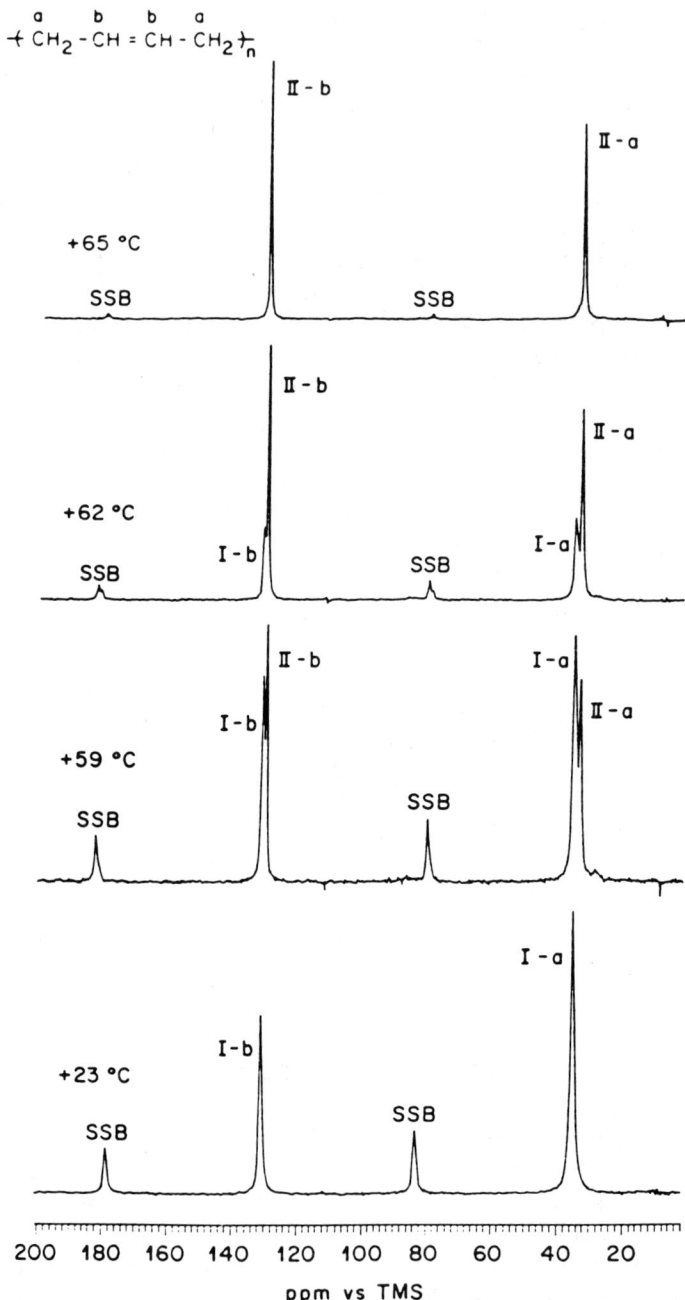

FIGURE 4.16 The 59-MHz carbon spectra of 1,4-*trans*-polybutadiene as a function of temperature showing the form I to form II transition. The spinning sidebands are marked. Adapted with permission (26).

4.2. Chain Conformation in the Solid State

tude of the γ-*gauche* effect was directly measured in an NMR/X-ray study of several poly(butylene terephthalate) model compounds, and Fig. 4.17 shows the conformational results determined from the crystal structure and the chemical shifts measured by NMR (29). From these model studies, it is expected that the all-*trans* conformation would appear at about 3.5 ppm downfield relative to the *gauche* conformation. The spectra of the α and β phases of poly(butylene terephthalate) have been recorded at high temperature, and the results are shown in Fig. 4.18 (30). The spectra of the β phase were recorded by placing a spool under tension in a magic angle spinning

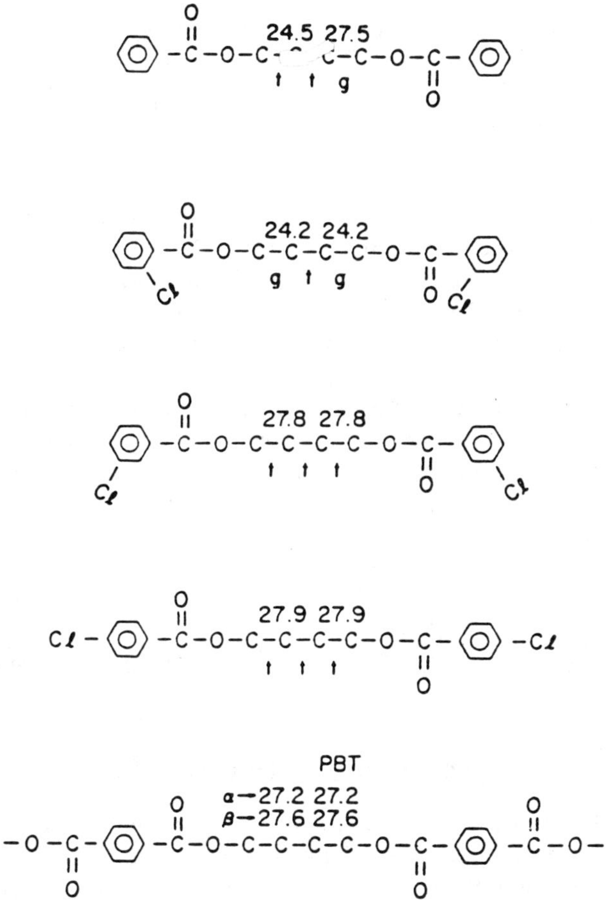

⟨O⟩–C(=O)–O–C–C–C–C–O–C(=O)–⟨O⟩
           24.5 27.5
           t t g

Cl–⟨O⟩–C(=O)–O–C–C–C–C–O–C(=O)–⟨O⟩–Cl
           24.2 24.2
           g t g

Cl–⟨O⟩–C(=O)–O–C–C–C–C–O–C(=O)–⟨O⟩–Cl
           27.8 27.8
           t t t

Cl–⟨O⟩–C(=O)–O–C–C–C–C–O–C(=O)–⟨O⟩–Cl
           27.9 27.9
           t t t

PBT
α — 27.2 27.2
β — 27.6 27.6

–O–C(=O)–⟨O⟩–C(=O)–O–C–C–C–C–O–C(=O)–⟨O⟩–C(=O)–O–

FIGURE 4.17 The model compounds, chemical shifts, and conformations for the compounds used as models to study the conformation of poly(butylene terephthalate) (PBT). Reproduced with permission (30).

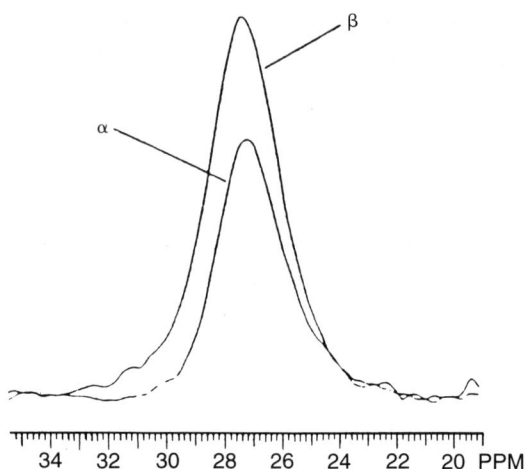

FIGURE 4.18 The 50-MHz carbon NMR spectra of poly(butylene terephthalate) in the α and β phases measured at 105°C. Adapted with permission (30).

rotor, and it was reported that higher resolution spectra could be obtained at a higher temperature (105°C), where it is easier to distinguish between the crystalline and the amorphous phases. The results showed that the chemical shifts were nearly equivalent for both forms, and the chemical shifts were close to those expected for the all-*trans* conformation. On the basis of this observation it was suggested that the methylenes are in the *trans* conformation in both phases, and the 10% increase in fiber repeat distance could be related to a change in the conformation of the ester group. It should be noted, however, that the data as presented here have not been universally accepted (31).

$$\left[-O-\overset{O}{\underset{\|}{C}}-\underset{}{\bigcirc}-\overset{O}{\underset{\|}{C}}-O-CH_2-CH_2-CH_2-CH_2-\right]$$

Poly(butylene terephthalate)

Poly(diacetylenes) are materials of interest because of their electrical and optical properties. Large single-crystal disubstituted diacetylenes can be obtained by topochemical polymerization of the monomers, and thermochromic transitions have been observed in many of these materials (32). NMR is one of the methods of choice for studying these polymers because it is expected that the change in

4.2. Chain Conformation in the Solid State

conformation giving rise to the change in optical properties should also affect the carbon chemical shifts.

Figure 4.19 shows the CPMAS spectra of one such polymer, poly(ETCD), that undergoes a transition from a blue phase to a red phase at 115°C (33). The NMR results show that only small changes in the CPMAS spectra of the polymer are observed upon the phase transition. The largest change is the 4-ppm downfield shift of the acetylinic carbon and the 2-ppm downfield shift of the side-chain β and γ carbons. A study of related poly(diacetylenes) showed that the change in chemical shift of the acetylenic carbons is common to all the related polymers that undergo the blue-to-red phase transition (34). These NMR data are consistent with a transition in which the polymer undergoes slight rotations along the backbone away from a planar structure in the blue-to-red-phase transition, and the side chains are more extended in the high-temperature red phase. The fact that no change is observed in the chemical shift of the carbonyl

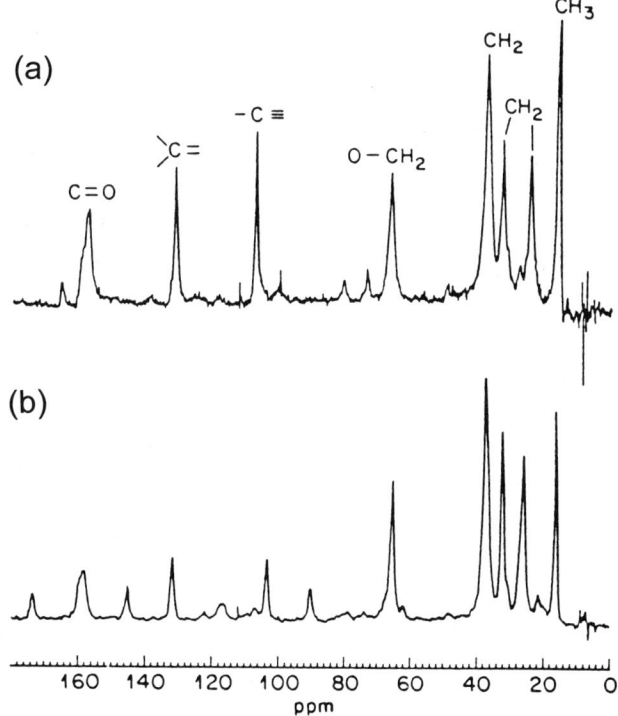

FIGURE 4.19 The 50-MHz carbon CPMAS spectra of the (a) blue and (b) red phases of poly(ETCD). Adapted with permission (33).

carbon shows that the hydrogen bonding pattern between side chains is maintained in both the blue and the red phases.

$$\left[=\underset{R}{\overset{R}{C}}-C\equiv C-\underset{R}{\overset{}{C}}= \right]$$

Poly(ETCD)

$R= CH_2\text{-}CH_2\text{-}CH_2\text{-}CH_2\text{-}O\text{-}\overset{\overset{O}{\|}}{C}\text{-}NH\text{-}CH_2\text{-}CH_3$

We noted earlier that polysilanes exist in several possible crystalline conformations that depend on the identity of the side group (Section 4.2.1). Figure 4.20 shows the solid-solid phase transition for

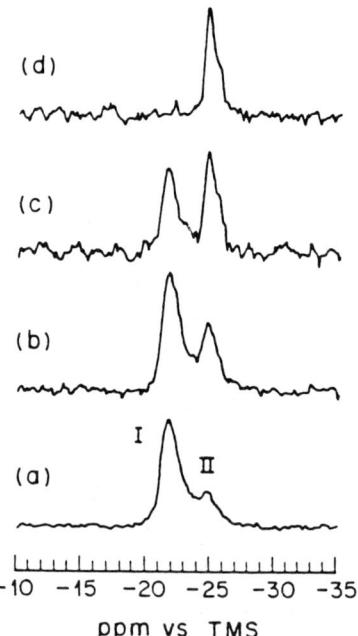

FIGURE 4.20 The 39.75-MHz ^{29}Si CPMAS spectra of poly(di-n-hexyl silane) acquired at (a) 25°C, (b) 39.5°C, (c) 41.5°C, and (d) 44.3°C. The transition midpoint is 42°C. Adapted with permission (9).

poly(di-*n*-hexyl silane) that can be observed by ^{29}Si NMR as a function of temperature (9). In this case the two phases correspond to an all-*trans* backbone conformation in which the *n*-hexyl side chains are packed in ordered arrays (I) and a disordered state (II). This transition can also be monitored optically and has a midpoint around 42°C.

These same NMR methods can be used to monitor other types of phase transitions in solids, including those such as the crystal-to-liquid crystal transitions. Polyphosphazines are in a group of polymers that are known to undergo such transitions before melting, and Fig. 4.21 shows the effect of temperature on the ^{31}P NMR spectra of poly[bis(4-ethylphenoxy)phosphazine] (35). This polymer undergoes the crystal-to-liquid crystal transition at 100°C, and above this temperature a single, relatively sharp line is observed.

Poly[bis(4-ethylphenoxy)phosphazine]

4.2.4 OTHER STUDIES

Solid-state NMR methods are used to study a wide variety of materials in addition to the crystalline and amorphous polymers, including gels, inclusion compounds, cross-linked polymers, polymers above T_g, melts, and ionomers. The methods used in these studies depend to a large degree on the molecular dynamics of the material under study. Since many of these materials are more mobile than rigid solids, cross polarization and magic angle spinning may not be the methods of choice.

Many polymers at temperatures far above T_g have a mobility that is intermediate between those observed for solids and solutions. While the lines are often broader than those observed for solutions, these materials are not rigid enough to be effectively cross polarized.

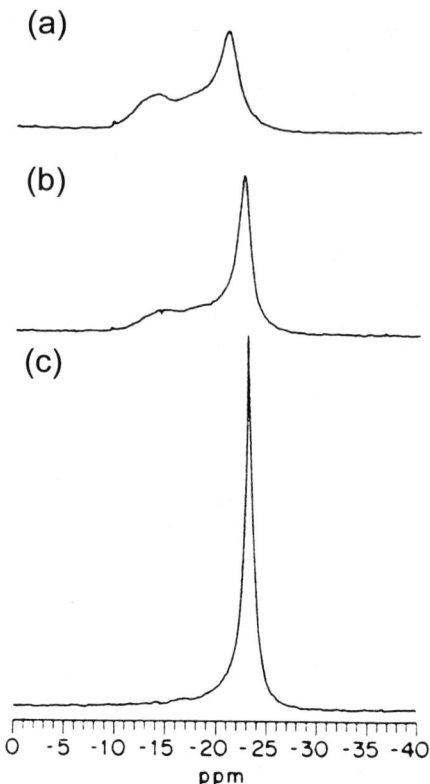

FIGURE 4.21 The 80-MHz ^{31}P NMR spectra of poly[bis(4-ethylphenoxy)phosphazine] at (a) 23°C, (b) 80°C, and (c) 120°C. The lowest temperature spectrum has contributions from the crystalline, interfacial, and amorphous phases. Adapted with permission (35).

In such cases, chain motion nearly averages the chemical shift anisotropy and the dipolar interactions, so the observed linewidths can be on the order of a kilohertz, compared with the 40-kHz lines observed in rigid solids.

This behavior is illustrated in the proton spectra of cis-1,4-polybutadiene shown in Fig. 4.22 (36). The top spectra show that two groups of signals can be resolved in the static proton spectra, and that each line has a width of ca. 300 Hz. The residual broadening from dipolar couplings and chemical shift anisotropy can be further averaged with magic angle spinning to 31 Hz. The linewidth observed with magic angle spinning is only slightly larger than that observed for many polymers in solution. Once the lines have been narrowed by this method, these polymers can be analyzed by any of the solution NMR methods presented in Chapter 3. In a similar way, Fig. 4.23

4.2. Chain Conformation in the Solid State

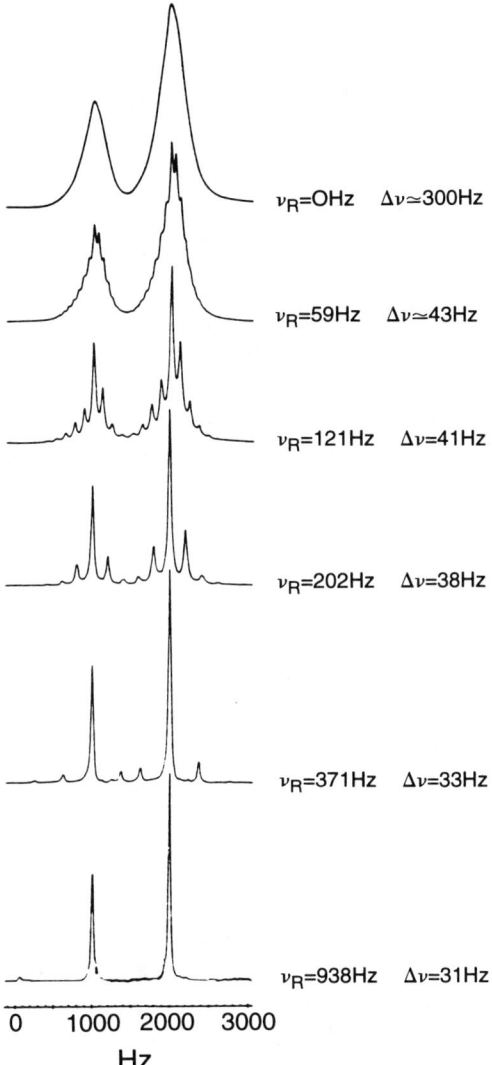

FIGURE 4.22 The 300-HMz magic angle spinning proton spectra of *cis*-1,4-polybutadiene at several spinning speeds. Adapted with permission (36).

shows that the carbon linewidths can also be effectively reduced with magic angle spinning (36).

Another class of materials that have chain mobilities intermediate between those of solids and liquids is swollen cross-linked polymer gels, such as the polystyrene derivatives that are often used for solid-phase synthesis of peptides and other biomolecules. Figure 4.24 compares the static and magic angle spinning proton NMR spectra of

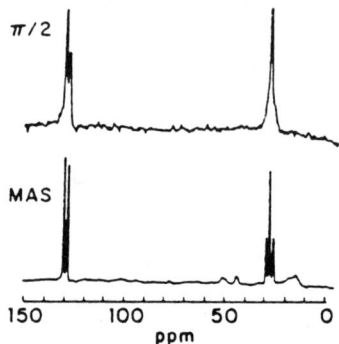

FIGURE 4.23 The 75-MHz static and magic angle spinning carbon spectra of *cis*-1,4-polybutadiene. Adapted with permission (36).

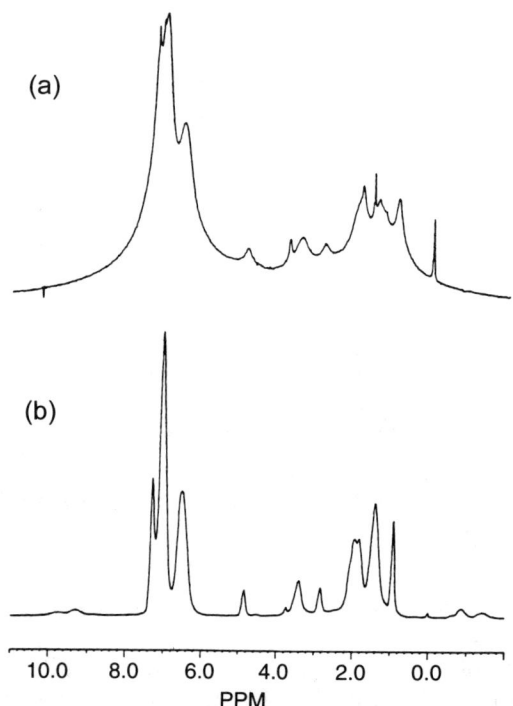

FIGURE 4.24 The (a) static and (b) magic angle spinning proton NMR spectra of a derivatized poly(styrene-divinyl benzene) swollen in chloroform. Adapted with permission (37).

4.2. Chain Conformation in the Solid State

a derivatized poly(styrene-divinyl benzene) cross-linked gel swollen in chloroform (37). Again the lines are effectively narrowed by magic angle spinning and the resolution is comparable to that obtained for polystyrene in solution without cross linking. A similar sharpening in the carbon spectra was also reported (37).

The inclusion compounds formed between some polymers and certain matrix materials are another example of a class of materials with mobilities intermediate between those of solutions and rigid solids. It has been observed that some small molecules, such as perhydrotriphenylene or urea, form stable crystal structures into which long-chain polymers may be included in narrow channels. Figure 4.25 shows a schematic diagram of an isolated polymer chain in the center of a 5.5-Å channel formed from perhydrotriphenylene (38). Solid-state carbon NMR has been used to study a number of such inclusion compounds, including those formed between perhydrotriphenylene and polyethylene (38), *trans*-1,4-polybutadiene (39), and *trans*-1,4-polypentadiene (40). Figure 4.26 shows the CPMAS spectrum of the perhydrotriphenylene/polyethylene inclusion compound (38). As noted above (Section 4.2.1), polyethylene exists in an all-*trans* conformation in the crystal structure and the methylene carbon resonance is shifted by 2.5 ppm from the amorphous material that

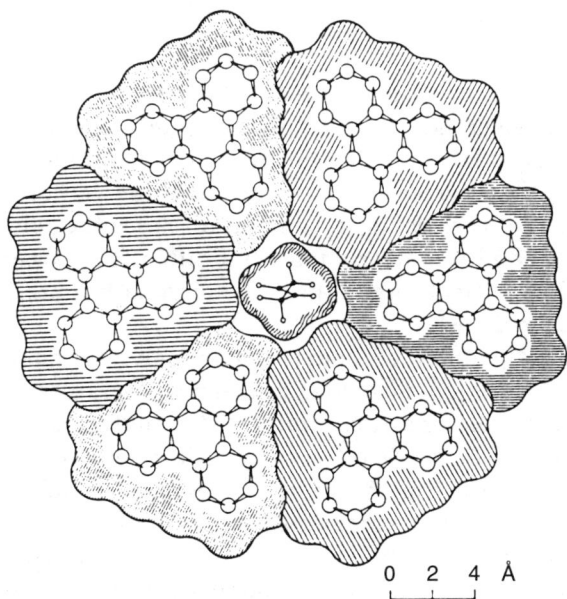

FIGURE 4.25 A schematic representation of an isolated polymer chain in the channel of a perhydrotriphenylene matrix. Adapted with permission (38).

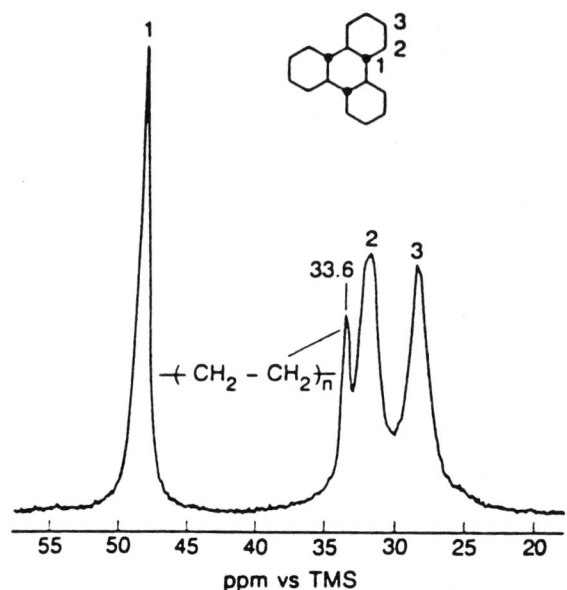

FIGURE 4.26 The 100-MHz carbon NMR spectrum of the polyethylene/perhydrotriphenylene complex obtained with magic angle spinning and dipolar decoupling. The peaks labeled 1, 2, and 3 are assigned to the matrix. Adapted with permission (38).

contains about 40% of the bonds in the *gauche* conformation. The chemical shift observed for the polyethylene inclusion compound is nearly identical (33.6 ppm) to that of the crystalline phase, demonstrating that the polymer exists in an all-*trans* conformation in the narrow channel of the inclusion compound. Since the relaxation times for the polymer and the matrix material differ, it is possible to choose experimental conditions to enhance the signals from either the polymer or the matrix (38).

Inclusion compounds of some linear polymer chains, such as polyethylene, poly(ethylene oxide), and some polyesters, can be formed in the 5.5-Å channels in a urea matrix. This matrix and the inclusion compounds have been studied by X-ray diffraction and other methods, and a schematic drawing of the urea inclusion complex is shown in Fig. 4.27 (41). Figure 4.28 compares the CPMAS spectra of crystalline poly(ε-caprolactone), which exists in an extended all-*trans* conformation, with the urea inclusion compound (42). As in the case of the polyethylene/perhydrotriphenylene complex, the poly(ε-caprolactone) appears to adopt an all-*trans* conformation in the narrow channel.

4.2. Chain Conformation in the Solid State 273

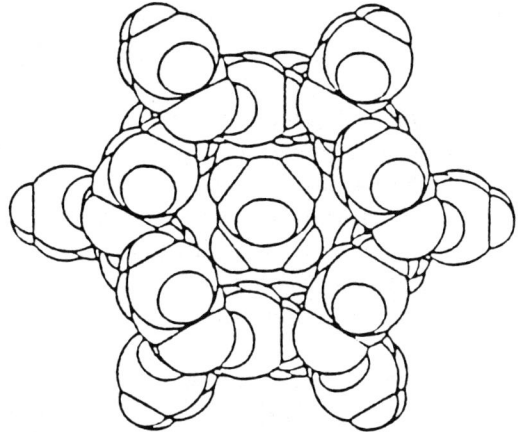

FIGURE 4.27 A schematic representation of the inclusion compound formed by urea and n-hexadecane. Adapted with permission (41).

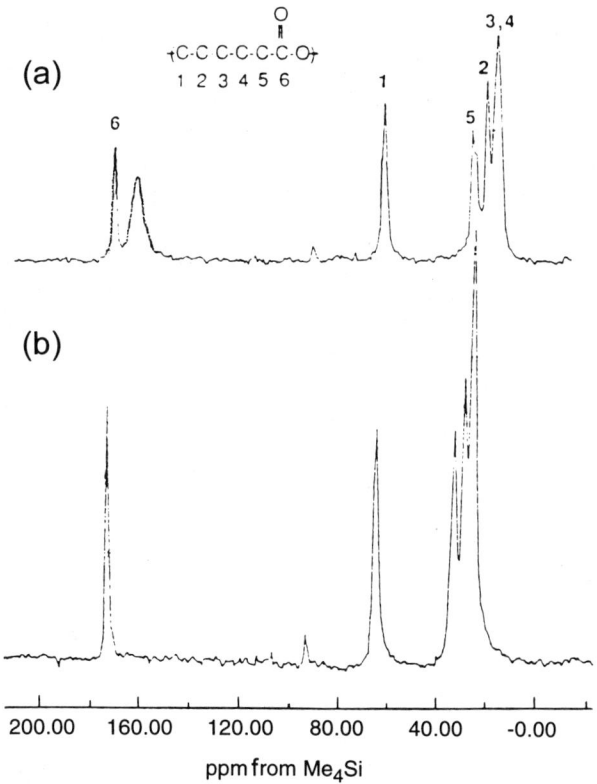

FIGURE 4.28 The CPMAS spectra of (a) the poly(ε-caprolactone)/urea inclusion compound and (b) poly(ε-caprolactone). Adapted with permission (42).

4.3 THE ORGANIZATION OF POLYMERS IN THE SOLID STATE

NMR studies have been valuable not only for studying the local structure or conformation of polymer chains, but also for measuring the long-range order. Such measurements are made primarily by selectively exciting one portion of the sample (i.e., the rigid or mobile phase) and measuring the rate of spin diffusion to the other phase. If the intrinsic spin diffusion rate is known, then the average distance between the phases can be calculated for a given morphological model.

In most cases the selective excitation of a particular portion of the sample is based on a difference in the relaxation rates between the phases. In semicrystalline polymers, for example, there is typically a large difference in the spin–spin relaxation rates between the rigid crystalline and the more mobile amorphous phase. Figure 4.29 shows the Goldman–Shen pulse sequence that is frequently used to study spin diffusion based on such differences in T_2 (43). Following the initial 90° pulse the magnetization from the rigid phases decay to zero due to T_2 relaxation during the delay period τ_0, while the magnetization from the mobile phase does not. The magnetization from the mobile phase is returned to the $+z$ axis with a pulse that is opposite the phase of the first one, creating a nonequilibrium spin state in which the mobile phase is close to equilibrium, while the rigid phase is saturated. If the phases are in close proximity (i.e., the separation is smaller than the distance the magnetization can diffuse) then the rigid phase can relax by the transfer of magnetization to the mobile phase, and the rigid signals will increase in intensity while mobile signals are decreased by spin diffusion from the rigid phase. After the

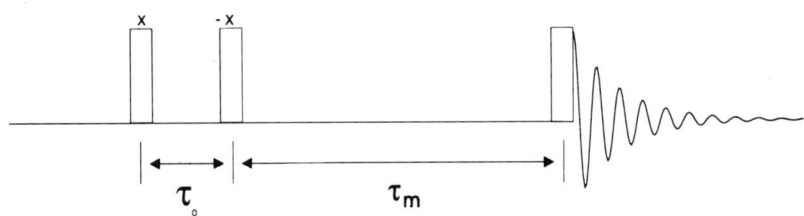

FIGURE 4.29 The Goldman–Shen pulse sequence (43). After the initial 90^0_x pulse the magnetization from the rigid phase decays during the τ_0 delay while that from the mobile phase does not. The second 90^0_{-x} pulse restores the magnetization from the mobile phase to the z axis. The magnetization is sampled with a third pulse after a spin diffusion period.

4.3. The Organization of Polymers in the Solid State

delay time t the intensity of all the signals is measured by the final 90° pulse.

Information about the domain structure of polymers is obtained by solving the diffusion equation

$$\dot{m}(r,t) = D\nabla^2 m(r,t), \quad (4.1)$$

where D is the diffusion constant (in units of cm^2/s) and $m(r,t)$ is the local magnetization density. The solutions to this equation are expressed in terms of the response function $R(t)$, which can be obtained from the measured intensities as a function of time (44),

$$R(t) = 1 - \frac{M(t) - M(\infty)}{M(0) - M(\infty)}, \quad (4.2)$$

where M is the intensity of the rigid portion of the material. The expression for $R(t)$ depends on the boundary conditions and on the dimensionality. The simplest cases are for diffusion in one, two, and three dimensions, where $\bar{b}$ is taken to be the average domain dimension. For diffusion in one dimension $R(t)$ is given by

$$R(t) = 1 - \varphi(t) \quad (4.3)$$

where

$$\varphi(t) = e^{Dt/\bar{b}^2} \text{erfc}\left(Dt/\bar{b}^2\right) \quad (4.4)$$

and efrc is the complement of the error function (44). This expression can be reduced to

$$\begin{aligned} R(t) &= 1 - \phi(t) \\ &= \left(\frac{2}{\sqrt{\pi}}\right)\left(\frac{Dt}{\bar{b}^2}\right)^2 \quad \text{for } t \ll \frac{\bar{b}^2}{D} \\ &= 1 - \sqrt{\pi}\left(\frac{\bar{b}^2}{Dt}\right) \quad \text{for } t \gg \frac{\bar{b}^2}{D}. \end{aligned} \quad (4.5)$$

For diffusion in two dimensions the expression for $R(t)$ is

$$\begin{aligned} R(t) &= 1 - \phi(t)^2 \\ &= (3\sqrt{\pi})\left(\frac{Dt}{\bar{b}^2}\right)^{1/2} \quad \text{for } t \ll \frac{\bar{b}^2}{D} \\ &= 1 - \sqrt{\pi}\left(\frac{\bar{b}^2}{\pi Dt}\right) \quad \text{for } t \gg \frac{\bar{b}^2}{D}, \end{aligned} \quad (4.6)$$

and for three-dimensional diffusion the result is

$$R(t) = 1 - \phi(t)^3$$
$$= (6\sqrt{\pi})\left(\frac{Dt}{\overline{b}^2}\right)^{1/2} \quad \text{for } t \ll \frac{\overline{b}^2}{D}$$
$$= 1 - \left(\frac{\overline{b}^2}{\pi Dt}\right)^{3/2} \quad \text{for } t \gg \frac{\overline{b}^2}{D}. \tag{4.7}$$

Physically, the one-dimensional diffusion applies to layer-like or lamellar structures, two-dimensional diffusion is most applicable to rod-like structures, and three-dimensional diffusion applies to spheres or cube-like structures. Figure 4.30 shows a plot of the response function plotted against $\sqrt{t}$, which is equal to $\sqrt{Dt}/\overline{b}$, for one-, two-, and three-dimensional diffusion (44).

Spin diffusion is an important tool for measuring the length scale of phase separation in semicrystalline polymers, blends, and a wide variety of phase-separated materials. Such studies have contributed

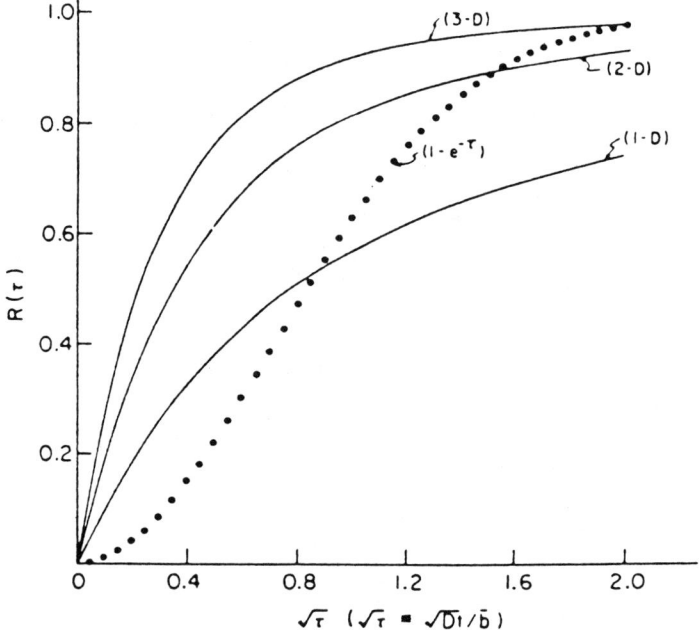

FIGURE 4.30 The response function $R(\tau)$ for the recovery of magnetization in the Goldman–Shen experiment for one-, two-, and three-dimensional diffusion as a function of the diffusion time. Adapted with permission (44).

4.3. The Organization of Polymers in the Solid State

greatly to our understanding of the structure of polymers, but it should also be noted that such studies are somewhat approximate because of uncertainties in measuring or estimating the diffusion constants, and because the morphology of the polymer under consideration may not exactly correspond to the simple one-, two-, or three-dimensional morphologies described above. It is sometimes necessary to use more sophisticated morphological models that include an interface that has a rate constant for spin diffusion which differs from both the rigid and the amorphous phases (45).

Determination of the diffusion constant used to calculate the length scales in the one-, two-, and three-dimensional models shown above is an important consideration. In most experiments the rigid phase is saturated during the τ_0 delay in the Goldman–Shen experiment shown in Fig. 4.29, and the rate-limiting step in spin diffusion is the spread of magnetization in the mobile phase. In such cases the spin diffusion depends only on the rate of spin diffusion in the mobile phase.

In the earliest studies the rate constants for spin diffusion were estimated from T_2 relaxation rates for the more mobile phase. The spin diffusion rate constant can be estimated as (46)

$$D = \frac{2r_0^2}{T_2}, \tag{4.8}$$

where r_0 is the radius of the hydrogen atom. This estimate was validated in a study of phase separation in polyurethanes by NMR and X-ray scattering (46). The upper bound for D can also be estimated as (44)

$$D = \frac{13a^2}{T_2}, \tag{4.9}$$

where a^2 and T_2 are the average distance between adjacent protons and the T_2 in the less mobile domain. This estimate was validated in a study of semicrystalline polymers (44).

In certain favorable cases it is possible to measure rather than estimate the spin diffusion rate constants. One approach is to measure the rate of spin diffusion from high-resolution proton spectra obtained using CRAMPS (20). If, as in the case of polystyrene, separate signals can be measured for the aromatic and aliphatic resonances, then the interchain rate of spin diffusion can be measured by monitoring the intensity following selective excitation of one of the resonances (47). This is of course more difficult in semicrystalline polymers where the spectra of the phases cannot be separately measured. Alternatively, the diffusion constants could be determined by measuring spin diffusion in a system where the length scales have been well determined by some other experiment. In one study the spin diffusion rates were measured for polystyrene–polybutadiene block copolymers where the domain dimensions were known from electric microscopy (43). The results showed that for very mobile polymers, such as polybutadiene, the estimate of D based on Eq. (4.8) and Eq. (4.9) did not provide a good estimate of the spin diffusion rate constant over the entire temperature range.

The Goldman–Shen pulse sequence [Fig. 4.29 (43)] was initially used to measure domain spacings in semicrystalline polymers, blends, and mixtures by acquiring the wide-line free induction decay and fitting it to two decay processes that were assigned to the rigid and mobile phases. More recently many experiments have been developed to measure spin diffusion in high-resolution spectra. The simplest application is to follow the spin diffusion period with a cross-polarization pulse sequence to detect the high-resolution carbon spectrum, as shown in the pulse sequence of Fig. 4.31. The intensity of the proton magnetization is directly related to the intensity of the cross-polarized carbon signal. Alternatively, other pulse sequences such as the *dipolar filter* (22) *chemical shift filter* (48), or *chemical shift gradients* (49) can be used to create a difference in spin temperature for the two components under study. Selective excitation (47) and 2D CRAMPS

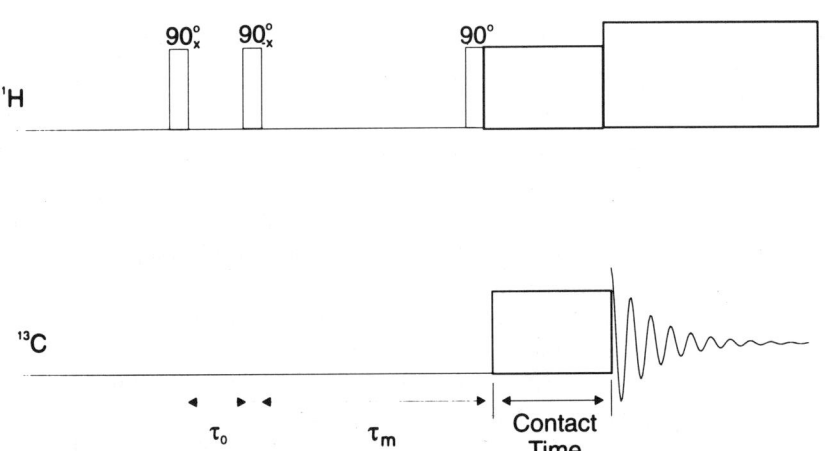

FIGURE 4.31 The pulse sequence for cross-polarized carbon detection of the Goldman–Shen pulse sequence. As in Fig. 4.29, τ_0 and τ_m are the periods for magnetization decay for the rigid phase and the spin diffusion period, and the cross polarization occurs during the contact time.

4.3. The Organization of Polymers in the Solid State

NMR (50) have also been used for these studies, as have various types of solid-state carbon–proton 2D NMR methods. Several of these methods will be discussed in the following sections.

4.3.1 THE ORGANIZATION OF SEMICRYSTALLINE POLYMERS

Solid-state NMR methods have made important contributions to our understanding of the organization of semicrystalline polymers, in part because the length scales that can be measured by spin diffusion and other techniques are smaller than those that can be measured by microscopy. These methods are also nondestructive and are relatively easy to perform.

The domain size of the crystalline and amorphous fractions of polyethylene has been characterized by several spin diffusion methods. Figure 4.32 shows the results from an early study of polyethylene structure using the Goldman–Shen pulse sequence (44). The difference in spin polarization between the crystalline and the amorphous fractions was produced with a 42-μs delay between the first two 90° pulses, and spin diffusion was monitored via the intensity of the

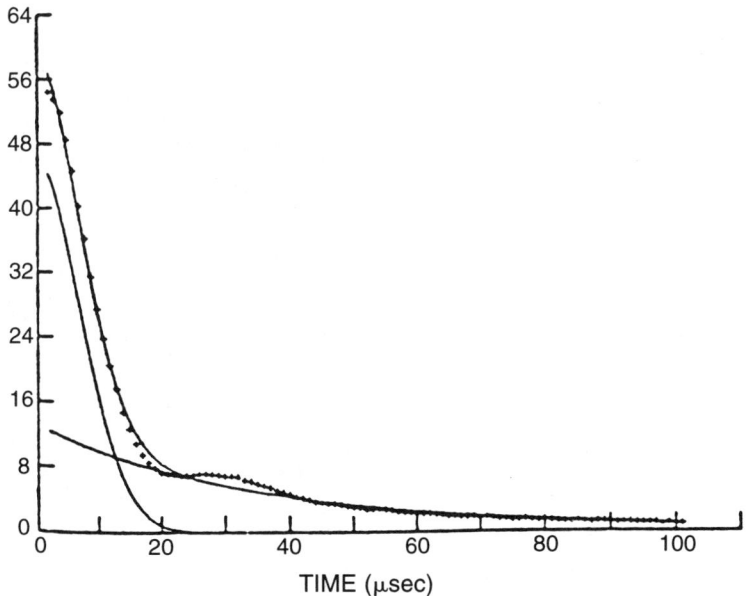

FIGURE 4.32 The proton free induction decay for polyethylene. The data (+) was fit to a Gaussian and Lorentzian component as shown by the solid lines. The relative intensity of the two components as a function of the spin diffusion time was used to measure the response function $R(t)$. Adapted with permission (44).

Gaussian and Lorentzian components of the free induction decay that were assigned to the crystalline and amorphous phases. The recovery factor $R(t)$ is plotted along with the fits to the 1D, 2D, and 3D morphological models in Fig. 4.33. Although polyethylene is known to have a lamellar morphology, the spin diffusion data for the noncrystalline domains appear to be better described by a 2D or 3D model. Using a diffusion coefficient of 8.3×10^{-12} cm^2/s, the average domain spacing for the 2D and 3D models was calculated to be 90 and 144 Å, respectively. As discussed below, more complex morphological models that incorporate interfacial regions may provide a more realistic model for spin diffusion in lamellar semicrystalline polymers.

The use of the Goldman–Shen pulse sequence to measure domain sizes via spin diffusion requires a large difference in the spin–spin relaxation rates of the phases under study. This is often a serious limitation since the relaxation times for the crystalline, interfacial, and amorphous phases are often too close to be easily separated. For this reason, other methods have been developed to create a difference in polarization between the phases. In some materials there are large differences in the proton $T_{1\rho}$'s that can be used to create the polarization difference (51), but this method is not ideal since the rate of spin diffusion is reduced only by a factor of two during the spin-lock

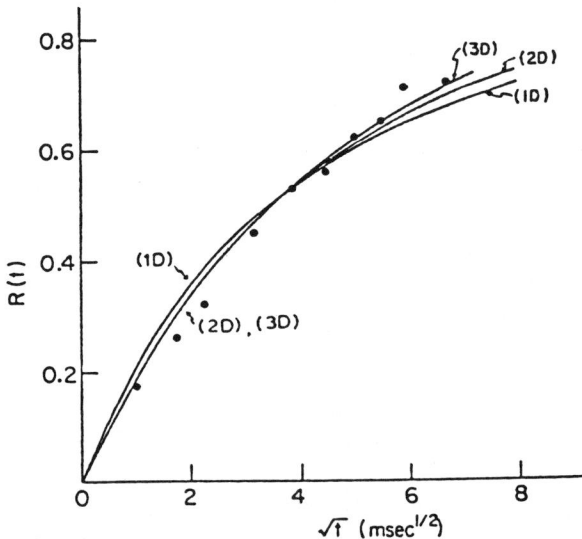

FIGURE 4.33 The response function $R(t)$ plotted as a function of $\sqrt{t}$ for polyethylene. The solid lines are calculated for one-, two-, and three-dimensional diffusion. Adapted with permission (44).

4.3. The Organization of Polymers in the Solid State

period. A better approach is to create the difference in spin polarization using multipulse NMR methods that quench spin diffusion during the preparation period. Such methods have been used to investigate the phase structure of poly(ethylene terephthalate) (52) and polyethylene (53).

The structures of other polyethylene samples have been more recently investigated using the dipolar filter pulse sequence to create the polarization difference between the crystalline and the amorphous fractions, which can then be monitored by the carbon spectrum using magic angle spinning and cross polarization (22). The pulse sequence used in these investigations is shown in Fig. 4.34. The dipolar filter part of the pulse sequence is the multipulse sequence designed to average the dipolar couplings. As used here, however, the delay time τ between pulses is rather long (7–20 μs), so the dipolar couplings for the rigid phase are not effectively averaged, and these signals are saturated by applying the sequence several times, while

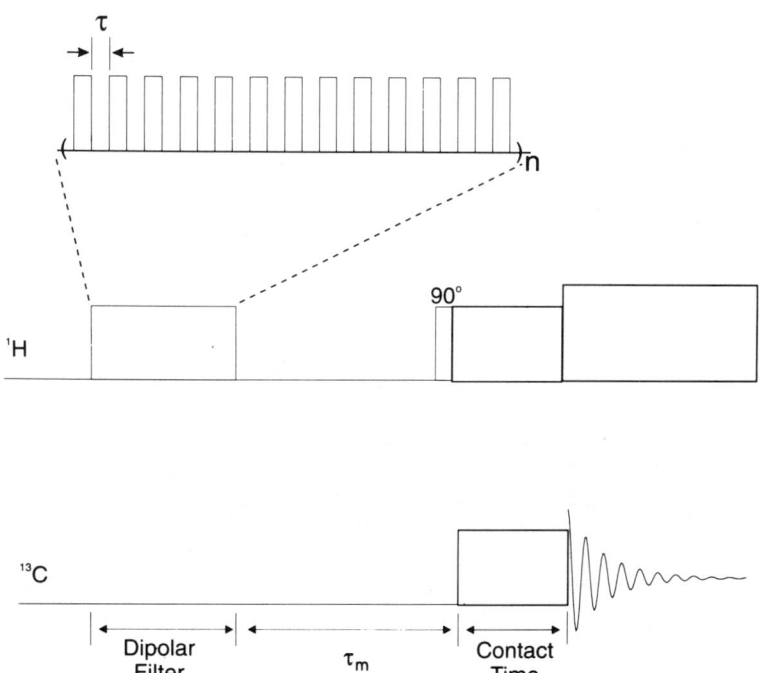

FIGURE 4.34 The pulse sequence diagram for the dipolar filtering experiment detected by cross polarization. The dipolar filter is a multipulse sequence that averages the dipolar couplings of the mobile phase while saturating the signals of the rigid phase. After a spin diffusion period (τ_m) the magnetization is measured using cross polarization.

282   4. The Solid-State NMR of Polymers

the signals from the mobile phase are perturbed to a much smaller degree. Following the spin diffusion time, the proton magnetization is sampled with the cross-polarization pulse sequence. This approach has the advantage that the carbon spectrum is observed, where the resolution is typically much greater than that in the proton spectrum.

Both low- and high-density polyethylenes were studied by this method, and Fig. 4.35 shows a plot of the spin diffusion data for low-density polyethylene (53). The data could not be adequately fitted

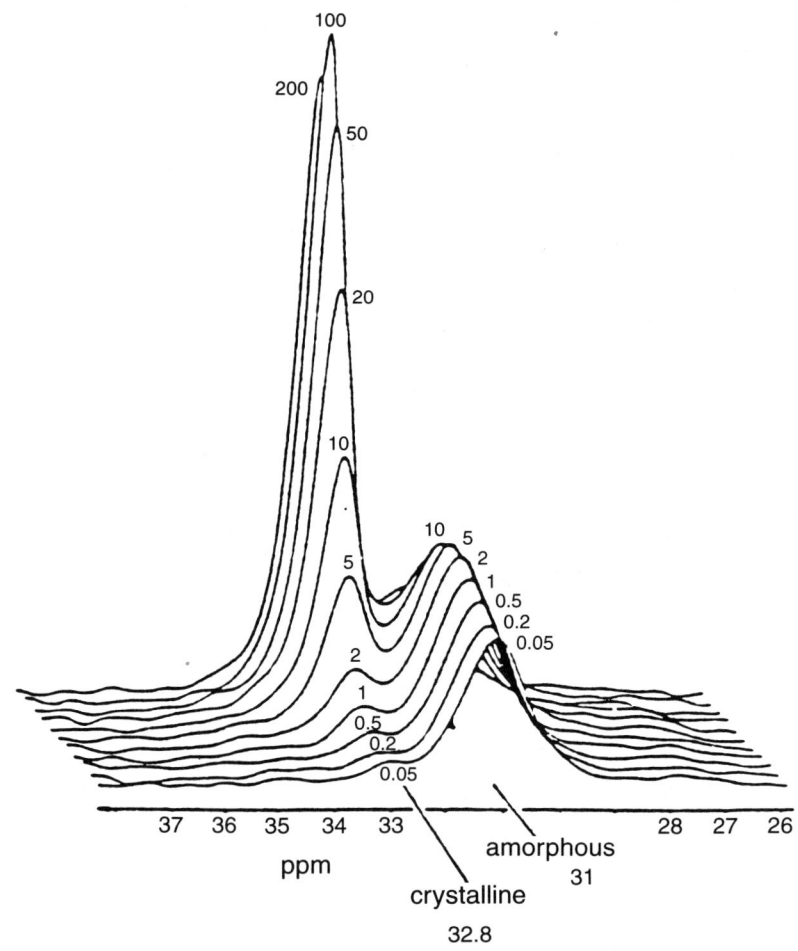

FIGURE 4.35 Spin diffusion in low-density polyethylene measured by the dipolar filter. Note the rapid diffusion from the more mobile to the less mobile amorphous materials in the first 5 ms followed by a slower spin diffusion to the crystalline region. Adapted with permission (53).

to a simple two-phase model, indicating that there may be a substantial interfacial component that was not detected in the earlier experiments. Direct evidence for an interfacial layer was obtained from the intensity of the amorphous component as a function of the spin diffusion time. Following saturation of the rigid protons with the dipolar filter pulse sequence, the intensity of the amorphous phase reached a maximum after 5 ms, indicating that there is spin diffusion between the mobile amorphous material to a less mobile amorphous material, the interface. At a longer spin diffusion time (10 ms) there is extensive spin diffusion to the crystalline phase. The diffusion behavior was modeled as a three-domain lamellar structure containing crystalline, amorphous, and interfacial domains. Despite the large differences in the crystallite thickness for the high (40 ± 10 nm) and low (9 ± 2.5 nm) density polyethylenes, the thickness of the interfacial layer was shown to be 2.2 ± 0.5 nm in both samples.

The spin diffusion between protons is most frequently used to study the domain structure of polymers because protons are abundant in polymers, the rates of spin diffusion are in a measurable range, and they are relatively easy to measure. However, other nuclei can also be used to measure spin diffusion in polymers. Some, like carbon or nitrogen, have a low natural abundance and require isotopic labeling, while others, such as phosphorus or fluorine, do not.

Phosphorus spin diffusion has been used to study the domain structures of semicrystalline polyphosphazines, such as poly[bis(3-methylphenoxy)-phosphazine] (54) and poly[bis(phenoxy)phosphazine] (55). Figure 4.36 shows the variable-temperature magic angle spinning spectra of poly[bis(3-methylphenoxy)phosphazine] acquired as a Bloch decay, rather than with cross polarization (54). Phosphorus has a large chemical shift anisotropy, so many spinning sidebands are observed. The lines narrow with increasing temperature, and the two main signals at -10.7 and -19.4 ppm are assigned to the crystalline and mesomorphic phases. Spin diffusion between the two phases is demonstrated in Fig. 4.37, which shows the 2D spin exchange spectrum in which cross peaks are observed between the chemical shifts assigned to the two phases. The spin exchange spectrum results from the solid-state implementation of the NOESY pulse sequence (Fig. 1.62), and cross peaks are observed if magnetization exchanges between the phases on the time scale of the mixing time, which in these experiments was 2 s. As we will see later (Section 5.3), cross peaks can also be observed if there is chain diffusion between the phases, but chain diffusion and spin diffusion can be distinguished since chain diffusion has a strong temperature dependence which spin diffusion does not (56). The exchange of magnetization

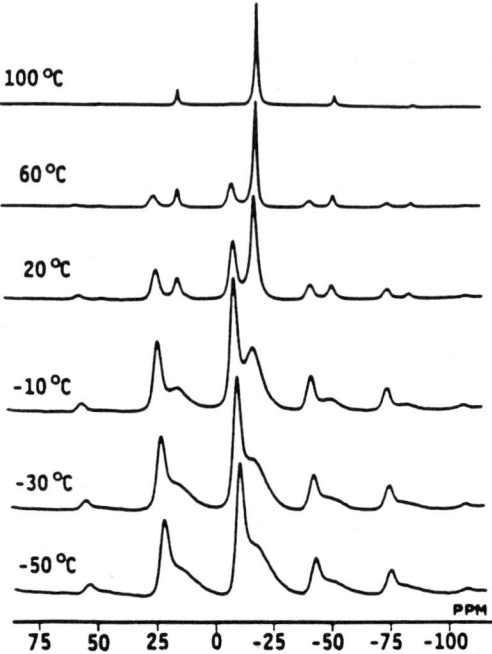

FIGURE 4.36 The variable-temperature ^{31}P NMR spectra of poly[bis(3-methoxyphenyl)phosphazine] acquired with magic angle spinning. The peaks at −10.7 and −19.4 ppm are assigned to the crystalline and mesomorphic phases. Adapted with permission (54).

between the phases of poly[bis(3-methylphenoxy)phosphazine] was shown to be independent of temperature (54).

In cases where resonances for the phases are well resolved, the one-dimensional analogue of the 2D NOESY NMR experiment can be used to study the spin diffusion, as shown in Fig. 4.38 (55). In such an experiment, a t_1 period is inserted that is long enough for the spin from the two phases to precess 180° out of phase. The 90° pulse then creates a difference in Z magnetization that is monitored by the final pulse after a spin diffusion period. Figure 4.39 shows the one-dimensional exchange spectra for poly[bis(3-methylphenoxy)phosphazine] for spin diffusion times between 0 and 4 s (55). The intensities of the two main peaks are fit to a set of coupled differential equations from which the spin–lattice relaxation rates and the rate of spin diffusion between the phases are obtained. To obtain an estimate of the domain spacing from these measurements, the spin diffusion constant D must be estimated. Using the approximate density of protons in

4.3. *The Organization of Polymers in the Solid State* 285

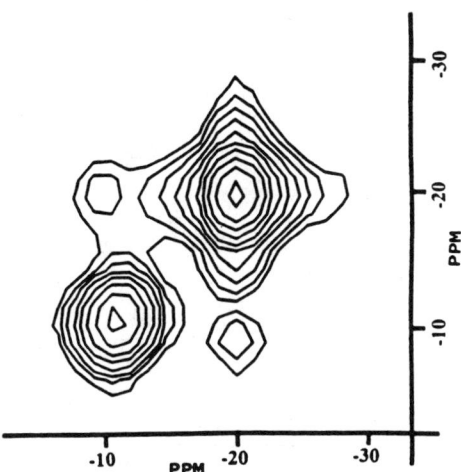

FIGURE 4.37 The 2D spin exchange spectrum of poly[bis(3-methoxyphenyl)phosphazine] acquired with a 2-s mixing time. Note the cross peaks between the crystalline and the mesomorphic phases. Adapted with permission (54).

poly[bis(3-methylphenoxy)phosphazine] to scale the D value estimated for polyethylene (6.2×10^{-12} cm^2/s), it was estimated that the proton spin diffusion coefficient was 4.3×10^{-12} cm^2/s. Correcting for the phosphorus spin density and the difference between the magnetogyric ratios of protons and phosphorus yielded the phosphorus spin diffusion coefficient of 4.8×10^{-14} cm^2/s. From this it may be calculated that the phosphorus–phosphorus spin diffusion rate of ca. 0.3 s^{-1} corresponds to a domains spacing of 4.5 nm, which is

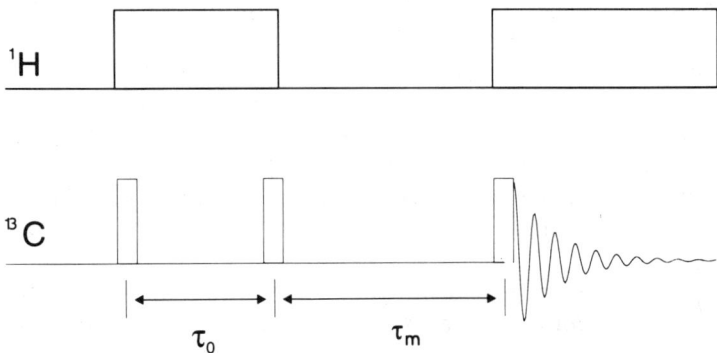

FIGURE 4.38 A schematic diagram of the 1D version of the 2D pulse sequence used to measure spin diffusion. Instead of measuring the spectra of many values of t_1, a single t_1 time is chosen that allows the signals for two phases to precess 180° out of phase.

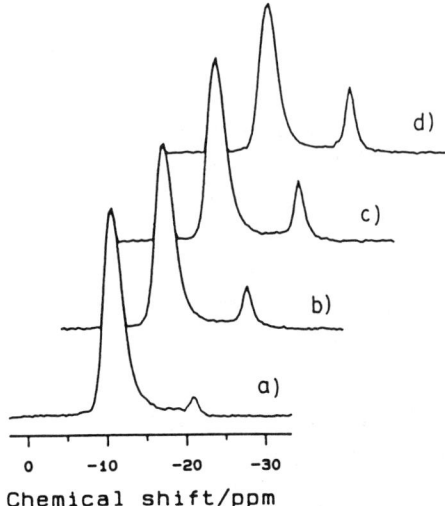

FIGURE 4.39 The 1D spin exchange spectra for poly[bis(phenoxy)phosphazine] for spin diffusion times of (a) 0.2, (b) 1, (c) 2, and (d) 4 s. Adapted with permission (55).

interpreted as a lamellar thickness of 9 nm. A similar analysis has been reported for poly[bis(phenoxy)phosphazine], where the results were interpreted in terms of a three-domain structure that included a crystalline, amorphous, and interfacial phase (55).

In certain favorable cases, NMR measurements can be used to determine the crystallinity in semicrystalline polymers. The biggest drawbacks in most studies are the long spin–lattice relaxation times for the crystalline phase that make it difficult to acquire spectra with a high signal-to-noise ratio. In polyethylene at high field strengths, for example, separate signals can be observed for the crystalline and amorphous signals, but it is difficult to obtain quantitative Bloch decay spectra because the T_1's for the crystalline phase may be as long as 1500 s (15). Cross polarization is usually used to acquire these spectra since the delay between pulses is governed by the proton T_1's, which are typically on the order of a few seconds. We noted earlier (Section 1.8.5.4) that it is difficult to measure intensities quantitatively using cross polarization.

The differences in relaxation times for the crystalline, interfacial, and amorphous materials can in favorable cases be used to measure the fraction of polymer in each phase. Figure 4.40 shows a plot of the intensity for a polyethylene sample obtained with the cross-polarization T_1 experiment (Fig. 1.40) (14). The insities are obtained over a wide range of delay times and fitted to three exponentials correspond-

4.3. The Organization of Polymers in the Solid State 287

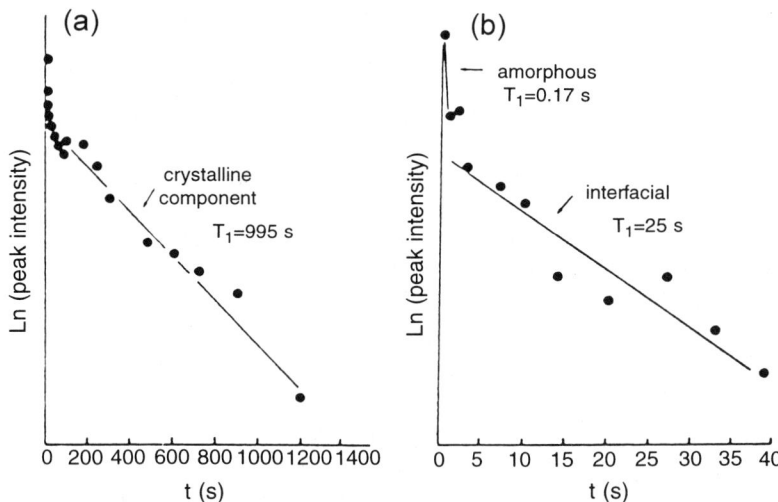

FIGURE 4.40 A plot of the signal intensity vs delay time for the cross-polarization T_1 experiment on linear polyethylene. Drawing (a) shows the decay over the entire range of relaxation delay times and (b) shows the initial relaxation behavior. Adapted with permission (14).

ing to the three phases. The spin–lattice relaxation times for the amorphous, interfacial, and crystalline components are 0.17, 25, and 995 s, and the crystallinity is calculated from the amplitude of the slowest relaxing fraction. The fraction crystallinity is often difficult to measure, and it is frequently observed that the crystallinities measured by various methods (density, differential scanning calorimetry, X-ray diffraction, Raman spectroscopy, etc.) are not all consistent with each other. The crystallinity values measured for polyethylene by NMR are in close agreement with those measured by Raman spectroscopy (15).

The differences in the relaxation times for polyethylene are due to a difference in the molecular dynamics of the different phases, as will be discussed in Section 5.3. In some polymers the relaxation may proceed by a pathway such that there is not a large difference between the crystalline and the amorphous phases, and the above methods cannot be used to determine the crystallinity. This behavior was observed in poly(methyl methacrylate) where only a small difference in the relaxation times was observed for the crystalline and amorphous phases (57). For poly(methyl methacrylate) it has been suggested that the T_1's are determined by the rapidly rotating methyl groups, and thus there is not a significant difference between the crystalline and the amorphous phases.

Nuclei other than carbons can also be used to measure the fraction crystallinity from the multicomponent spin–lattice relaxation. Figure 4.41 shows the recovery curves measured by deuterium NMR for poly(ethylene oxide)-d_4 for the solution crystallized, annealed, and quenched samples. The crystallinity depends on the sample history, and the fraction crystalline is obtained by extrapolating the long time relaxation behavior to zero time.

For most polymers the relaxation times are determined by the local constraints on chain motion, in both the crystalline and the amorphous states. One exception to this behavior is polyethylene, where the T_1's depend on the morphology of the crystals and the size of the crystallites. This is illustrated in Fig. 4.42, which shows a plot of the T_1's for polyethylene vs the crystallite thickness [L (Å)] determined by laser Raman acoustical mode spectroscopy (15). For thicknesses greater than 100 Å, the T_1 increases linearly with the thickness, while for thicknesses less than 100 Å there is only a rough increase with increasing thickness. Furthermore, if the interfacial structure is removed by oxidation, the T_1's increase dramatically. These results are interpreted with a model in which the T_1's for the crystalline fraction are shortened by contact with the interfacial material, so

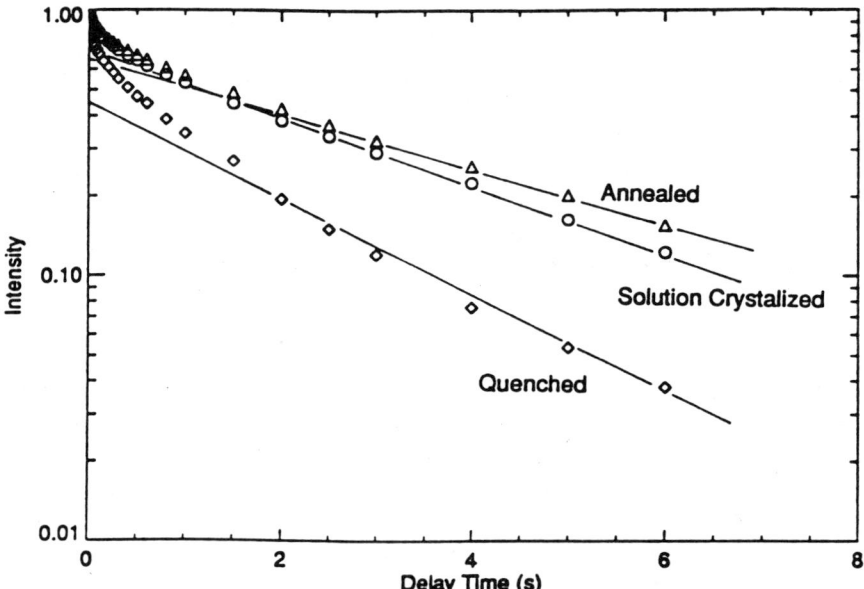

FIGURE 4.41 The spin–lattice relaxation of poly(ethylene oxide)-d_4 measured by deuterium NMR for solution crystallized, annealed, and quenched samples. The crystallinity is calculated from the amplitude of the longest decay process.

4.3. The Organization of Polymers in the Solid State

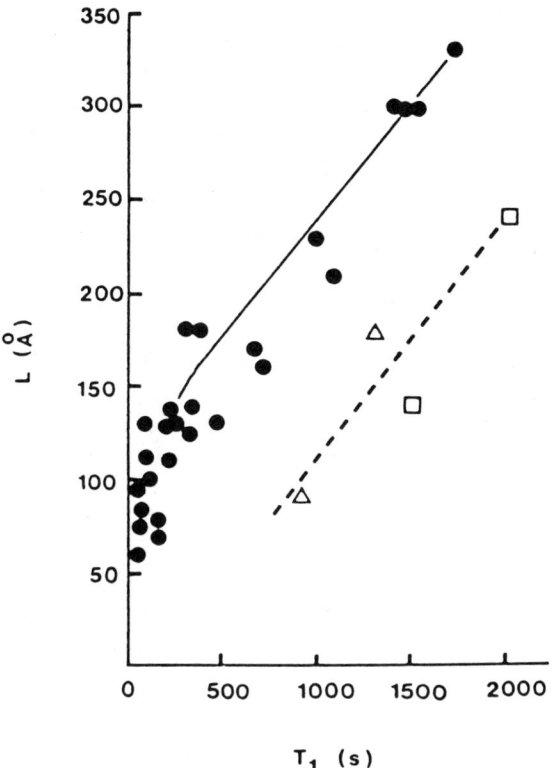

FIGURE 4.42 A plot of crystallite thickness vs the carbon T_1 for polyethylene. The open symbols show the relaxation times for polyethylene from which the interfacial material has been removed by oxidation. Adapted with permission (15).

longer relaxation times are associated with thicker crystallites (see also Section 5.3.2).

4.3.2 MULTIPHASE POLYMERS

Solid-state NMR is also a useful method for the investigation of the structure of phase-separated polymers. Among the key questions in such systems are how large are the phases and what is the degree of phase mixing. These questions can be answered using the methods introduced in Section 4.3 to measure the domain sizes, and by observing differences in the relaxation times for the material in the different phases.

Polyurethane is an example of a commercially important material that has been extensively investigated by solid-state NMR. Several polyurethanes were investigated in an early study using the Goldman–Shen pulse sequence, and the recovery curve for the rigid phase of one such material is shown in Fig. 4.43 (46). The fit to the data points in Fig. 4.43 is calculated for a lamellar model that accounts for some spin diffusion during the preparation period (75 μs) and is in excellent agreement with the domain spacings measured by small-angle X-ray scattering. The observation that the T_2 of the mobile phase changes with the spin diffusion time was taken as evidence for the presence of an interfacial region (46).

The NMR studies of polyurethanes have been used not only to measure the length scale of phase separation, but also to measure the degree to which the blocks are separated into the "hard" or "soft" domains. Deuterium NMR is an effective way to make such measurements because the lineshapes for the material in the soft domains are much sharper than those for the hard domains, and the contributions of the hard and soft material to the deuterium wide-line spectra can be easily quantified. Figure 4.44 shows the 55-MHz deuterium NMR spectra for a series of polyurethanes containing 50, 60, and 70% hard segments that have been specifically labeled in the butanediol section of the hard segment (58). The relative contribution of the deuterated butanediol dissolved in the hard and soft domains is obtained by

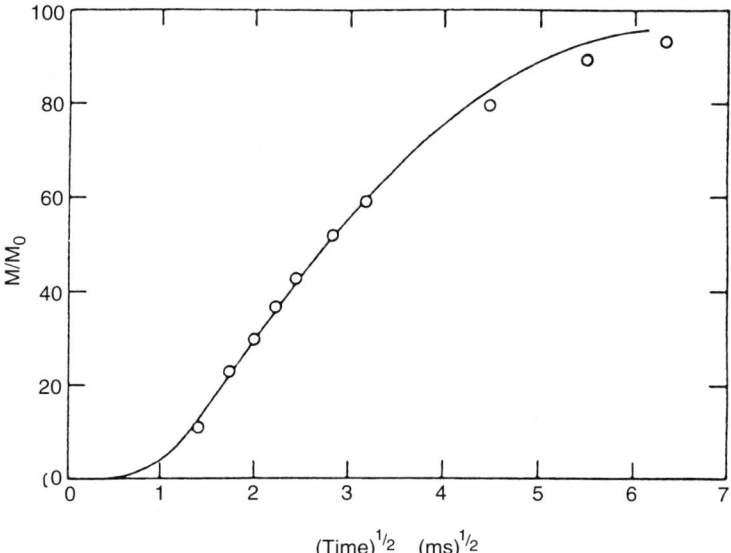

FIGURE 4.43 The recovery curve for the rigid phase in polyurethanes measured by the Goldman–Shen experiment. The solid line is a fit to a lamellar model. Adapted with permission (46).

4.3. The Organization of Polymers in the Solid State

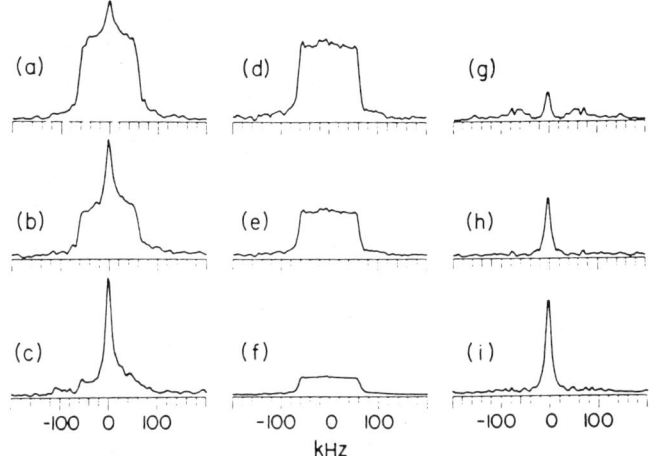

FIGURE 4.44 The 55-MHz ^{2}H spectra of hard-segment labeled polyurethanes. Spectra (a)–(c) show the spectra for polymers containing 70, 60, and 50% hard segments. Spectra (d)–(f) show the fraction of hard segments contributing to spectra (a)–(c), and spectra (g)–(i) show the result of subtracting the hard-segment spectra from the observed spectra. Adapted with permission (58).

subtracting the spectra for a polymer containing only the hard segments from the observed spectra, and the material with the motionally averaged lineshape is assigned to the amorphous and interfacial material. Table 4.1 shows that the NMR results compare favorably with those measured by small-angle X-ray scattering for the material containing 60 and 70 wt% hard segments. Since the X-ray experiments were performed on similar materials but not on the same samples, it was suggested that such differences might arise from the differences in synthesis and molding procedures used to produce the samples. Similar results were reported for model polyurethanes con-

TABLE 4.1
NMR and Small-Angle X-Ray (SAX) Determination of Hard Segments in the Interfacial Region and Dissolved in Soft Segments for Deuterium-Labeled Polyurethanes (58)

wt% hard segment	% dissolved and interfacial (SAX)	% dissolved and interfacial (NMR)
70	15.6	14
60	18.4	20
50	23.7	50

taining deuterated piperazine rings in the hard segment and poly(tetramethylene oxide) ($M_n \approx 2000$) for the soft segment, where 85% of the labeled rings were observed in the hard domains (59).

$$\left[\begin{array}{c} \text{O} \\ \parallel \\ \text{C}-\text{NH}-\bigcirc-\text{CH}_2-\bigcirc-\text{NH}-\text{C}-\text{O}-\text{CH}_2-\text{CD}_2-\text{CD}_2-\text{CH}_2-\text{O} \end{array} \right]$$

Hard segment

$$\left[\begin{array}{c} \text{CH}_3 \\ | \\ -\text{CH}-\text{CH}_2-\text{O}- \end{array} \right]$$

Soft segment

Novel 2D NMR experiments have been developed to measure the length scale of phase separation in multiphase polymers, and these experiments have been used to measure the degree of mixing of the phases in polyurethanes. Figure 4.45(a) shows the pulse sequence for the 2D wide-line separation (WISE) NMR spectroscopy (60). The WISE experiment starts with a 90° pulse to the protons and evolution of the proton chemical shifts followed by cross polarization and detection of the carbon spectrum. The purpose of this experiment is to obtain a correlation of the carbon and proton chemical shifts. Since no decoupling is applied during the proton evolution, the high-resolution carbon spectrum is correlated with the wide-line proton spectrum, and this 2D experiment can distinguish between carbons attached to protons in a mobile environment in which the proton linewidths are motionally averaged vs those in a rigid environment that have an extremely broad proton spectrum.

Figure 4.46 shows the application of this experiment to polyurethanes, where the goal is to distinguish between the soft-segment poly(propylene glycol) dissolved in the soft domain and that dissolved in the hard domain (61). Cross sections through the 2D data show that the proton linewidths for the methylene carbons in the poly(propylene glycol) are sharp (ca. 3.2 kHz) compared to those expected for soft segments dissolved in a rigid matrix (ca. 50 kHz). From these data it is concluded that the phase separation is nearly complete, and that very little of the soft segments is dissolved in the hard domains.

The structure of phase-separated diblock copolymers has been extensively investigated by spin diffusion using the Goldman–Shen (43) and dipolar filtering (22) pulse sequences. Spin diffusion can also be monitored using a variation of the WISE experiment shown in Fig.

4.3. The Organization of Polymers in the Solid State

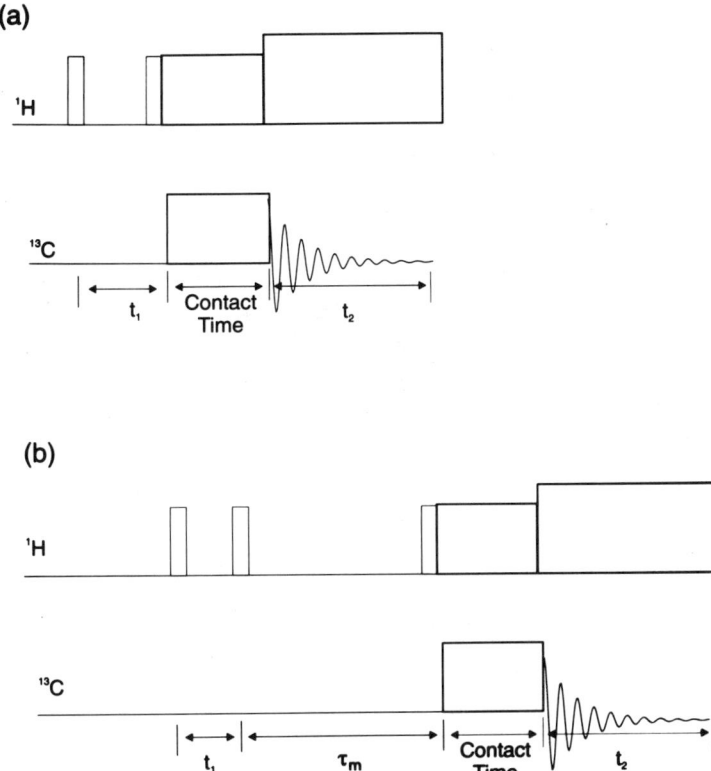

FIGURE 4.45 The pulse sequence diagram for the *wi*de-line *se*paration (WISE) experiment (a) without and (b) with a spin diffusion period. The wide-line proton spectra are correlated with the high-resolution carbon spectra (60).

4.45(b), in which a spin diffusion period is inserted into the pulse sequence between the t_1 period and the cross polarization. Figure 4.47 shows the application to a 50:50 poly(styrene-b-dimethyl siloxane) copolymer (60). At the shortest mixing times the carbons are correlated with their directly bonded protons and the proton linewidths are determined by the molecular mobility. Since the dimethyl siloxane block is far above its T_g and extremely mobile, its carbon signal at 0 ppm is correlated with a sharp proton line. The polystyrene block is more rigid and is correlated with a wide line in the proton dimension. The polystyrene linewidth is similar to that expected for the polystyrene homopolymer, and this experiment demonstrates that the presence of the mobile block does not appreciably affect the mobility of the polystyrene block. However, if 20 ms is allowed for proton spin diffusion, the proton lineshapes for the

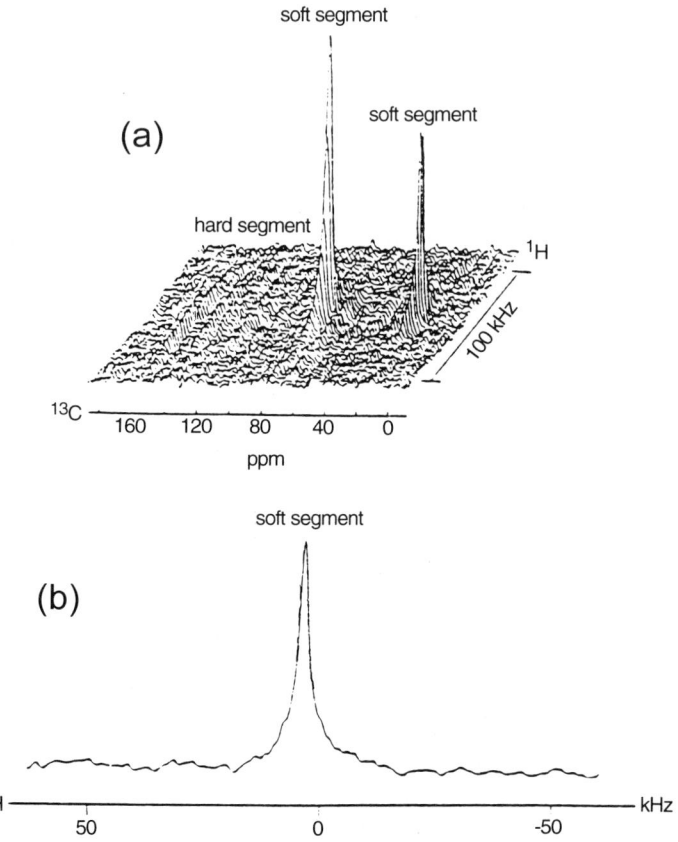

FIGURE 4.46 The (a) 2D WISE spectra and (b) a cross section through the data at the frequency of the methylene carbon for a polyurethane. The narrow linewidth for the methylene carbon shows that few of the soft segments dissolved in the hard-segment domains. Adapted with permission (61).

polystyrene appear as a combination of the sharp and broad resonances, indicating that there is substantial spin diffusion on this time scale. After 200 ms there is more extensive spin diffusion and the system is approaching equilibrium.

The observation that there is no sharp component to the polystyrene proton lineshape in the absence of spin diffusion demonstrates that the interface between the polystyrene and the poly(dimethyl siloxane) blocks is extremely sharp, and that very little of the two polymers has a mobility that is intermediate between the rigid and the mobile phases. Figure 4.48 shows the 2D WISE spectrum of poly(styrene-b-methyl phenyl siloxane) (86:14 mol%) in which different behavior is observed (60). In this case the lineshape for the methyl group on the siloxane appears as a composite of a broad and a narrow proton

4.3. The Organization of Polymers in the Solid State

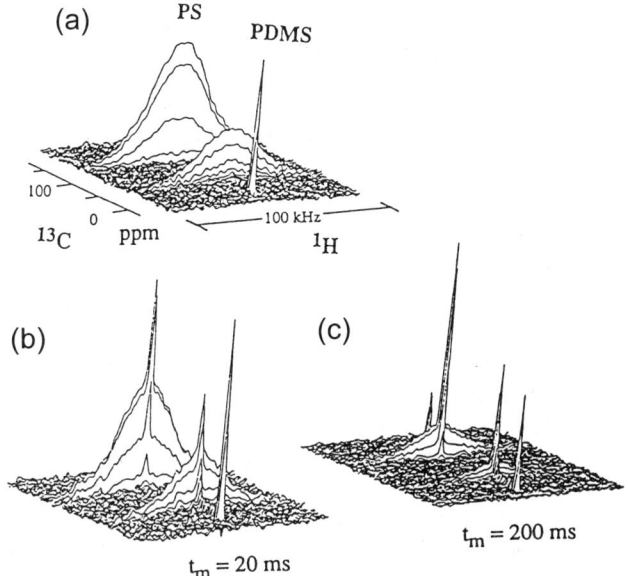

FIGURE 4.47 The 2D WISE NMR spectra for the 50:50 poly(styrene-b-dimethyl siloxane) copolymer with spin diffusion mixing times of (a) 0, (b) 20, and (c) 200 ms. Adapted with permission (60).

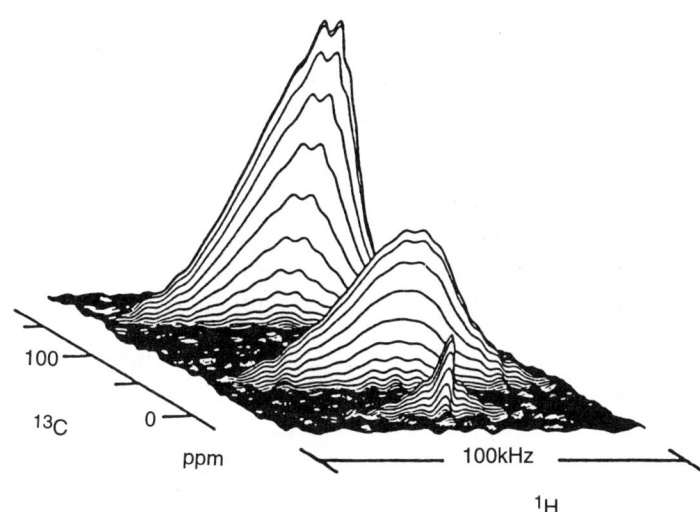

FIGURE 4.48 The 2D WISE NMR spectrum of the 86:14 poly(styrene-b-methyl phenyl siloxane) copolymer. Adapted with permission (60).

lineshape, indicating that about 50% of these groups have a significantly reduced mobility as a consequence of being in close contact with the polystyrene block.

The structure of a series of the poly(styrene-b-methyl phenyl siloxane) diblock copolymers with different molecular weights, compositions, and morphologies has been investigated by spin diffusion methods. Spin diffusion was measured with the dipolar filter pulse sequence, and some of the results are shown in Fig. 4.49 and Table 4.2 (45). The spin diffusion curves shown in Fig. 4.49 for three of the polymers show that clear differences in the morphology can be inferred from the raw data. It is well known that the morphology of block copolymers depends on the block molecular weights and compositions, and by varying these parameters several morphologies may be obtained for a single system. The data were modeled as a three-phase system: a polystyrene-rich phase, a poly(methyl phenyl siloxane)-rich phase, and an interface in which diffusivity changes linearly between the values of the two blocks. The dimensions of the phases and the morphologies for six samples are compiled in Table 4.2. The data show close agreement with the domain sizes measured by small-angle X-ray scattering (45).

WISE NMR has also been used to study the structure of poly(ethylene-co-methacrylic acid) copolymers alone and in mixtures with n-hexadecane. The results provide a measure of the role of the methacrylic acid groups, the size of the crystalline and amorphous domains, and the distribution of n-hexadecane dissolved in the copolymer. Figure 4.50 shows the 50-MHz 2D WISE spectra of the poly(ethylene-co-methacrylic acid) copolymer containing 14 wt% methacrylic acid at spin diffusion times of 0 and 0.02 s (62). The quaternary carbon from the methacrylic acid monomer at 46 ppm is correlated with a broad line in the proton dimension, showing that its mobility is greatly restricted. Since this same resonance is correlated with a narrow proton linewidth in a polymer where the acid groups have been esterified, these data show that the mobility is restricted by intermolecular hydrogen bond formation. Furthermore, the spin diffusion data show that these groups are in close proximity to the amorphous phase of the polyethylene and that n-hexadecane imbibed in the copolymer is localized in the amorphous domains.

$$\left[-CH_2-CH_2-\right]\left[-CH_2-\underset{CH_3}{\overset{COOH}{C}}-\right]$$

Poly(ethylene-co-methacrylic acid)

4.3. The Organization of Polymers in the Solid State

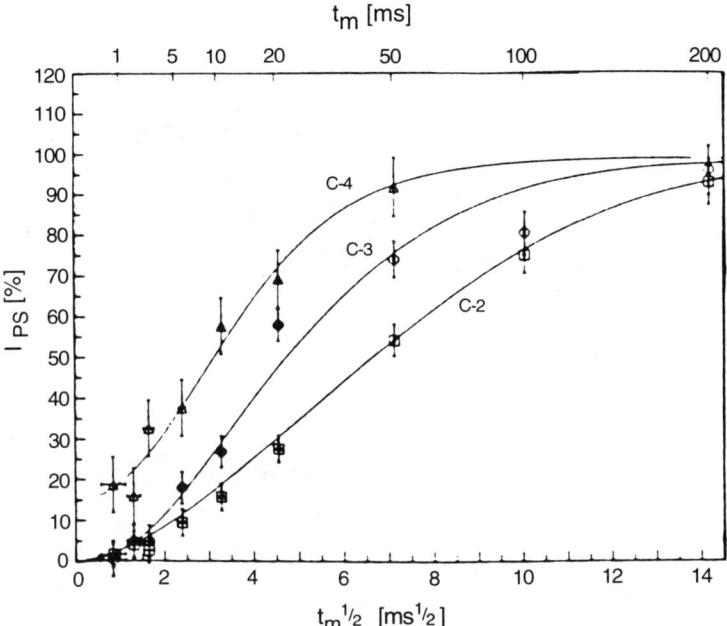

FIGURE 4.49 Magnetization recovery curves for poly(styrene-b-methyl phenyl siloxane) copolymers containing (C-2) 37%, (C-3) 21%, and (C-4) 16% methyl phenyl siloxane. The solid lines are fits to the morphological models shown in Table 4.2. Adapted with permission (45).

TABLE 4.2
Microphase Domain Sizes of Poly(styrene-b-methyl phenyl siloxane) Diblock Copolymers Determined by NMR and Small-Angle X-Ray Scattering (45)

Sample	$\Phi_{PMPS}{}^a$	Morphology[b]	$d_{PMPS}{}^c$ (nm)	$d_{interface}{}^c$ (nm)	$d_{PS}{}^c$ (nm)	$L_{NMR}{}^d$ (nm)	$L_{SAX}{}^d$ (nm)
C1	41	OBDD	8	4	3	14–24	24–27
C2	37	Lamellar	5	4	9.5	19–27	19–25
C3	21	Spherical	6	4	3	13–21	18–20
C4	16	Spherical	3.5	4	2.5	9–18	
C5	26	Spherical	1.5	1.5	0.5	2–8	
C6	45	Lamellar	2.5	1.5	2.5	6–12	

[a] The mole fraction of methyl phenyl siloxane in the block copolymer.
[b] OBDD is the ordered bicontinuous double diamond morphology.
[c] The domain sizes for poly(methyl phenyl siloxane), the interface, and the polystyrene.
[d] The long period ($L = d_{PMPS} + 2d_{interface} + d_{PS}$).

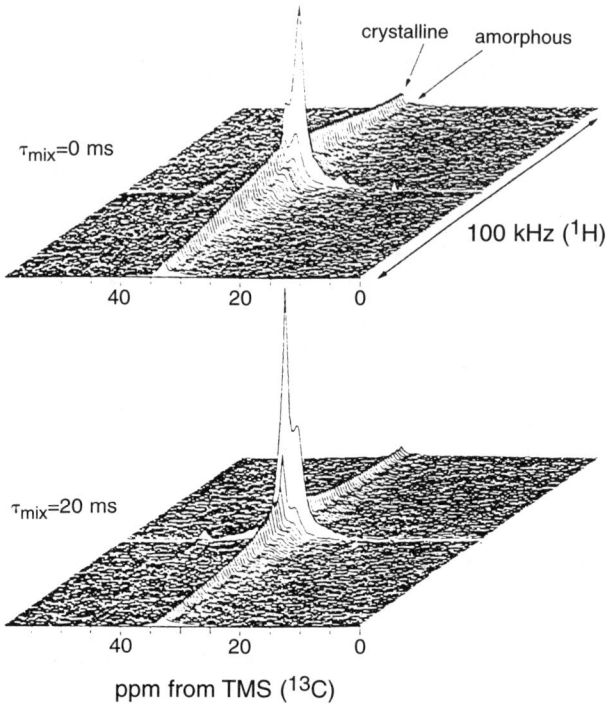

FIGURE 4.50 The 2D WISE NMR spectra of the poly(ethyl-co-methacrylic acid) copolymer with spin diffusion mixing times of 0 and 0.02 s. Adapted with permission (62).

In the experiments discussed thus far, the difference polarization between the phases is created by differences in the relaxation times that are related to differences in the molecular dynamics. In some cases of interest, however, the differences in dynamics may not be large enough to study these materials by spin diffusion. One novel way to study such materials is the *chemical shift filter*, which selects signals for cross polarization based on their proton chemical shift in a multipulse line-narrowing experiment (48). This method has been used to study a series of blends, blocks, and statistical copolymers of styrene and methyl methacrylate, and several of the spin diffusion curves are shown in Fig. 4.51 (63). Polystyrene and poly(methyl methacrylate) are immiscible and the block copolymers with equal block lengths are known to form lamellar structures. The domain sizes calculated from the spin diffusion data span a range between 0.5

4.3. The Organization of Polymers in the Solid State

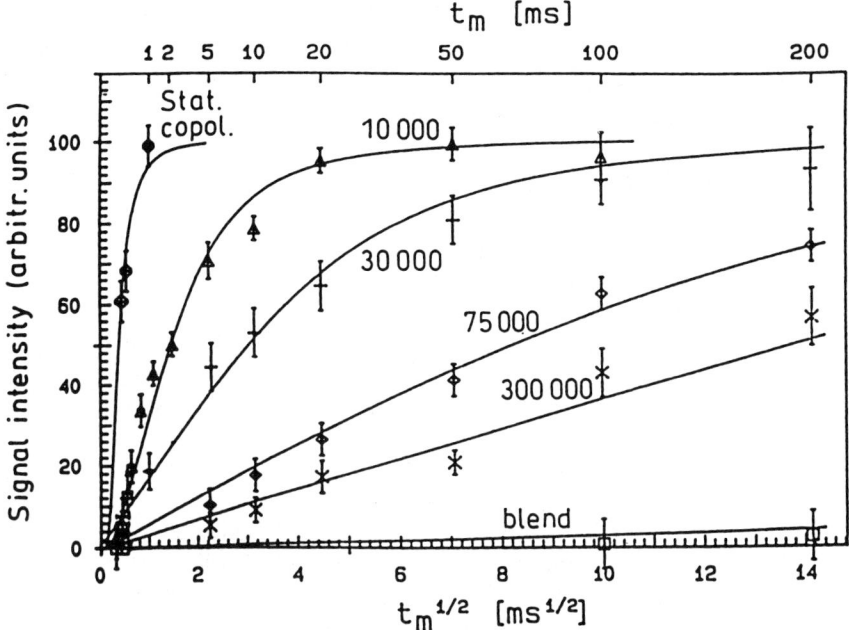

FIGURE 4.51 The spin diffusion recovery curves for polystyrene and poly(methyl methacrylate) copolymers and blends. The data are shown for the statistical copolymer; the equal length block copolymers of 10,000, 30,000, 75,000, and 300,000; and the blend of polystyrene and poly(methyl methacrylate). Adapted with permission (63).

and 100 nm and are in close agreement with those expected theoretically.

Spin diffusion methods have been used to study not only semicrystalline and phase-separated polymers, but also the formation of domains in homopolymers near or above the glass transition temperature. The presence of mobile domains in a polymer near T_g has been inferred by the observation of a composite lineshape in the wide-line proton NMR spectrum of chloral polycarbonate, or in polycarbonate with deuterated methyl groups (64). The difference in mobility between the rigid and the mobile domains is sufficiently large that the Goldman–Shen pulse sequence can be used to study the domain sizes. The results of these studies show that the mobile material composes about 15% of the total at 40–60° above the T_g, the domain size is on the order of 30 Å, and the mobile material occurs in isolated islands surrounded by less mobile material (65). More recently this system has been studied in the presence of 10 wt% trioctyl phosphate, where a domain size of 10 Å was measured (66). The mobile domains were modeled with a statistical lattice model in which there are two diluent molecules for every 11 polymer repeat units.

4.3.3 POLYMER BLENDS

Solid-state NMR has been an important tool for the characterization of polymer blends and mixtures. Among the most important questions in this field to which solid-state NMR has made important contributions are what is the length scale of mixing in polymer blends and what are the specific intermolecular interactions that lead to miscibility. NMR is one of the methods of choice for the determination of miscibility because the length scale measured using spin diffusion is much smaller than that inferred from DSC or other studies. With the introduction of new 2D NMR methods and isotopic labeling, it is now possible to measure the local structure in polymer blends. From such studies it is possible to identify the interacting groups in miscible blends and to gain a deeper insight into the factors that determine miscibility.

The molecular-level mixing of polymer chains causes changes in the structure of these materials that can be measured via the NMR spectra and the relaxation rates. On the most local level, strong intermolecular interactions due to hydrogen bonding, charge transfer complex formation, or ionic interactions can cause chemical shift changes that are large enough to be measured in the solid-state spectra. On a longer length scale spin diffusion experiments can be used to measure the average distances between the polymer chains. Blend formation is also known to affect the molecular dynamics (Section 5.3.4).

Most binary mixtures of polymer are immiscible because the entropy of mixing is extremely small for high-molecular-weight polymers, so miscibility is typically observed when there are favorable intermolecular interactions (67). If these interactions are strong enough it is sometimes possible to observe such effects directly in the

4.3. *The Organization of Polymers in the Solid State* 301

solid-state NMR spectra. One such example is shown in Fig. 4.52, which compares the 50-MHz carbon CPMAS spectra of two polymers labeled with electric donor and acceptor moieties with the miscible blend (68). The miscibility is due to a charge transfer complex formation between the aromatic rings on the donor and acceptor polymers, and the formation of the complex causes chemical shift changes due to intermolecular ring current effects (Section 1.5.1). While there is extensive overlap between the signals for the donor and acceptor carbons, several of the resonances are resolved in the spectra obtained with dipolar dephasing to visualize the nonprotonated carbons. The largest chemical shift differences (ca. 1.2 ppm) are observed for

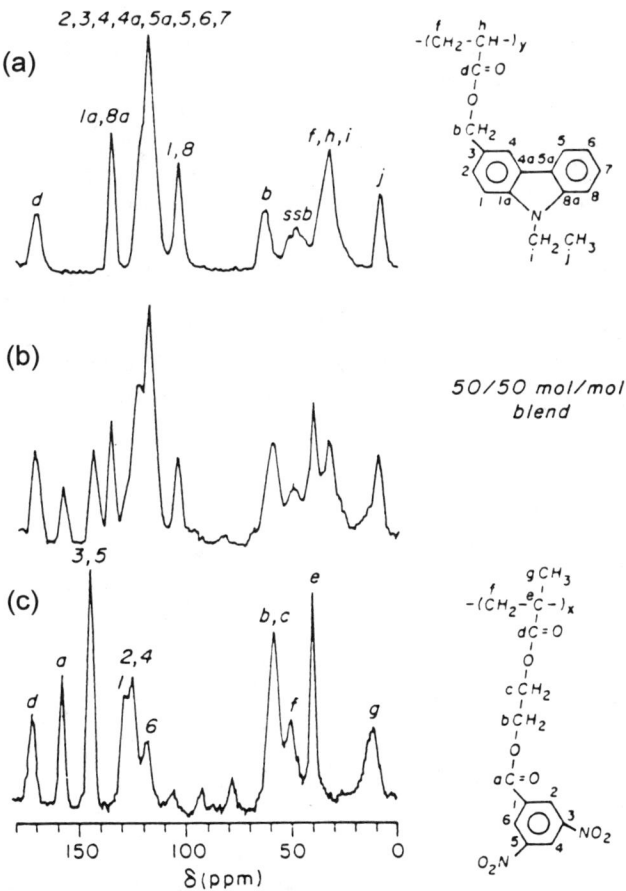

FIGURE 4.52 The carbon NMR spectra of (a) pNECMA, (b) the 50/50 blend of pNECMA and pDNBEM, and (c) pDNBEM. Adapted with permission (68).

carbons 1, 3, and 5, of the poly[2-[(3,5-dinitrobenzoyl)oxy] ethyl methacrylate] moiety.

Ionic interactions are also among those interactions strong enough to influence the chemical shifts of polymers in blends. This behavior is illustrated in Fig. 4.53, which compares the carbonyl region of the carbon CPMAS spectra of a zinc-neutralized copolymer of ethylene and methacrylic acid containing 15 mol% methacrylic acid monomers, 60% of which are neutralized, and the blends formed with poly(2-vinyl pyridine) and poly(4-vinyl pyridine) (69). Two peaks are observed in the spectra of the copolymer that are assigned to the zinc-neutralized and acidic carbonyl carbons at 189 and 185 ppm. Complex formation with poly(4-vinyl pyridine) leads to the appearance of a new peak at ca. 179.5 ppm, while such a peak is not observed with poly(2-vinyl pyridine). The shifted peak in the blend with poly(4-vinyl pyridine) is assigned to the carbonyl in the zinc-neutralized methacrylic acid that is complexed with the 4-vinyl pyridine since this peak is missing in the spectra of the acidic complex that has not been neutralized with zinc.

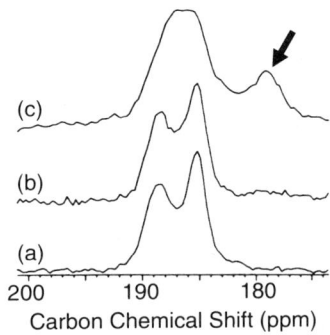

FIGURE 4.53 The carbonyl region of the CPMAS spectrum of (a) the zinc-neutralized ethylene methacrylic acid copolymer and the complexes with (b) poly(2-vinyl pyridine) and (c) poly(4-vinyl pyridine). The arrow in (c) shows the appearance of the new resonance. Adapted with permission (69).

4.3. The Organization of Polymers in the Solid State

Hydrogen bonding is another strong interaction that can affect the chemical shifts of the carbons on the donor and acceptor groups. However, in many cases the resonances are not well resolved, so the carbonyl signals are shifted and broadened. This effect is illustrated in Fig. 4.54, which compares the carbonyl region of the CPMAS spectra of a blend of a polybenzimidazole and a polyimide and a mechanical mixture (70). The blend is miscible as a consequence of hydrogen bonding, and the carbonyl resonance appears broadened and shifted.

Polybenzimidazole

XU 218

The above experiment shows some of the cases in which blend formation driven by strong intermolecular interaction leads to an easily observable change in the solid-state carbon NMR spectra. It is much more common, however, that blend formation is driven by weak intermolecular interactions and that the chemical shift changes induced by blend formation are much smaller than the solid-state linewidths, and are therefore undetectable. In such cases the length scale of chain mixing is measured by spin diffusion. The experiments for measuring spin diffusion include those discussed in previous sections and several other experiments that take advantage of the fact that the length scale of mixing is much smaller in polymer blends than in semicrystalline polymers and multiphase polymer systems.

Intermolecular cross polarization is perhaps the most straightforward method for determining molecular-level mixing in polymer blends. In such an experiment one of the polymers is deuterated and mixed with a protonated polymer (71). The idea is that the deuterated polymer has no directly bonded protons and cannot be cross polarized, while if the blends are mixed on a length scale of less than 20 Å, then

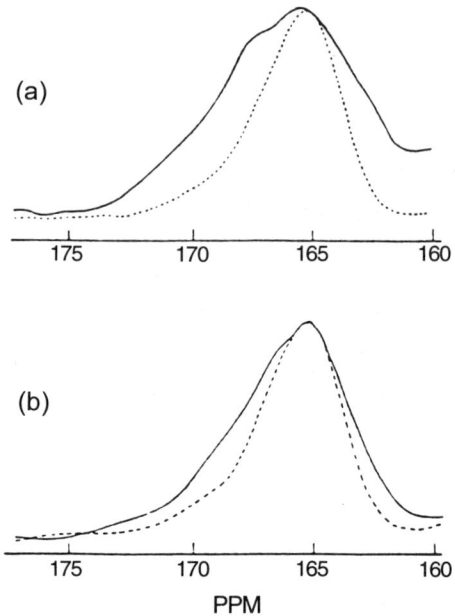

FIGURE 4.54 The carbonyl region of the CPMAS spectra of the (a) 75/25 wt% and (b) 50/50 wt% polybenzimidazole/XU 218 blends and mixtures. The blends are shown by the solid line, and the mechanical mixture of the two polymers by the dotted line. Adapted with permission (70).

the deuterated polymer can be cross polarized from the protons on the other polymer chain. This effect is illustrated in Fig. 4.55, which compares the cross-polarization behavior of the blend of deuterated polystyrene and protonated poly(methyl vinyl ether) with that of the mechanical mixture (72). The polystyrene aromatic signals at 125 ppm are only weakly cross polarized with a 2-ms cross polarization time in the mechanical mixture while they are strongly cross polarized in the blend. The introduction of the deuterated polymer is also a useful way to simplify the spectra of the blend when there are many overlapping signals. Since the deuterated polymer has no directly attached protons it can be observed with the nonprotonated carbons using the dipolar dephasing experiment (73).

The length scale of polymer mixing is most commonly measured via the T_1 and $T_{1\rho}$ relaxation rate measurements (74). If the chains are closer than the length scale of spin diffusion, then magnetization transfer from the slower to the more rapidly relaxing chains is a very effective relaxation mechanism. In such a case the relaxation times of

4.3. The Organization of Polymers in the Solid State

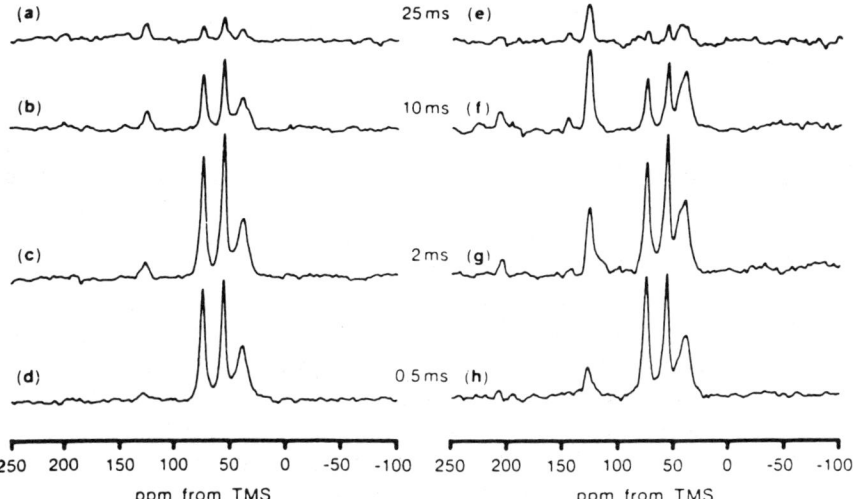

FIGURE 4.55 The carbon CPMAS spectra of the mechanical mixture and the blend of polystyrene-d_8 and poly(methyl vinyl ether). Spectra (a)–(d) show the effect of contact time on the cross-polarized intensity of the mechanical mixture. Spectra (e)–(h) show the same experiment for the blend. Adapted with permission (72).

both chains are equal to a weighted average of the values for the individual chains and are given by

$$k = k_a \frac{N_a \varphi_a}{N_a + N_b} + k_b \frac{N_b \varphi_b}{N_a + N_b}, \quad (4.10)$$

where k, k_a, and k_b are the relaxation rates (either T_1 or $T_{1\rho}$) for the blend and the pure polymers a and b; N_a and N_b are the total number of protons contained in the polymers a and b; and φ_a and φ_b are the mole fractions. In those cases where the length scale of mixing is longer than the spin diffusion length scale, the measured relaxation rates will be the same as those for the pure polymers. In cases where there is partial mixing the relaxation behavior may be more complex, and multiexponential relaxation may be observed (71). One limitation of this approach to measuring polymer mixing is that the relaxation rates for the pure polymers must differ by more than a factor of two to accurately measure the averaging of the relaxation rates.

The length scale of spin diffusion L measured by such experiments depends on the relaxation time and the diffusion coefficient D discussed in Section 4.3, and is given by

$$L = \sqrt{\frac{6D}{k}}, \qquad (4.11)$$

where, as above, k is the relaxation rate (74). Assuming a diffusion coefficient of ca. 10^{-12} cm^2 s^{-1}, the length scale of spin diffusion for a polymer with a proton spin–lattice relaxation time of 0.5 s corresponds to ca. 170 Å. Rotating frame relaxation rates are typically much faster and are therefore affected by a shorter spin diffusion length scale. For a proton $T_{1\rho}$ value of 5 ms, the length scale of chain mixing that can be detected by such measurements is on the order of 17 Å. Exact measurement of the length scale of phase separation is somewhat uncertain because of the uncertainties in estimating D.

Figure 4.56 shows the effect of blending on the proton $T_{1\rho}$ relaxation times for poly(dimethyl phenylene oxide) and poly(4-methyl styrene) (75). These polymers are ideal candidates for the study of polymer mixing because the relaxation times for the pure polymers differ from each other by an order of magnitude, so changes in the relaxation times due to the effects of blend formation can be easily

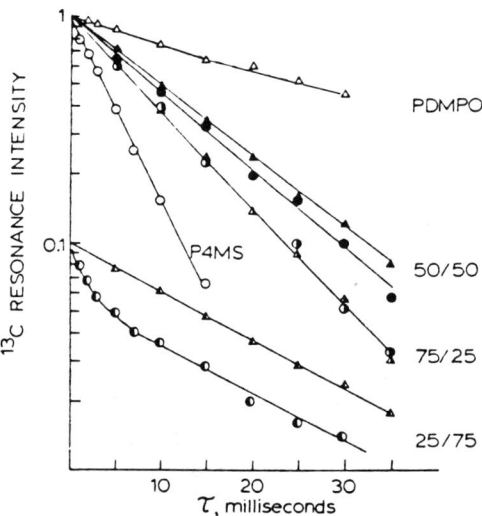

FIGURE 4.56 The proton rotating frame relaxation data for poly(2,6-dimethyl phenylene oxide), poly(4-methyl styrene), and the 50/50, 75/25, and 25/75 blends. The data for the 25/75 blend are offset from the other data to show the nonexponential behavior. The triangles represent measurements for carbons on poly(2,6-dimethyl phenylene oxide) and the circles for poly(4-methyl styrene). Adapted with permission (75).

visualized. These data show that the relaxation rates measured for the 75:25 mixture are identical for both chains, indicating homogeneous mixing. In the samples containing a 50:50 mixture, single-exponential relaxation rates are again observed, but the relaxation rates for poly(4-methyl styrene) and poly(dimethyl phenylene oxide) are no longer identical. At a ratio of 27:75, multiexponential relaxation behavior is observed for poly(dimethyl phenylene oxide). These data show that the mixtures of poly(4-methyl styrene) and poly(dimethyl phenylene oxide) are not homogeneous on the length scale measured by $T_{1\rho}$ (30 Å) for the samples prepared with a 50:50 or 25:75 ratio. This approach was initially used to study blends of polystyrene and poly(dimethyl phenylene oxide), where it was reported that these polymers do not form homogeneous mixtures (73).

The miscibility of poly(styrene-co-acrylonitrile) and poly(methyl methacrylate) blends has been studied using this method as a function of composition, and Fig. 4.57 shows a plot of the proton spin–lattice relaxation rate as a function of poly(methyl methacrylate) composition (74). The plot is linear, demonstrating that the chains are mixed on the length scale of less than 170 Å. Figure 4.58 shows a plot of the rotating frame relaxation rate for the same materials. In this case there appears to be a systematic deviation from the linear relationship between the relaxation rate and the composition, suggesting that these polymers are not miscible on the length scale of spin diffusion measured by the $T_{1\rho}$ (17 Å). Further evidence for inhomogeneous mixing on this length scale was obtained from the nonexponential relaxation of the sample with 78 mol% poly(methacrylate), where nonexponential relaxation was observed with a lower powered spin-locking field.

The miscibility of polystyrene, and poly(methyl vinyl ether) has been extensively investigated by NMR methods and illustrates the wide variety of methods that can be used to study polymer blends. It was noted in the early studies that the solvent has a large affect on the appearance of the films, where optically clear films are obtained for the films cast from toluene, while cloudy films are obtained from chloroform solutions (76). These and many following studies have shown the polymers are phase separated in the films cast from chloroform.

We have already discussed two methods that have been used to study the intermolecular interactions in polystyrene and poly(methyl vinyl ether). In Section 3.4.2 solution 2D NOESY experiments were used to show intermolecular interactions between the phenyl groups of polystyrene and the methoxy group of poly(methyl vinyl ether) that are believed to be the driving force for miscibility (77). In this section intermolecular cross polarization (Fig. 4.55) between deuterated

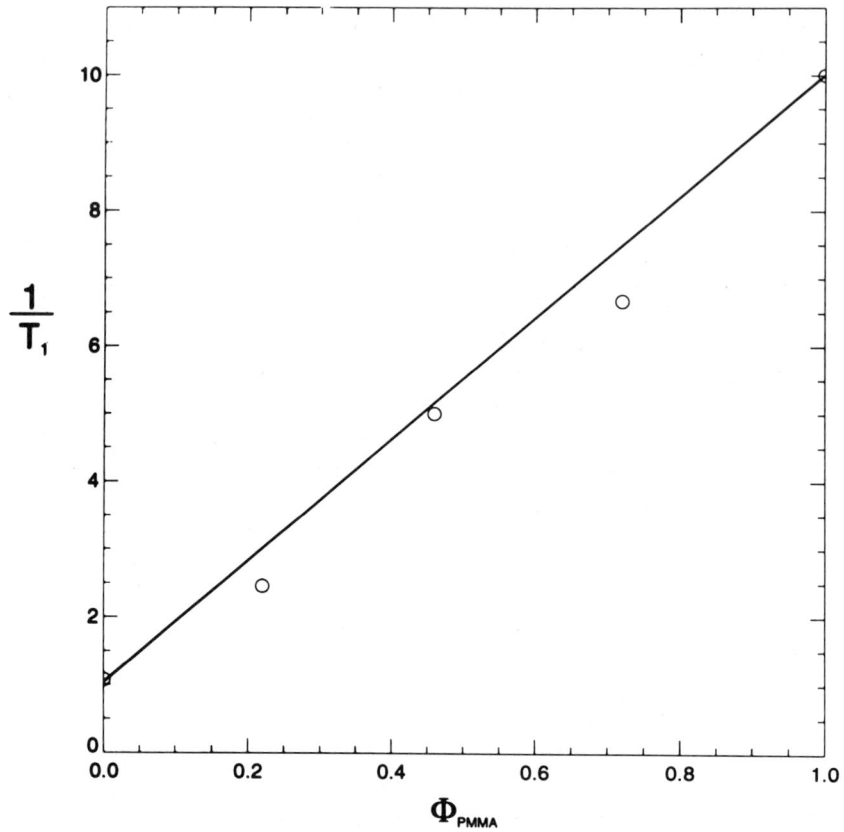

FIGURE 4.57 A plot of the inverse proton spin–lattice relaxation time as a function of composition for blends of poly(styrene-co-acrylonitrile) and poly(methyl methacrylate). Adapted with permission (74).

polystyrene and protonated poly(methyl vinyl ether) was used to show that the chains are in close contact. Other solid-state methods have been used not only to show that the chains are in close contact, but also to examine the limits of the miscibility.

In the early studies it was demonstrated that the blend cast from toluene had a single proton spin–lattice relaxation time for all compositions, showing that the chains are mixed on the length scale probed by such measurements (170 Å) (76). However, if the samples were heated above 140°C and quenched in ice water, two phases were observed by the T_1 measurements. Since the relaxation rates for the two phases were not the same as those for the pure polymers, these data show that the phase separation is incomplete. A nonlinear relationship between the $T_{1\rho}$ and the blend composition has been

4.3. The Organization of Polymers in the Solid State

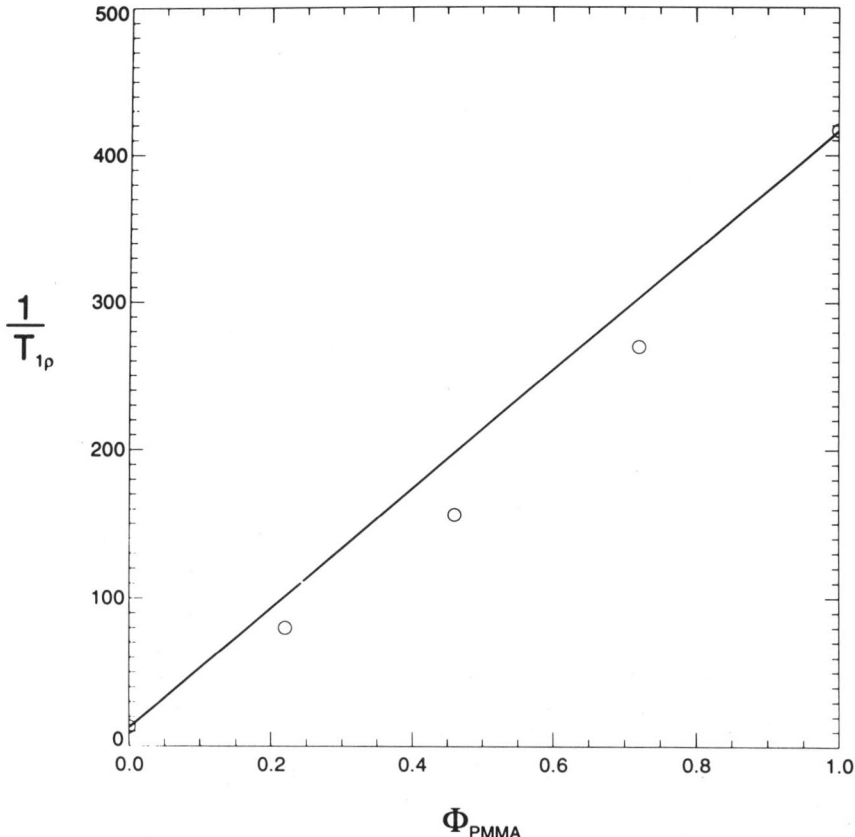

FIGURE 4.58 A plot of the inverse proton rotating frame spin–lattice relaxation time as a function of composition for blends of poly(styrene-co-acrylonitrile) and poly(methyl methacrylate). Adapted with permission (74).

reported for the films cast from toluene, suggesting that the blends are not completely mixed on the length scale probed by the $T_{1\rho}$ measurements (17 Å) (78).

The solid-state equivalent of the proton 2D NOESY pulse sequence has been used to study the mixing of the polystyrene and poly(methyl vinyl ether) blends from toluene and chloroform. The broad lines in the proton spectra that result from the dipolar interactions require the use of both magic angle spinning and multipulse decoupling during both the t_1 and the t_2 periods, and the NOESY pulse sequence as modified for use in solids is shown in Fig. 4.59 (50). The resulting 1D spectra and the 2D correlation maps for the blends are shown in

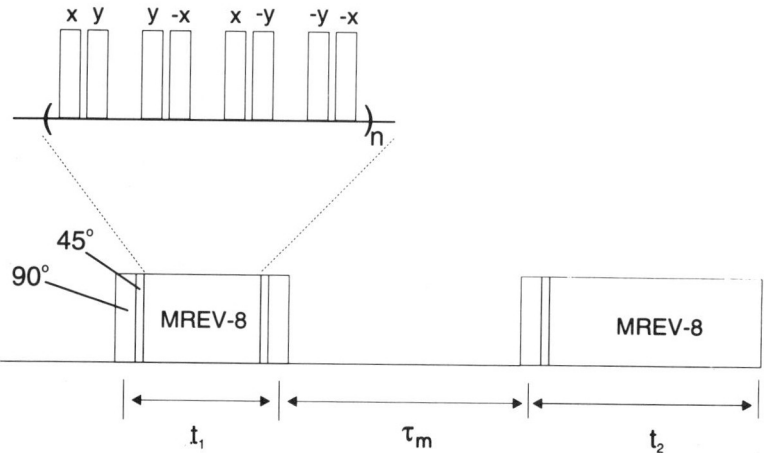

FIGURE 4.59 A pulse sequence diagram for the solid-state implementation of the 2D NOESY pulse sequence. Proton multipulse decoupling is applied in both the t_1 and the t_2 dimensions (50).

Fig. 4.60 (50). Three groups of signals can be observed in the high-resolution proton NMR spectra that are assigned to the polystyrene aromatic protons at 7 ppm, the poly(methyl vinyl ether) methine and methoxy protons at 3.5 ppm, and a band containing the remaining aliphatic protons at higher field. Two sets of cross peaks are observed in the 2D spectra acquired for the chloroform blend that can be assigned to the intramolecular polystyrene aromatic to aliphatic interactions and intramolecular methine/methoxy to aliphatic interactions in poly(methyl vinyl ether). In the toluene blends, additional cross peaks are observed between the polystyrene aromatic protons and the poly(methyl vinyl ether) methine/methoxy peak. Since the only mechanism for magnetization transfer in these experiments is through-space dipolar interactions, these spectra demonstrate that a large fraction of the chains are in molecular contact (50).

The rate of intermolecular spin diffusion and the fraction of polymer in the mixed phase for the polystyrene and poly(methyl vinyl ether) blends can in principle be calculated from the intensities of the diagonal and cross peaks as a function of mixing time (50). A more practical way to make such measurements, however, is to use selective excitation followed by detection of the proton signals using CRAMPS (47). The pulse sequences for making such measurements can be either of the selective inversion or of the saturation-transfer type, as shown in Fig. 4.61. In a selective inversion experiment a group of protons, such as the aromatic protons of polystyrene, are

4.3. The Organization of Polymers in the Solid State

inverted and the signals are measured after a spin diffusion period. In a saturation-transfer experiment the selective pulses are applied several times to drive the spin system toward equilibrium.

Figure 4.62 shows how the selective inversion pulse sequence can be used to measure the rate of intermolecular spin diffusion for the polystyrene and poly(methyl vinyl ether) blends from chloroform and toluene (47). In this experiment the polystyrene aromatic signals were inverted and the intensity of the poly(methyl vinyl ether) methine/methoxy peak was measured as a function of the mixing time. Biexponential behavior is observed for the blend cast from toluene, in which the signal initially rises due to magnetization transfer from the aromatic protons of polystyrene followed by decay due to spin–lattice relaxation. In the chloroform blend no such initial rise in magnetization is observed, as expected for phase-separated mixtures. From the

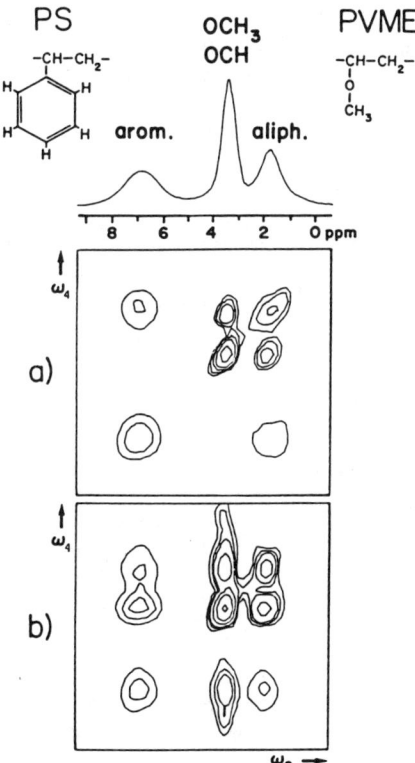

FIGURE 4.60 The solid-state proton 1D NMR spectra of a polystyrene/poly(methyl vinyl ether) blend and the 2D spin exchange spectra for the blends from (a) chloroform and (b) toluene. Adapted with permission (50).

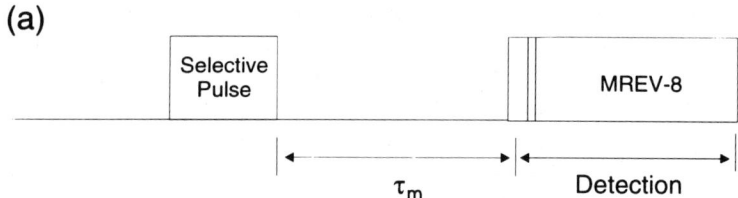

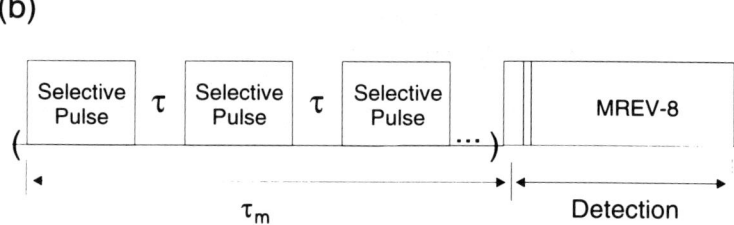

FIGURE 4.61 The pulse sequence diagrams for (a) selective inversion and (b) saturation-transfer NMR experiments for protons in the solid state (47).

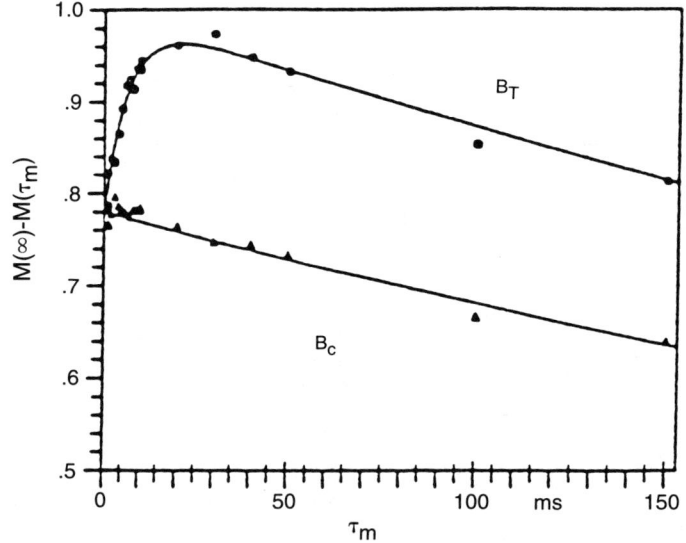

FIGURE 4.62 The intensity of the methine/methoxy peak in blends of polystyrene/poly(methyl vinyl ether) cast from (a) toluene and (b) chloroform following selective inversion of the polystyrene aromatic signals. Adapted with permission (47).

4.3. The Organization of Polymers in the Solid State

fits to Fig. 4.62 and similar experiments with the pure polymers it is reported that the rates for intramolecular spin diffusion for polystyrene and poly(methyl vinyl ether) are $7.0 \pm 0.8 \times 10^3$ and 38 ± 6 s^{-1}, and the rate of intermolecular spin diffusion in the toluene blend is 154 ± 17 s^{-1}. The intermolecular spin diffusion is an order of magnitude slower than that for pure polystyrene and four times faster than that for pure poly(methyl vinyl ether), suggesting that the mobility of the poly(methyl vinyl ether) is not substantially changed in the blend. The degree of chain mixing in the polystyrene/poly(methyl vinyl ether) blends from toluene was investigated in this same study using the saturation-transfer experiment [Fig. 4.61(b)] (47). After correction for experimental imperfections, such as the partial saturation of the poly(methyl vinyl ether) methine/methoxy peak by the selective pulse applied to the polystyrene aromatic protons, the data showed that $79 \pm 0.3\%$ of the total polymer mass was contained in the mixed phase.

The domain size in the blends cast from toluene has also been studied by 2D WISE spectroscopy, as shown in Fig. 4.63 (60). The spectrum acquired with the shortest mixing time shows that even in the mixed phase there is a substantial difference in the mobility of the two components, in agreement with the spin diffusion results presented above (47). These data show that at 320 K a spin diffusion time of ca. 5 ms is required to equilibrate the proton lineshapes for the polystyrene and poly(methyl vinyl ether) chains. It is estimated that the average length scale for spin diffusion $\langle |x| \rangle$ is given by

$$\langle |x| \rangle = \sqrt{\frac{D_{\text{eff}} t_m \pi}{4}}, \qquad (4.12)$$

where D_{eff} is the effective spin diffusion coefficient for the blend that is given by

$$\sqrt{D_{\text{eff}}} = \frac{\sqrt{D_{\text{PS}} D_{\text{PVME}}}}{\sqrt{D_{\text{PS}}/2} + \sqrt{D_{\text{PVME}}/2}} \qquad (4.13)$$

and t_m is the mixing time. Using values of $0.5 \pm 0.2 \times 10^{-12}$ cm^2/s and $0.1 \pm 0.05 \times 10^{-12}$ cm^2/s for the spin diffusion coefficients for polystyrene and poly(methyl vinyl ether), it is estimated the the blend must be heterogeneous on the length scale of 3.5 ± 1.5 nm at 320 K.

Solid-state NMR has also been used to examine the intermolecular interaction in the polystyrene and poly(methyl vinyl ether) blends using the solid-state NOEs (79). The pulse sequence is similar to that used to measure carbon–proton NOEs in solution and consists of a series of pulses that saturate the protons followed by a carbon ob-

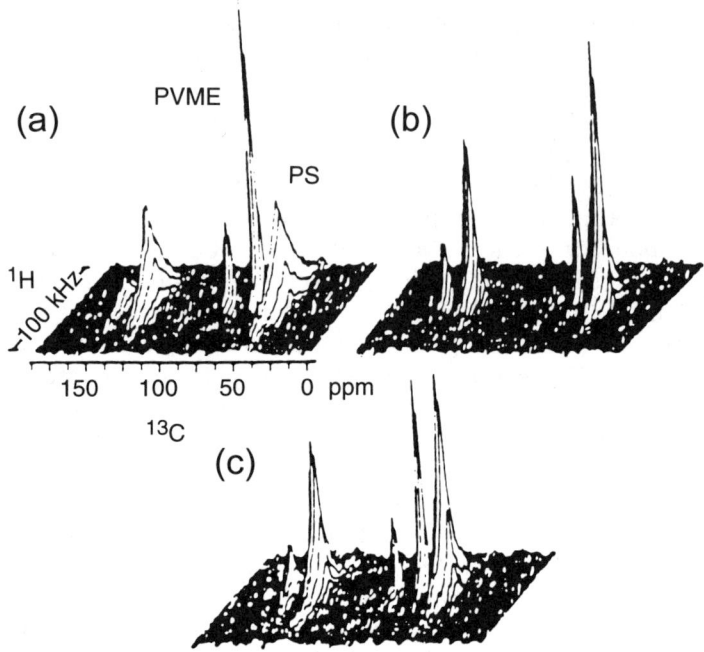

FIGURE 4.63 The 2D WISE spectra of 50:50 blends of polystyrene and poly(methyl vinyl ether) with spin diffusion mixing times of (a) 0.2 ms, (b) 5 ms, and (c) 1 ms. Adapted with permission (60).

serve pulse as shown in Fig. 4.64. Due to the constraints imposed by the polymer lattice, there is typically not enough molecular motion in solids for the observation of NOEs. One exception to this rule is the rotating methyl groups, which are often rapidly rotating even in polymers at temperatures well below the glass transition tempera-

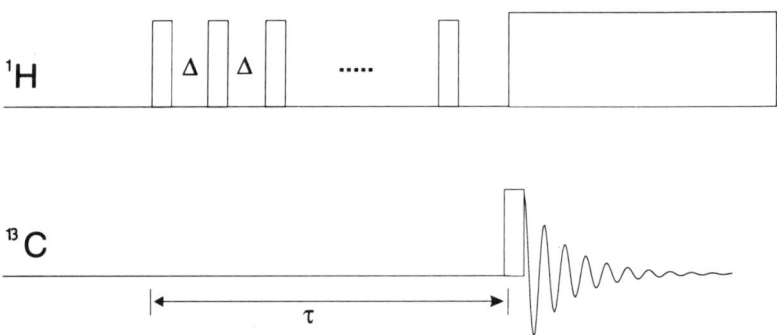

FIGURE 4.64 The pulse sequence diagram for measuring solid-state carbon nuclear Overhauser effects following saturation by a train of proton pulses.

4.3. *The Organization of Polymers in the Solid State* 315

ture. Thus, NOEs can be observed to carbons that are in close proximity to these methyl groups.

Figures 4.65 and 4.66 show the NOE difference spectra for polystyrene/poly(methyl vinyl ether) blends cast from toluene and chloroform (79). The largest peaks in the difference spectra are those

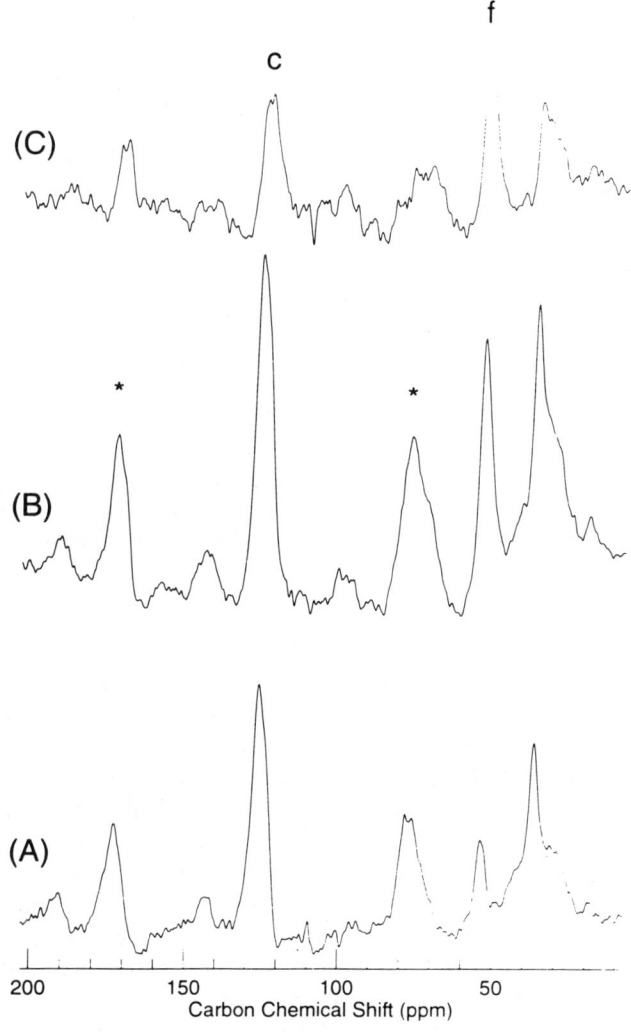

FIGURE 4.65 The solid-state NOE difference spectra for a polystyrene/poly(methyl vinyl ether) blend cast from toluene. The spectra show (a) the Bloch decay spectra, (b) the NOE enhanced, and (c) the difference spectra. Adapted with permission (79).

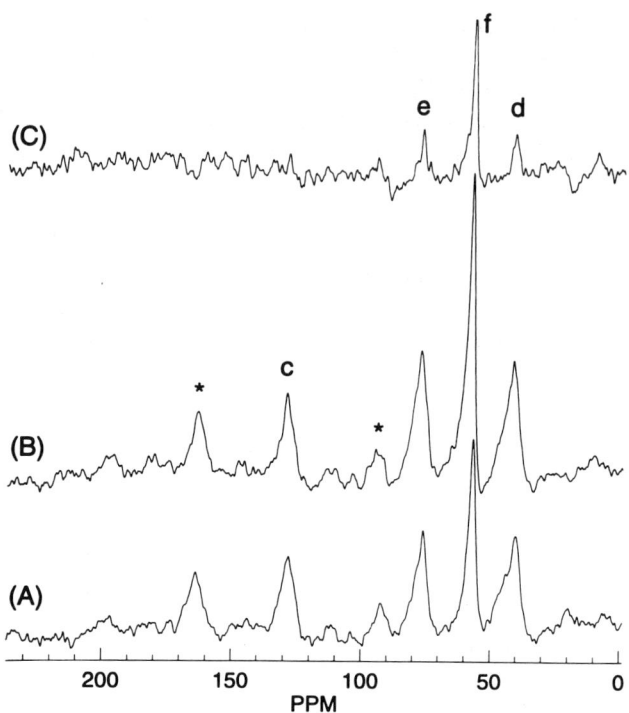

FIGURE 4.66 The solid-state NOE difference spectra for a polystyrene/poly(methyl vinyl ether) blend cast from chloroform. The spectra show (a) the Bloch decay spectra, and (b) the NOE enhanced and (c) the difference spectra. Adapted with permission (79).

from the methyl carbons, which experience the largest NOEs. The feature to note about the difference spectra for the toluene blend is that NOEs are observed not only for the poly(methyl vinyl ether) carbons, but also for the polystyrene aromatic carbons. This can arise only if these groups are in close contact (ca. 5 Å) and supports the proposal based on solution-state NOE studies of blends in concentrated solutions that miscibility is a consequence of the intermolecular interactions between the polystyrene aromatic group and the poly(methyl vinyl ether) methoxy group (77,80). No such intermolecular NOEs are observed in the blend cast from chloroform, which other studies have shown to be phase separated.

A method similar to the 2D CRAMPS experiment (Fig. 4.59) has been used to examine chain mixing in polybenzimididazole and polyetherimide blends (49). The pulse sequence is the same as that in the 2D experiment, but this experiment is performed in the 1D mode.

4.3. The Organization of Polymers in the Solid State

In such an experiment a *chemical-shift-based polarization gradient* is created by the choice of the spectrometer offset and by fixing the number of times the MREV-8 pulse sequence is applied during the preparation period of the experiment (81). Following the creation of the gradient, a period is allowed for spin diffusion followed by observation using the normal CRAMPS methods.

Polybenzimidazole

Poly(ether imide)

The chemical-shift-based polarization gradient can be used to study spin diffusion in a wide variety of polymers even though there is extensive overlap of the signals from the two polymers in the solid-state proton spectra. Figure 4.67 shows the solid-state proton spectra for the two component polymers and the blend (49). Since the polymers contain overlapping signals it is not possible to selectively excite the resonances from one chain. Instead this experiment takes advantage of the fact that the fraction of the signals in the aromatic and aliphatic regions differs for the two polymers. For the poly(ether imide) the fractions of signals in the aromatic and aliphatic regions are 0.75 and 0.25, while for the polybenzimidazole the fractions are 0.92 and 0.08. If the aliphatic protons are excited there will be rapid intramolecular spin diffusion followed by a slower intermolecular spin diffusion. Since the fraction of aromatic and aliphatic protons differs for the two polymers, the intermolecular spin diffusion can be directly measured.

Figure 4.68 compares the chemical-shift-based spin diffusion experiment for two blends of polybenzimidazole and polyetherimide that have been subjected to different thermal histories at long and short spin diffusion times (49). The differences between these spectra are due to differences in the degree of mixing, since heating to 400°C for an hour is known to cause phase separation. Following the initial polarization gradient, the return to equilibrium is very different for the two samples. The domain size can be calculated from the spin

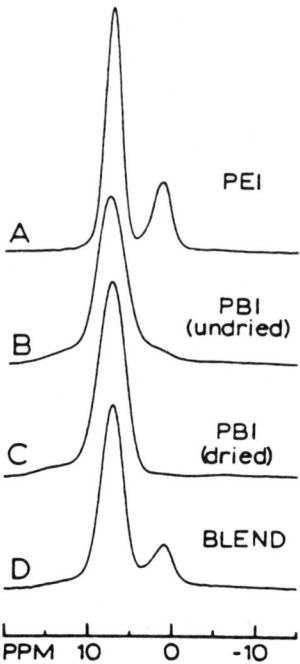

FIGURE 4.67 The multipulse proton NMR spectra of (a) poly(ether imide), (b) undried polybenzimidazole, (c) dried polybenzimidazole, and (d) the blend of poly(ether imide) and polybenzimidazole. Adapted with permission (81).

diffusion time constant that is obtained from a plot of the reduced intensity as a function of $\sqrt{t}$ as shown in Fig. 4.69 (47). Accounting for the density of proton in such a blend, these results reveal a domain size of 2.4 nm. Such an estimate is considered reasonable for such a blend because of the large size of the monomers. An additional advantage of this method is that the fraction of polymer in the mixed phase can also be estimated.

The miscibility of poly(vinylidine fluoride) and poly(methyl methacrylate) has been investigated using ^1H-^{19}F-^{13}C triple-resonance cross-polarization methods with magic angle spinning (82). These experiments take advantage of the high natural abundance and sensitivity of fluorine, and utilize ^{13}C-^{19}F cross polarization, double ^{19}F-^1H-^{13}C cross polarization, and ^1H-^{19}F-^{13}C cross depolarization to show that a fraction of the blend is mixed on the molecular level and that there are isolated domains of both poly(vinylidine fluoride) and poly(methyl methacrylate). The pulse sequences for the triple-resonance cross-polarization experiments are shown in Fig.

4.70. Figure 4.71 shows the ^{19}F–^{13}C cross-polarization spectrum and the ^{19}F–^{1}H–^{13}C double cross-polarization spectra for the 60/40 poly(methyl methacrylate)/poly(vinylidine fluoride) blend (82). The feature to note about these spectra is that in addition to the poly(vinylidine fluoride) signals at 50 and 125 ppm, signals from the poly(methyl methacrylate) are also observed, including the carbonyl peak at 176 ppm and the signals from the quaternary carbon and the methine and methylene carbons in the range of 45–55 ppm. Peaks that appear in the ^{19}F–^{13}C cross-polarization spectrum arise from those carbons in close proximity to the fluorines of poly(vinylidine fluoride), and can arise only from molecular-level mixing of the chains. The peaks that appear in the ^{19}F–^{1}H–^{13}C double cross-polarization spectrum arise from carbons that are near protons that can be cross polarized from the poly(vinylidine fluoride) fluorines and also show chain mixing. However, because of the short ^{19}F $T_{1\rho}$, it is not possible to determine the fraction of poly(methyl methacrylate) chains that are in close contact to poly(vinylidine fluoride) from such experiments.

A quantitative measure of the fraction of chains in the mixed phase can be determined from the ^{19}F–^{1}H–^{13}C cross-depolarization experiment shown in Fig. 4.70(b). In the case of poly(vinylidine fluoride) the

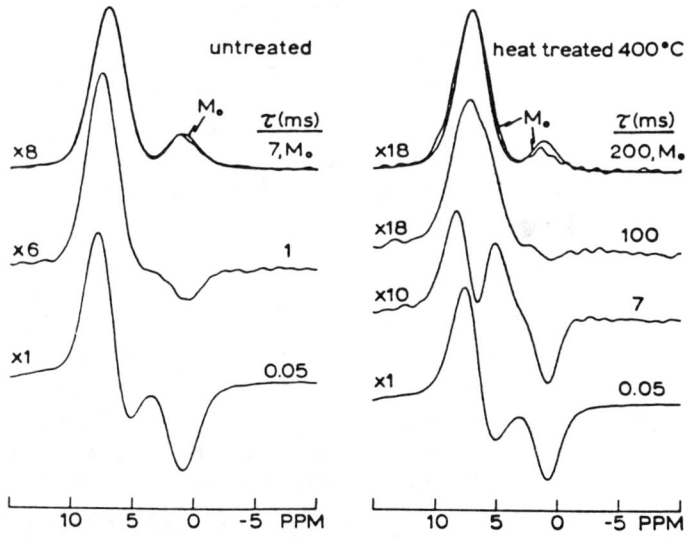

FIGURE 4.68 Spin diffusion in (a) untreated and (b) heat treated blends of polyetherimide and polybenzimidazole. The expansion factor for the plotting of each spectra is shown to the left. Adapted with permission (81).

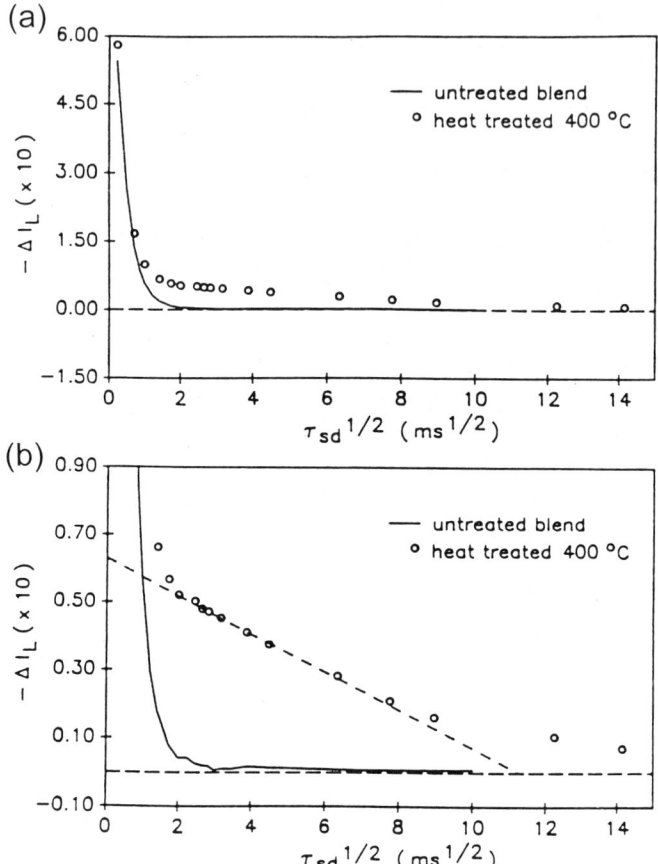

FIGURE 4.69 The scaled intensity data as a function of the square root of the spin diffusion time for the untreated (—) and heat treated (○) blends of polyetherimide and polybenzimidazole. Adapted with permission (81).

^{19}F $T_{1\rho}$ is extremely short (0.8 ms) so that the ^{19}F–^{1}H cross polarization leads to a loss of magnetization for those protons in close contact with the poly(vinylidine fluoride). Thus the magnetization obtained from the ^{1}H–^{13}C cross polarization can arise only from poly(methyl methacrylate) chains isolated from the fluorines of poly(vinylidine fluoride). This experiment can be used to study the domain size since proton spin diffusion occurs during the spin-locking period. Data from such an experiment are shown in Fig. 4.72 for poly(vinylidine fluoride) and the 60/40 poly(methyl methacrylate)/poly(vinylidine fluoride) blend (82). These data are plotted as the ratio of the intensities without and with ^{1}H–F^{19} cross polarization to account for the loss of

4.3. The Organization of Polymers in the Solid State

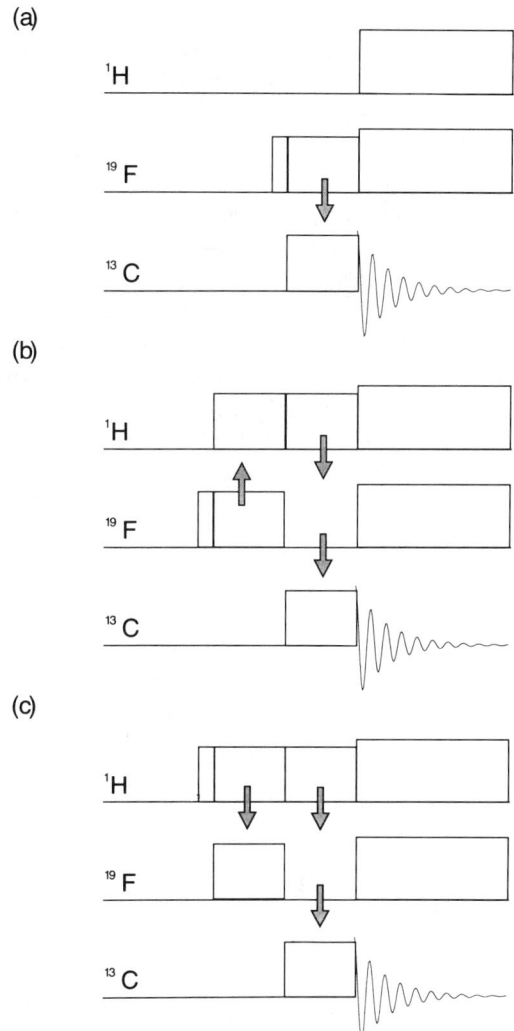

FIGURE 4.70 Pulse sequence diagrams for (a) ^{19}F–^{13}C cross polarization, (b) ^{1}H–^{19}F–^{13}C double cross polarization, and (c) ^{1}H–^{19}F–^{13}C cross depolarization (82).

signals due to the proton $T_{1\rho}$ relaxation. In poly(vinylidine fluoride) the signals rapidly decay to zero since all of the protons are in close proximity to fluorines while the signals for the blend decay more slowly. These data are fit with a model that has isolated poly(vinylidine fluoride), a mixed phase, poly(methyl methacrylate) near the mixed phase, and poly(methyl methacrylate). The results show that the domain size for the mixed region is on the order of 6 Å, and the size of the domain made up of the mixed phase and the poly(methyl methacrylate) near the mixed phase is on the order of 12–17 Å. Furthermore, the fractions of unmixed poly(methyl methacrylate) and poly(vinylidine fluoride) are about 30 and 15% for the 60/40 poly(methyl methacrylate)/poly(vinylidine fluoride) blend.

The miscibility of polymer blends has been more recently investigated using the 2D heteronuclear correlation experiment shown in Fig. 4.73 (83,84). The original purpose of this experiment was to correlate the carbons and protons that have strong dipolar couplings. During the t_1 period the multipulse decoupling is applied to the protons and carbons with the BLEW-24 and BB-24 pulse sequence to remove both the proton–proton and the carbon–proton dipolar couplings, allowing the carbon chemical shifts to be correlated with the high-resolution proton spectra (85). Magnetization is transferred from protons to carbons with the WIM-24 pulse sequences, which quenches proton–proton spin diffusion during the cross polarization (83). This

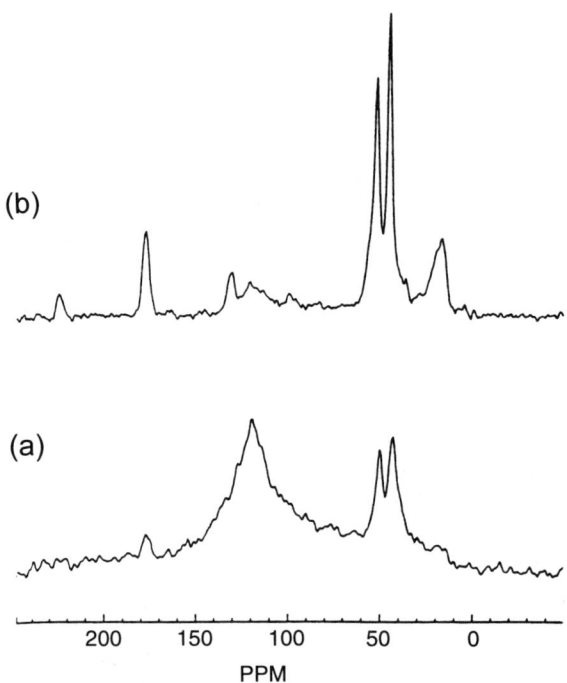

FIGURE 4.71 The (a) $^{19}F-^{13}C$ and (b) $^{19}F-^{1}H-^{13}C$ double cross-polarization spectra of the 60/40 poly(methyl methacrylate)/poly(vinylidine fluoride) blend obtained with a contact time of 2 ms. The $^{1}H-^{19}F$ contact time was 0.4 ms. Adapted with permission (82).

4.3. The Organization of Polymers in the Solid State   323

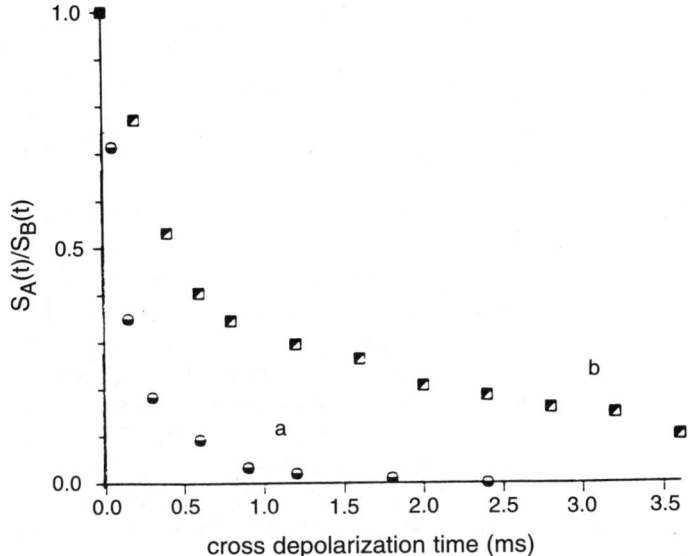

FIGURE 4.72 Plots of signal intensity for the ^{1}H–^{19}F–^{13}C cross-depolarization experiment for (a) poly(vinylidine fluoride) (circles) and (b) the 60/40 poly(methyl methacrylate)/poly(vinylidine fluoride) blend (squares). The data are a ratio of the intensities without and with the ^{1}H–^{19}F cross polarization to account for $T_{1\rho}$ relaxation. Adapted with permission (82).

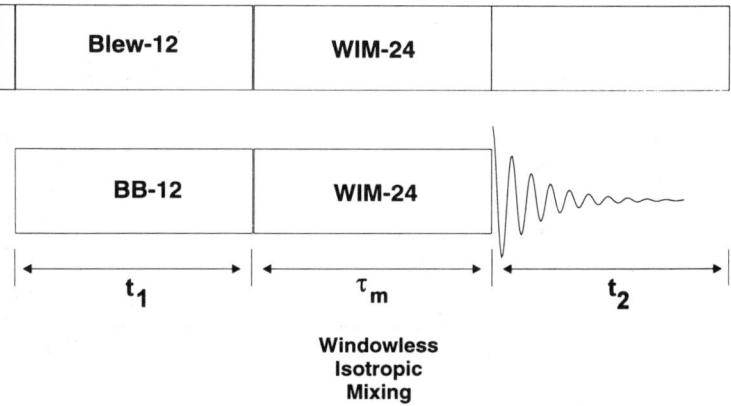

FIGURE 4.73 The pulse sequence diagram for 2D heteronuclear correlation in the solid state. Multiphase decoupling is applied to both the carbons and the protons during the t_1 period, and selective cross polarization without spin diffusion occurs during the 24-pulse WIM mixing period (82).

is in contrast to the normal cross polarization by spin locking, where proton spin diffusion proceeds at half the rate observed in the absence of the RF field. If a spin diffusion delay is inserted between the evolution and the mixing periods, correlations are observed not only between nearby protons and carbons, but also between carbon and more distant protons via the spin diffusion.

This effect is illustrated in the heteronuclear correlation spectra of polycarbonate and the aromatic diamine (TPD) that is used as a charge transport layer in photoconductors (86). Figure 4.74 shows the heteronuclear correlation spectra of the blend acquired with no spin diffusion delay period, where the peaks show the direct correlation between the carbons and the protons. The peaks of greatest interest are in the methyl region where well-resolved correlations are observed for the methyl groups of polycarbonate and TPD. Figure 4.75 shows this section of the 2D spectra as the spin diffusion mixing time is increased from 20 μs to 2 ms. The horizontal lines show the chemical shifts of the polycarbonate and TPD methyl groups, which are partially resolved along the proton dimension. For long spin diffusion times both intra- and intermolecular spin diffusion occurs

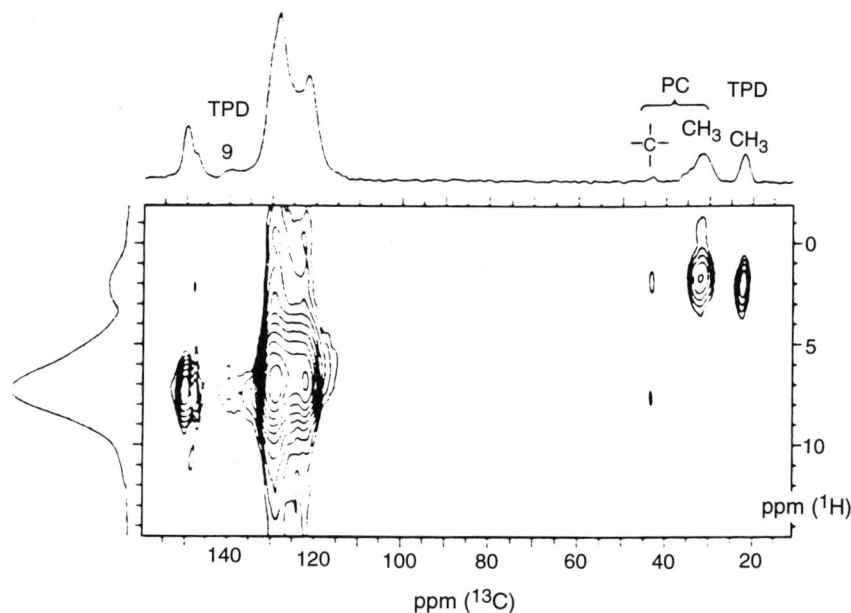

FIGURE 4.74 The solid-state 2D heteronuclear correlation spectra of the 50/50 mixture of polycarbonate and TPD without a spin diffusion mixing period. The proton spectrum appears along the left axis and the carbon spectrum on the top axis. Adapted with permission (86).

and magnetization is transferred from the methyl protons to the aromatic protons and the chemical shift difference between the methyl groups disappears as the spin system approached equilibrium. An analysis of the peak intensities as a function of mixing time gives a spin diffusion time constant of ~ 1 ms, from which it is calculated that the polymers must be mixed on a length scale of ~ 1.6 nm (86). Since this dimension is approximately the size of the monomers, this mixture is miscible at the molecular level.

TPD

This same method has been used to study short-range order in blends of ring deuterated polystyrene and poly(2,6-dimethyl phenylene oxide) (85), as illustrated in the heteronuclear correlation spectra with short and long spin diffusion periods shown in Fig. 4.76. The important feature to note about the spectrum obtained with a short spin diffusion delay [Fig. 4.76(a)] is that there is no correlation between the polystyrene aliphatic carbon resonances (peak 3, 35–50 ppm) and the protons in the aromatic region because the ring protons of polystyrene have been replaced with deuterons. When such a correlation is noted in the spectrum with a long spin diffusion delay it must arise from intermolecular spin diffusion. The results from this study show a complete equilibration of the magnetization within 10 ms, demonstrating that the chains must be mixed on a length scale of less than 2.1 nm. These data were compared to lattice model calculations that suggested that while the components are mixed on the molecular length scale the polystyrene segments can form clusters with a radii of 0.8–1.5 nm (87). These results can be compared with an earlier study of the miscibility by $T_{1\rho}$ measurements where it was shown that the blend is not completely miscible over a spin diffusion time period of 25 ms (71).

In certain favorable circumstances it is possible to use the heteronuclear correlations to measure the geometry of the intermolecular contacts in miscible blends. Such an example is illustrated in the heteronuclear correlation spectra of the blend of poly(vinyl phenol)

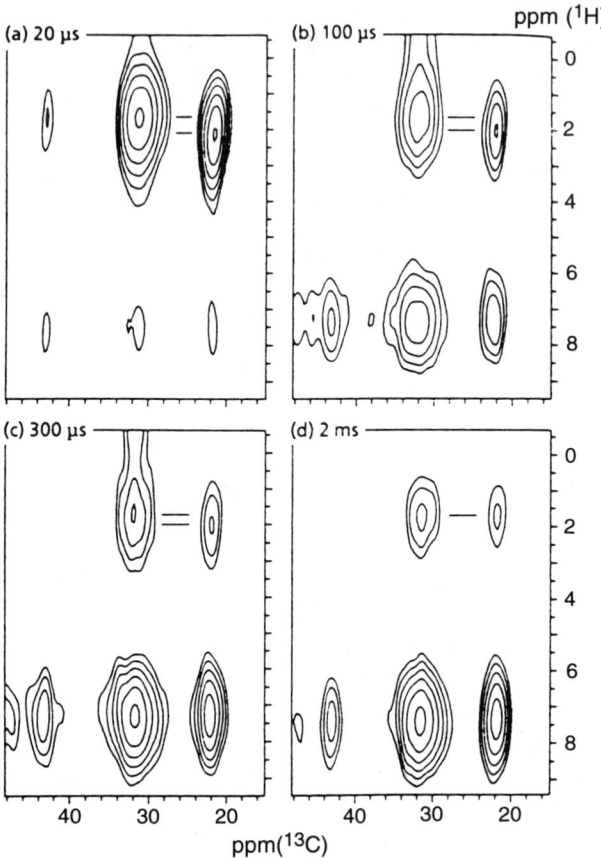

FIGURE 4.75 The aliphatic region of the heteronuclear correlation spectra of polycarbonate and TPD with a spin diffusion mixing time of (a) 20 μs, (b) 100 μs, (c) 300 μs, and (d) 2 ms. The horizontal lines between the methyl cross peaks show how the chemical shifts converge at longer mixing times due to spin diffusion. Adapted with permission (86).

and poly(methyl acrylate) shown in Fig. 4.77 (88). The strongest peaks in the spectrum arise from the correlation of the carbons with their directly bonded protons, while weaker correlations are noted between the nonprotonated carbons and the nearby protons. Figure 4.77 shows a weak correlation between the carbonyl carbon of poly(methyl acrylate) and protons resonating at ca. 5 ppm. Since this correlation is not observed in a spectrum with a poly(vinyl phenol) sample in which the hydroxyl protons have been exchanged for deuterons, this correlation must arise from the close contact between

4.3. The Organization of Polymers in the Solid State   327

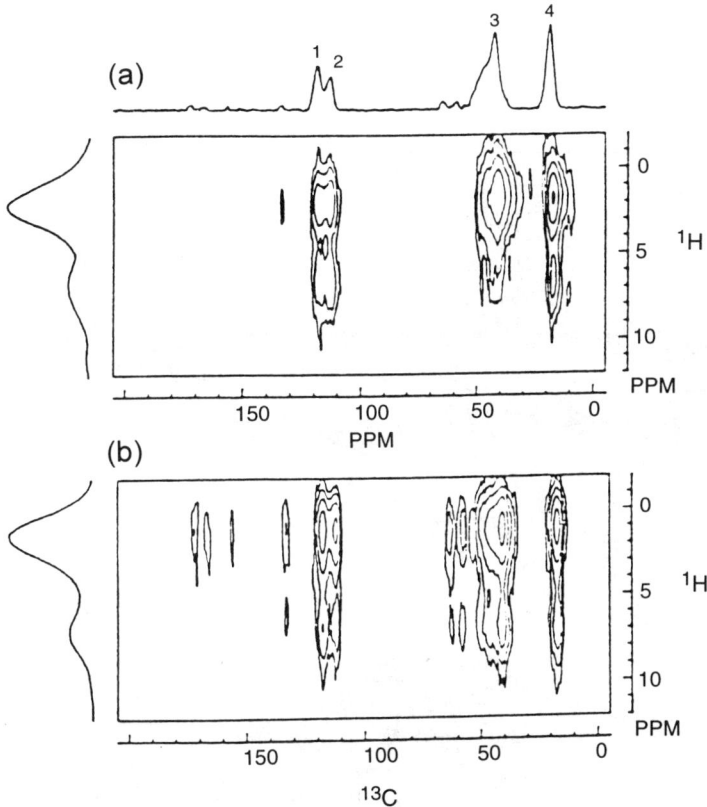

FIGURE 4.76 The heteronuclear correlation spectra of the 50/50 wt% blend of poly(2,6-dimethyl phenylene oxide) and polystyrene-d_5 with spin diffusion mixing times of (a) 0.3 ms and (b) 10 ms. Adapted with permission (87).

the carbonyl of poly(methyl acrylate) and the hydroxyl protons of poly(vinyl phenol) due to intermolecular hydrogen bonding.

This method can be used to measure not only strong interactions such as hydrogen bonding, but also the weaker interactions that determine miscibility for a wide variety of polymers. Figure 4.78 shows the heteronuclear correlation spectrum for the blend of polycarbonate and polycaprolactone (89). This blend is known to exhibit complex phase behavior with the coexistence of two crystalline phases and a mixed amorphous phase (90), and it has been suggested that miscibility arises as a consequence of n–π complex formation between the polycaprolactone carbonyl and the aromatic rings of polycarbonate (91). The carbonyl resonance of polycaprolactone shows three correlations in the heteronuclear correlation spectrum shown in Fig. 4.78, including intramolecular correlations to the nearby methy-

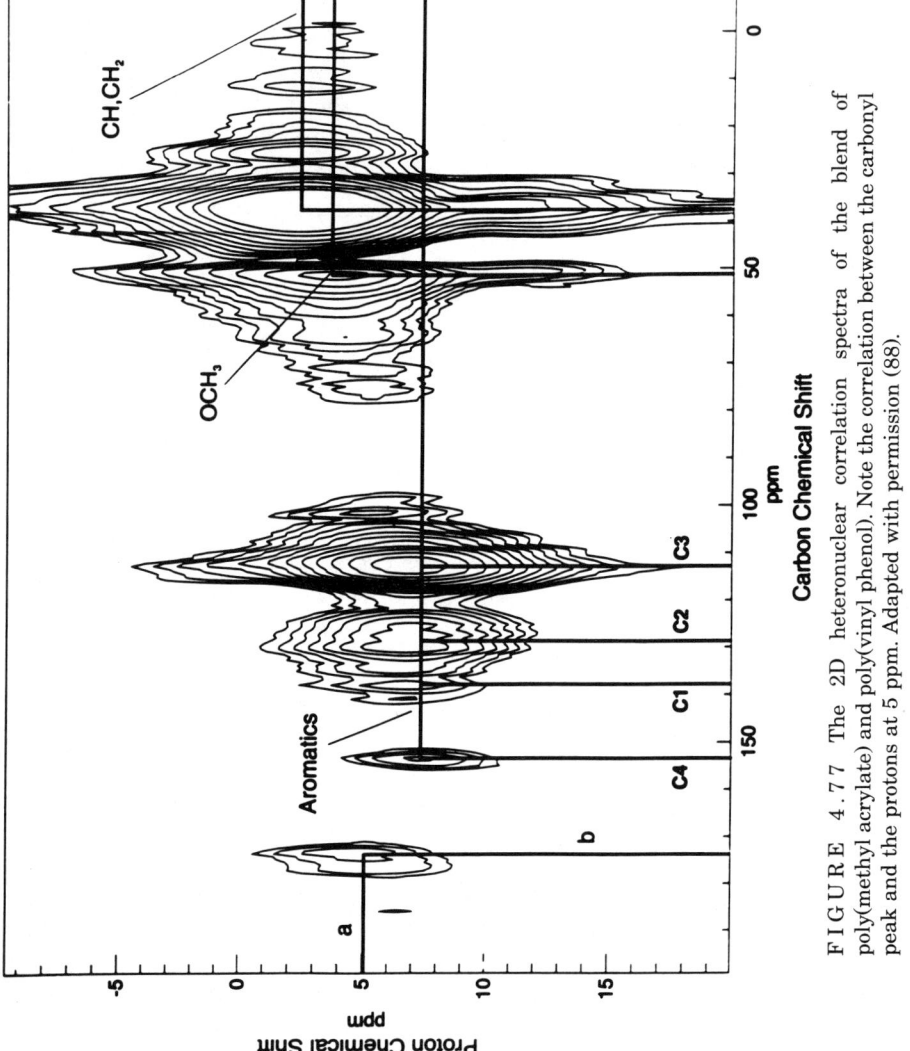

FIGURE 4.77 The 2D heteronuclear correlation spectra of the blend of poly(methyl acrylate) and poly(vinyl phenol). Note the correlation between the carbonyl peak and the protons at 5 ppm. Adapted with permission (88).

4.3. The Organization of Polymers in the Solid State

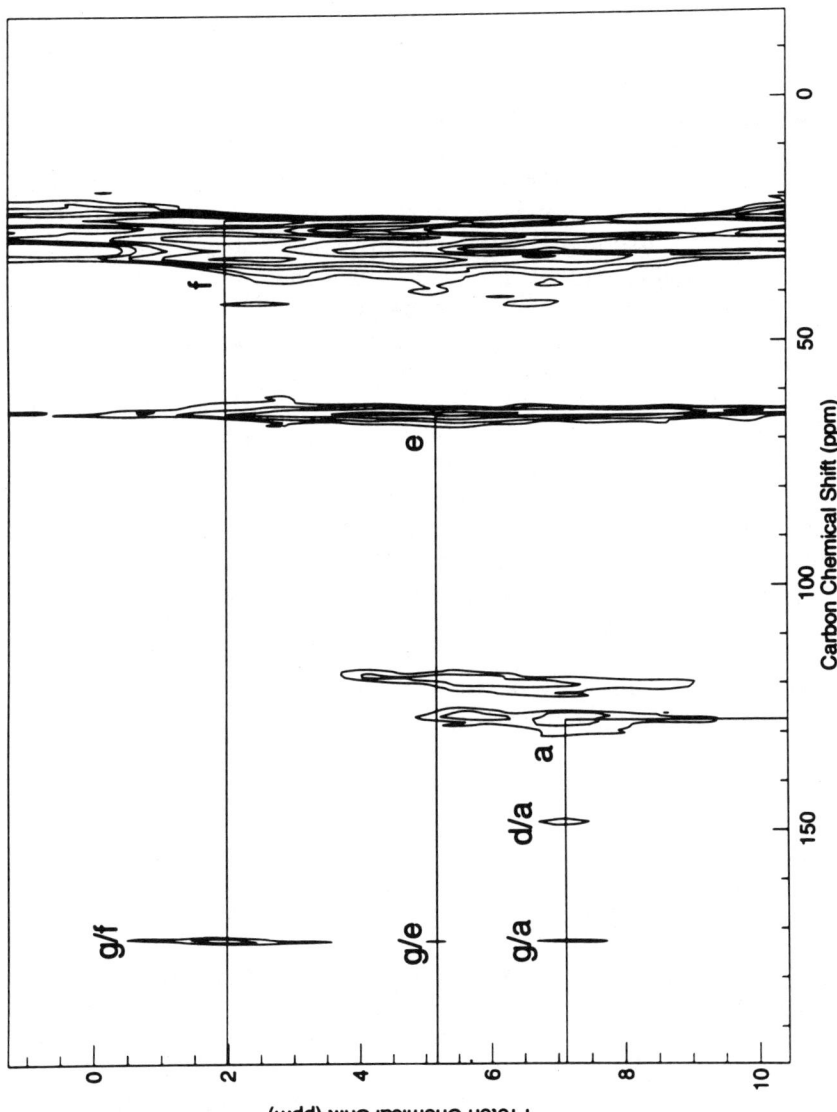

FIGURE 4.78 The 2D heteronuclear correlation spectrum of the blend of poly-caprolactone and polycarbonate. Adapted with permission (89).

lene groups (89). The important peak to note is the correlation between the polycaprolactone carbonyl and the aromatic protons of polycarbonate that supports the suggestion that miscibility is a consequence of the intermolecular interaction between the polycaprolactone carbonyl and the aromatic rings of polycarbonate.

Proton–proton spin diffusion is most commonly used to study the structure of solid polymer blends because the protons are the most abundant nuclei and they have a high sensitivity, although fluorine and phosphorus and fluorine can also be used in favorable circumstances. The study of polymer structure using the spin diffusion between nuclei with lower natural abundances is also possible with isotopic labeling. Carbon–carbon spin diffusion is typically not observed in natural abundance samples because the natural abundance of ^{13}C is 1.1%, so the probability of having two ^{13}C's next to each other is very small. However, the structure of polymer blends can be studied by carbon–carbon spin diffusion if one of the polymers is isotopically enriched. This approach was used to study the miscibility of ^{13}C-labeled poly(ethylene terephthalate) with bisphenol-A polycarbonate (92). The same 2D and 1D experiments used to study phosphorus spin diffusion in polyphosphazines (54,55) (Figs. 4.37 and 4.39) were also used to study the ^{13}C–^{13}C spin diffusion. Figure 4.79 shows the solid-state 2D NOESY spectrum and cross sections for the 75:25 poly(ethylene terephthalate):bisphenol-A polycarbonate blend obtained with a 0.4-s mixing time and a 45-μs delay for dipolar dephasing to reduce the intensity of the protonated carbon signals (92). The data show correlations between the polymers, indicating that the chains must be mixed on the molecular length scale.

Poly(ethylene terephthalate)

Polycarbonate

This same approach has been used to study specific intermolecular interactions in blends of poly(2,6-dimethyl oxide) ^{13}C labeled at the methyl group and polystyrene (93). In this case it was observed that intermolecular ^{13}C–^{13}C spin diffusion was fastest between the methyl group of poly(2,6-dimethyl phenylene oxide) and the protonated aromatic carbons of polystyrene, suggesting that there are specific interaction that drive the miscibility in this blend. There was good agreement between the observed spin diffusion rates and those calculated from a lattice model. This methodology was also used to study the interactions between bisphenol-A polycarbonate and the plasticizer di-n-butyl phthalate labeled with ^{13}C at the carbonyl position (94).

4.3. *The Organization of Polymers in the Solid State* 331

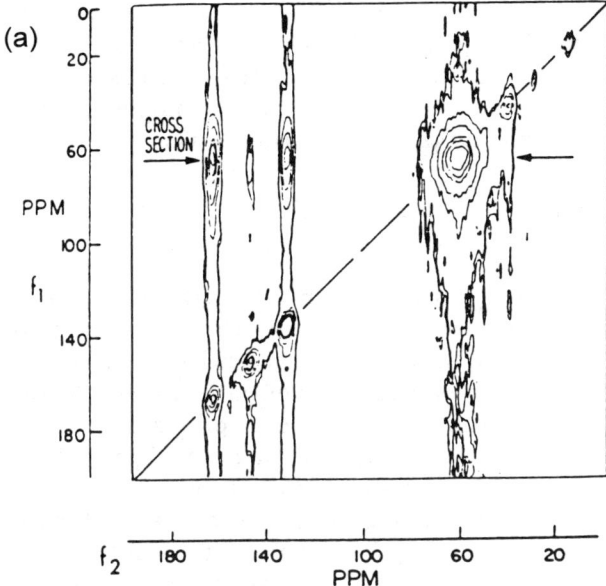

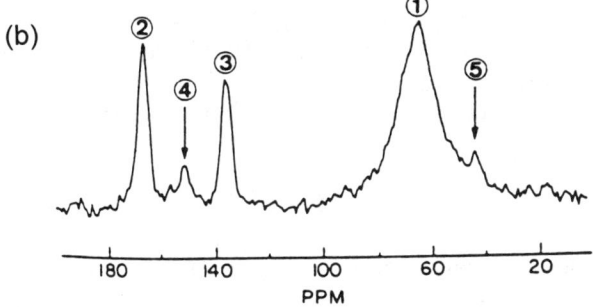

FIGURE 4.79 The (a) 2D carbon spin diffusion spectrum of the 75/25 blend of poly(ethylene terephthalate) and polycarbonate, where ethylene carbons are labeled with ^{13}C, and (b) a cross section through the 2D spectra showing intra- and intermolecular cross peaks. Adapted with permission (92).

4.3.4 NMR STUDIES OF ORIENTED POLYMERS

Several solid-state NMR methods have been used to characterize the distribution of polymer chains in anisotropic materials. Such materials are of interest because they may possess superior mechanical, electrical, or optical properties, and because polymers may become oriented as a consequence of their processing.

Polymers can become oriented as a consequence of mechanical stress, as shown in Fig. 4.80. The application of such a stress causes the polymer chains to tilt toward the stress (or draw) axis, and when the stress is removed the sample retains some permanent elongation that is characterized by the draw ratio λ given by

$$\lambda = \frac{L}{L_0}, \qquad (4.14)$$

where L_0 is the original length and L is the stretched length. The polymer chains are now anisotropic and many of the NMR parameters may be sensitive to the orientational distribution function. In Section 1.8.5.3 it was seen that powder patterns were observed for the chemical shift anisotropy, the dipolar couplings, and the quadrupolar couplings as a consequence of averaging over all possible orientations. If the distribution of polymer chains is anisotropic, then the lineshapes observed for such samples may be dramatically different than those for isotropic samples. In the most favorable cases, the NMR spectra can be used to measure the average orientation angle and the distribution of orientations.

The position of a chain segment in an oriented sample can be described by a set of Euler angles relative to some reference frame. Where the deformation is uniaxial, the distribution function has a cylindrical symmetry and reduces to a function of one variable, $N(\theta)$. Because of this symmetry the distribution function can be expanded as a series of even ordered Legendre polynomial P_l's of $\cos \theta$ as

$$N(\theta) = \sum_{i=0}^{\infty} a_l P_l(\cos \theta), \qquad (4.15)$$

where the Legendre polynomials are given by

$$P_2(\cos \theta) = \tfrac{1}{2}(3 \cos \theta - 1)$$

$$P_4(\cos \theta) = \tfrac{1}{8}(35 \cos^4 \theta - 30 \cos^2 \theta + 3), \qquad (4.16)$$

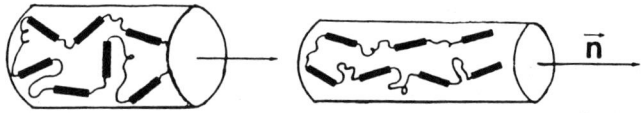

FIGURE 4.80 A schematic diagram of polymer orientation by the application of uniaxial stress that causes the chains to align along the draw direction. Adapted with permission (A. J. Brandolini and C. Dybowski (eds.), "High Resolution NMR Studies of Oriented Polymers," p. 283. VCH, Newark, New Jersey, 1986).

etc. As a general strategy the spectrum is compared with one calculated for a model distribution function, and typically includes only a few terms.

^{19}F NMR has been used to study anisotropy in compressed and drawn samples of polytetrafluoroethylene, as shown in Fig. 4.81 (95). These data compare the spectra for samples under tension ($\lambda = 1.40$) and compression ($\lambda = 0.80$) where the magnetic field is aligned normal to the stress direction, and illustrate the large effect that stress can have on the polymer orientation and the NMR spectra. In the absence of mechanical stress the ^{19}F spectrum appears as a superposition of the chemical shift anisotropy pattern for the rigid crystalline material and a motionally narrowed line for the amorphous polytetrafluoroethylene. By studying the spectrum as a function of the angle between the draw axis to the magnetic field, the distribution for polytetrafluoroethylene chains in the drawn sample has been determined (95).

The quadrupolar couplings in ^{2}H NMR have been extensively used to study orientation in polymer samples. As shown in Section 5.3.1.1., the quadrupolar coupling depends on the angle between the C–D bond vector and the magnetic field, and the typical Pake-type powder patterns are observed by averaging over all possible orientations in isotropic samples. These patterns are distorted considerably in anisotropic samples, and the orientational distribution function for the chain orientation can be determined by comparing the wide-line

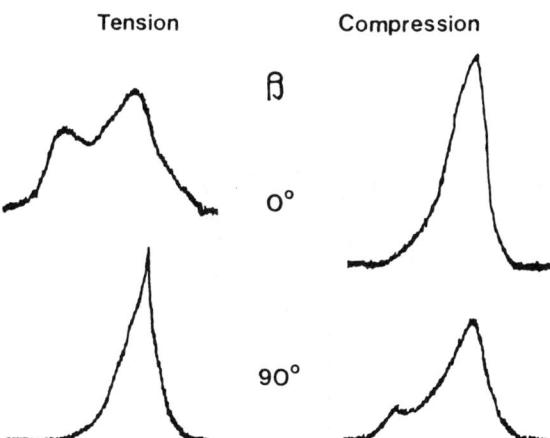

FIGURE 4.81 The ^{19}F spectra of polytetrafluoroethylene under tensile ($\lambda = 1.40$) and compressive ($\lambda = 0.8$) stress oriented parallel and perpendicular to the magnetic field. Adapted with permission (95).

^{2}H spectra with spectra calculated for various distributions as a function of the angle between the draw direction and the magnetic field.

Figure 4.82 compares the experimental and simulated wide-line ^{2}H spectra for a drawn sample of polyethylene-d_4 ($\lambda \approx 9$) as a function of the angle between the draw direction and magnetic field (96). These data show that the lineshape is extremely sensitive to the angle, demonstrating that there is a high degree of alignment in the drawn sample. The close agreement between the observed and the simulated spectra was obtained for a model in which there is a Gaussian distribution of chain angles about the preferred direction that has a width of 6.6 ± 0.5°.

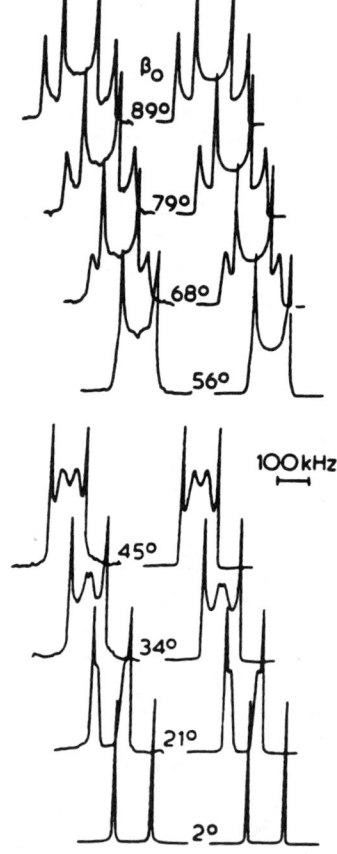

FIGURE 4.82 Comparison of the observed (left) and calculated (right) ^{2}H NMR lineshape of drawn polyethylene-d_4 ($\lambda \approx 9$) for several values of β, the angle between the draw direction and the magnetic field. Adapted with permission (96).

4.3. The Organization of Polymers in the Solid State

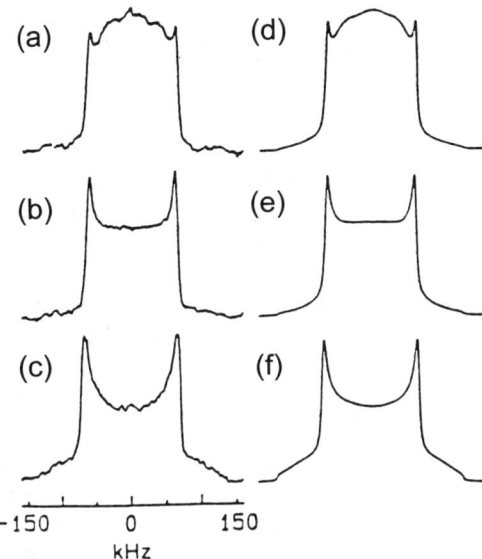

FIGURE 4.83 Comparison of the observed [(a)–(c)] and simulated [(d)–(f)] quadrupolar echo ^{2}H NMR spectra for stretched and aligned films of ring deuterated poly(phenylene vinylene) at (a) $\beta = 0°$, (b) $\beta = 15°$, and (c) $\beta = 90°$. Adapted with permission (97).

Figure 4.83 shows a similar analysis as applied to stretched films of ring deuterated poly(phenylene vinylene) at angles of 0°, 15°, 90° (97). In this case the data are well modeled with an orientation distribution consisting of a well-oriented component with a Gaussian width of 10° and a more poorly oriented component with a width of 30°. These results were obtained at low temperatures where the averaging of the quadrupole lineshape by molecular motion does not contribute to the spectra.

Poly(phenylene vinylene)

Information about the orientational distribution functions can also be obtained from the magic angle spinning spectra using 2D NMR methods with the pulse sequence shown in Fig. 4.84 (98). This pulse sequence differs from many of the others presented here in that the start of the 2D experiment is triggered by an optical signal from the spinning rotor. This experiment takes advantage of the fact that in spinning a partially ordered sample, phase and intensity variations are observed for the center band and sidebands that depend on the position of the rotor at the time the spectrum is excited, as long as the orientation axis is not parallel to the rotor axis. Therefore, by synchronizing the data acquisition with the rotor position a 2D spectrum can be obtained in which the intensity and phase of the sidebands vary depending on the angle between the orientation axis and the rotor axis (β), the degree of orientation, and the orientation of the chemical shift tensor principle axis system in the frame of reference of the orientation axis.

Figure 4.85 compares the observed and simulated cross sections through the 2D spectrum of highly oriented polyethylene ($\lambda = 20$, $\beta = 30°$) (98). The first point to note about these spectra is that significant intensity is observed in the cross sections through the 2D spectra at the frequency of the higher-order sidebands. This can occur only for oriented samples. The observation of such signals can therefore be taken as evidence for orientational ordering. The numerical studies show that the intensity pattern for the sidebands is extremely sensitive to the chain orientation and that this method is an effective

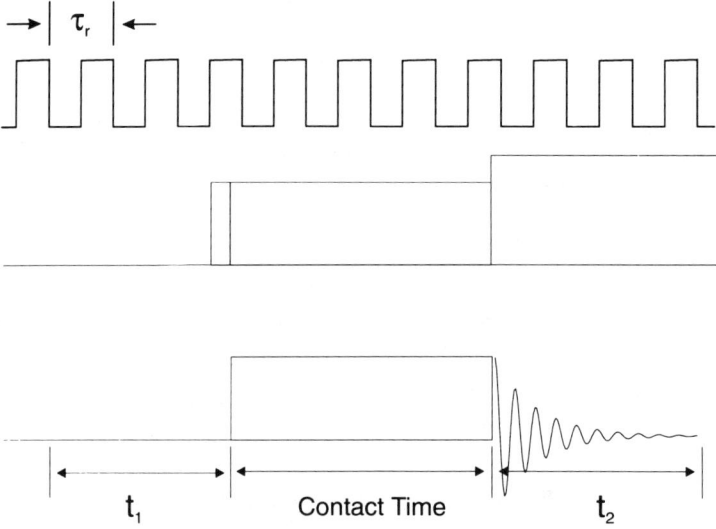

FIGURE 4.84 The pulse sequence diagram for the 2D NMR experiment to study orientational order in magic angle spinning samples (96).

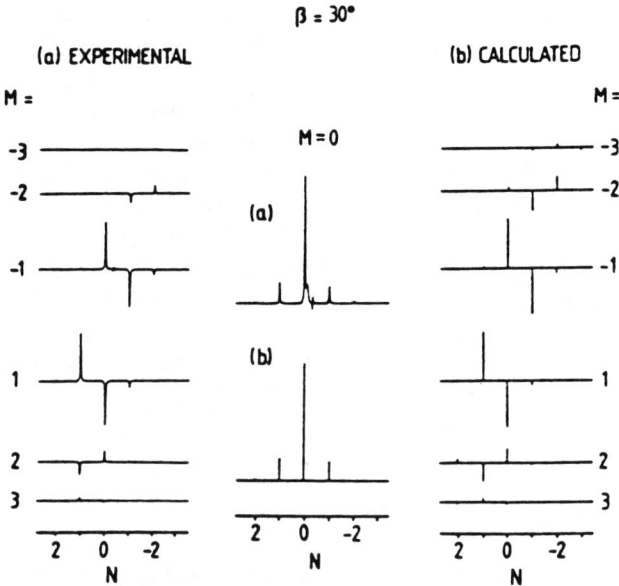

FIGURE 4.85 Comparison of the (a) observed and (b) simulated cross sections through the 2D magic angle spinning spectra of highly oriented polyethylene ($\lambda = 20$) with an angle of 30° between the magnetic field at the draw direction of the sample. Adapted with permission (98).

way to measure the orientational distribution function in both highly ordered and less highly ordered systems (98,99). Furthermore, the chemical shift tensors are larger for carbonyl and aromatic carbons so the orientational distribution function can in principle be characterized more completely in these systems. This analysis has also been applied to oriented poly(oxy methylene) (100), poly(ethylene terephthalate) (99), and a high modulus fiber based on the liquid crystalline copolymer of hydroxy benzoic acid and hydroxynapthalene (61).

4.4 NMR IMAGING OF POLYMERS

NMR imaging [or magnetic resonance imaging (MRI)] is a promising method that has been extensively used in the medical community, and is now being used to study the structure of polymers. It has the advantage of being nondestructive and it is not limited to surface analysis or optically clear materials, but it has some severe spatial resolution problems in comparison to the imaging of biological samples (101). The resolution in imaging experiments is limited by the spin–spin relaxation times and linewidths, which are much greater for polymers than for the water or fat that is typically imaged in biological samples. While the linewidths are a limitation in polymer imaging, new methods are being developed to obtain high-resolution images of a wide variety of polymeric materials.

MRI differs from most other NMR methods in that strong magnetic field gradients are used. This is in contrast to most high-resolution NMR studies where the field is made as homogeneous as possible to obtain the highest resolution. In MRI magnetic field gradients are applied so that the resonance frequency for the same atoms can be distinguished if they are at different points in space.

The spatial resolution is obtained by applying field gradients that are deliberately inhomogeneous along the different field directions. The gradient is the spatial derivative of the field and is given by

$$G_l = \frac{\partial B_l}{\partial_l}, \tag{4.17}$$

where $l = x, y, z$. As we noted in Chapter 1, the resonance frequency for a spin in the magnetic field is given by

$$\omega = -\gamma |(1 - \sigma)B_0|, \tag{4.18}$$

where B_0 is the magnetic field, σ is the isotropic chemical shift, and γ is the magnetogyric ratio. In the presence of a gradient the frequency becomes a function of spatial position $[\mathbf{r} = f(x, y, z)]$ and is given in vector notation by

$$\omega = -\gamma |(1 - \sigma)B_0 + G \cdot r|. \tag{4.19}$$

This equation shows how the frequency of a resonance in a gradient is modified by its spatial position. This behavior is shown schematically in Fig. 4.86, which compares the spectra for two samples in the presence and absence of a field gradient (101). The resolution in MRI is determined by the linewidth and the gradient strength. Two points in space can be resolved if the linewidth $\Delta \omega$ is less than the product of the magnetogyric ratio, the gradient field strength, and the spatial position:

$$\Delta \omega < \gamma \cdot G \cdot r. \tag{4.20}$$

The success of biological imaging is due in large part to the fact that water or fat is imaged, and the intrinsic linewidths for these resonances are much smaller than those observed in polymers, where the linewidths due to proton dipolar interactions may be as large as 50 kHz. A number of strategies have been developed to deal with the linewidth problems in polymers and to obtain useful images. MRI, for

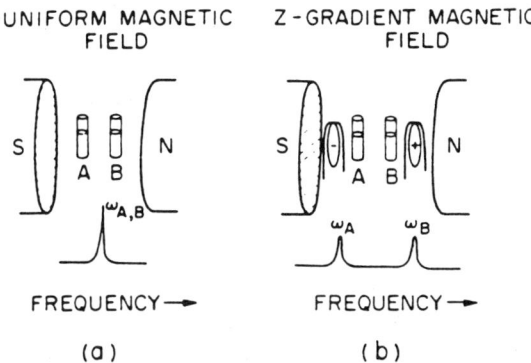

FIGURE 4.86 A schematic representation and the NMR spectra of (a) two capillaries of water in a uniform magnetic field and (b) a gradient magnetic field. Adapted with permission (F. Bovey, "Nuclear Magnetic Resonance Spectroscopy," 2nd ed. Academic Press, New York, 1988).

example, has been extremely useful for studying solvent ingress into polymers, where a liquid with a small linewidth can be imaged (102). MRI has also been successfully applied to elastomers above T_g, where the lines are narrowed relative to rigid solids by rapid chain motion (103). More complex methods using multipulse NMR and magic angle spinning have been used to obtain images of rigid solid polymers (101).

A variety of pulse sequences have been developed that attempt to overcome the intrinsic imaging limitations in solid materials. The choice of the pulse sequence depends to a large degree on the characteristics of the sample and its NMR properties. Contrast is frequently an issue, and methods have been developed, usually based on differences in the relaxation times, to enhance the signals from a section of the materials of interest (101).

Imaging in most materials is performed using two-dimensional NMR methods. The major differences are that MRI uses field gradients during the t_1 period and the acquisition time, and the final spectrum is a plot of the intensity as a function of two spatial coordinates. It is possible to obtain 3D images, but more typically selective pulses and gradients are combined to image a plane through the sample.

Figure 4.87 shows a pulse sequence that can be used to obtain a planar image through a materials sample. In addition to the normal control over the RF levels and the NMR receiver, the field gradients along the x, y, and z directions are carefully controlled. The initial pulse and z gradient are used to select a plane through the sample

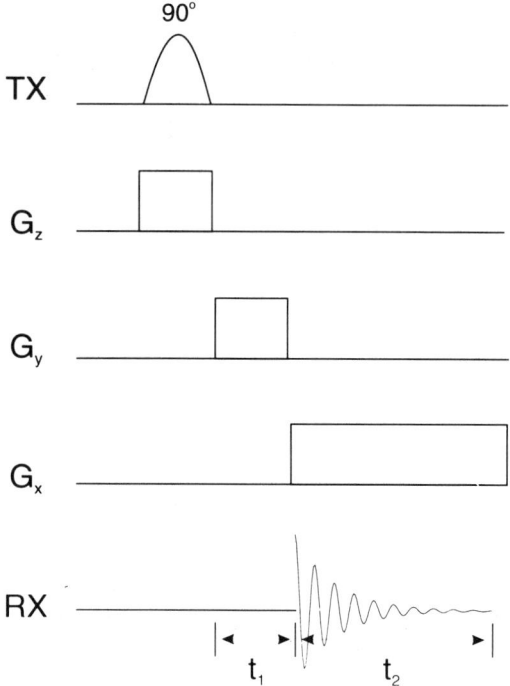

FIGURE 4.87 The pulse sequence diagram for materials imaging. In addition to the status lines for the transmitter (TX) and receiver (RX), the status lines for the x, y, and z gradients (G_x, G_y, and G_z) are also shown.

that can be imaged in the x and the y dimensions. In the presence of the z gradient, the chemical shift depends on the position along the z axis (Fig. 4.85), and slices through the sample are obtained using narrow frequency (or selective) RF pulses. The signal-to-noise ratio of the image depends on the number of nuclei excited by the selective pulse, so the slice thickness is typically on the order of a few millimeters. Resolution in the y direction is obtained as the sample evolves under the influence of the y gradient (the t_1 period), and resolution in the x direction is obtained as the signal is acquired in the presence of the x gradient (the t_2 period). Fourier transformation along the t_1 and t_2 dimensions gives rise to the final image, which is a plot of the intensity as a function of the x and y positions.

As noted above, the linewidths in solid polymers are much larger than those for the fat or water signals that are used in biological imaging, and one of the important areas of research is finding new methods for imaging in materials with larger linewidths (104). There are many instances, however, where the linewidths of polymers are sufficiently small that the methods used for biological imaging can be

4.4. NMR Imaging of Polymers

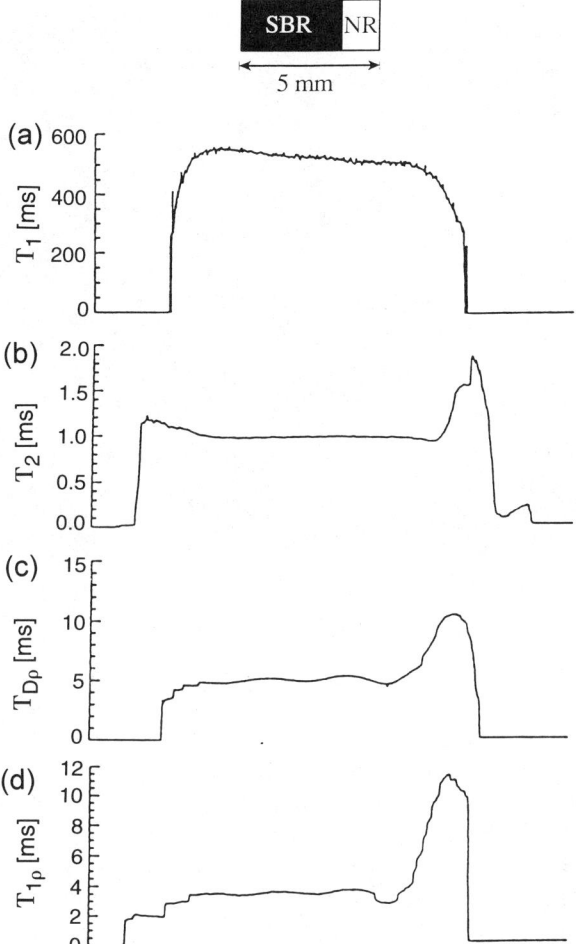

FIGURE 4.89 Cross sections through an image of styrene–butadiene rubber and natural rubber obtained with different filtering pulse sequences applied before the imaging pulse sequence. The data are shown for the (a) T_1 filter, (b) the T_2 filter, (c) the dipolar filter, and (d) the $T_{1\rho}$ filter. Adapted with permission (106).

The resolution obtainable in such studies was estimated by imaging of a phantom of *cis*-polybutadiene laminated with spacers of decreasing thickness, as shown in Fig. 4.91 (103). From such images it is estimated that features as small as 0.07 mm can be visualized in a polymer with a 150-Hz linewidth.

NMR imaging of elastomers has been used to investigate the morphology, composition, and defects in composite materials such as non-steel belted radial tires. Figure 4.92 shows an image of a cross

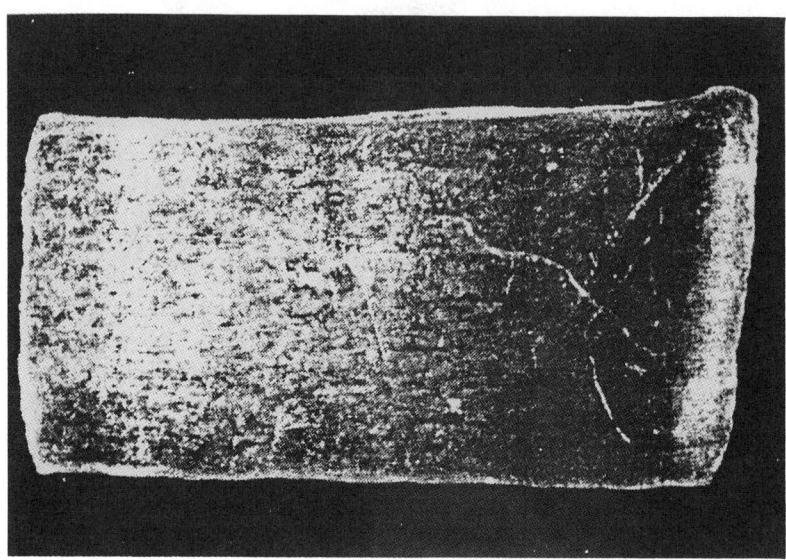

FIGURE 4.90 A T_2-weighted NMR image of a block of *cis*-polybutadiene. The dimensions of the block are 120 × 123 mm. Adapted with permission (103).

section through a tire in which the elastomer can be distinguished from the reinforcing fibers that have a much shorter T_2 and appear as dark circles in the T_2-weighted image (107). The resolution in this experiment is ca. 200 µm and the image has some distortions due to the magnetic susceptibility variations associated with the voids and filler particles. It was also shown that the variations in intensity of the elastomeric portions are associated with composition variations in the rubber, and it is estimated that a 20% variation in the composition of the *cis*-polybutadiene/styrene–butadiene rubber can be identified in the T_2-weighted image.

The application of strain to elastomers can induce changes in the free volume and causes crystallization. Both of these factors cause changes in the chain dynamics of the polymers and therefore affect the intensities in the T_2-weighted image. This effect is illustrated in Fig. 4.93, which shows an image of a polysiloxane strip with a cut created from a spatially resolved T_2 measurement (108). The changes in T_2 are correlated with stress, and this correlation has been used to create a stress image of this material where the stresses range from 0 to 2.4 MPa.

MRI has added considerably to our understanding of the diffusion of organic solvents into polymers. The diffusion of solvents into

4.4. NMR Imaging of Polymers

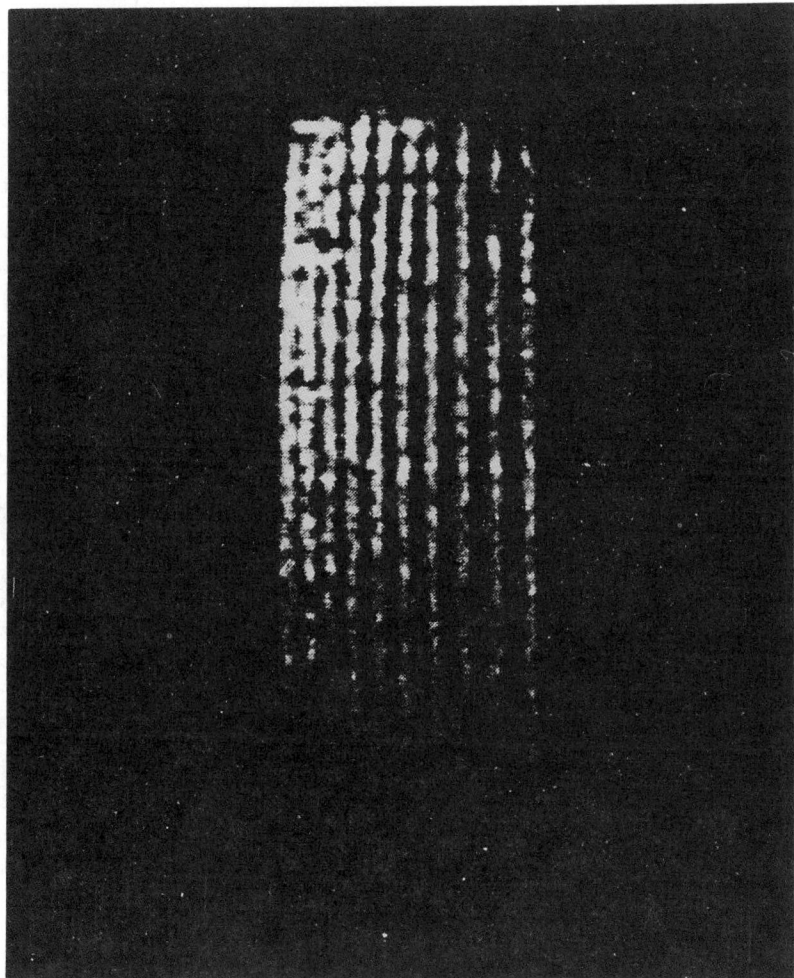

FIGURE 4.91 A T_2-weighted image of 10 pieces of carbon-black filled *cis*-polybutadiene separated by spacers ranging in thickness from 0.07 to 1.3 mm. Adapted with permission (103).

polymers is frequently a factor affecting the lifetime of the polymers, since solvents can cause crazing or stress cracking. MRI is an excellent way to study this process since it is nondestructive and relatively high-resolution images can be obtained since the sharp signals from the diffusing solvent are imaged. These studies often complement the more common studies of weight gain for polymers immersed in solvents.

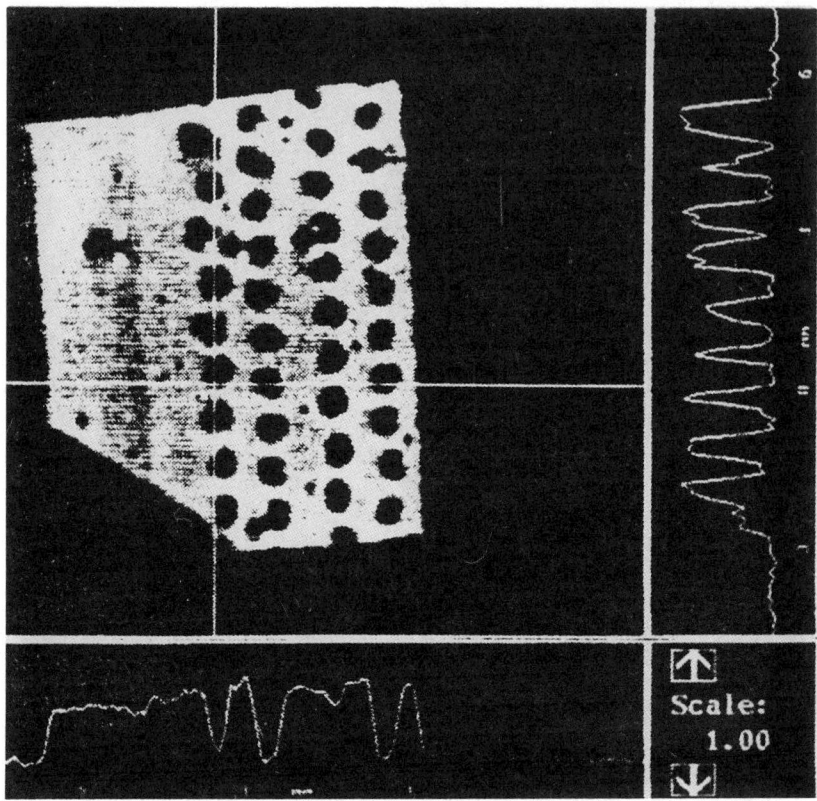

FIGURE 4.92 An NMR image of a non-steel belted radial tire section. The bright portions of the image are due to the elastomer. Adapted with permission (107).

1 mm
⊢⊣

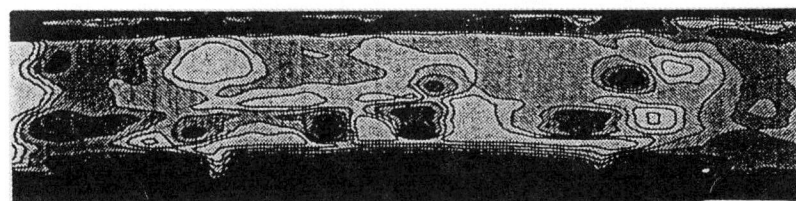

FIGURE 4.93 A stress image of a stretched polysiloxane band. The gray scale indicates local stress in the range from 0 to 2.4 MPa. Adapted with permission (108).

4.4. NMR Imaging of Polymers

Polycarbonate is a tough, high-clarity polymer that is affected by organic solvents such as acetone and methanol. These solvents have different effects on polycarbonate that can be investigated by MRI. Acetone diffuses rapidly into polycarbonate and causes stress cracking, while methanol dissolves more slowly and does not. The use of MRI to study diffusion of solvents is illustrated in Fig. 4.94, which shows images of polycarbonate rods in the presence of an acetone/methanol-d_4 mixture at 2.25, 10.5, 22.25, and 30.33 hr (109). Since the solvent contained deuterated methanol, the bright rings in Fig. 4.94 show the ingress of acetone into the polycarbonate rods. Ingress of the solvents into the polycarbonate rods was demonstrated to be Fickian, since it depends on $\sqrt{t}$, and the diffusion coefficients could be independently measured for both acetone and methanol in experiments where either one or the other solvents were deuterated. Similar methods were used to study the cyclic sorption–desorption of water and methanol into poly(methyl methacrylate) rods, where more complex diffusion behavior was observed as a function of the number of cycles (110).

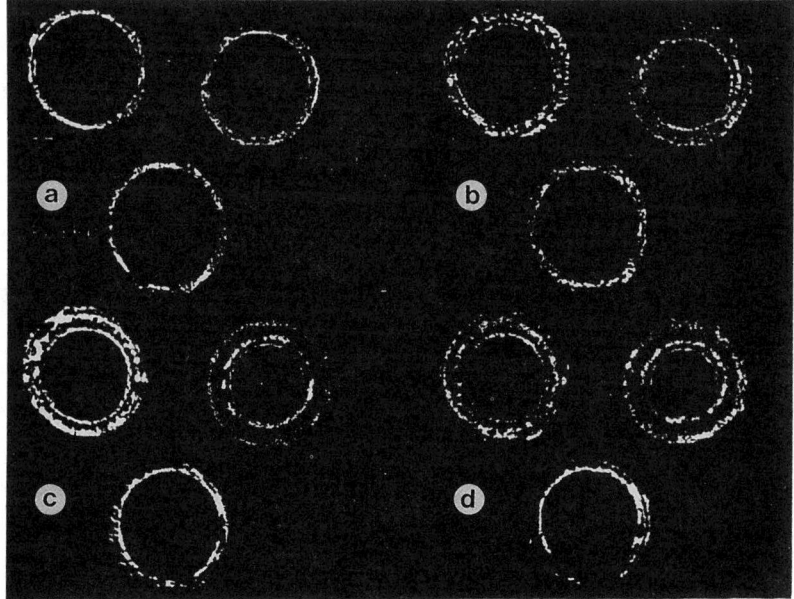

FIGURE 4.94 Proton NMR images of the diffusion of acetone/methanol-d_4 mixtures into polycarbonate rods with times of (a) 2.25 hr, (b) 10.5 hr, (c) 22.25 hr, and (d) 30.33 hr. Adapted with permission (109).

348   4. The Solid-State NMR of Polymers

MRI can also be used to follow the progress of polymerization, since the signals associated with the monomers disappear as polymerization proceeds to completion. One application of this technology is shown in Fig. 4.95, which shows two images of a block of wood and a poly(vinyl acetate) emulsion (111). The bright areas in such an image are due to the more mobile adhesive, and the dark areas to the wood. At the shortest times (10 min) a thin glue line is noted between the pieces of wood. After 4 hr the glue line has disappeared, but a large

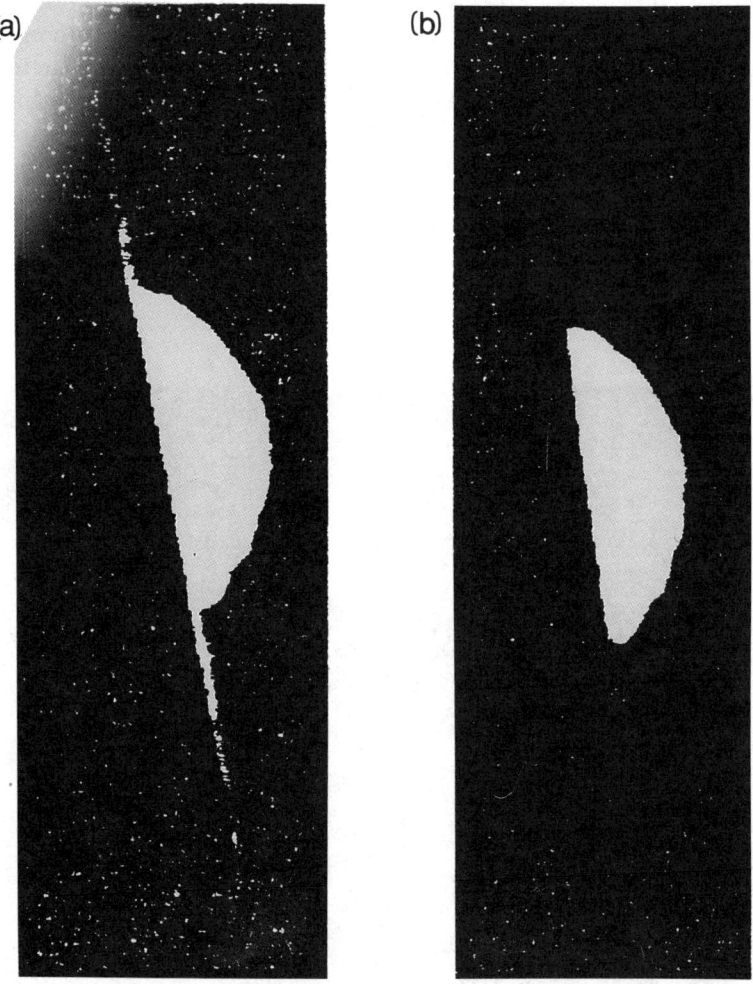

FIGURE 4.95 A materials image of the drying of a plasticized poly(vinyl acetate) emulsion between two blocks of wood with drying times of (a) 10 min and (b) 4 hr. Adapted with permission (111).

portion of the glue is still mobile and shows up as a bright spot on the image. It is estimated that the resolution in such a study is 100 μm. MRI has also been used to monitor the progress of polymerization reactions in poly(methyl methacrylate) (112).

NMR imaging in solid polymers is considerably more challenging than the experiments described above, and requires equipment that is not standard on most NMR spectrometers. The NMR imaging of solids is an area of intense investigation, and results have been reported for both static (113,114) and magic angle spinning experiments (104,105).

REFERENCES

1. W. L. Earl and D. L. Vander Hart, *Macromolecules* **12**, 762 (1979).
2. V. J. McBrierty and K. J. Packer, Nuclear Magnetic Resonance in Solid Polymers, Cambridge University Press, Cambridge, 1993.
3. A. Tonelli, "NMR Spectroscopy and Polymer Microstructure: The Conformational Connection." VCH, New York, 1989.
4. D. L. Vander Hart, *J. Magn. Reson.* **44**, 117 (1981).
5. A. Bunn, E. Cudby, R. Harris, K. Packer, and B. Say, *Chem. Commun.* **15**, 15 (1981).
6. A. E. Tonelli, M. A. Gomez, H. Tanaka, and M. H. Cozine (eds.), *in* "Solid State 13C NMR Studies of the Structures, Conformations, and Dynamics of Semicrystalline Polymers." Plenum Press, New York, 1991.
7. M. A. Gomez, H. Tanaka, and A. E. Tonelli, *Polymer* **28**, 2227 (1987).
8. L. J. Mathias, "Solid State NMR of Polymers." Plenum, New York, 1991.
9. F. Schilling, F. Bovey, A. Lovinger, and J. Zeigler, *Macromolecules* **19**, 2660 (1986).
10. F. A. Bovey and F. C. Schilling (eds.), "The Solid State ^{29}Si and ^{13}C NMR of poly(di-n-alkyl silanes)," p. 295. Plenum, New York, 1991.
11. V. G. Damaschun, *Kolloid-Z* **180**, 64 (1962).
12. F. C. Schilling, M. A. Gomez, and A. E. Tonelli, *Macromolecules* **24**, 6552 (1991).
13. D. Powell, A. Sikes, and L. Mathias, *Macromolecules* **21**, 1533 (1988).
14. D. E. Axelson, "Carbon-13 Solid State NMR of Semicrystalline Polymers." VCH, Deerfield Beach, Florida, 1986.
15. D. E. Axelson, L. Mandelkern, R. Popli, and P. Mathieu, *J. Polym. Sci. Polym. Phys. Ed.* **21**, 2319 (1983).
16. S. J. Opella and M. H. Frey, *J. Am. Chem. Soc.* **101**, 5854 (1979).
17. A. Choli, W. Ritchey, J. L. Koenig, *Spectrosc. Lett.* **16**, 21 (1983).
18. J. J. Dechter, *J. Polym. Sci. Poly. Lett. Ed.* **23**, 261 (1985).
19. L. M. Ryan, R. E. Taylor, A. J. Paff, and B. C. Gerstein, *J. Chem. Phys.* **72**, 508 (1980).
20. B. C. Gerstein, R. G. Pembleton, R. C. Wilson, and L. M. Ryan, *J. Chem. Phys.* **66**, 361 (1977).
21. J. Schaefer, E. O. Stejskal, and R. Buchdahl, *Macromolecules* **10**, 384 (1977).
22. N. Egger, K. Schmidt-Rohr, B. Blumich, W. D. Domke, and B. Stapp, *J. Appl. Polym. Sci.* **44**, 289 (1992).
23. C. A. Fyfe, A. Rudin, and W. J. Tchir, *Macromolecules* **13**, 1322 (1980).

24. P. D. Murphy, R. A. De Pietro, C. J. Lund, and W. D. Weber, *Macromolecules* **27**, 279 (1994).
25. G. Natta and P. Corradini, *Nuovo Cimento Suppl.* **1**, 40 (1960).
26. F. C. Schilling, M. A. Gomez, A. E. Tonelli, F. A. Bovey, and A. E. Woodward, *Macromolecules* **20**, 2954 (1987).
27. J. Suehiro and M. Takayangagi, *Macromol. Sci.* **B4**, 39 (1970).
28. C. De Rosa, R. Napolitano, and B. Pirozzi, *Polymer* **26**, 2039 (1986).
29. M. F. Grenier-Loustalot and G. Bocelli, *Eur. Polym. J.* **20**, 957 (1984).
30. M. A. Gomez, M. H. Cozine, and A. E. Tonelli, *Macromolecules* **21**, 388 (1988).
31. B. C. Perry, R. P. Grasso, J. L. Koenig, J. B. Lando, *Macromolecules*, **22**, 2014 (1989).
32. G. Wegner, *Discuss Faraday Soc.* **68**, 494 (1980).
33. H. Tanaka, M. A. Gomez, A. E. Tonelli, and M. Takur, *Macromolecules* **22**, 1208 (1989).
34. H. Tanaka, M. Thakur, M. A. Gomez, and A. E. Tonelli, *Macromolecules* **22**, 2427 (1987).
35. H. Tanaka, M. Gomez, A. Tonelli, S. V. Chichester-Hicks, and R. C. Haddon, *Macromolecules* **22**, 1031 (1989).
36. A. D. English, *Macromolecules* **18**, 178 (1985).
37. H. D. H. Stover and J. M. J. Frechet, *Macromolecules* **22**, 1574 (1989).
38. P. Sozzani, F. A. Bovey, and F. C. Schilling, *Macromolecules* **24**, 67641 (1991).
39. P. Sozzani, R. W. Behling, F. C. Schilling, S. Bruckner, E. Helfand, F. A. Bovey, and L. W. Jelinski, *Macromolecules* **22**, 3318 (1989).
40. S. Bruckner, P. Sozzani, C. Boeffel, S. Destri, and G. Di Silvestro, *Macromolecules* **22**, 607 (1989).
41. M. Farina, (ed.) "Inclusion Compounds," Vol. 3, p. 297. Academic Press, New York, 1984.
42. N. Vasanthan, I. I. D. Shin, and A. E. Tonelli, *Macromolecules* **27**, 6515 (1994).
43. M. Goldman and L. Shen, *Phys. Rev.* **144**, 321 (1966).
44. T. Cheung and B. Gerstein, *J. Appl. Phys.* **52**, 5517 (1981).
45. W. Z. Cai, K. Schmidt-Rohr, N. Egger, B. Gerharz, and H. W. Spiess, *Polymer* **34**, 267 (1993).
46. R. Assink, *Macromolecules* **11**, 1233 (1978).
47. P. Caravatti, P. Neuenschwander, and R. Ernst, *Macromolecules* **19**, 1889 (1986).
48. K. Schmidt-Rohr, J. Clauss, B. Blumich, and H. W. Spiess, *Magn. Reson. Chem.* **28**, S3 (1990).
49. D. L. Vanderhart G. C. Campbell, and R. M. Briber, *Macromolecules* **25**, 4734 (1992).
50. P. Caravatti, P. Neuenschwander, and R. Ernst, *Macromolecules* **18**, 119 (1985).
51. T. T. P. Cheung, B. C. Gerstein, L. M. Ryan, R. E. Taylor, and C. R. Dybowski, *J. Chem. Phys.* **73**, 6059 (1980).
52. R. R. Havens and D. L. Vander Hart, *Macromolecules* **18**, 1663 (1985).
53. B. Blumich, A. Hagemeyer, D. Schaefer, K. Schmidt-Rohr, and H. W. Spiess, *Adv. Mater.* **2**, 72 (1990).
54. S. A. Taylor, J. L. White, N. C. Elbaum, R. C. Crosby, G. C. Campbell, J. F. Haw, and G. R. Hatfield, *Macromolecules* **25**, 3369 (1992).
55. K. Takegoshi, I. Tanaka, K. Hikichi, and S. Higashida, *Macromolecules* **25**, 3392 (1992).
56. K. Schmidt-Rohr and H. Spiess, *Macromolecules* **24**, 5288 (1991).
57. W. S. Veeman and E. M. Menger, *Bull. Magn. Reson.* **2**, 77 (1980).
58. J. J. Dumais, L. W. Jelinski, L. M. Leung, I. Gancarz, A. Galambos, and J. T. Koberstein, *Macromolecules* **18**, 116 (1985).

59. J. A. Kornfeld, H. W. Spiess, H. Nefzger, H. Hayden, and C. D. Eisenbach, *Macromolecules* **24**, 4787 (1991).
60. K. Schmidt-Rohr, J. Clauss, and H. Spiess, *Macromolecules* **25**, 3273 (1992).
61. H. J. Tao, D. M. Rice, W. J. MacKnight, and S. L. Hsu, *Macromolecules* **28**, 4036 (1995).
62. Y. H. Chin and S. Kaplan, *Magn. Reson. Chem.* **32**, S53 (1994).
63. H. Spiess, *Chem. Rev.* **91**, 1321 (1991).
64. P. T. Inglefeld, R. A. Amici, J. F. O'Gara, C. C. Jung, and A. A. Jones, *Macromolecules* **16**, 1552 (1983).
65. K. L. Li, A. A. Jones, P. T. Inglefield, and A. D. English, *Macromolecules* **22**, 4198 (1989).
66. Y. Liu, P. T. Inglefield, A. A. Jones, and R. P. Kambour, *Magn. Reson. Chem.* **32**, S18 (1994).
67. D. Walsh and S. Rostami, *Adv. Polym. Sci.* **70**, 119 (1985).
68. A. Simmons and A. Natansohn, *Macromolecules* **25**, 3881 (1992).
69. L. A. Belfiore, T. J. Lutz, and C. Cheng (eds.), "Solid State NMR Detection of Molecular Level Mixing Phenomena in Strongly Interacting Polymer Blends and Phase Separated Copolymers," Plenum, New York, 1991.
70. J. Grobelny, D. Rice, F. Karasz, and W. MacKnight, *Macromolecules* **23**, 2139 (1990).
71. E. Stejskal, J. Schaefer, M. Sefcik, and R. McKay, *Macromolecules* **14**, 275 (1981).
72. G. Gobbi, R. Silvestri, R. Thomas, J. Lyerla, W. Flemming, and T. Nishi, *J. Polym. Sci. Part C. Polym. Lett.* **25**, 61 (1987).
73. J. Schaefer, M. D. Sefcik, E. O. Stejskal, and R. A. McKay, *Macromolecules* **14**, 188 (1981).
74. V. McBrierty, D. Douglass, and T. Kwei, *Macromolecules* **11**, 1265 (1978).
75. L. Dickinson, H. Yang, C. Chu, R. Stein, and J. Chein, *Macromolecules* **20**, 1757 (1987).
76. T. Kwei, T. Nishi, and R. Roberts, *Macromolecules* **7**, 667 (1974).
77. P. Mirau, H. Tanaka, and F. Bovey, *Macromolecules* **21**, 2929 (1988).
78. C. W. Chu, L. C. Dickinson, and J. C. W. Chien, *J. Appl. Polym. Sci.* **41**, 2311 (1990).
79. J. L. White and P. A. Mirau, *Macromolecules* **26**, 3049 (1993).
80. P. Mirau and F. Bovey, *Macromolecules* **23**, 4482 (1990).
81. G. C. Campbell and D. L. Vander Hart, *J. Magn. Reson.* **96**, 69 (1992).
82. W. E. J. R. Maas, W. A. C. van der Heijden, W. S. Weeman, J. M. J. Vankan, and G. H. W. Buning, *J. Chem. Phys.* **95**, 4698 (1991).
83. P. Caravatti, G. Bodenhausen, and R. R. Ernst, *Chem. Phys. Lett.* **89**, 363 (1982).
84. P. Caravatti, L. Braunschweiler, and R. R. Ernst, *Chem. Phys. Lett.* **100**, 305 (1983).
85. D. P. Burum and A. Bielecki, *J. Magn. Reson.* **94**, 645 (1991).
86. S. Kaplan, *Macromolecules* **26**, 1060 (1993).
87. S. Li, D. M. Rice, and F. E. Karasz, *Macromolecules* **27**, 2211 (1994).
88. J. L. White and P. A. Mirau, *Macromolecules* **27**, 1648 (1994).
89. P. A. Mirau and J. W. White, *Magn. Reson. Chem.* **32**, S32 (1994).
90. Y. W. Cheung and R. S. Stein, *Macromolecules* **27**, 2512 (1994).
91. M. M. Coleman and P. C. Painter, *J. Appl. Spectrosc. Rev.* **20**, 837 (1984).
92. M. Linder, P. M. Henrichs, J. M. Hewitt, and D. J. Massa, *J. Chem. Phys.* **82**, 1585 (1985).
93. P. Wang, A. A. Jones, P. T. Inglefield, D. M. White, and J. T. Bendler, *New Polym. Mater.* **2**, 221 (1990).

94. A. K. Roy, P. T. Inglefield, J. H. Shibata, and A. A. Jones, *Macromolecules* **20**, 1434 (1987).
95. A. J. Brandolini, M. D. Alvey, and C. Dybowski, *J. Polym. Sci., Polym. Phys. Ed.* **21**, 2511 (1983).
96. R. Hentschel, H. Sillescu, and H. W. Spiess, *Polymer* **22**, 1516 (1981).
97. J. H. Simpson, D. M. Rice, and F. E. Karasz, *Macromolecules* **25**, 2099 (1992).
98. G. S. Harbison and H. W. Spiess, *Chem. Phys. Lett.* **124**, 128 (1986).
99. G. S. Harbison, V. Vogt, and H. W. Spiess, *J. Chem. Phys.* **86**, 1206 (1987).
100. A. Hagemeyer, K. Schmidt-Rohr, and H. Spiess, *Adv. Magn. Reson.* **13**, 85 (1989).
101. B. Blümich, *Makromol. Chem.* **194**, 2133 (1993).
102. A. Webb and L. Hall, *Polymer*, **32**, 2926 (1991).
103. C. Chang and R. Komoroski, *Macromolecules*, **22**, 600 (1989).
104. D. J. Cory, J. C. de Boer, and W. S. Veeman, *Macromolecules* **22**, 1618 (1989).
105. D. G. Cory, J. B. Miller, A. N. Garroway, *Makromol. Symp.* **86**, 259 (1994).
106. P. Blumler and B. Blumich, *Magn. Reson. Imaging* **10**, 779 (1992).
107. S. N. Sarker, R. Komoroski, *Macromolecules*, **25**, 1420 (1992).
108. P. Blumer, B. Blümich, *Acta Polym.*, **44**, 125 (1993).
109. R. A. Grinsted and J. L. Koenig, *Macromolecules* **25**, 1229 (1992).
110. R. A. Grinsted, L. Clark, and J. L. Koenig, *Macromolecules* **25**, 1235 (1992).
111. A. O. K. Nieminen and J. L. Koenig, *J. Edhes. Technol.* **2**, 407 (1988).
112. B. J. Balcom, A. Carpenter, and L. D. Hall, *Macromolecules* **25**, 6818 (1992).
113. E. Gunther, B. Blumich, and H. W. Spiess, *Macromolecules* **25**, 3315 (1992).
114. B. Blumich, P. Blumler, E. Gunther, G. Schauss, and H. W. Spiess, *Makromol. Chem. Macromol. Symp.* **44**, 37 (1991).

5

THE DYNAMICS OF MACROMOLECULES

5.1 INTRODUCTION

Nuclear magnetic resonance has long been used to observe molecular motion and has been extensively applied to the complex dynamic problems presented by polymer chains. Such studies have customarily been divided into two regimes, polymer solutions, in which the chain motion is relatively fast, and polymer in the solid state, where the molecular motion is slower and may vary over many orders of magnitude—in some crystalline and glassy polymers this may reach the extreme of complete immobility. Since many synthetic polymers are useful because of their physical and mechanical properties in the solid state, which are ultimately related to molecular-level dynamics, the study of the molecular dynamics of polymers is a topic of great importance. The studies in solution primarily reveal information about *intramolecular* forces, while the molecular dynamics in the solid state are determined by the combination of *intra-* and *intermolecular* forces.

The molecular dynamics of polymers in solution are typically measured via relaxation rate measurements, such as the T_1, T_2, and NOE measurements introduced in Section 1.8. These same measurements are used in the solid state along with several others, including $T_{1\rho}$, T_{CH}, and some recently developed 2D NMR methods for studying the low-frequency molecular dynamics in polymers. The choice of experiment depends to a large degree on the time scale of the polymer molecular dynamics, since each of these methods is sensitive to the dynamics over a range of correlation times.

Nuclear relaxation is caused by fluctuating local fields from nearby nuclei through several possible mechanisms. It is the motions of atoms that cause such fluctuations in the local magnetic fields over a broad range of frequencies, and to the extent that these fluctuations have components at the resonance frequency for the nuclei of interest, they will cause relaxation. The distribution of motional frequencies is given by the *spectral density function*, which is defined by

$$J(\omega) = \frac{1}{2} \int_{-\infty}^{+\infty} G(\tau) e^{-i\omega\tau} d\tau, \qquad (5.1)$$

where $G(\tau)$ is the autocorrelation function of the time-dependent relation expressing the orientation of the internuclear vector—such as a ^{13}C–^{1}H vector—in the laboratory frame and is given by

$$G(\tau) = \langle F(t) F^*(t+\tau) \rangle. \qquad (5.2)$$

The brackets denote the ensemble average over a collection of nuclei, and F represents a function, related to spherical harmonics, describing the position and motion of the molecule. $J(\omega)$ may be thought of as expressing the power available at frequency ω to relax the spins in question. The spectral densities and autocorrelation functions are Fourier inverses of each other in the time and frequency domains, respectively. If $G(\tau)$ decays to zero in a short time, this corresponds to a short *correlation time* τ_c, which means that molecular motion is rapid and that the molecules have only a short memory of their previous state of motion.

In order to give these ideas specific meaning, we must adopt a dynamical model for the polymer molecule. The simplest model views it as a rigid sphere immersed in a viscous continuum that is reoriented in small diffusive steps. The correlation time can be thought of as the interval between these alterations in the state of motion of the molecule. For such a molecule the loss of memory of the previous motional state is exponential with the time constant τ_c,

$$G(\tau) = e^{-(\tau/\tau_c)}, \qquad (5.3)$$

and the spectral density function becomes

$$J(\omega) = \frac{\tau_c}{1 + \omega^2 \tau_c^2}. \qquad (5.4)$$

Figure 5.1 shows a plot of the spectral density function as a function of frequency for three values of τ_c. A large value for τ_c corresponds to the molecular motion of a large molecule, a stiff chain, or a small

5.1. Introduction

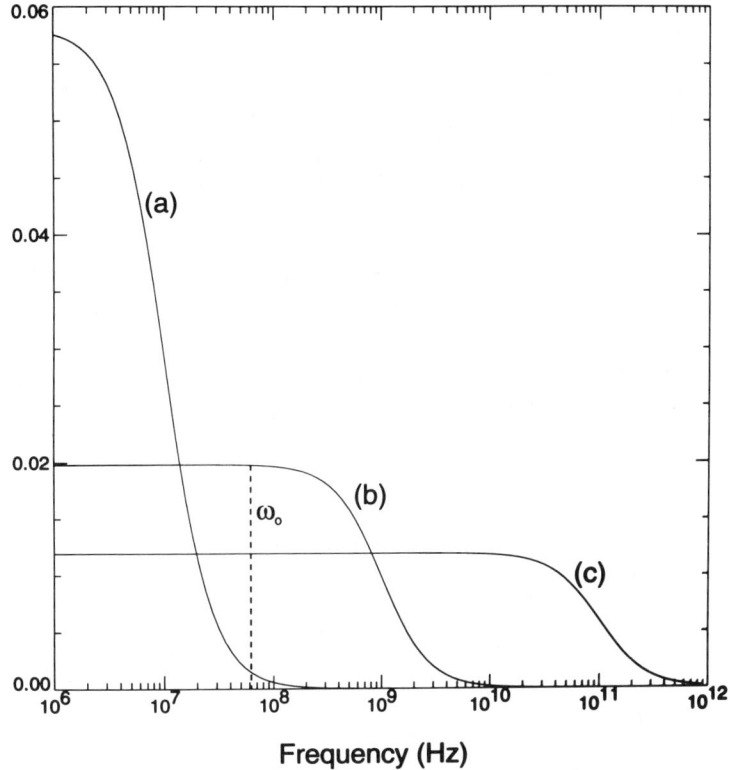

FIGURE 5.1 A plot of the spectral density functions as a function of frequency for (a) short, (b) intermediate, and (c) long correlation times. The dotted line at ω_0 shows the power of the spectral density at the spectrometer frequency.

molecule in a viscous medium, and a small value for τ_c corresponds to the rapid motion of a small molecule or a very flexible polymer chain. Also shown in Fig. 5.1 is the plot for an intermediate case, when $\tau_c \omega_0 \cong 1$, where ω_0 is the resonant frequency of the observed nuclei. The areas under the curves are the same for all three correlation times. This means that the molecular power available to cause relaxation is the same, and only the distribution varies with τ_c. For long correlation times the component at ω_0 is weak, and for short τ_c the frequency spectrum is broad, so no one component, particularly that at ω_0, is very intense. At some intermediate value of τ_c the intensity at ω_0 will be at a maximum. Since the relaxation rate depends on the component at ω_0, it will be at a maximum for the intermediate value of τ_c, and the relaxation rates will be much slower for very fast or very slow motion. The exact dependence of the relaxation rate on the correlation time will depend on which experiment is performed and which relaxation rate is measured. A critical part of measuring the molecular dynamics of polymers is choosing the proper model for the correlation functions and spectral densities. It is well established that a single exponential correlation function is not a good model for the molecular motion of an object as complex as a polymer chain (1).

Understanding the relaxation mechanism for the nuclei of interest is an important first step in the study of the molecular dynamics of polymers. Depending on the nucleus and its chemical environment, there can be several contributions to the observed relaxation rate, including those from dipolar interactions, chemical shift anisotropy, or quadrupolar relaxation, and the observed relaxation rate is given by the sum of all of these contributions,

$$\frac{1}{T_1} = \frac{1}{T_1^{DD}} + \frac{1}{T_1^{CSA}} + \frac{1}{T_1^{QUAD}}. \tag{5.5}$$

The dipolar contribution to the spin–lattice and spin–spin relaxation of carbons by protons is given by

$$\frac{1}{T_1^{DD}} = \frac{n}{10}\left(\frac{\mu_0}{4\pi}\right)^2 \frac{\gamma_C^2 \gamma_H^2 \hbar^2}{r^6} \{J(\omega_H - \omega_C) + 3J(\omega_C) + 6J(\omega_H + \omega_C)\} \tag{5.6}$$

and

$$\frac{1}{T_2^{DD}} = \frac{n}{20}\left(\frac{\mu_0}{4\pi}\right)^2 \frac{\gamma_C^2 \gamma_H^2 \hbar^2}{r^6}$$
$$\times \{4J(0) + J(\omega_H - \omega_C) + 3J(\omega_C) + 6J(\omega_C)$$
$$+ 6J(\omega_H) + 6J(\omega_H + \omega_C)\}, \tag{5.7}$$

where n is the number of attached protons, μ_0 is the vacuum magnetic permeability, γ_H and γ_C are the magnetogyric ratios for protons and carbons, ω_H and ω_C are the resonant frequencies for protons and carbons, and r is the internuclear distance, which for carbons with directly attached protons is taken as 1.08 Å for methine and methylene carbons and 1.09 Å for aromatic carbons. The nuclear Overhauser enhancement is also due to dipolar interactions and is given by

$$\text{NOE} = 1 + \frac{\gamma_H}{\gamma_C} \frac{\{6J(\omega_H + \omega_C) - J(\omega_H - \omega_C)\}}{\{J(\omega_H - \omega_C) + 3J(\omega_C) + 6J(\omega_H + \omega_C)\}}. \tag{5.8}$$

The relaxation of carbon atoms with a large chemical shift anisotropy, such as aromatic and double-bonded carbons, may also have contribu-

5.1. Introduction

tions from chemical shift anisotropy, which for axially symmetry tensors is given by

$$\frac{1}{T_1^{CSA}} = \frac{2}{15} \omega_C^2 \Delta\sigma^2 J(\omega_C) \tag{5.9}$$

and

$$\frac{1}{T_2^{CSA}} = \frac{1}{45} \omega_C^2 \Delta\sigma^2 \{4J(0) + J(3\omega_C)\}, \tag{5.10}$$

where

$$\Delta\sigma = \sigma_{33} + \tfrac{1}{2}(\sigma_{11} + \sigma_{22}). \tag{5.11}$$

It should be noted that in some cases the relaxation by chemical shift anisotropy provides an alternate relaxation pathway that competes with dipolar interactions and may affect the measured NOE value. It should also be noted that a signature of relaxation by chemical shift anisotropy is that it depends on the square of the field strength.

In polymers containing quadrupolar atoms, mainly deuterium, the relaxation due to quadrupolar interactions is given by

$$\frac{1}{T_1^{QUAD}} = \frac{3\pi^2}{10} \left(\frac{e^2 qQ}{h}\right)^2 (1+\xi)\{J(\omega_D) + 4J(2\omega_D)\} \tag{5.12}$$

and

$$\frac{1}{T_2^{QUAD}} = \frac{3\pi^2}{20} \left(\frac{e^2 qQ}{h}\right)^2 (1+\xi)\{3J(0) + 5J(\omega_D) + 2J(2\omega_D)\}, \tag{5.13}$$

where ξ is the asymmetry of the electric field gradient, which is taken as zero for deuterons, and (e^2qQ/h) is the quadrupolar coupling constant.

For the study of the molecular dynamics of polymers, it is frequently advantageous to choose a nucleus for which the relaxation mechanism is well understood and well characterized. For many solutions studies, ^{13}C is often the nucleus of choice, because for unsaturated carbons with directly bonded protons the relaxation is due exclusively to dipolar interactions. Aromatic, carbonyl, and double-bonded carbons are more complex as they may have contributions from relaxation by chemical shift anisotropy. Quadrupolar nuclei, such as deuterium, can also be used since the relaxation for these nuclei is dominated by quadrupolar interactions. Molecular dynamics are often difficult to study by proton NMR relaxation because the rates depend on the inverse sixth power of the internuclear distances and these values are neither fixed nor known in most polymers.

As a general rule, it is difficult to prove that a particular motional model for $G(\tau)$ is responsible for the relaxation of a polymer chain. The most common strategy is to make as many different types of relaxation measurements as possible over a range of temperatures and viscosities to demonstrate that the data are consistent with a particular motional model. This approach is necessary because the relaxation measurements only sample the spectral densities at a few points. Since the relaxation parameters such as the T_1, T_2, and NOEs have a different dependence on the spectral densities, several types of relaxation data can be used to argue for a particular model.

NMR studies of the molecular dynamics of solid polymers utilize these same types of measurements in addition to several others. Solids differ from solutions in that there is not enough molecular motion to average the dipolar interactions, the chemical shift anisotropy, or the quadrupolar interactions, and information about the molecular dynamics can be extracted from the lineshapes. One dramatic manifestation of this effect is for polymers above and below the glass transition temperature, where there is often a large change in lineshape at temperatures above T_g (2). Other types of motions that are fast on the time scale of chemical shift anisotropy or quadrupolar couplings, such as methyl rotation, lead to characteristic changes in the lineshape. In deuterium NMR, for example, the quadrupolar lineshapes resulting from aromatic ring flips or methyl group rotations can be easily recognized (3).

More recently, 2D NMR experiments analogous to solution 2D NOESY experiments have been developed to measure the *ultraslow* molecular dynamics of polymers (4). If a polymer undergoes molecular motion during the mixing time (milliseconds to seconds) then a characteristic off-diagonal ridge pattern develops in the solid-state deuterium or carbon spectra. The pattern of ridges depends on both the rate and the angle of reorientation, and the molecular dynamics can be mapped out to high precision directly from the experimental data.

5.2 POLYMER DYNAMICS IN SOLUTION

5.2.1 INTRODUCTION

It was observed quite early on that most synthetic polymers give fairly well-resolved proton spectra, particularly at moderately elevated temperatures, and thus must enjoy considerable segmental motion. It might have been anticipated that polymers would give broad, unresolved spectra since the correlation time for the rotational

5.2. Polymer Dynamics in Solution

diffusion of a random coil chain is given by

$$\tau_r = \frac{2M[\eta]\eta_0}{3RT}, \qquad (5.14)$$

where M is the molecular weight, $[\eta]$ is the intrinsic viscosity of the polymer, and η_0 is the solvent viscosity. Thus, in the absence of segmental dynamics that are rapid on the time scale of τ_r, broad lines would be expected for even moderate molecular weight polymers. Furthermore, it is frequently observed that the linewidths for the side-chain resonances are sharper than those for the main chain, demonstrating that the side chains experience additional motion relative to the main chain, presumably from rotation about the carbon–carbon bonds. Finally, in some polymers, such as the polybutadienes, different relaxation times are noted for neighboring atoms in the main chain (5). Any successful model must simultaneously account for all of these observations.

The contribution of segmental motion to the relaxation of polymers can be experimentally demonstrated by measuring the relaxation times as a function of molecular weight, as shown in Fig. 5.2 for polystyrene (6). At low molecular weights the relaxation times are sensitive to the degree of polymerization, but little change is observed in the T_1's for polymers with molecular weights greater than 10,000. At molecular weights above 10,000 the relaxation must therefore be due exclusively to segmental motion.

In addition, it was observed that the linewidths and relaxation times do not appear to depend on the macroscopic viscosity, and high-resolution spectra could be obtained for solution concentrations of 30–50%, where the solutions were too viscous to be poured from the NMR tube. Despite these characteristics, however, polymer spectra usually appear more poorly resolved than their small-molecule counterparts. Part of this broadening is inhomogeneous and attributed to the envelope of chemical shifts that results from the microstructure of polymers, including branches, defects, and stereochemistry. In many cases the linewidths for polymers are on the order of 2–5 Hz. Typically, the linewidths are larger than the value calculated from the T_2 [Eq. (1.16)] due to unresolved long-range J couplings.

5.2.2 Modeling the Molecular Dynamics of Polymers

Although a useful approximation, the isotropic rotational diffusion (single correlation time) model does not describe in detail any actual random coil polymer chain in solution. Polymers are characterized by

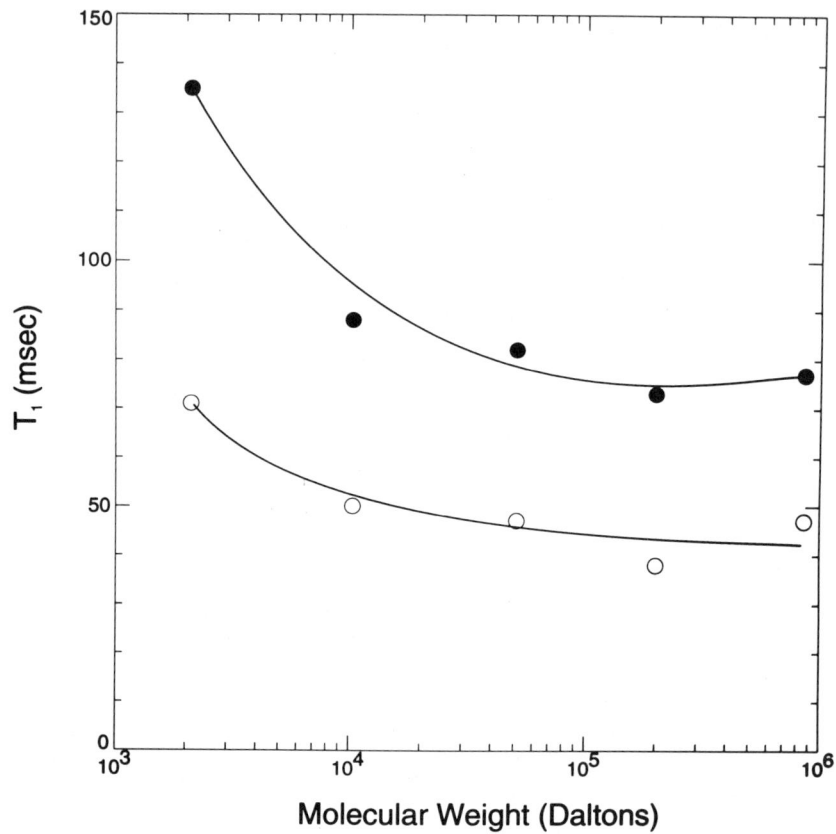

FIGURE 5.2 The effect of molecular weight on the ^{13}C NMR spin–lattice relaxation times for the (●) methine and (○) methylene carbons of polystyrene (6).

many types of molecular motion having widely differing rates, including overall tumbling, motions of large portions of the chain, and localized segmental motions such as *gauche–trans* isomerizations, ring flips, and methyl group rotations. They can therefore be better dealt with in terms of multiple correlation times, distributions of correlation times such as those used in describing dielectric and dynamic mechanical relaxation, lattice models, or conformational transition models.

The relaxation behavior of polymers in solution is frequently interpreted using a model with a distribution of correlation times. While critics of such models argue that they are not interpretable in terms of the molecular structure of polymers, a distribution of correlation times might be expected for long-chain macromolecules. As noted above, several different types of isolated and collective motions can

5.2. Polymer Dynamics in Solution

contribute to segmental reorientation in polymers. Because of the different types of motion (*gauche–trans* isomerization vs librations, for example) the reorientations of the internuclear vectors are expected to occur over a range of time scales that can be described by a distribution of correlation times. A fitting of the data to these models, however, provides little insight into the mechanism of reorientation.

Several different distributions of correlation times have been used for the interpretation of NMR relaxation data in polymers. To distinguish between the various models for molecular motion it is frequently necessary to measure several relaxation parameters (spin–lattice and spin–spin relaxation times and NOE) as a function of temperature and spectrometer frequency. Among the distribution functions that have been successfully used to interpret dielectric and dynamic mechanical spectra as well as NMR spectra are the Cole–Cole (7) and Fuoss–Kirkwood (8) distributions.

The equations for $J(\omega)$ can be easily modified to incorporate a distribution of correlation times (1). If τ_0 is the center of the distribution of correlation times and we define

$$s = \ln\left(\frac{\tau_c}{\tau_0}\right), \tag{5.15}$$

and the distribution function is expressed as a density function $F(s)$, where

$$F(s) = \int_{-\infty}^{\infty} F(s)\,ds = 1, \tag{5.16}$$

the correlation function can be written as

$$G(\tau) = \int_{-\infty}^{\infty} F(s) e^{|\tau|/\tau_0 e^s}\,ds, \tag{5.17}$$

which gives the spectral density function

$$J(\omega) = \int_{-\infty}^{\infty} \frac{F(s)\tau_0 e^s}{1 + \omega^2 \tau_0^2 e^{2s}}\,ds. \tag{5.18}$$

The Cole–Cole distribution function is (7)

$$F(s) = \frac{1}{2\pi} \frac{\sin(\varepsilon\pi)}{\cosh(\varepsilon s) + \cos(\varepsilon\pi)}, \tag{5.19}$$

which upon Fourier transformation gives the spectral density function

$$J(\omega) = \left(\frac{1}{\omega}\right) \frac{\cos[(1-\varepsilon)(\pi/2)]}{\cosh[\varepsilon \ln(\omega\tau_0)] + \sin[(1-\varepsilon)(\pi/2)]}. \tag{5.20}$$

The Fuoss–Kirkwood distribution function is (8)

$$F(s) = \left(\frac{\beta}{\pi}\right) \frac{\cos(\beta\pi/2)\cosh(\beta s)}{\cos^2(\beta\pi/2) + \sinh^2(\beta s)}, \quad (5.21)$$

giving

$$J(\omega) = \frac{2\beta}{\omega} \frac{(\omega\tau_0)^\beta}{1 + (\omega\tau_0)^{2\beta}}, \quad (5.22)$$

where ε and β are adjustable width parameters in the range $0 < (\varepsilon, \beta) < 1$. A smaller value for ε or β corresponds to a wider distribution in τ_c, and as ε and β approach unity the spectral densities approach those of the isotropic reorientation model. Figures 5.3 and 5.4 show the distribution functions as a function of ε or β.

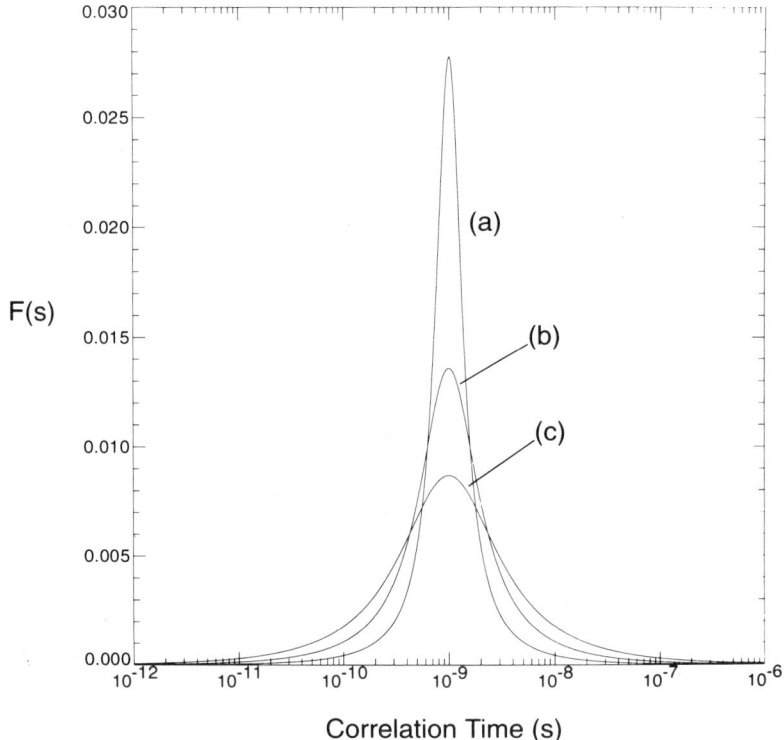

FIGURE 5.3 A plot of the Cole–Cole distribution function (7) as a function of the width parameter ε. The data are shown for ε values of (a) 0.9, (b) 0.8, and (c) 0.7.

5.2. Polymer Dynamics in Solution

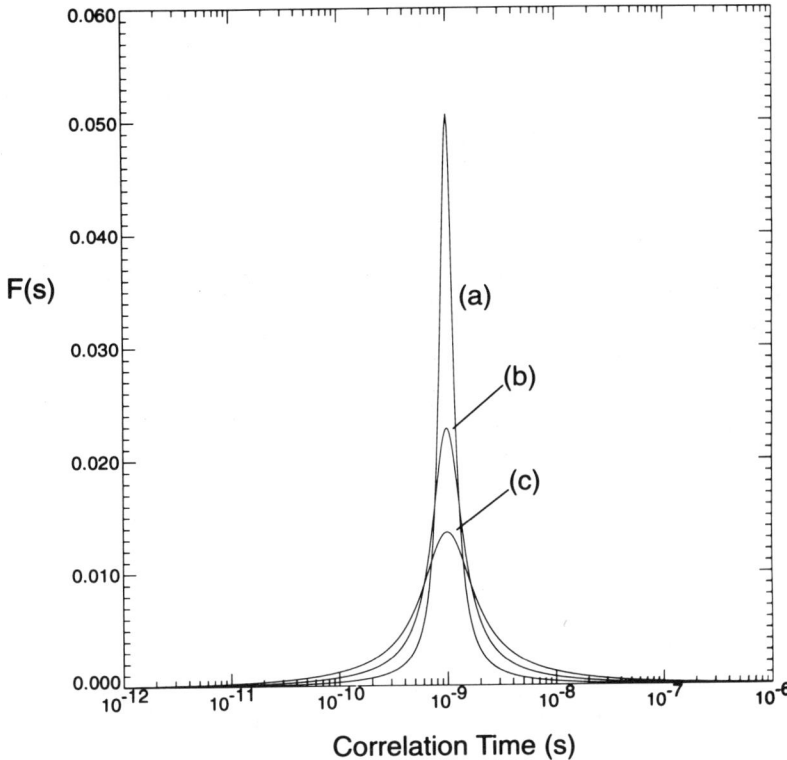

FIGURE 5.4 A plot of the Fuoss–Kirkwood distribution function (8) as a function of the width parameter β. The data are shown for β values of (a) 0.9, (b) 0.8, and (c) 0.7.

It may be noted that the distribution functions for the Cole–Cole and Fuoss–Kirkwood functions are symmetrical about the mean value τ_0. It has been argued that an asymmetric distribution weighted toward the long correlation time limit is more appropriate for polymers, and the log-χ^2 distribution has been suggested for this purpose (9). Defining

$$s = \log_b\left[1 + (b-1)\frac{\tau}{\tau_0}\right], \qquad (5.23)$$

we obtain the density function

$$F(s) = \frac{p}{\Gamma(p)}(ps)^{p-1}e^{-psd}, \qquad (5.24)$$

where b and p are adjustable parameters and $\Gamma(p)$ is the gamma function. The spectral density functions are obtained by Fourier transformation to give

$$J(\omega) = 2\int_0^\infty \frac{\tau_0 F(s)[b^s - 1]}{(b-1)\{1 + \omega^2\tau_0^2[(b^s - 1)/(b-1)]^2\}}\, ds, \quad (5.25)$$

but in this case must be evaluated numerically. It should be noted that b and p are not independent fitting parameters, and typically b is held at 1000 while the value of p is adjusted. Figure 5.5 shows the effect of the width parameter on the log χ^2 distribution function. For $p = 100$ the results are identical with the model for isotropic motion. While models that utilize a distribution of correlation times do not provide an insight into the nature of the motion that leads to relax-

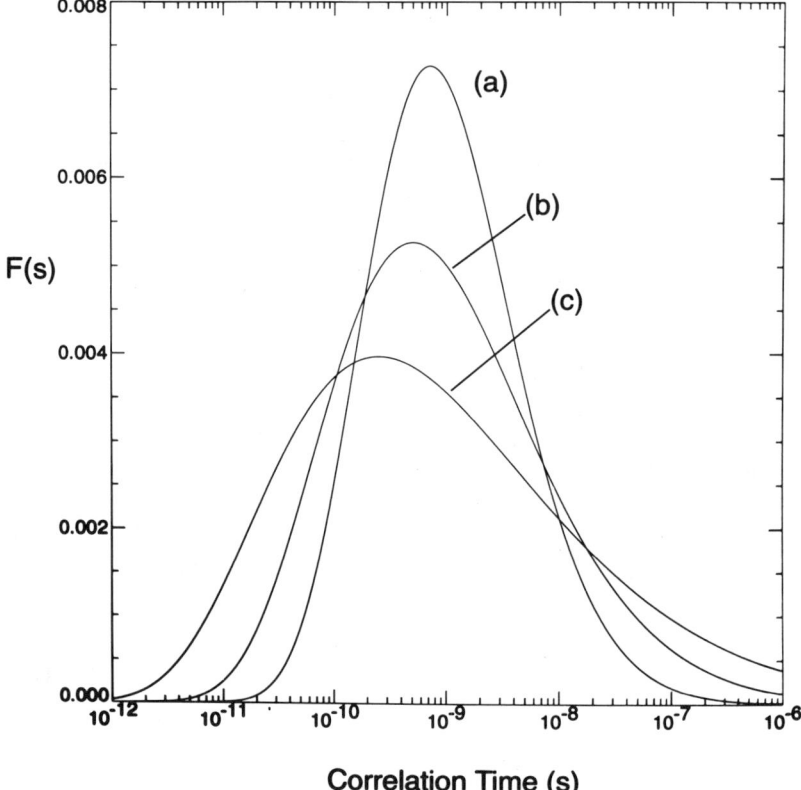

FIGURE 5.5 A plot of the log χ^2 distribution function as a function (9) of the width parameter p with $b = 1000$. The data are shown for a mean correlation time of 1 ns and p values of (a) 20, (b) 10, and (c) 5.

5.2. Polymer Dynamics in Solution

ation, they are extremely useful for comparing the molecular dynamics of a series of polymers.

The relaxation of polymers can also be interpreted using the so-called "model-free" approach to fitting of the relaxation data (10,11). In this approach it is assumed that the correlation function can be separated into a correlation function for slow reorientation (G_o) and a correlation function (G_i) for faster localized reorientation as

$$G(t) = G_o(t)G_i(t). \tag{5.26}$$

Implicit in this separation of correlation functions are the assumptions that the correlation times for the overall and internal motions are separated by orders of magnitude, the motions are uncoupled, and the internal motion is in the fast motion limit. This separation simplifies the relaxation analysis and a variety of motional models may be used for the molecular reorientation (isotropic, anisotropic, etc.) and the internal dynamics (isotropic, two-site jump, diffusion in a cone, etc.). This model has been used in several studies, including the successful fitting of the relaxation data as a function of temperature for poly(n-butyl methacrylate) (10,11). As with the models using distributions of correlation times, this interpretation provides little insight into molecular details of polymer dynamics.

Another approach to relating the NMR relaxation rates to polymer dynamics is to use models that specifically account for the relaxation due to conformational transitions in long-chain macromolecules. The advantage of this approach is that the relaxation times are interpreted with a physically reasonable model for the dynamics. Such models include *gauche–trans* isomerizations and cooperative three- and four-bond transitions. Cooperative motions must be taken into account since the entire chain does not move during a conformational transition. The disadvantage of these models is that they are complex and many parameters may be required to fit the limited amount of relaxation data that can be acquired. Also, there may be several models that also fit the relaxation data.

One class of successful models used for the interpretation of relaxation data is based on the diamond lattice model, in which conformational transitions are modeled as jumps between sites on a tetrahedral lattice (12,13). These models include three- and four-bond jumps with no conformational bias. The correlation function for the diamond lattice model is (14)

$$G(t) = e^{-t/\theta}e^{-t/\rho}\text{erfc}(t/\rho)^{1/2}, \tag{5.27}$$

where $\theta = (4.74w_4)^{-1}$ and $\rho = \{K_D^2(3.16w_3 - 9.16w_4)\}^{-1}$, where K_D is a lattice constant and w_3 and w_4 are the transition rates for the three- and four-bond jumps. The corresponding spectral densities are given by

$$J(\omega) = \frac{2\theta\rho(\theta-\rho)}{(\theta-\rho)^2 + \omega^2\theta^2\rho^2} \left\{ \left(\frac{\theta}{2\rho}\right)^{1/2} \left[\frac{(1+\omega^2\theta^2)^{1/2}+1}{1+\omega^2\theta^2}\right]^{1/2} \right.$$

$$\left. + \left(\frac{\theta}{2\rho}\right)^{1/2} \left[\frac{\omega\rho\theta}{(\theta-\rho)}\right] \left[\frac{(1+\omega^2\theta^2)^{1/2}-1}{1+\omega^2\theta^2}\right]^{1/2} - 1 \right\}. \quad (5.28)$$

Another approach is to model conformational transitions in polymers as a damped diffusional process along the chemical sequence. Several types of transitions are classified in Hall–Helfand (HH) theory (15), depending on the displacement on the polymer tails. Some transitions (crankshaft motions) lead to no change in the polymer tails while others do. The autocorrelation function is given by

$$G(t) = e^{t/\tau_2} e^{t/\tau_1} I_0(t/\tau_1), \quad (5.29)$$

where I_0 is the modified zeroth-order Bessel function, τ_1 is the correlation time for correlated jumps responsible for orientation diffusion along the chain, and τ_2 is the correlation time corresponding to damping from either nonpropagative specific motions or distortions of the chain with respect to its most stable local conformations. The spectral density is obtained by Fourier transformation and is given by

$$J(\omega) = \text{Re}\left[\frac{1}{(\alpha+i\beta)^{1/2}}\right], \quad (5.30)$$

where $\alpha = 1/\tau_2^2 + 2/\tau_1\tau_2 - \omega^2$ and $\beta = -2\omega(1/\tau_1 + 1/\tau_2)$. Modifications to this model have been introduced that account for correlated motions for nonneighboring bonds, but these models have a large number of parameters. The use of this model with the first-order modified Bessel functions has been used to account for the higher-order correlations while reducing the number of parameters by Voivy, Monnerie, and Brochon (VMB) (14).

These models have been used for the interpretation of the relaxation data in a variety of polymer systems (16,17). As might be expected, the unrealistic isotropic motion model gives a poor fit to the data, and better fits are obtained with the HH and VMB models, although the fit is not perfect over the entire temperature range.

These data led to the proposal that fast anisotropic motion also contributes to the relaxation. An excellent fit to the data was obtained by introducing an additional term for fast anisotropic reorientation of the CH vector in a cone of half-angle θ about the rest position (16,18),

$$G(t) = (1-a)e^{-t/\tau_1}e^{-t/\tau_2}I_0(t/\tau_1) + ae^{-t/\tau_0}e^{-t/\tau_1}e^{-t/\tau_2}I_0(t/\tau_1), \quad (5.31)$$

where τ_0 is the correlation time associated with the local anisotropic reorientation and

$$1 - a = \left[\frac{\cos\theta - \cos^3\theta}{2(1-\cos\theta)}\right]^2. \quad (5.32)$$

Good fits were obtained to the relaxation data for several polymers in bulk and in solution, and the authors suggest that such local anisotropic fluctuations are a general feature of polymer relaxation (16,17).

5.2.3 Observation of Polymer Relaxation in Solution

NMR spectroscopy, primarily carbon NMR, has been extensively used to study the molecular dynamics of polymers in solution and in polymers at temperatures above T_g, where the polymer lines are partially narrowed by molecular motion and solution-like spectra can be obtained. As a general rule, the relaxation is measured over the largest range of temperature and magnetic field available and fitted to a variety of relaxation models that range from the very simple, like isotropic motion, to the more complex. These data are fit either to models with a distribution or correlation times or to more physically meaningful models.

The molecular dynamics of polymers in solution have been investigated as a function of temperature, solvent, and magnetic field, and it has been generally reported that the isotropic motion correlation function [Eq. (5.3)] is a poor model for describing the complex dynamics of polymers. Later in this chapter we will examine some of the models for spectral densities that provide a better description of the relaxation in polymers. However, even without such an analysis, certain of the general principles governing the relaxation in polymers can be gleaned from an analysis of the raw data and the correlation times calculated assuming a single exponential correlation function. Some of these data are compiled in Table 5.1. For the purposes of comparison, all of the data were gathered for polymers in the fast motion limit ($\omega\tau_c \ll 1$) with the exception of poly(butene-1-sulfone).

TABLE 5.1
^{13}C Spin–Lattice Relaxation Times and Approximate Correlated Times Calculated from the Isotropic Diffusion Model[a]

Polymer	Solvent[b]	Temperature (°C)	Concentration (wt%)	τ_c (ns)
Polyethylene	TCB	110	33	0.019
	ODCB	100	25	0.018
	ODCB	30	25	0.040
Polypropylene	ODCB	100	25	0.044
		30	25	0.13
Polyisobutylene	CDCl$_3$	30	5	0.16
Polybutene-1	CCl$_3$CHCl$_2$	100	10	0.14
1,4-Polybutadiene				
cis	CDCl$_3$	54	20	0.016
trans	CDCl$_3$	54	20	0.021
Poly(vinyl chloride)	TCB	107	10	0.15
Poly(vinylidine chloride)	HMPA-d_{18}	40	15	0.63
		89	15	0.20
Poly(vinylidine fluoride)	DMF	41	20	0.079
Poly(vinyl alcohol)	DMSO	30	20	0.55
Polyacrylonitrile	DMSO	50	20	0.33
Polystyrene				
Atactic	Toluene-d_8	30	15	0.49
Isotactic	Toluene-d_8	30	15	0.55
Poly(methyl methacrylate)				
Isotactic	CDCl$_3$	38	10	0.39
Syndiotactic	CDCl$_3$	38	10	0.62
Atactic	DMF	41	20	0.81
Poly(α-methyl styrene)	CDCl$_3$	30	10	0.49
Poly(oxymethylene)	HFIP	30	3	0.082
Poly(ethylene oxide)	C$_6$D$_6$	30	5	0.018
Poly(propylene oxide)	CDCl$_3$		5	0.049
Poly(methyl thiirane)	CDCl$_3$	30	10	0.057
Poly(phenyl thiirane)	CDCl$_3$	25	15	0.25
Poly(styrene peroxide)	CHCl$_3$	23	22	0.21
Poly(butene-1-sulfone)	CDCl$_3$	40	25	23

[a] F. Bovey and L. Jelinski, *J. Phys. Chem.* **89**, 571 (1985).
[b] Solvent abbreviations: TCB, 1,2,4-trichlorobenzene; ODCB, o-dichlorbenzene; HMPA, hexa(methyl-d_3) phosphoramide; DMF, dimethylformamide; DMSO, dimethyl sulfoxide; HFIP, hexafluoro-2-propanol.

From these rough data we can reach the following conclusions:

1. *Effect of side chains.* Comparison of polyethylene with polypropylene, polybutene-1, and poly(vinyl chloride) shows that the presence of side chains impedes chain mobility and that the larger the side chain the slower the motion. A methyl group slows the chain by a factor of 2–3 and a chlorine is comparable, while ethyl and

phenyl groups have a further impeding effect. The main-chain conformational barrier may be regarded as composed of a symmetrical threefold torsional potential on which is superimposed nonbonded interactions between the side groups, the latter increasing with the size of the side groups. The above findings are readily interpretable in these terms. It may at first seem surprising that the presence of two α-substituents, as in polyisobutylene, poly(vinylidine chloride), and poly(methyl methacrylate), does not impede chain motions significantly more than a single substituent. This may be attributed to the fact that two substituents increase the crowding in *all* rotational states, including the eclipsed states, without necessarily increasing the barrier between them.

2. *Effect of heteroatoms.* The chains of poly(ethylene oxide) are among the most flexible in Table 5.1, with a correlation time at 30°C comparable to that of polystyrene at 100°C. Poly(propylene oxide) shows an analogous relationship to polypropylene. Poly(methyl thiirane), poly(propylene sulfide), and poly(phenyl thiirane) show the liberating influence of a sulfur atom, and poly(styrene peroxide) shows the comparable effect of a peroxy link. These findings are consistent with the known low rotation barriers of heteroatoms, but it is not evident why poly(methylene oxide), with an oxygen atom inserted at every carbon atom, should be less flexible than even polyethylene. This has been rationalized on the basis of the energetic preference for *gauche* conformers in poly(methylene oxide), which tends to inhibit chain motions requiring a succession of both *trans* and *gauche* bonds intermixed. The addition of sulfone groups, in contrast to sulfur atoms alone, stiffens the chain markedly, as in poly(butene-1-sulfone). These chains show other unusual motional behavior.

3. *Effect of double bonds.* Although the double bonds prevent rotation at the olefinic carbons, the reduced barriers at the allyic bonds more than compensate for this and, as a result, both *cis*- and *trans*-1,4-polybutadiene are substantially more flexible than polyethylene.

4. *Effect of stereochemistry.* A dependence of chain mobility on stereochemical configuration has been observed in vinyl polymers, both for entire chains and for steric sequences in atactic chains. Syndiotactic sequences commonly appear to be more restricted than isotactic sequences. This has been reported for poly(methyl methacrylate), polypropylene, polybutene-1, and poly(vinyl chloride), while the reverse is true for polystyrene.

In Table 5.1 we compared the T_1's and correlation times for a variety of polymers by assuming that the relaxation data could be fit to the single correlation time isotropic rotation model (1). While such a method is useful for gross comparisons, such a simple model typically does not fit the relaxation data over a range of temperatures and frequencies. When the relaxation is measured over a range of temperatures, a T_1 minimum is often observed, and the minimum values calculated from the isotropic motion model for polystyrene and poly(methyl methacrylate) are 50% smaller than the measured values (19). Similar observations have been made for poly(n-butyl methacrylate) and poly(n-hexyl methacrylate) (20), as well as for a variety of other polymers (17,21). The failure of the isotropic model for most polymers necessitates the use of more complex models.

The relaxation of polymer solutions can be well described by a distribution of relaxations times using the Cole–Cole (7), Fuoss–Kirkwood (8), or log χ^2 models (9). Table 5.2 compares the relaxation distributions for polystyrene, poly(methyl methacrylate), and poly(propylene oxide) at several temperatures and concentrations for relaxation data interpreted with the log χ^2 and Cole–Cole distribution functions (19). Examination of Table 5.2 shows several trends with respect to temperature and concentration. The width of the distribution (ε) tends to increase with increasing temperature, an observation consistent with slower motions having a higher activation energy. The increase is less obvious with more concentrated solutions, perhaps because of entanglements. It is also observed near ambient

TABLE 5.2
The Interpretation of NMR Relaxation Times Using the Log χ^2 and Cole–Cole Distribution Functions (19)

Polymer	Solvent[a]	Temperature (°C)	Concentration (wt%)	Log χ^2		Cole–Cole	
				$-\log(\tau_0)$	p	$-\log(\tau_0)$	ε
Polystyrene	CCl$_3$CHCl$_2$	30	5.5	8.35	20	8.35	0.72
		100	5.5	9.3	30	9.28	0.80
		30	19	8.0	20	8.15	0.72
		100	19	9.15	20	9.1	0.72
Poly(methyl methacrylate)	ODCB	40	7	8.8	9	9	0.57
		100	7	9.6	20	9.64	0.65
		40	24	8.0	8	8.2	0.55
		100	24	9.05	12	9.17	0.6
Poly(propylene oxide)	CDCl$_3$	−45	7	9.82	25	9.79	0.75
		5	7	10.38	40	10.27	0.91
		−45	21	9.4	20	9.36	0.7
		5	21	10.44	20	10.25	0.8

[a] ODCB, o-dichlorobenzene.

temperature that the distribution width for poly(propylene oxide) is much smaller than those for the stiffer polystyrene and poly(methyl methacrylate). However, as the temperature is decreased to $-45°C$ the width of the distribution approaches those observed for the stiffer polymers even though the correlation time is much faster. Thus, there appears to be a general relationship between the correlation time and the width of the distribution.

Figure 5.6 shows a plot of the relaxation times at three frequencies along with a fit to the Fuoss–Kirkwood distribution function for the protons and carbons of poly(4-vinyl pyridine) (22). A good fit is obtained over the entire temperature range by adjusting the value of β, the distribution width parameter, between 0.96 and 0.7. The value of β varies linearly with inverse temperature, and τ_0 varies with temperature according to $\tau_0 = Ae^{-E_a/RT}$, where $A = 1.3 \times 10^{-12}$ s and $E_a = 4$ kcal/mol. One important advantage of using such distributions for the interpretation of NMR relaxation data is that reasonable values of the activation energies E_a are obtained from the temperature studies.

While reasonable fits to the relaxation data can often be obtained using the distribution of correlation times for many polymers, such models are often not employed because they do not provide an insight into the dynamics on the molecular level. Also, in some cases the data on the low-frequency side of the T_1 minimum are not well fit by such models (1,19). The use of such models also assumes that the shape of the correlation function is temperature dependent, with the slower motions having higher activation energies than the more rapid motions. An experimental method has recently been suggested to determine if this is a valid assumption in the vicinity of $1/\omega_0$ (23). If the shape of the correlation function is temperature independent, then the relaxation times measured at different spectrometer frequencies can be superposed on a plot of ω_0/T_1 vs $\log[\omega_0\tau_c(T)]$, as shown in Fig. 5.7 for the proton relaxation rates of molten polybutadiene measured at 32, 60, and 100 MHz. Similar plots have shown that temperature-independent correlation functions are obtained for polyisoprene (21), deuterated 1,2-polybutadiene (24), and poly(1-naphthal acrylates) (25). With the restriction that the shape of the correlation function does not change with temperature, the distributions of correlation time models often provide a poor fit of the experimental data over the entire range of temperature and frequency (21).

In Section 5.2.2 several generalized models for relating molecular motion to the relaxation rates measured by NMR, including the diamond lattice (13), the Hall–Helfand (15), the Viovy–Monnerie–Brochon (14), and the de la Batie–Laupretre–Monnerie (17) models, were presented. Such models were introduced to give the

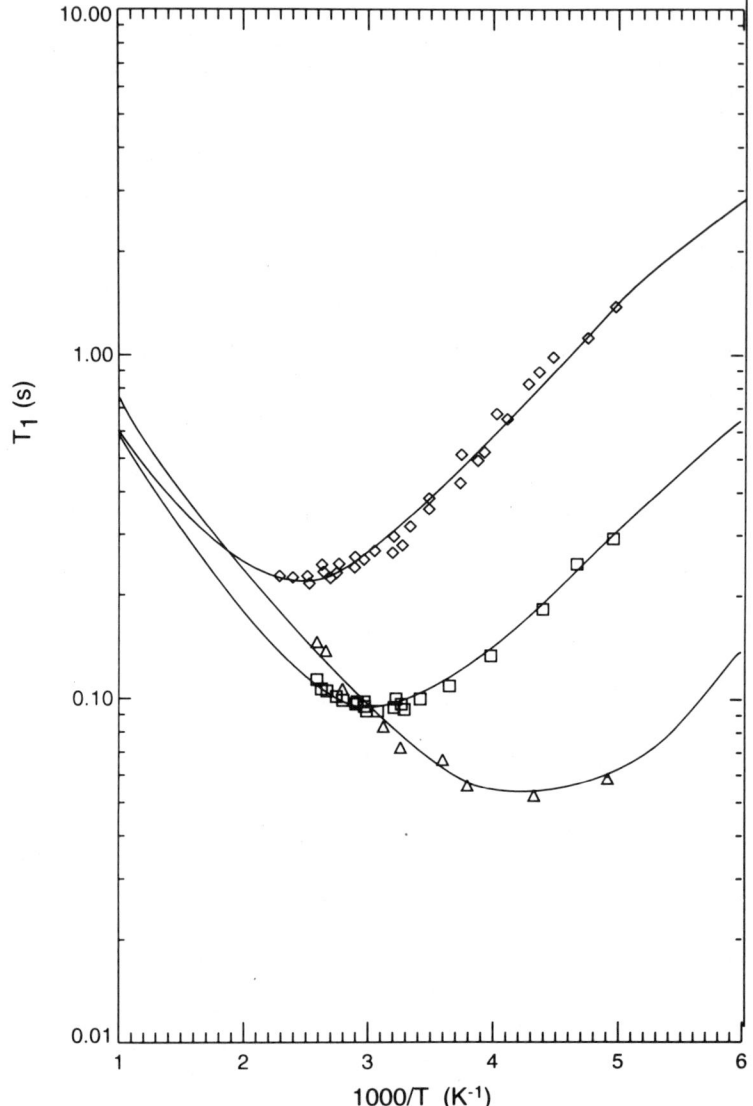

FIGURE 5.6 Plots of the spin–lattice relaxation time as a function of the inverse temperature for the main-chain carbons and protons of poly(4-vinyl pyridine) in methanol. The data are shown for protons at (◇) 250 MHz and (□) 100 MHz, and for (△) carbons at 25 MHz. The solid lines are the fits to the Fouss–Kirkwood distribution function. Adapted with permission (22).

interpretation of relaxation rates a physical basis, and because some observations suggest that the molecular motion is very closely related to the chain structure. In a ^{13}C NMR study it was observed that the relaxation rates for the methine and methylene carbons of 1,4-polybutadiene do not scale with the number of attached protons (5).

5.2. Polymer Dynamics in Solution 373

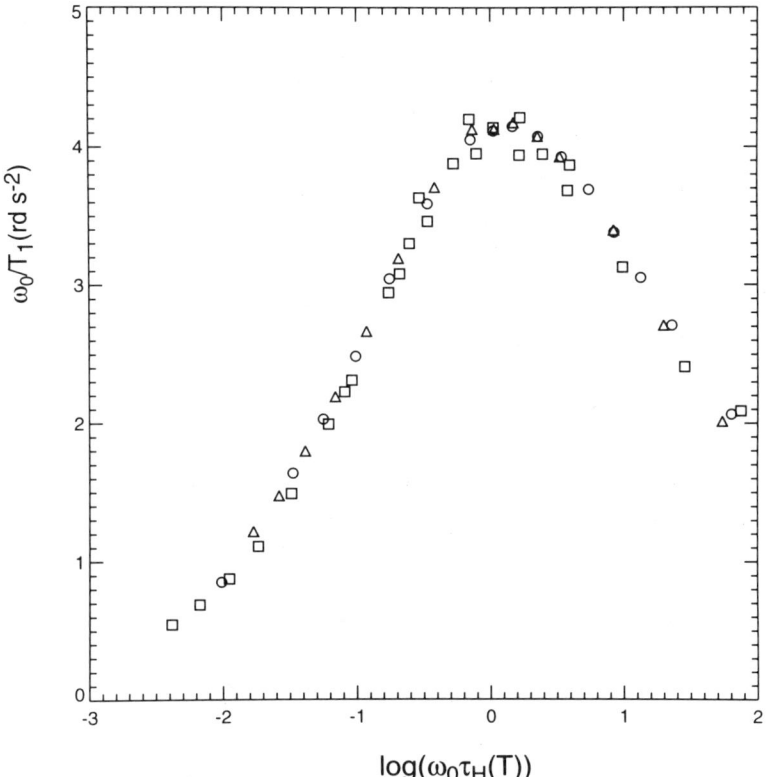

FIGURE 5.7 The normalized relaxation rate of the proton spin system of molten polybutadiene as a function of the reduced variable $\omega \tau_c(T)$. The symbols correspond to data gathered at (□) 32 MHz, (○) 60 MHz, and (△) 100 MHz. Adapted with permission (23).

Instead of the expected ratio of 2, the measured value was in the range of 1.5–1.7, suggesting that the molecular motion in polybutadiene was such that reorientation of the CH bond vectors in the methylene and methine groups was very different. Models for interpretation of the relaxation rates must therefore contain enough molecular detail to explain such observations.

The dynamics of several polymers, including bulk and solution-state polyisoprene, polybutadiene, poly(methyl vinyl ether), polyacrylates, polystyrenes, and poly(ethylene oxide), have been investigated by NMR relaxation and fitted to the models presented in this section. In the earlier studies these data were fitted to the diamond lattice model (13), which is characterized by the correlation times ρ and θ that are related to the probabilities for three- and four-bond jumps. Typical data for the ^{13}C T_1 and NOE are shown in Fig. 5.8 along with the fits

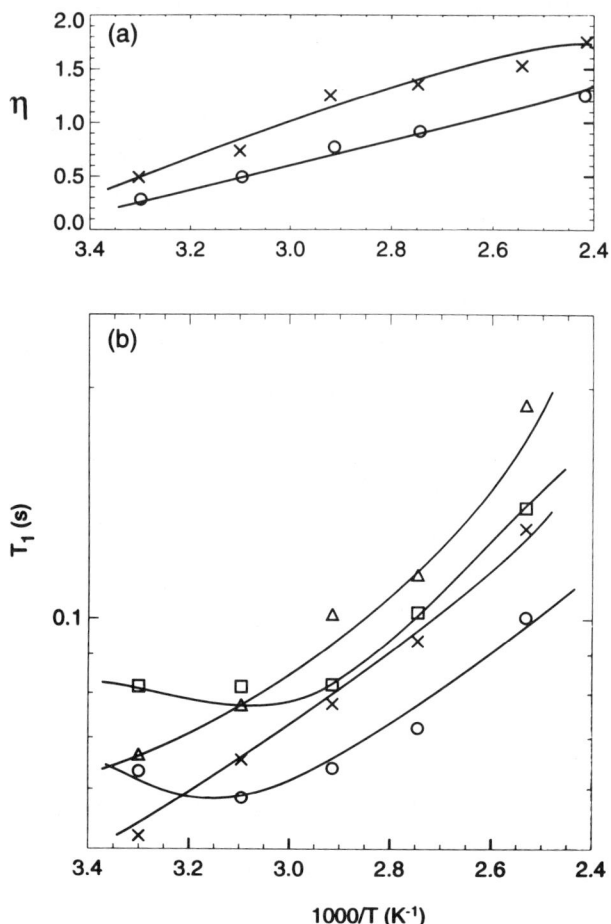

FIGURE 5.8 The (a) carbon–proton NOEs and (b) spin–lattice relaxation rates for polystyrene (MW = 110,000) as a function of temperature. The data are shown for methylene and methine carbons at (○) 400 mg/ml and (×)100 mg/ml, and the *ortho* and *meta* carbons at (□) 400 mg/ml and (△) 100 mg/ml. Adapted with permission (1).

using the diamond lattice model, and a summary of the correlation times as a function of concentration and molecular weight is given in Table 5.3 (19). Note that the values for ρ are similar to those measured using the distribution of correlation times (Table 5.2) and are not very sensitive to molecular weight. The dependence of the ratio of ρ/θ is due to the molecular weight dependence of θ. It should be noted that these data were derived from measurements of only two parameters at a single magnetic field and may not be a unique fit to the molecular dynamics.

5.2. Polymer Dynamics in Solution

TABLE 5.3
The Diamond Lattice Model Correlation Times for Polystyrene in Pentachloroethane Determined from ^{13}C NMR Relaxation Rate Measurements (19)

Rate[a]	MW	Conc. (wt%)	Temperature (K)				
			303	323	345	370	400
$-\log \rho$	110,000	5.5	(8.4)	8.7	8.92	9.09	9.32
ρ/θ	110,000	5.5	0.4	0.4	0.4	0.4	0.4
$-\log \rho$	110,000	19	(8.10)	(8.4)	8.7	9.13	9.49
ρ/θ	110,000	19	0.4	0.4	0.4	0.2	0.1
$-\log \rho$	10,000	5.5	8.3	8.57	8.80	9.20	9.55
ρ/θ	10,000	5.5	1.5	1.0	0.8	0.4	0.25
$-\log \rho$	10,000	19	(7.87)	(8.23)	8.41	8.73	9.17
ρ/θ	10,000	19	1.0	1.0	1.0	0.7	0.4

[a] The values in parentheses are subject to considerable uncertainty.

More recently the molecular dynamics of polymers in the solution and in the molten state have been successfully studied using a modified version of the Hall–Helfand model that includes an additional term to account for high-frequency libration (17). In a study of the molecular dynamics of poly(methyl vinyl ether) it was reported that the isotropic rotation and the Hall–Helfand and Viovy–Monnerie–Brochon autocorrelation functions provided a poor fit of the experimental data at several temperatures and frequencies, and underestimated the observed T_1 minimum. Inclusion of the extra term in the spectral densities to account for high-frequency librational motions [Eq. (5.31)] led to a good fit for the experimental data when $\tau_1/\tau_2 = 2$, $\tau_1/\tau_0 = 400$, and $a = 0.4$, where τ_1 is the correlation time for correlated jumps leading to orientational diffusion along the chain, τ_2 is the correlation time corresponding to nonpropagative motions or distortions, τ_0 is the correlation time for the librational motion, and a is the fraction of the relaxation due to the high-frequency librations. These parameters provide a good fit to the relaxation of molten poly(methyl vinyl ether) and poly(methyl vinyl ether) dissolved in chloroform (17).

Several of the above models were compared in a study of the local dynamics of polyisoprene (21). The relaxation times were measured at three frequencies over a range of temperatures, and composed a complete data set for testing the various models. Plots of the temperature–frequency superposition were used to show that the correlation function does not change shape as a function of temperature in the vicinity of $1/\omega_0$. Using this constraint it was not possible to fit all the relaxation data, particularly near the T_1 minimum, using the single exponential, the Cole–Cole, the Fuoss–Kirkwood, or the log χ^2 distribution functions. The Hall–Helfand model also provided a poor fit of the data. However, several of the models that had two well-separated time constants provided a good fit to the data. These included a biexponential correlation function, a Cole–Cole distribution modified to include a term for rapid libration, and the model combining the Hall–Helfand correlation function with a term for high-frequency libration. The best fit was obtained with the biexponential correlation function with $\tau_1/\tau_0 = 1000$ and $a = 0.49$. The best fits with the other models were obtained with a narrow width parameter and a separation of correlation times similar to that used for the biexponential correlation function.

The model combining the Hall–Helfand correlation function with a term for high-frequency libration has been successfully used to study a variety of polymers, and the fractions of the relaxation attributable to librations are compiled in Table 5.4 for polymers in solution and for bulk polymers above T_g (24). The trends in these data can be understood from simple steric arguments. Since double bonds are rigid, CH vectors attached to unsaturated carbons are expected to have a smaller range of librations. This explains why the relaxations for neighboring methine and methylene groups in polyisoprene and 1,4-polybutadiene have different NT_1 values (5). The side groups in vinyl polymers are expected to restrict the dynamics, and polymers with similarly sized groups, such as poly(methyl vinyl ether) and 1,2-polybutadiene, have similar librational contributions to the relaxation. Polymers with large side groups, such as poly(1-naphthylalkyl acrylates) have a smaller librational contributions to the relaxation. Only

TABLE 5.4
The Percentage of the Decay of the Correlation Function Due to Libration Motions in Several Polymers (24)

Polymer	Bond	%
Polyisoprene	CH	41
Polyisoprene	CH_2	48
Polyisobutylene	CH_2	21
1,2-Polybutadiene	CD/CD_2	42
Poly(methyl vinyl ether)	CH/CH_2	40
Poly(1-napthylacrylate)	CH	22
Poly(1-napthyl methyl acrylate)	CH	15
Poly(1-napthyl ethyl acrylate)	CH	16
Bulk polysiobutylene	CH_2	21
Bulk poly(methyl vinyl ether)	CH/CH_2	40

small differences are observed in the librational contributions to the relaxation for polymers in bulk and in solution.

The relaxation for polymers with longer side chains can be interpreted using either generalized motional models or the model-free approach. In a study of the relaxation of poly(n-alkyl acrylates) it was reported that the data could be fit with a model that incorporates a distribution of correlation times for the main chain and multiple internal rotations about the carbon–carbon bonds in the side chains (20). This is a rather complex correlation function with many parameters, only some of which can be easily related to the molecular structure. These same data were fit using the model-free approach using a double exponential correlation function for the main-chain resonances (11). After fitting the main-chain data, the generalized order parameter S^2 was calculated for each of the side-chain atoms by assuming that the motions of the side chain are in the fast motion limit and are independent of the main chain. As expected the generalized order parameter decreased and the internal motion correlation time increased with increasing distance from the main chain. It should be noted, however, that the correlation times for the terminal methyl group in poly(n-hexyl acrylate) were not physically reasonable. Although the model-free approach lacks the physical significance of the more generalized models, it provides a good basis for comparing the molecular dynamics of polymers. The model-free approach has also been used to study the dynamics of polymers with hindered side chains, such as poly[(S)-4-methyl-1-hexane] and poly(methyl vinyl ether) (26).

The NMR studies of the molecular dynamics of polymers have shown that local segmental motions make an important contribution to the relaxation of polymers. One manifestation of this observation is that the relaxation times are insensitive to molecular weight for high-molecular-weight polymers and do not scale with the macroscopic viscosity. However, the molecular dynamics are expected to depend on the solution viscosity. Assuming that the correlation time measured by NMR is inversely proportional to the rate constant for *gauche–trans* isomerization, the temperature and viscosity dependence of the correlation time is given by Kramer's (27) theory as

$$\tau_c = A\eta e^{E_a/RT}, \qquad (5.33)$$

where the prefactor A is a constant, η is the solution viscosity, and E_a is the activation energy for isomerization. The molecular dynamics of poly(ethylene oxide) (28,29) and polyisoprene (30) have been studied as a function of solution viscosity and it has been observed that the correlation times do not follow this simple relationship, but are instead given by a power law dependence

$$\tau_c = A\eta^\alpha e^{E_a/RT}, \qquad (5.34)$$

where the coefficient α is determined empirically. From the study of polyisoprene in 10 solvents covering a factor of 70 in viscosity, the value of α was determined to be 0.41 ± 0.02 at constant temperature (30), and Fig. 5.9 shows the superposition of these data over a range of temperatures. The breakdown of Kramer's theory as applied to polymer dynamics measured by NMR may be due to the fact that there is no clear separation in the time scales of the polymer motion and of the solvent.

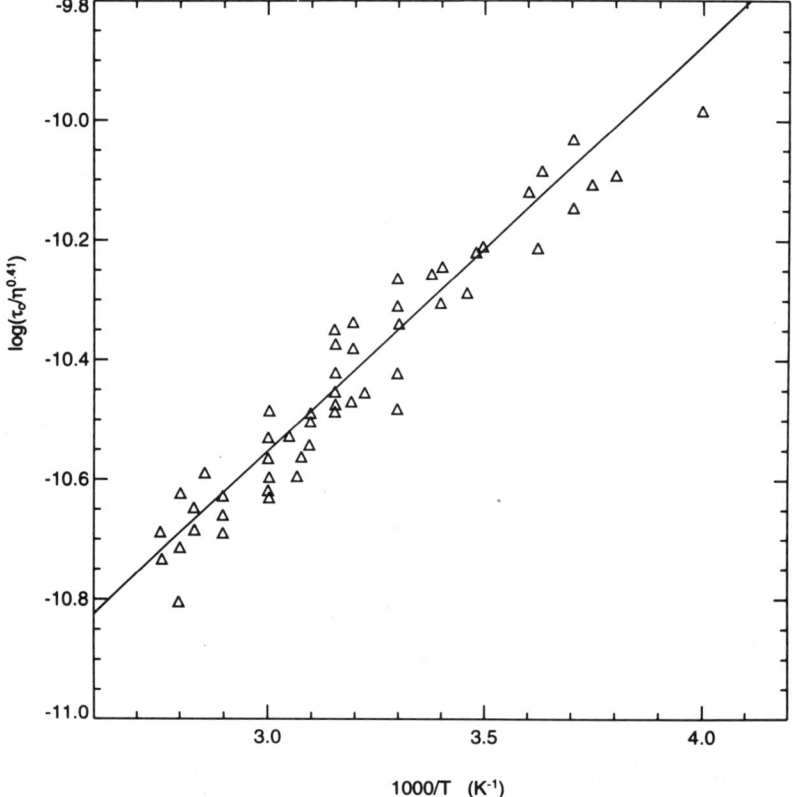

FIGURE 5.9 The temperature dependence of the reduced correlation time for the methylene carbons of polyisoprene measured in 10 solvents ranging in viscosity from 0.33 to 15.4 cP. Adapted with permission (30).

5.3 POLYMER DYNAMICS IN THE SOLID STATE

5.3.1 INTRODUCTION

The molecular dynamics of solid polymers have been extensively studied in order to understand the structure-function relationships in semicrystalline, amorphous, and rubbery materials. The intense interest in the molecular dynamics in solids is driven by the fact that most applications for polymers are in the solid state. The molecular dynamics are known to occur over a wide range of time scales, ranging from nanoseconds to seconds, and combinations of NMR methods have been used to study the dynamics over this wide range.

The study of the polymer dynamics in the solid state is different in many ways from the study of the solution dynamics because of the differences between solids and solutions. Chain motion is restricted in the solid state by neighboring chains so the chemical shift anisotropy and dipolar interactions are not averaged to near zero as they are in solution. The broadening due to chemical shift anisotropy and dipolar couplings makes it difficult to observe high-resolution spectra, and the spectra are usually acquired with magic angle spinning and high-power decoupling. However, if a site on the polymer is isotropically labeled or if the resonance is well resolved, such as a carbonyl resonance, the broadening due to chemical shift anisotropy can be observed. Molecular motion that is fast on the time scale of the chemical shift anisotropy causes a characteristic change in the lineshape that can be modeled, and when the motion becomes very fast, as for polymers near T_g, the lineshape collapses and a high-resolution spectrum is observed. The dipolar interactions are more difficult to observe directly from the spectra, but can be studied by 2D NMR methods. As with chemical shift anisotropy, molecular motion averages the dipolar lineshape in a way that depends on the geometry of the motion and its rate. Deuterium NMR has also been extensively used to study the dynamics of polymers. The quadrupolar coupling is on the order of 130 kHz, and the lineshape is sensitive to motions on this time scale. The lineshape again depends on the geometry of molecular motion, and lineshape changes due to aromatic ring flips or methyl rotation are easily recognizable.

The methods for studying the dynamics of polymers over a wide time scale can be classified as lineshape methods, relaxation rate measurements, or ultraslow exchange measured by 2D NMR. Each of these methods is sensitive to the dynamics over a range of time scales that depend on the type of interactions, and Fig. 5.10 shows a plot of the time scale over which each of the methods is most sensitive. The fastest motions can be measured most efficiently by the relaxation

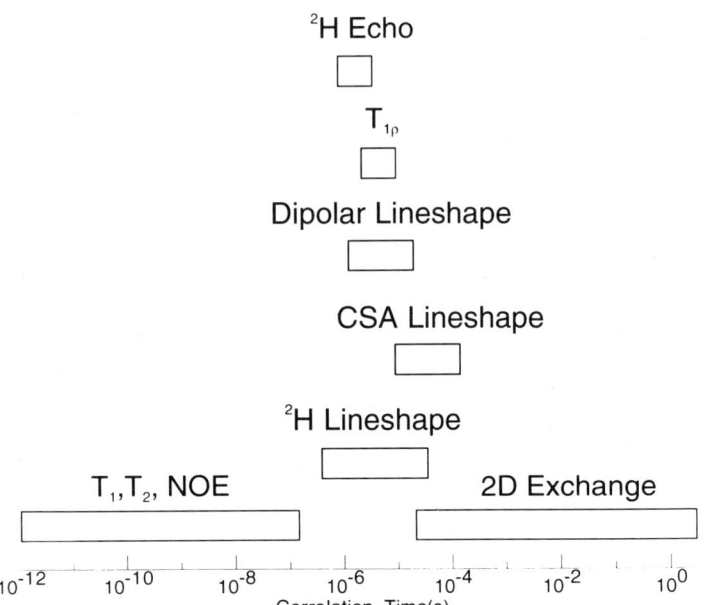

FIGURE 5.10 A plot showing the dynamics that can be measured by lineshape measurements, relaxation rate measurements, and ultraslow exchange.

times T_1 and T_2. Motions in the intermediate time scale ($10^{-7} > \tau_c > 10^{-3}$) can be measured using a variety of methods including changes in the deuterium, dipolar or chemical shift anisotropy lineshape, the $T_{1\rho}$, or the echo intensity in deuterium NMR. Motions in the slowest regime can be measured using 2D real-time exchange measurements of the deuterium or chemical shift anisotropy lineshape. The upper limit on the dynamics that can be studied by the 2D spin alignment or spin exchange is the T_1 of the material under study.

5.3.1.1 Polymer Dynamics and NMR Lineshapes

The observation of NMR lineshapes, particularly those due to quadrupolar coupling in deuterated polymers or the chemical shift anisotropy in the carbon, silicon, phosphorus, or nitrogen spectra is a rich source of information about the molecular dynamics of polymers. The lineshapes have a characteristic appearance in the absence of molecular motion, and the changes in lineshape due to the molecular dynamics depend on the type of molecular motion and the time scale over which the motion occurs. If the motion has a large amplitude and a nearly isotropic angular distribution, as for polymers above T_g, then solution-like spectra are obtained.

5.3. Polymer Dynamics in the Solid State

Deuterium NMR is often used to study polymer dynamics because the relaxation is dominated by the quadrupolar coupling, so the relaxation mechanism is well characterized (3). Deuterium has a low sensitivity, so polymers with site-specific labels are required. This is a disadvantage in that new polymers must be synthesized, but an advantage in the sense that the labels are incorporated into the polymer at well-defined sites, such as the main-chain or the side-chain aromatic rings. In the absence of molecular motion the frequency of a given deuteron is given by

$$\omega = \omega_0 \pm \delta(3\cos^2\theta - 1 - \eta\sin^2\theta\cos 2\phi), \qquad (5.35)$$

where ω_0 is the resonance frequency, $\delta = 3e^2qQ/8\hbar$ ($e^2qQ/\hbar$ is the quadrupolar coupling constant), η is the asymmetry parameter, and the orientation of the magnetic field in the principal axis of the electric field gradient tensor is specified by the angles θ and ϕ. For rigid solids $\delta/2\pi = 62.5$ kHz and $\eta \approx 0$ for C–D bonds. Thus, two lines are observed corresponding to the transitions for each deuteron. In isotropic samples, averaging over all possible orientations gives rise to the well-known "Pake" spectrum (31) shown in Fig. 5.11.

Molecular motion of the polymer will cause the lineshapes to change in a way that depends on the geometry and time scale of the molecular motion. If the motion is rapid on the time scale defined by the coupling constant, $1/\delta$, it is said to be in the fast motion limit ($\tau_c < 10^{-7}$ s). This leads to a characteristic change in lineshape given by

$$\omega = \omega_0 \pm \bar{\delta}(3\cos^2\theta - 1 - \bar{\eta}\sin^2\theta\cos 2\phi), \qquad (5.36)$$

where $\bar{\delta}$ and $\bar{\eta}$ are the coupling constant and asymmetry parameter for the *averaged* electric field gradient tensor. It is important to note that $\bar{\eta}$ may be different from zero even though $\eta = 0$. This is illustrated in Fig. 5.12 for kink 3-bond rotation, crankshaft 5-bond motion, and aromatic ring flips (3). Such motions give rise to $\bar{\eta}$ values of 1, 0, and 0.6, and illustrate how different types of molecular motion result in characteristic deuterium lineshapes.

^{2}H NMR is a powerful method for the study of the molecular dynamics of polymers, but caution must be used since it is possible for different motions to give rise to the same values for $\bar{\delta}$ and $\bar{\eta}$, and the same lineshape. For example, the same $\bar{\eta} = 1$ pattern is obtained for a two-site jump model with a jump angle of 90° when the population of site two is twice that of site one, for a two-site jump with equal populations and an angle of 109°, or with restricted diffusion about a tetrahedral axis with a distribution width of 90° (32). Such models can be distinguished via the anisotropic relaxation of the lineshape

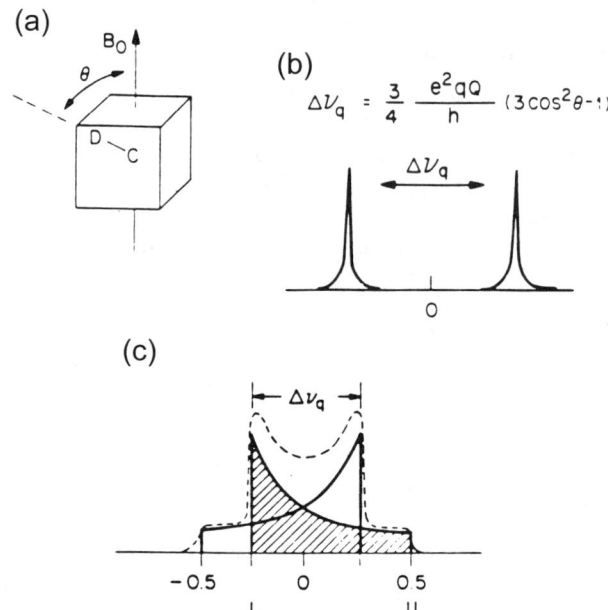

FIGURE 5.11 The deuterium NMR lineshape. For a given C–D bond vector (a) with an orientation of θ with respect to the magnetic field, two transitions are observed (b) with a splitting given by $\frac{3}{4}(e^2qQ/h)(3\cos^2\theta - 1)$. For isotropic samples the observed spectrum (c) results from averaging over all possible values of θ. Adapted with permission (F. Bovey, "Nuclear Magnetic Resonance Spectroscopy," 2nd ed. Academic Press, New York, 1988).

using the Jeener–Broekaert pulse sequence shown in Fig. 5.13 to measure T_{1Q}, the relaxation of the spin-aligned state (33). The T_{1Q}-distorted lineshape is given by

$$I(\nu, \tau_2) = \frac{1}{4\pi} \int_0^{2\pi} \int_0^{\pi} \{W[\nu - \nu_+^*(\theta, \phi)] + W[\nu - \nu_-^*(\phi, \theta)]\} e^{-\tau_2/T_{1Q}(\theta, \phi)} \sin\theta\, d\theta\, d\phi, \quad (5.37)$$

where I is the intensity as a function of frequency and the delay time τ_2, θ and φ are the spherical coordinates of the magnetic field in the molecular frame, $W(\nu)$ is the broadening function accounting for unresolved dipolar interactions and inhomogeneous magnetic fields, and $\nu_\pm^*$ is the rapid exchange frequency of either of the two spectral lines for each orientation which are given by the motionally averaged quadrupole tensor. The T_{1Q} relaxation time is given by

$$\frac{1}{T_{1Q}(\theta, \phi)} = \frac{9}{4}\pi^2 \nu_Q^2 J(\omega_0), \quad (5.38)$$

5.3. Polymer Dynamics in the Solid State

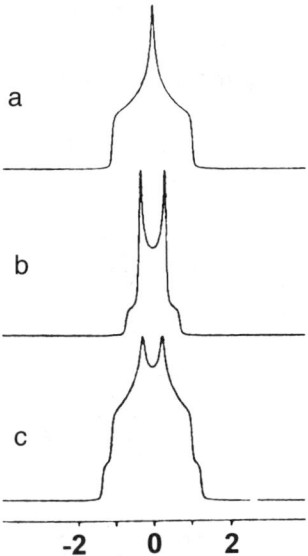

FIGURE 5.12 Theoretical deuterium NMR lineshapes for deuterons undergoing (a) kink 3-bond motion, (b) crankshaft 5-bond motion, and (c) 180° flips, as in aromatic rings. The rate of molecular motion is assumed to be in the fast motion limit. Adapted with permission (3).

where ν_Q is the quadrupolar coupling constant and $J(\omega_0)$ is the spectral density calculated for the particular model of molecular motion (34). For a two-site jump the spectral density is given by

$$J(\omega_0) = \frac{3(1 - \sin^2\theta \cos^2\phi - \sin^2 2\theta \sin^2\phi) P_1 P_2 \tau_c \sin^2\alpha}{1 + \omega_0^2 \tau_c^2}, \quad (5.39)$$

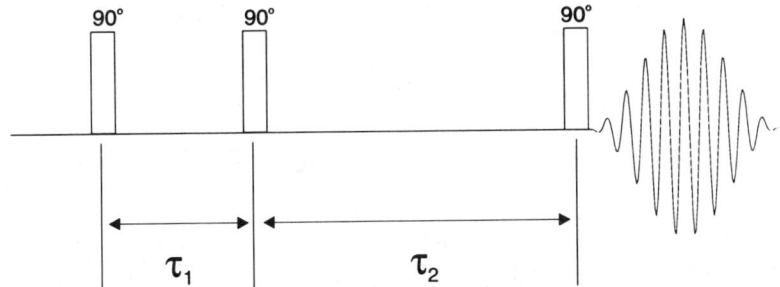

FIGURE 5.13 The pulse sequence diagram for the Jeener–Broekaert pulse sequence used to measure the T_{1Q}-distorted lineshapes. The lineshape is measured as a function of the delay times τ_1 and τ_2.

where α is the jump angle and the correlation time τ_c is given by the inverse of the sum of the rate constants for jumping between sites one and two $(k_{12} + k_{21})^{-1}$. Figure 5.14 compares the T_{1Q}-distorted lineshapes for the three models of molecular motion detailed above as a function of the delay time τ_2 in the Jeener–Broekaert pulse sequence.

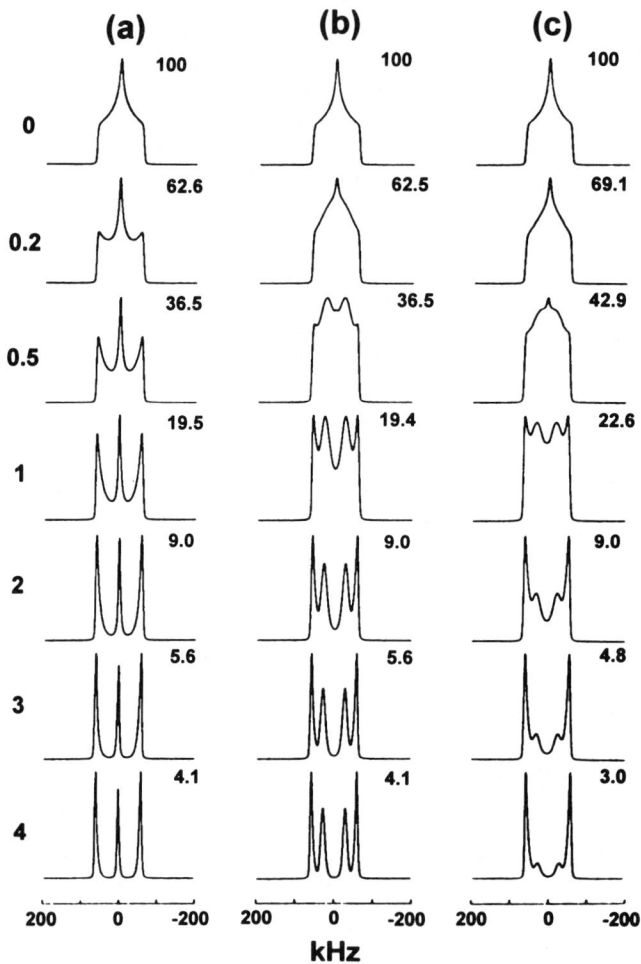

FIGURE 5.14 The T_{1Q}-distorted lineshapes calculated for (a) the two-site jump ($\alpha = 90°$, $P_1 = 2P_2$), (b) another two-site jump model ($\alpha = 109.5°$, $P_1 = P_2$), and (c) a restricted diffusion model ($\beta = 109.5°$, $\psi = 70°$). Note that identical lineshapes are observed for short τ_2 delays, but that differences in lineshape become obvious at longer values. The values to the left of the spectra are the delay times τ_2 and the values to the right are the integrated intensities. Adapted with permission (32).

5.3. Polymer Dynamics in the Solid State

It is clear from this comparison that the motional models giving rise to the same deuterium lineshape can be distinguished via the T_{1Q}-distorted lineshapes.

In those cases where the molecular dynamics are on the intermediate time scale, $\tau_c \approx 1/\delta$, the lineshapes cannot be so easily calculated. However, information about the molecular dynamics can be obtained using echo methods. Deuterium spectra are typically very broad and cannot be acquired using the 90° pulse–acquire pulse sequences because much of the data is lost during the period after the pulse during which the electronics recover after the strong RF pulse. These data are usually acquired with the quadrupolar echo pulse sequence shown in Fig. 5.15. Data collection begins at the top of the quadrupolar echo, which avoids collecting data in the period immediately following the pulses. However, if molecular motion occurs during the echo delay times, it will not be refocused and changes in the lineshape and intensity may be observed as a function of the echo delay time (3).

If the motion is in the slow motion limit, $\tau_c < 10^{-4}$ s, then the lineshape observed with the quadrupolar echo pulse sequence will be for the most part insensitive to the molecular motion. If the correlation time is less than the spin–lattice relaxation time then characteristic changes in the lineshape can be observed with pulse sequences that resemble the 2D NOESY pulse sequence but which differ significantly in the details (Fig. 5.16). As in the NOESY spectra, the frequencies of the deuterons in the t_1 dimension are correlated with those in the t_2 dimension. If no molecular motion occurs during the mixing time then the signals for a particular deuteron are the same in both dimensions and the signals appear along the diagonal. Thus,

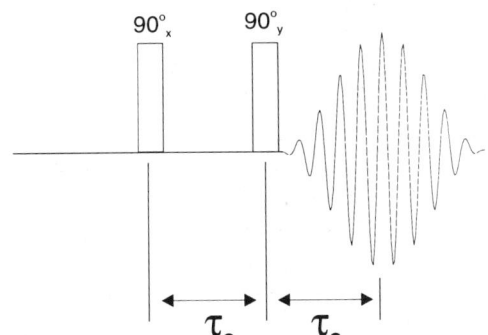

FIGURE 5.15 The pulse sequence diagram for the quadrupolar echo pulse sequence. Data acquisition begins after the delay time τ following the second pulse.

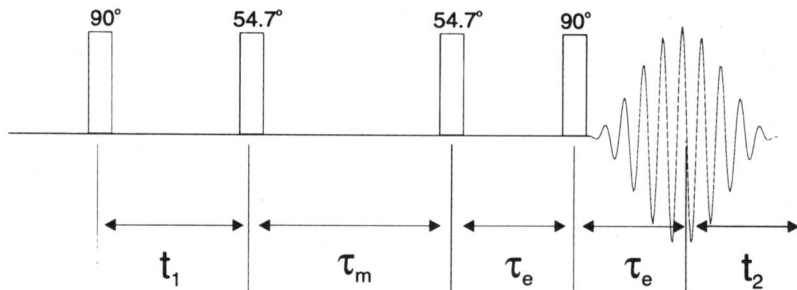

FIGURE 5.16 The pulse sequence diagram for measuring ultraslow exchange in polymers. As in other 2D NMR experiments, many spectra are acquired as a function of the t_1 delay and Fourier transformed to give the spectrum with two frequency axes. Off-diagonal intensity is generated by molecular motion during the mixing time.

for mixing times that are short compared to the correlation time a normal Pake pattern appears along the diagonal. If the deuterons reorient during the mixing time then they have different frequencies in the t_1 and t_2 dimensions and a ridge is observed in the 2D plane that depends on the angle of reorientation. Such a ridge pattern is observed because the frequency depends on the orientation of the deuteron in the magnetic field. Figure 5.17 shows the 2D deuterium exchange spectrum of deuterated dimethyl sulfone, which is known to undergo a two-site jump in the solid state (35). The jump angle is determined from the off-diagonal ellipses as

$$|\tan \theta| = \frac{a}{b}, \qquad (5.40)$$

where a and b are the principal axes of the ellipses, parallel and perpendicular to the diagonal. The data for dimethyl sulfone yield a jump angle of 106°, which compares favorably with the value of 103° measured from X-ray experiments. This analysis shows how the jump angle can be measured directly from the experimental data. Such an analysis is appropriate for dimethyl sulfone and crystalline polymers, but not for amorphous materials where the jumps between sites are less well defined. To study the dynamics of amorphous polymers the experimental spectra are compared with spectra calculated for a distribution of correlation times and jump angles.

The study of the molecular dynamics using the chemical shift anisotropy of carbon, nitrogen, phosphorus, or silicon is analogous to the deuterium studies. The chemical shift depends on the orientation

5.3. Polymer Dynamics in the Solid State

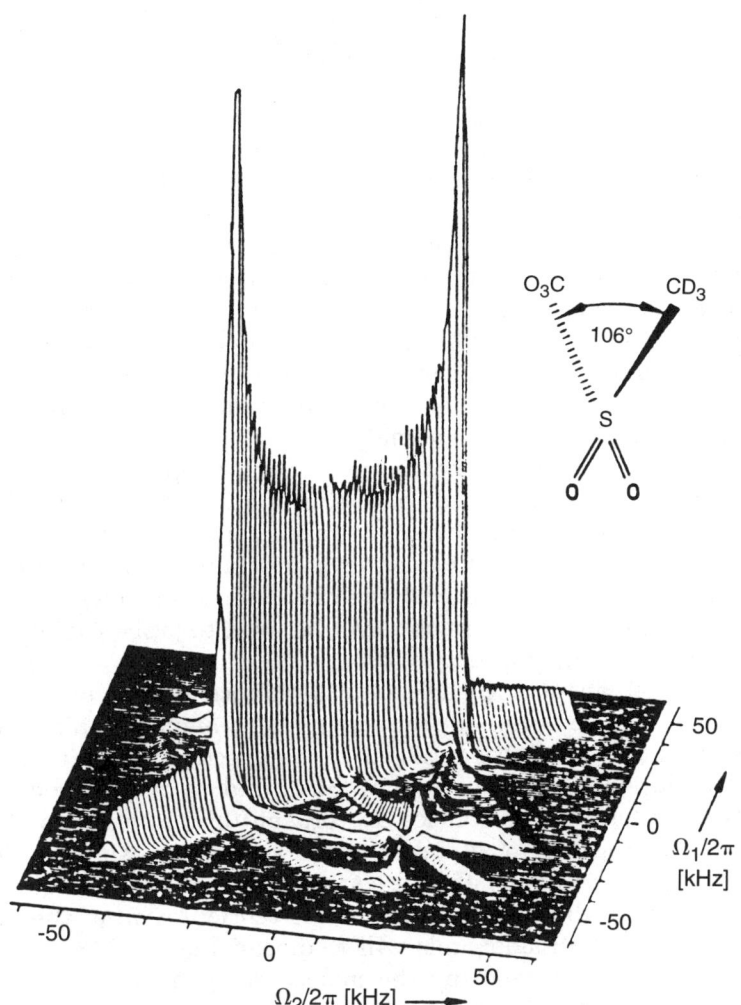

FIGURE 5.17 The 2D deuterium exchange spectrum of dimethyl sulfone. Note the off-diagonal ridges that arise from the two-site jump during the mixing time. Adapted with permission (92).

of the principal axis of the chemical tensor in the magnetic field and is given by

$$\omega = \omega_0 + \tfrac{1}{2}\delta(3\cos^2\theta - 1 - \eta\sin^2\theta\cos 2\phi), \quad (5.41)$$

where δ is the strength of the anisotropic coupling given by $\delta = \tfrac{2}{3}\Delta\sigma\gamma B_0$. The chemical shift anisotropy is typically much smaller than the quadrupolar coupling with deuterium (0-4 kHz vs 62 kHz) and therefore is sensitive to molecular motions on a different time scale.

The procedure for using the chemical shift anisotropy to study the dynamics is similar to that of the deuterium studies. However, since the shielding varies from atom to atom, a spectrum first must be acquired at a low enough temperature to quench the molecular motion so that the anisotropy in the absence of motion can be measured. Site-specific isotopic labeling is often used for low-sensitivity nuclei like carbon and nitrogen. Also, since anisotropy patterns tend to overlap in the carbon spectra it is important to study well-resolved resonances, such as carbonyls or methyl groups, or to isotopically enrich the carbon of interest so its signals are much stronger than the natural abundance signals.

As with deuterium, if the motion is anisotropic but in the fast motion limit on the time scale of chemical shift anisotropy, then the resonance position is given by an averaged value of the chemical shift anisotropy as in Eq. (5.36) for the deuterium lineshape. Complex lineshapes are observed for exchange in the intermediate time scale, and 2D NOESY-like experiments can be used to probe the dynamics in the slow exchange limit.

Information about the molecular dynamics in the slow motion regime can also be obtained from slow magic angle spinning experiments. As shown in Section 1.8.5.3, if the spectrum is acquired with magic angle spinning that is slow compared to the chemical shift anisotropy, then the resonances are broken up into a series of sidebands. It is possible to suppress these sidebands using the TOSS pulse sequence that consists of a series of 180° pulses (36). However, if there is molecular motion that causes reorientation of the atoms, then the suppressed sidebands will reappear. Figure 5.18 shows the pulse sequence for measuring the molecular dynamics via the reappearance of sidebands (37). During the preparation period the sidebands are suppressed with the TOSS pulse sequence and the resulting magnetization is stored along the $-z$ axis during the mixing time, which is an integral number of rotor periods, and the signal is detected with the final pulse. The rate and geometry of reorientation can be determined from the appearance of sidebands as a function of the mixing time. The 2D version of this experiment has also been used to study the molecular dynamics of polymers (38,39).

Information about the molecular dynamics of polymers can also be obtained from the heteronuclear dipolar couplings in 1D or 2D NMR experiments. In high-resolution cross-polarization and magic angle spinning studies the carbon spectra are usually acquired with high-powered (ca. 50 kHz) proton decoupling. However, if the molecular

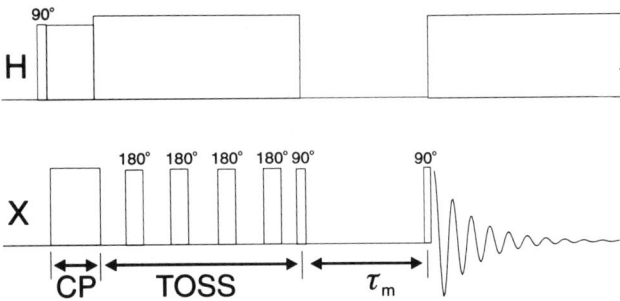

FIGURE 5.18 The pulse sequence diagram for measuring slow exchange via the reappearance of TOSS-suppressed spinning sidebands (37). After cross polarization and sideband suppression with the TOSS pulse sequence, the magnetization is stored along the $-z$ axis during the mixing time. Molecular motion leads to the reappearance of the spinning sidebands.

motion is on the same time scale as the decoupling field, interference with the decoupling will be observed and the spin–spin relaxation rate will be given by (40)

$$\frac{1}{T_2} = \frac{\gamma_C^2 \gamma_H^2 \hbar^2}{5r^6} \left(\frac{\tau_c}{1 + \omega_2^2 \tau_c^2} \right), \quad (5.42)$$

where ω_2 is the decoupler field strength. When $\omega_2^2 \tau_c^2 \approx 1$ the linewidth will be at a maximum and the signal will be broadened beyond detection.

The heteronuclear lineshapes can be measured in a 2D NMR experiment using rotational dipolar spin echoes with the pulse sequence shown in Fig. 5.19 (41). In such an experiment the magnetization is sampled after two rotor periods with a carbon 180° pulse to refocus the magnetization after one rotor period. Proton–carbon decoupling is applied during the first rotor period, and the t_1 period is incremented by including an increasing number of multipulse proton decoupling WAHUHA pulse cycles during the second rotor period. The spinning speed is chosen such that 16 WAHUHA pulse cycles fit evenly into one rotor cycle. The effect of the WAHUHA pulse cycles is to decouple the protons from each other such that the signal is modulated by the $^{13}C-^1H$ coupling alone. Fourier transformation of the signal intensities as a function of the number of WAHUHA pulse cycles gives the lineshape from the $^{13}C-^1H$ dipolar coupling. As for the deuterium and chemical shift anisotropy lineshapes, molecular motion averages the lineshape in a way that depends on the rate and amplitude of the molecular motion.

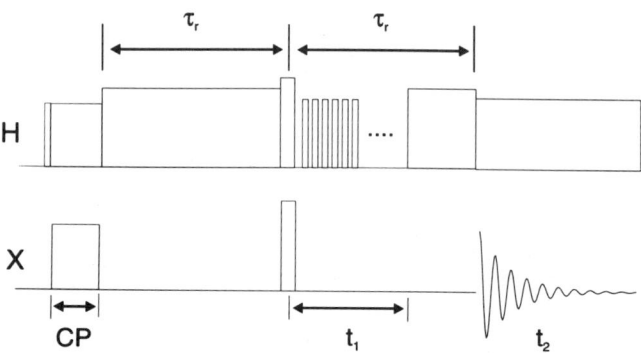

FIGURE 5.19 The pulse sequence diagram for the rotational dipolar spin echo experiment (41). The t_1 period is incremented by including increasing numbers of WAHUHA proton decoupling pulse cycles in the second rotor period.

5.3.1.2 Polymer Dynamics and Relaxation Rates

The NMR relaxation times can provide important information about the molecular dynamics of polymers in the solid state. Some of these measurements, such as T_1's, T_2's, and NOEs, are similar to those measured in solution, while others like $T_{1\rho}$ and T_{CH} utilize cross polarization, which is not effective for polymers in solution.

The carbon and proton spin–lattice relaxation times provide gross information about the molecular dynamics of polymers, but it is frequently difficult to interpret the relaxation times in terms of the correlation times, as has been effectively done for polymers in solution (Section 5.2). It is particular difficult for protons because proton spin diffusion is very efficient in the solid state, so the measured relaxation time represents an average for the entire spin system. If the polymer contains methyl groups, then methyl group rotation is the most effective mechanism to cause relaxation and will dominate the relaxation of the entire spin system, limiting the information about the molecular dynamics that can be obtained from such measurements. Proton T_1 measurements are most useful for studying the structure of multipulse polymers as discussed in Section 4.3.

Due to limited molecular motion in the solid, the carbon T_1's can be extremely long, especially for crystalline polymers, and the carbon T_1's are most efficiently measured using the cross-polarization T_1 pulse sequence shown in Fig. 1.40. The relaxation is often nonexponential, especially for semicrystalline polymers, and the long relaxation times are usually associated with the crystalline phase, and the shorter ones with the amorphous phase. In polyethylene a third component is observed that can be assigned to the interface (42). As

5.3. Polymer Dynamics in the Solid State

with proton relaxation, it is often observed that some mechanism other than the segmental chain dynamics is the main source of relaxation. This is illustrated in Fig. 5.20, which shows the carbon spin–lattice relaxation for the methine, methylene, and methyl carbons of isotactic polypropylene as a function of temperature (43). At temperatures around 100 K the methyl peak disappears from the spectrum because of the interference of methyl rotation with the high-powered decoupling [Eq. (5.41)]. The data in Fig. 5.20 show a minimum in the T_1's for the methine and methylene carbons at the same temperature as the proton T_1 minimum (44). From this observation it is concluded that segmental chain motion does not provide an efficient relaxation pathway and that the relaxation for all the carbons in polypropylene is attributable to methyl group rotation. Similar conclusions have been reported for poly(methyl methacrylate) of different tacticities and for samples containing plasticizers (45).

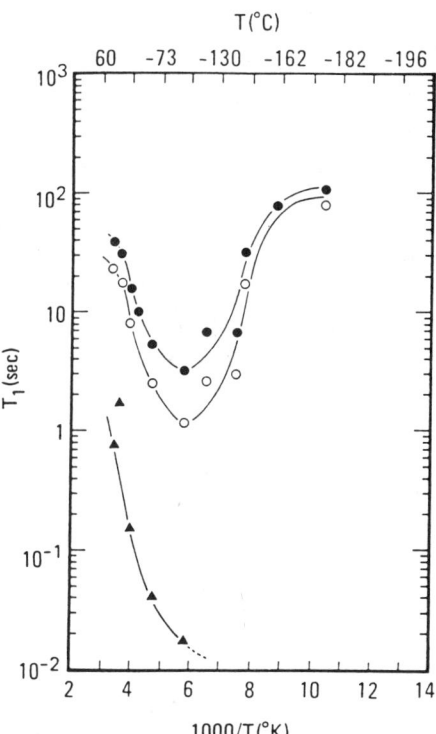

FIGURE 5.20 The ^{13}C NMR relaxation times for the (●) methylene, (○) methine, and (▲) methyl carbons of isotactic polypropylene. Adapted with permission (43).

Nuclear Overhauser effects can sometimes be observed for solid polymers, particularly for those near or above the glass transition temperature. Such measurements are also difficult to interpret in terms of the molecular motions because they are also mainly due to methyl group rotation (46). These measurements are most useful for studying the structure of polymer blends and other multiphase polymer mixtures (47).

Information about the molecular dynamics of polymers can also be obtained from the cross-polarization dynamics. As noted in Section 1.8.5.4, the time course of the magnetization in the cross-polarization experiment represents a compromise between the build-up due to proton–carbon cross relaxation (T_{CH}) in the spin-locking field and the decay due to carbon and proton rotating frame relaxation ($T_{1\rho}^C$ and $T_{1\rho}^H$) and is given by

$$M(t) = \frac{M_0[1 - e^{-\lambda t/T_{CH}}]e^{-t/T_{1\rho}^H}}{\lambda}, \qquad (5.43)$$

where $\lambda = 1 + T_{CH}/T_{1\rho}^C + T_{CH}/T_{1\rho}^H$. In most polymers $T_{1\rho}^H, T_{1\rho}^C \gg T_{CH}$, so the magnetization builds up due to carbon–proton cross relaxation and decays due to proton rotating frame relaxation. Slightly modified cross-polarization pulse sequences have been developed to measure each of these relaxation rates (48).

The proton rotating frame relaxation rates can be measured from the decay of cross-polarized magnetization or using the related pulse sequence shown in Fig. 1.43. However, effective spin diffusion during the spin-locking pulse sequence averages the relaxation times, making it difficult to relate the $T_{1\rho}^H$ relaxation times to the molecular dynamics. As shown in Section 4.3, the $T_{1\rho}^H$ relaxation times are very useful for measuring the length scale of phase separation in polymer blends and multiphase polymer systems.

The carbon–proton cross-relaxation time can be measured from the build-up of cross polarization or, more efficiently, using the modified cross-polarization pulse sequence shown in Fig. 1.43. The relaxation rate is given by

$$\frac{1}{T_{CH}} = \frac{1}{2}\sin^2\theta_C \sin^2\theta_H M_2^{CH}[J(\Delta\omega)], \qquad (5.44)$$

where θ_C and θ_H are the angles between the external field and the rotating frame fields of the carbons and protons, M_2^{CH} is the second moment of the heteronuclear dipolar coupling interaction, and J is the spectral density function that describes the modulation of the cross polarization by the degree of spin-lock mismatch and by molecu-

5.3. Polymer Dynamics in the Solid State

lar motions. T_{CH} is shortest for rigid materials and may be related to important materials properties such as the dynamic modulus (49).

In favorable cases the rotating frame carbon spin–lattice relaxation time $T_{1\rho}^C$ can be related to the molecular dynamics of polymers. The $T_{1\rho}^C$ is measured using the pulse sequence shown in Fig. 5.21 in which the carbon magnetization is spin locked for a variable contact time in the absence of the proton spin-locking field. This relaxation parameter is often of interest because the relaxation is due to molecular motions in the 15- to 100-kHz regime that may be of direct interest for understanding the mechanical properties of polymers. Interpretation of $T_{1\rho}^C$ relaxation is complicated by the possibility of two contributions to the decay, and the observed relaxation time $T_{1\rho}^*$ is given by

$$\frac{1}{T_{1\rho}^*} = \frac{1}{T_{1\rho}^C} + \frac{1}{T_H^D}, \qquad (5.45)$$

where $T_{1\rho}^C$ is the desired relaxation time and T_{CH}^D is the time constant for cross relaxation between the spin-locked carbons and the unlocked protons. The relaxation rates are given by

$$\frac{1}{T_{1\rho}^C} = \frac{n}{10}\left(\frac{\mu_0}{4\pi}\right)^2 \frac{\gamma_H^2 \gamma_C^2 \hbar^2}{r^6} \{4J(\omega_1) + J(\omega_H - \omega_C) + 3J(\omega_C)$$

$$+ 6J(\omega_H) + 6J(\omega_H + \omega_c)\} \quad (5.46)$$

and

$$\frac{1}{T_{CH}^D} = \frac{1}{2}\sin^2\theta M_{CH}^2 \{\pi\tau_D e^{-\omega_1 \tau_D}\}, \qquad (5.47)$$

where τ_D is the correlation time for spin fluctuations. The relative contributions of the two pathways to the relaxation of spin-locked

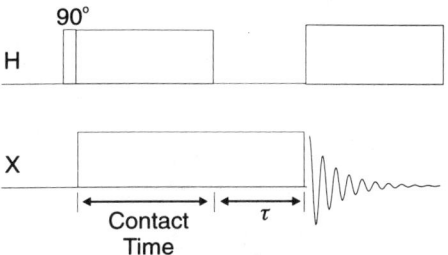

FIGURE 5.21 The pulse sequence diagram for measuring $T_{1\rho}^C$. The data are collected as a function of the carbon spin-locking time.

carbons can be estimated by measuring the relaxation time as a function of the spin-lock power, since the relaxation by the spin–spin pathway has a quadratic dependence on ω_1, while relaxation by T_{CH}^D has an exponential dependence. It has been reported that the relaxation in glassy polymers is primarily due to the $T_{1\rho}^C$ pathway and that the relaxation times can be interpreted in terms of chain dynamics over the 15- to 100-kHz time scale (50), while the relaxation in semicrystalline polymers has a substantial contribution from T_{CH}^D relaxation and cannot be so easily interpreted (43).

5.3.2 The Dynamics of Semicrystalline Polymers

A variety of NMR methods, ranging from relaxation studies to deuterium NMR and 2D NMR methods, have been used to investigate the molecular dynamics of semicrystalline polymers. Multiple signals are often observed in the NMR spectra that are assigned to the crystalline, amorphous, and interfacial materials. The measurement of the molecular dynamics plays an important role in the assignment of the signals and in understanding structure–property relationships.

Polymers in the crystalline phase are in a rigid environment and are most efficiently observed using cross polarization and magic angle spinning. They can often be distinguished from the amorphous and interfacial material by choosing a short cross-polarization contact time, or in the CPT_1 experiment (Fig. 1.40) by choosing a delay time longer than the relaxation time of the more mobile material.

Although the chain dynamics are restricted by the crystalline environment, dynamics can be observed over a range of time scales ranging from methyl group rotation to aromatic ring flips to the jumping of monomers between sites in the crystalline lattice. Methyl group rotation is fast except at very low temperatures and can affect the appearance of the spectra and influence the spin–lattice relaxation times. Such behavior is illustrated in Fig. 5.22, which shows the spectra for 70% crystalline isotactic polypropylene as a function of temperature (43). At 300 K the signals from the methine, methylene, and methyl carbons are well resolved, but as the temperature is lowered to 105 K the methyl signal disappears, and then reappears at temperatures below 77 K. At temperatures near 105 K the rate of methyl rotation is near the modulation period of the decoupler and the line is broadened beyond detection, as expected from Eq. (5.42).

Main-chain motion in some crystalline polymers is sufficiently rapid to interfere with either high-power decoupling or magic angle spinning. Such behavior is illustrated in Fig. 5.23, which shows the effect of the spinning rate on the linewidth observed for poly(oxymethylene), where broader lines are observed with higher

5.3. Polymer Dynamics in the Solid State

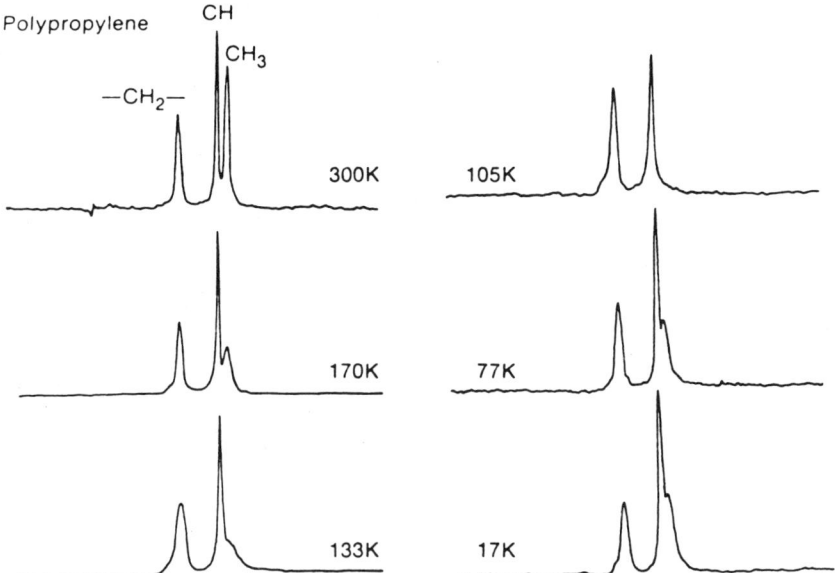

FIGURE 5.22 The ^{13}C cross-polarization spectra of polypropylene as a function of temperature. Note the disappearance of the methyl signal at 105 K. Adapted with permission (43).

spinning rates (51). Similar effects are observed in the crystalline phase of poly(ethylene oxide), where chain motion causes the $T_{1\rho}^{H}$ to be extremely short, making it difficult to acquire a spectrum using cross polarization and magic angle spinning (52,53).

Solid-state ^{2}H NMR has been used to study the molecular dynamics of the crystalline and amorphous phases of nylon 66 in polymers labeled with deuterium at several sites. This approach is illustrated in Fig. 5.24, which shows the wide-line deuterium spectrum at 97°C of polymers labeled at carbons C_1 and C_6, C_2 and C_5, and C_3 and C_4 of the diamine moiety (NY16NHME, NY25NHME, and NY34NHME) and at the C_2 and C_5, and C_3 and C_4 positions of the adipoyl moiety (NY25COME and NY34COME) (54). These spectra show that the molecular dynamics of the various methylenes are not identical under these conditions.

$$\overset{O}{\underset{1}{\overset{\parallel}{C}}}-\overset{2}{CH_2}-\overset{3}{CH_2}-\overset{4}{CH_2}-\overset{5}{CH_2}-\overset{O}{\underset{6}{\overset{\parallel}{C}}}-NH-\overset{1}{CH_2}-\overset{2}{CH_2}-\overset{3}{CH_2}-\overset{4}{CH_2}-\overset{5}{CH_2}-\overset{6}{CH_2}-NH-$$

Nylon 66

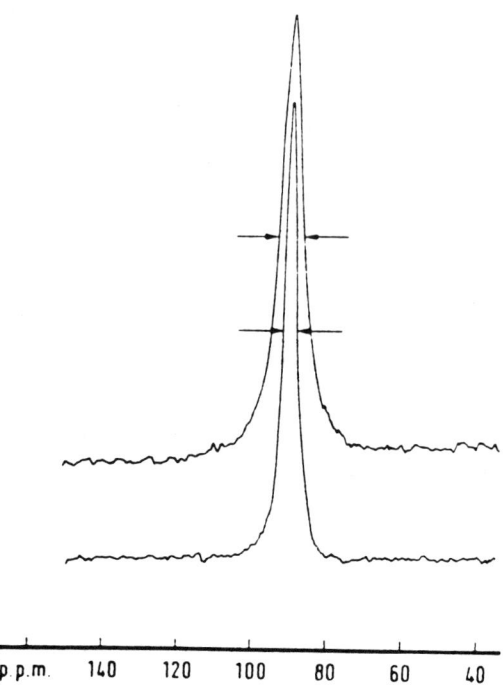

FIGURE 5.23 The increase of the ^{13}C linewidths of poly(oxymethylene) with the increased rate of magic angle spinning. The data are shown for spinning speeds of (a) 5.6 kHz and (b) 3.1 kHz. Adapted with permission (51).

The molecular dynamics of the crystalline and amorphous phases of nylon 66 can be individually measured since the differences in the dynamics between the crystalline and the amorphous phases give rise to large differences in the spin–lattice relaxation rates, as shown in Fig. 5.25 (55). For nylon 66 at very low temperatures ($T < -50°C$) there appears to be some mixing of the phases, perhaps because some of the amorphous material is constrained and has a T_1 like that of the crystalline phase. At higher temperatures it appears that the longer T_1 component has an amplitude near 40% and the shorter component has an amplitude of 60%. The long T_1 component is assigned to the crystalline phase and agrees well with the crystallinity measured by X-ray diffraction (35%) and DSC (39%).

Several models for the molecular dynamics of nylon 66 were examined to explain the temperature dependence of the lineshapes for the polymers labeled with deuterium at the diamine, adipoyl, and NH positions (55). The best fit was obtained for models incorporating libration motions that change with amplitude as a function of temper-

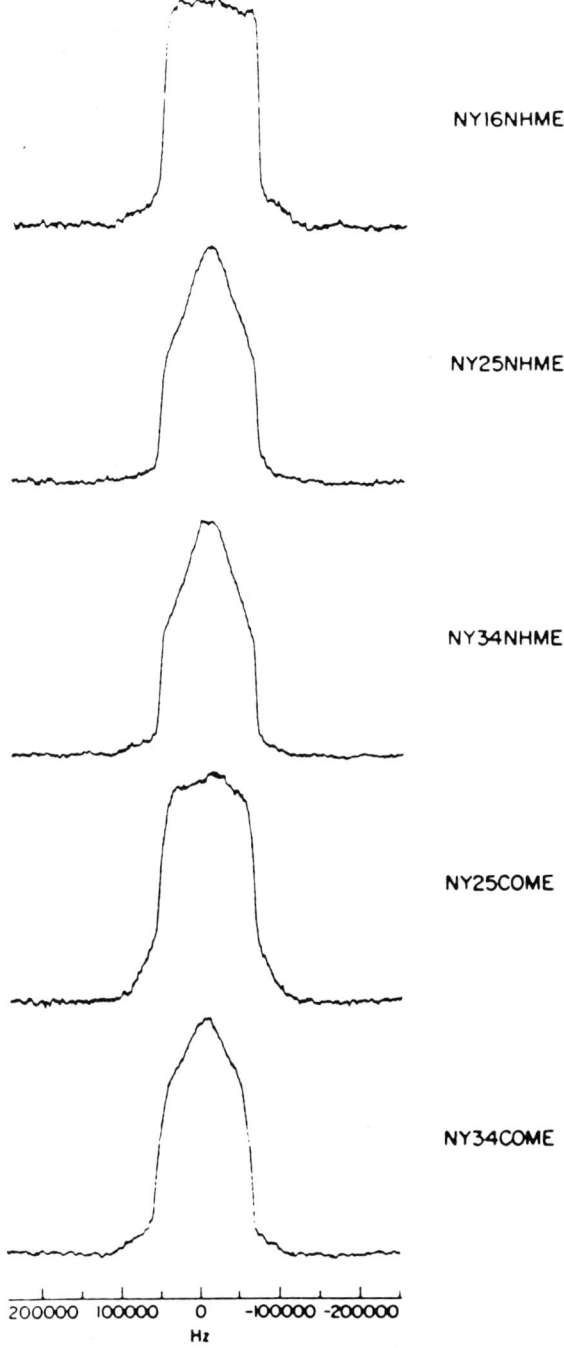

FIGURE 5.24 The fully relaxed ^{2}H NMR spectra at 97°C of nylon 66 polymers selectively deuterated on the diamine moiety at the C_1 and C_6 carbons (NY16NHME), the C_2 and C_5 carbons (NY25NHME), and the C_3 and C_4 carbons (N346NHME), and on the adipoyl moiety at the C_2 and C_5 carbons (NY25COME) and the C_3 and C_4 carbons (NY34COME). Adapted with permission (54).

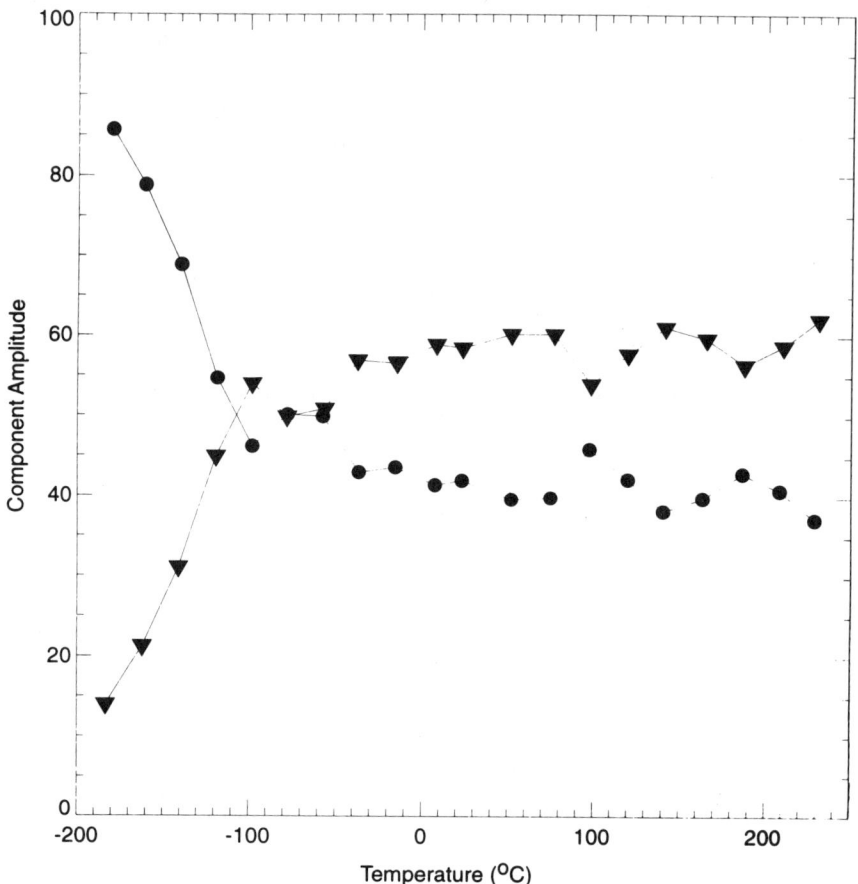

FIGURE 5.25 The relative amplitudes of the fast (▼) and slow (●) component of the deuterium T_1 relaxation for nylon 66 labeled with deuterium on the C_1 and C_6 positions of the diamine moiety. Adapted with permission (55).

ature, and Fig. 5.26 compares the experimental and simulated lineshapes for nylon 66 labeled with deuterium at the C_3 and C_4 positions of the adipoyl moiety along with the distribution of librational angles [$P(\Delta\theta)$] required to give a good fit between the experimental and the simulated data. Further evidence for the librational model was obtained from the T_{1Q}-distorted lineshape shown in Fig. 5.27. The lineshape could be modeled with either a two-site jump model or the librational model, but the T_{1Q}-distorted lineshape is much closer to the librational model, even though the signal-to-noise ratio is quite low in these experiments.

5.3. Polymer Dynamics in the Solid State

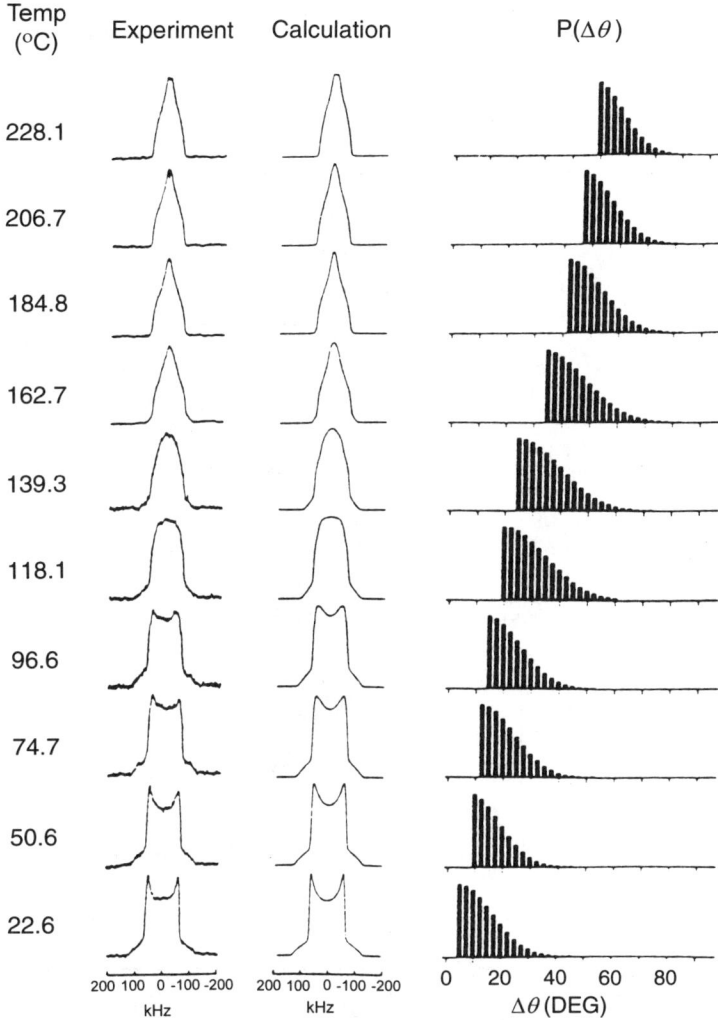

FIGURE 5.26 Comparison of the experimental and simulated lineshapes for nylon 66 labeled with deuterium on the C_3 and C_4 positions of the adipoyl moiety as a function of temperature. Also shown is the distribution of librational angles required to fit the spectra as a function of temperature. Adapted with permission (55).

The analysis of the deuterium relaxation for the nylon 66 polymers shows that the hydrogen bonded NH groups act as "pinning points" that restrict the mobility of the nylon chains (55,56). The mean correlation times for the ND groups vary between 200 and 500 ps over the temperature range 185–225°C. It is also reported that above 160°C the methylene groups in both the diamine and the adipoyl

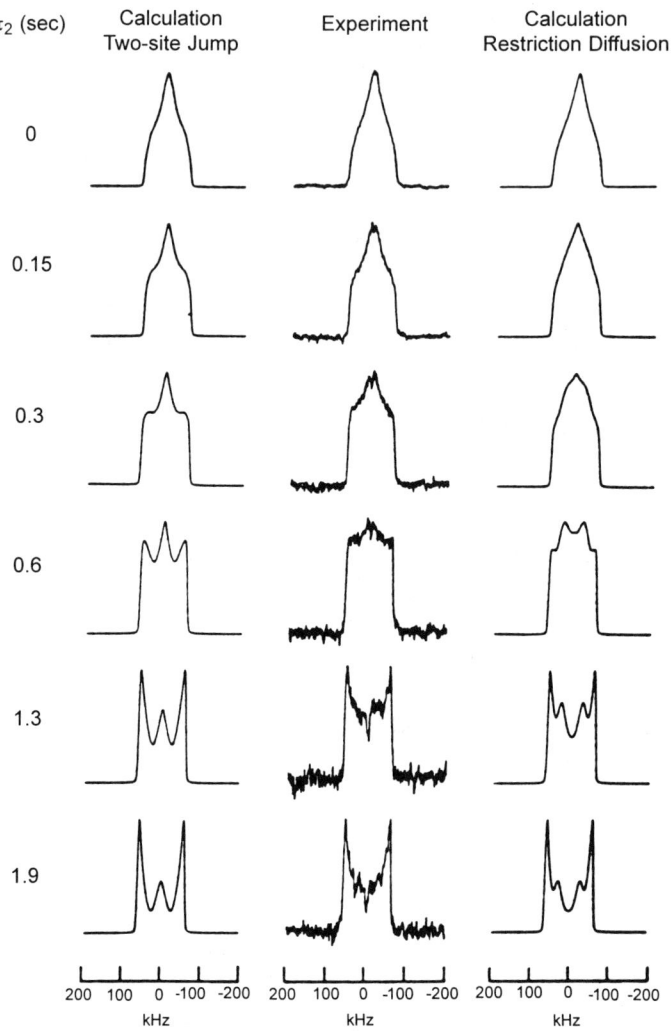

FIGURE 5.27 The T_{1Q}-distorted lineshape for nylon 66 labeled with deuterium on the C_3 and C_4 positions of the adipoyl moiety as a function of the delay time τ_2. The experimental data (center) are compared with calculated lineshapes for a two-site jump model and restricted diffusion. Adapted with permission (55).

moieties have similar correlation times ($\tau_c \approx 40$ ps) that are much shorter than those for the ND groups. Below this temperature, however, the adipoyl groups experience larger amplitude fluctuations than do the methylenes in the diamine moiety. These experimental data are in close agreement with the motions observed in molecular dynamics simulations of nylon 66 crystals (56).

5.3. Polymer Dynamics in the Solid State

The molecular dynamics of the amorphous phase of nylon 66 were studied by the same methods, where it was observed that the N–D sites exhibit little motion below T_g, but a fraction of the sites undergo isotropic motion above T_g (57). The C–D sites exist in two discrete environments below T_g, one that exhibits only librational motions and one that experiences both librational motions and bond rotations. Above T_g a fraction of the C–D sites also undergo isotropic motion.

Poly(phenylene terephthalamide) is a commercially important, highly crystalline aromatic polyamide that can exist in two crystalline modifications, depending on the precipitation conditions. The idealized structure of poly(phenylene terephthalamide) is shown in Fig. 5.28 and consists of stacks of hydrogen bonded aramide molecules organized into a pleated sheet structure. The dynamics of poly(phenylene terephthalamide) have been measured with site-specific deuterium labeling of the amide (58), the terephthalamide rings (59), and the phenylene diamine rings (60).

Although poly(phenylene terephthalamide) is a highly crystalline material, the deuterium NMR spectra and relaxation times are similar to those observed for semicrystalline polymers in that the relaxation is biexponential and the observed spectrum is a composite of a rigid and motionally averaged lineshape. This is illustrated in the ^{2}H NMR spectra of the N-deuterated polymer recorded at -184 and 228°C shown in Fig. 5.29 (58). The outer singularities are characteristic of a nearly rigid material while the inner singularities are characteristic of N–D bonds undergoing 180° reorientations. About 25% of

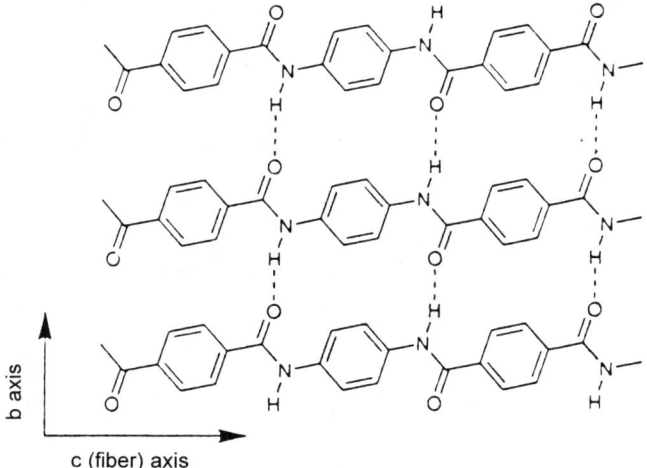

FIGURE 5.28 The idealized structure of hydrogen bonded sheets of poly(phenylene terephthalamide). Adapted with permission (59).

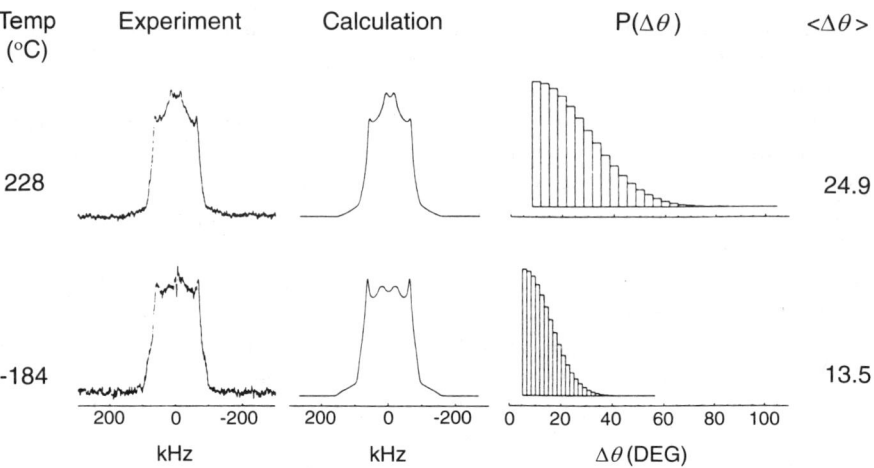

FIGURE 5.29 The experimental and simulated deuterium lineshapes for poly(phenylene terephthalamide) labeled with deuterium at the amide position at −184 and 228°C. Also shown is the distribution of librational motions required to fit the lineshape. Adapted with permission (60).

the signal is associated with the flipping N–D bonds, and the fraction of mobile material is temperature independent. These data are in accord with biexponential spin–lattice relaxation, where the fraction of slowly relaxing deuterons is approximately 75% and independent of temperature, even though the relaxation times change considerably over the temperature range. The best fit between the experimental and the simulated lineshapes was obtained by a lineshape from the more rigid population and a lineshape from a population of N–D bonds that are flipping rapidly ($\tau_c < 10^{-9}$ s) about the 1,4 axis of the terephthalamide rings. The only change with temperature was an increase in the amplitude of librational motion as the temperature increased.

Similar observations have been reported for poly(phenylene terephthalamide) labeled with deuterium at the terephthalamide rings (59) and the phenylene diamine rings (60). Again two dynamic populations are inferred from the biexponential spin–lattice relaxation and the lineshape that is a composite of the more rigid material and rapidly flipping aromatic rings. However, the fraction of flipping rings is temperature dependent, as shown in Fig. 5.30. From these data there appear to be two populations of flipping rings, with about 25% of the rings flipping in the low-temperature regime.

Since poly(phenylene terephthalamide) is a highly crystalline material, the dynamic heterogeneity observed in the deuterium NMR

5.3. Polymer Dynamics in the Solid State

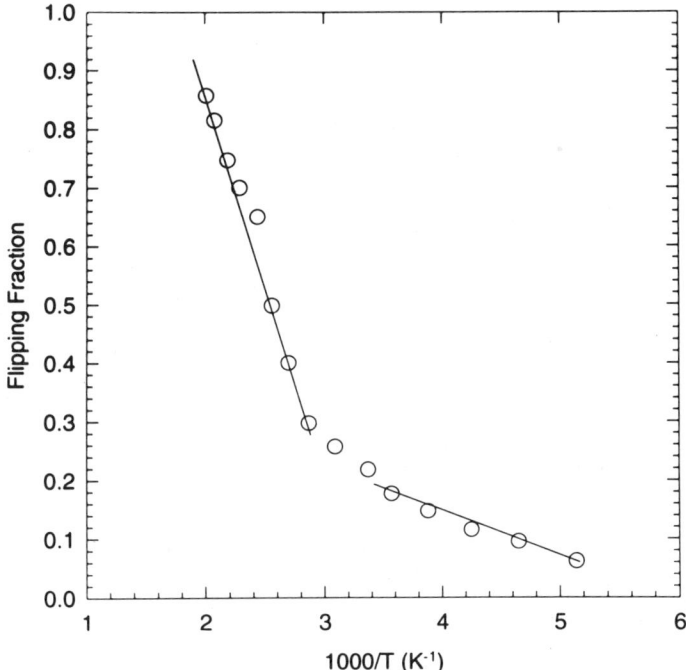

FIGURE 5.30 The temperature dependence of the fraction of rapidly flipping rings in poly(phenylene terephthalamide) labeled with deuterium on the terephthalamide rings. Adapted with permission (59).

spectra and relaxation times is associated with a small crystallite size and crystalline defects (59). X-ray diffraction shows that the crystallites are quite small, on the order of 30–35 Å. For such a crystallite about 40% of the chains reside on the surface and ca. 20% of all amides would be incapable of forming hydrogen bonds. Thus, the mobile population in poly(phenylene terephthalamide) is associated with these surface sites. The terephthalamide rings in form I have a slightly lower degree of mobility than do those in form II. The dynamic populations are not affected by fiber formation since the crystallite size is not affected by this process.

Poly(phenylene vinylene) is a highly crystalline polymer of current interest for possible use in light-emitting diodes. The molecular dynamics of poly(phenylene vinylene) films have been studied by deuterium NMR in polymers in which the ring protons have been replaced by deuterons, and typical data as a function of temperature are shown in Fig. 5.31 (61). At low temperatures the spectrum is characteristic of a rigid polymer. As the temperature is increased a

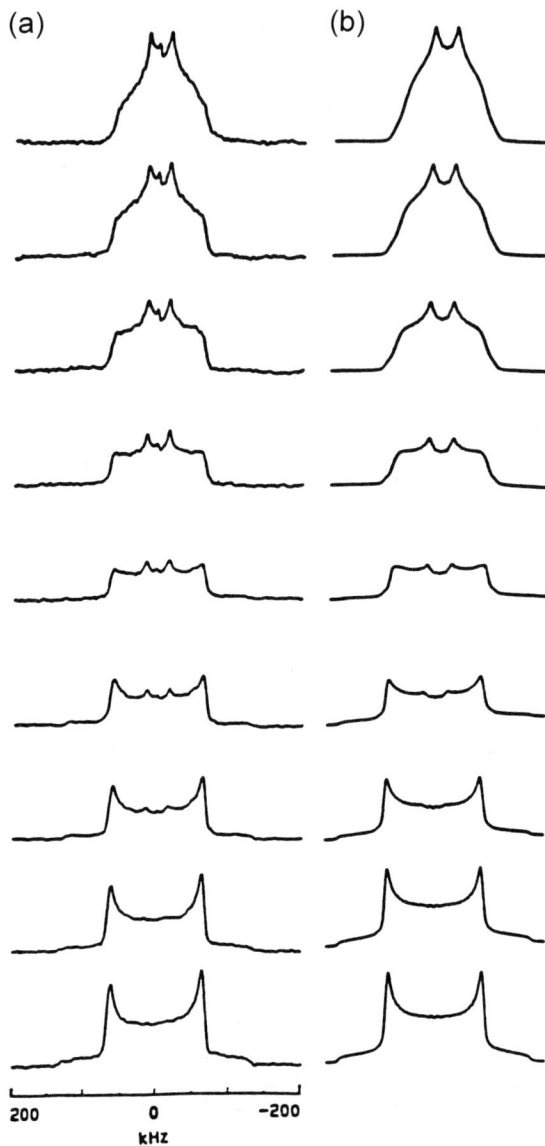

FIGURE 5.31 Comparison of the (a) experimental and (b) simulated quadrupolar echo ^{2}H NMR spectra of ring deuterated poly(phenylene vinylene) as a function of temperature. Adapted with permission (61).

central component to the lineshape appears that can be assigned to aromatic ring flips about the 1,4 axis of the phenylene rings, and at the highest temperature it appears that all of the aromatic rings undergo such 180° flips. In addition, there appears to be some librational motion which further narrows the lines at the highest tempera-

5.3. Polymer Dynamics in the Solid State

tures. The ring flips have an activation energy of 63 kJ mol^{-1} and a relatively narrow distribution of flipping rates (ca. two orders of magnitude). In samples doped with sulfuric acid a higher activation energy is observed (88 kJ mol^{-1}) that was attributed to an increase in the double bond character of the poly(phenylene vinylene) (62). A wider distribution of flipping rates (ca. four orders of magnitude) was also observed, indicating that there is more heterogeneity in the doped samples, presumably due to a lower degree of crystalline perfection in the doped samples.

Poly(phenylene vinylene)

Poly(oxy-1,4-phenyleneoxy-1,4-phenylenecarbonyl-1,4-phenylene), more commonly known as PEEK, is a high-performance material having only oxygens and carbonyl carbons connecting the main-chain aromatic rings. PEEK is insoluble in common organic solvents, but can be fabricated because it can be heated above its crystalline melting point without degradation. Figure 5.32 shows the cross-polarization and magic angle spinning spectra of PEEK samples that are 1 and 26% crystalline (63). Although there is some overlap in the carbon spectra, the lines can be assigned and the protonated and nonprotonated carbon signals can be separated by dipolar dephasing (64).

Poly(ether ether ketone)

The molecular dynamics of PEEK samples have been investigated using NMR relaxation methods and dipolar rotational spin echo methods for samples containing 1, 8, and 26% crystalline material (63). As noted in Section 5.3.1, rotational dipolar spin echo is a 2D experiment that correlates the carbon chemical shift in one dimension with the dipolar lineshape in the other dimension. The results from these experiments are shown as stick figure diagrams of the dipolar lineshape in Fig. 5.33 for the samples with 1, 8, and 26% crystallinity. The lineshape for the sample with 26% crystallinity is best matched by a two-component lineshape. About 20% of the lineshape is as-

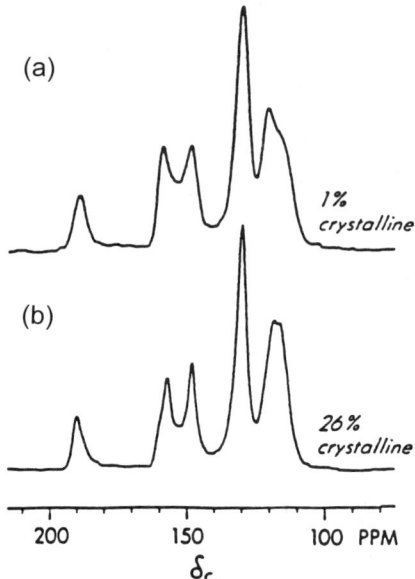

FIGURE 5.32 The 15.1-MHz cross-polarization and magic angle spinning spectra of PEEK samples with crystallinities of (a) 1% and (c) 26%. Adapted with permission (63).

signed to rings undergoing 180° ring flips superimposed on small-amplitude ring oscillations with wiggles, while the other 80% is essentially immobile. Since the sample is 26% crystalline, this means that the rings in the amorphous fraction are essentially immobile. These data are consistent with the observation of only 20–40% flipping rings in the sample with 1% crystallinity.

The lower frequency dynamics of crystalline polymers have also been investigated by solid-state NMR using methods related to the 2D NOESY experiments. In such "ultraslow" experiments the change in orientation of the atoms in the polymer chain is translated into a change in frequency during the mixing time of a 2D experiment, and the frequencies before and after the motion can be directly correlated. The longer time limit for the use of these studies is the decay of the spin system due to spin–lattice relaxation.

Polyethylene is a semicrystalline polymer that has been extensively studied by solid-state NMR. An important feature in the relaxation spectrum of polyethylene is the so-called α transition that has been attributed to solid-state chain diffusion. It is possible to directly observe this diffusion between the crystalline and the amorphous domains because their chemical shifts differ by ca. 2 ppm because of

5.3. Polymer Dynamics in the Solid State

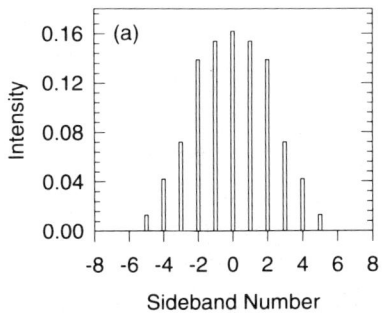

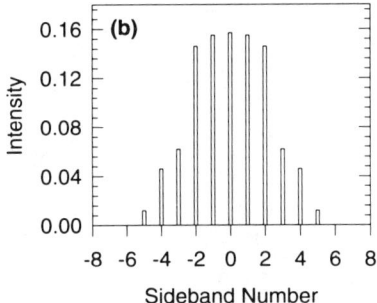

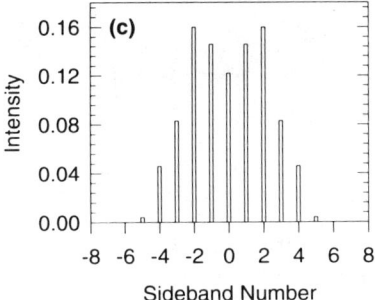

FIGURE 5.33 The dipolar lineshapes of the protonated carbons obtained from the rotational dipolar spin echo experiment for PEEK samples with crystallinities of (a) 1%, (b) 8%, and (c) 26%. Adapted with permission (63).

the differences in chain conformation (65). Figure 5.34 shows a solid-state 2D NOESY spectrum obtained with magic angle spinning that demonstrates chain diffusion between the crystalline and the amorphous material (66). These spectra clearly show the cross peaks (ca and ac) that arise from the transfer of material between the two phases. The jump rates and activation energies (105 kJ mol^{-1}) agree

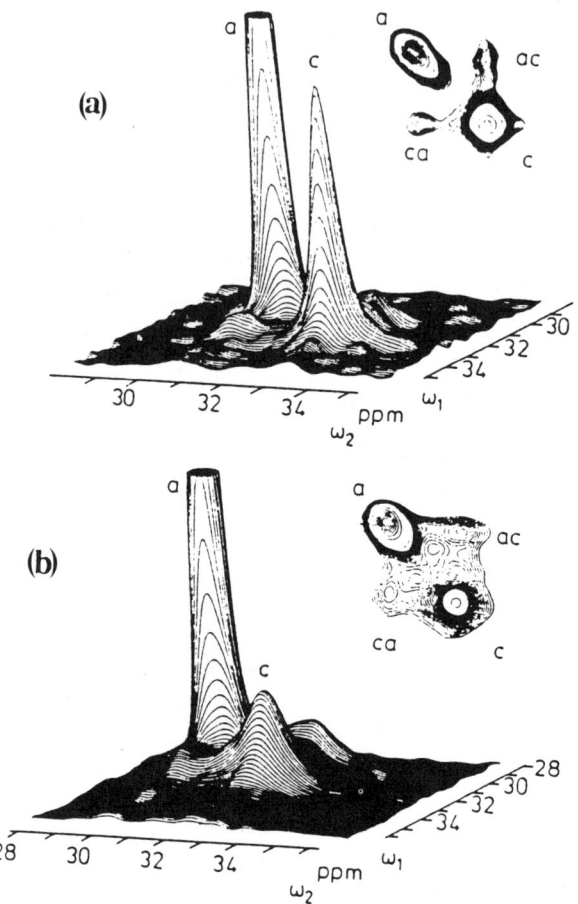

FIGURE 5.34 The 2D magic angle spinning exchange spectra for high-density polyethylene at (a) 363 K with a mixing time of 1 s and at (b) 383 K with a mixing time of 2 s. The cross peaks between the crystalline and the amorphous phases are labeled ca and ac. Adapted with permission (66).

well with the values previously reported for the α process and show that chain diffusivities between 10^{-17} and 10^{-21} m s^{-1} can be measured by such NMR methods. These data also explain the nuclear Overhauser effects observed for the crystalline and amorphous phases, the nonexponential spin–lattice relaxation, and the dependence of the relaxation times on the crystallite size.

Static and slow magic angle spinning NMR experiments have been used to study monomer dynamics in a variety of crystalline polymers such as poly(oxymethylene) (4,38,39), polypropylene (67), and

5.3. Polymer Dynamics in the Solid State

poly(vinylidine fluoride) (68). All of these studies take advantage of the facts that as a monomer changes orientation with respect to the magnetic field, the frequency [given by Eq. (5.35) or (5.36)] will also change, and that the two-site jump will give rise to well-defined ridge patterns in the 2D spectra (35).

Poly(oxymethylene) has been extensively studied by such methods. The crystalline conformation of poly(oxymethylene) is a 9_5 helix, and in the absence of magic angle spinning a chemical shift anisotropy pattern is observed with singularities at 67, 86, and 111 ppm. From the 2D dipolar correlation spectrum it has been determined that the orientation of the chemical shift anisotropy tensor is determined by the local symmetry of the CH_2O_2 unit and the z axis is oriented perpendicular to the O–C–O plane and makes an angle of 55° with the helix axis (69). The y axis lies in the O–C–O plane and bisects the O–C–O angle, and the x axis is perpendicular to the y and z axes.

The static 2D exchange spectra for poly(oxymethylene) as a function of temperature and mixing time are shown in Fig. 5.35 (39). At low temperature and short mixing time all the intensity is confined to the diagonal, indicating that there is little motion on the time scale of the mixing time. As the temperature is increased a ridge pattern is observed that increases in intensity as the temperature is increased or the mixing time is lengthened. From the geometry of the ridge pattern and the intensity as a function of mixing time it is determined that the helical motions can be described as a one-dimensional random walk with an elementary process of a 200° jump that corresponds to the jump of a monomer to a neighboring site of the crystalline helix. The rates and activation energies (75 ± 8 kJ mol^{-1}) measured by NMR are in close agreement with those measured for the α relaxation by mechanical and dielectric spectroscopy (70).

The slow molecular motion of poly(oxymethylene) has also been studied using a similar 2D NOESY-type pulse sequence, but with slow magic angle spinning (38). In such an experiment several spinning sidebands are observed and the mixing time is set to be an integral number of rotor periods. Cross peaks between the spinning sidebands will be observed if molecular motion occurs during the mixing time. Figure 5.36 shows the results of an experiment on poly(oxymethylene) with short and long mixing times with slow magic angle spinning. Information about the geometry of the molecular motion is obtained from the intensity of the diagonal and cross peaks, since they depend on the orientation of the chemical shift anisotropy tensor before and after the mixing time. The data in Fig. 5.36 are best matched by 200° jumps of the monomers between sites in a crystalline lattice. This experiment has also been used to study isotactic polypropylene in the absorption mode (39).

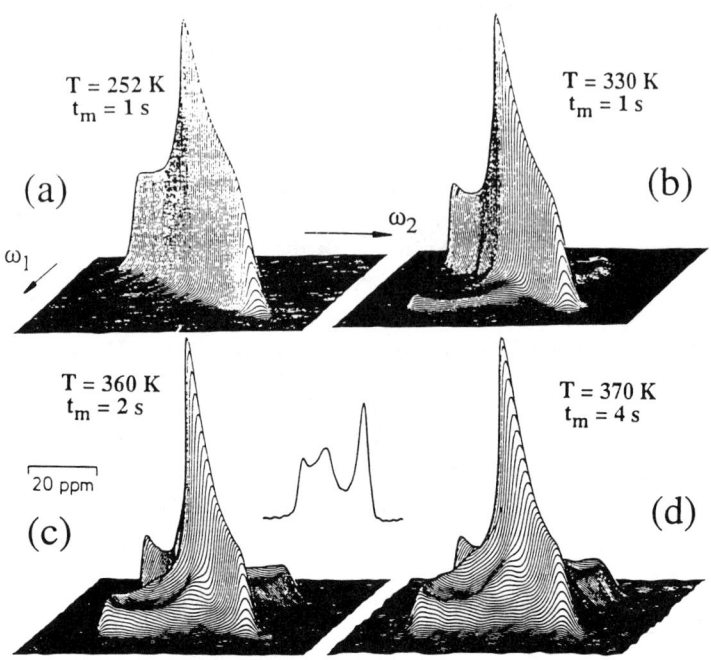

FIGURE 5.35 The 2D ^{13}C exchange spectra for poly(oxymethylene) at (a) 252 K with a mixing time of 1 s, (b) 330 K with a mixing time of 1 s, (c) 360 K with a mixing time of 2 s, and (d) 370 K with a mixing time of 4 s. Note the appearance of the ridge that arises from a two-site jump between positions in the crystalline lattice. Adapted with permission (39).

The 2D exchange studies provide important information about the geometry of the jumps and crystalline lattice, but do not provide information about the jump mechanism. Such information can be obtained from the 3D exchange experiment shown in Fig. 5.37. In order to distinguish between different jumping mechanisms it is necessary to determine the molecular orientation at least three times. In poly(oxymethylene), for example, it would be possible to distinguish between a 400° jump and two 200° jumps. Figure 5.38 shows the experimental and simulated 3D exchange spectra for highly oriented poly(oxymethylene), where it is found that the best fit between the experimental and the simulated data is for an elementary jump of 200° (4).

Deuteron 2D NMR methods have also been used to study the molecular dynamics of crystalline polymers using the pulse sequence shown in Fig. 5.16. Again the 1D spectrum appears along the diagonal and the off-diagonal intensity arises from exchange occurring during the mixing time, and since the frequency of a particular deuteron

5.3. Polymer Dynamics in the Solid State

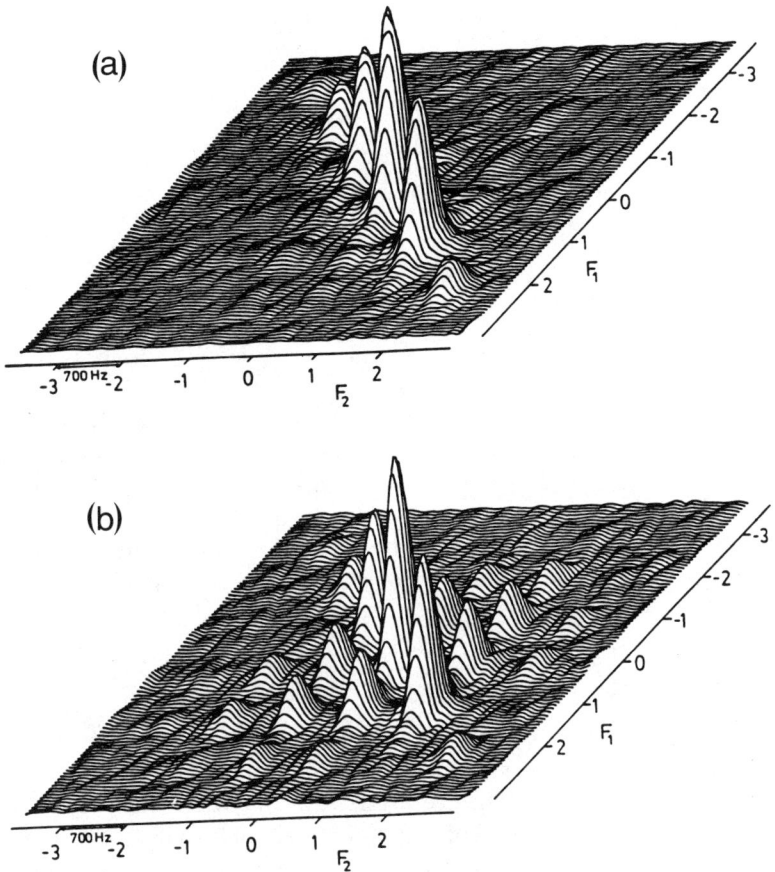

FIGURE 5.36 The 2D exchange spectra of poly(oxymethylene) acquired with slow magic angle spinning at (a) 316 K with a mixing time of 1 s and (b) 336 K with a mixing time of 4.5 s. The cross peaks arise from molecular motion during the mixing time. Adapted with permission (38).

depends on its orientation to the magnetic field before and after the mixing time, a characteristic ridge pattern is observed that depends on the geometry of the molecular motion.

Deuteron 2D exchange NMR has been used to study the ultraslow motion in the crystalline phase of poly(vinylidine fluoride) to understand the unusual α relaxation that occurs ca. 370 K. Figure 5.39 shows the experimental and simulated 2D spectra for a mixing time of 0.2 s. The best fit between the experimental and the simulated spectra is obtained for a model in which there are either 67° or 113° jumps between sites in the crystalline lattice. These data are consis-

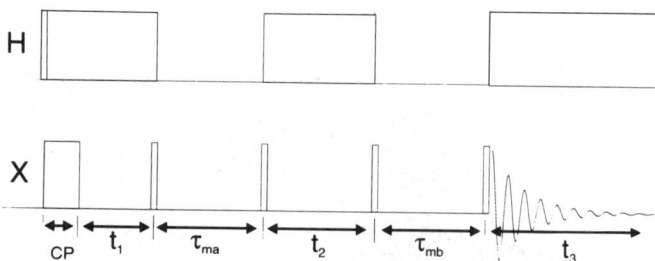

FIGURE 5.37 The pulse sequence diagram for the 3D spin exchange experiment. The two evolution times t_1 and t_2 are independently incremented, and exchange occurs during the two mixing times τ_{ma} and τ_{mb}.

tent only with one crystal structure of poly(vinylidine fluoride) (71), and from these data it can be concluded that chain motion is characterized by an electric dipole moment transition only along the molecular direction and by a conformational change between the *tgtg*$^-$ and the *g*$^-$*tgt* conformations. The same methods have been used to identify the 113° jumps of the monomer as the crystalline phase of isotactic polypropylene (67).

5.3.3 THE DYNAMICS OF AMORPHOUS POLYMERS

Solid-state NMR has been extensively used to study the molecular dynamics of amorphous polymers in an attempt to gain a molecular-level understanding of the relationship between the structure of polymers and their macroscopic properties. Included in these studies is the use of NMR relaxation to obtain molecular-level assignments

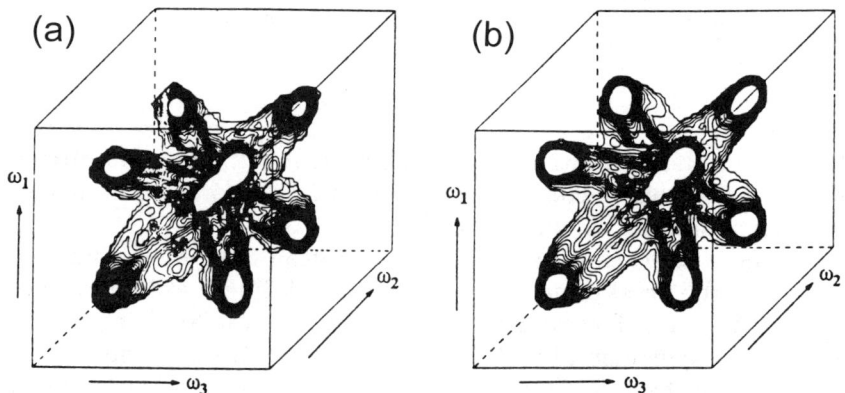

FIGURE 5.38 The (a) experimental and (b) simulated ^{13}C 3D exchange spectra of highly oriented poly(oxymethylene). Adapted with permission (4).

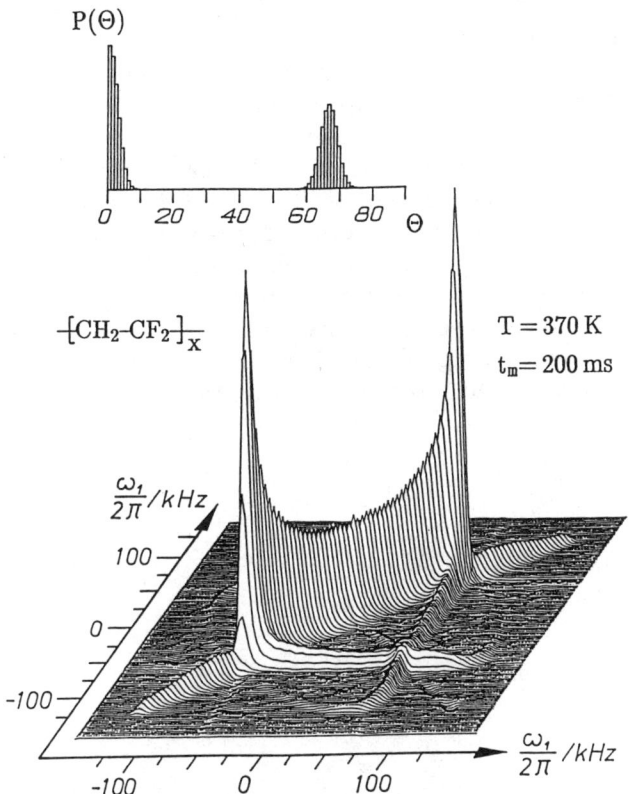

FIGURE 5.39 The (a) experimental wide-line ^{2}H exchange spectra for poly(vinylidine fluoride) at 370 K with a mixing time of 0.2 s. The spectrum was best fit with a two-site jump around an angle of 67 ± 3° as shown by the inset plot. Adapted with permission (68).

for the relaxation peaks observed in dielectric and dynamic mechanical spectroscopy and to understand the impact strength, gas permeability, and other macroscopic properties of amorphous polymers.

As in the case of crystalline polymers, the molecular dynamics of glassy polymers span a range of time scales from very rapid dynamics, like methyl group rotation, to ultraslow motions, like chain diffusion. Because amorphous polymers are by their very nature disordered, the relaxation times are often nonexponential because the chains exist in a variety of molecular environments. In Section 5.3.2 we saw that the chain reorientation in the crystalline lattice can be described by a two-site jump model. In amorphous polymers a distribution of correlation times spanning several orders of magnitude is typically a better description of the molecular dynamics below or near the glass transition temperature.

The most rapid motions observed in polymers are the localized dynamics, such as methyl group rotation, that interfere with dipolar decoupling and magic angle spinning. Such an effect is demonstrated in Fig. 5.40, which shows the effect of temperature on the magic angle spinning spectra of polycarbonate (2). At high temperatures the spectrum of polycarbonate shows resolved resonances for several of the peaks, including the protonated aromatic carbons, the quaternary carbon, and the methyl carbon. Similar, but slightly broadened, spectra are observed at lower temperatures, while at intermediate temperatures the methyl signal broadens and disappears from the spectra, presumably due to interference with the dipolar decoupling. The

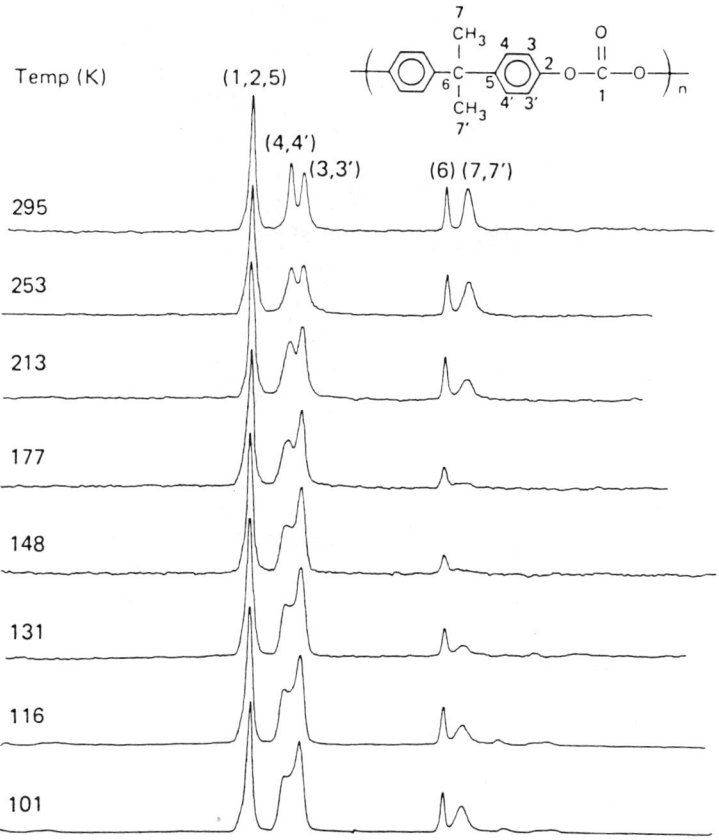

FIGURE 5.40 The ^{13}C cross-polarization and magic angle spinning spectra of polycarbonate as a function of temperature. Adapted with permission (2).

protonated aromatic carbons also show changes as a function of temperature. This lineshape results from aromatic carbons in a variety of magnetic environments that lead to inhomogeneous broadening from conformational and ring current effects. At 295 K there is a rapid averaging among these conformations and sharp lines are observed.

The relaxation of amorphous polymers has been studied for a variety of materials, mostly by carbon NMR, and it has been reported that carbon T_1's, $T_{1\rho}^C$'s, and T_{CH} relaxation times can be used to monitor the chain dynamics over a range of frequencies (50). An analysis of $T_{1\rho}^C$ relaxation in a variety of glassy polymers has shown that the relaxation is primarily due to spin–spin rather than spin–lattice relaxation. Therefore $T_{1\rho}^C$ is a valuable probe of chain dynamics in the 10- to 50-kHz region, while the T_1 relaxation provides information about chain motion in the 5- to 100-MHz regime.

The relaxation of a variety of glassy polymers, including poly(methyl methacrylate), polycarbonate, poly(ether sulfone), poly(2,6-dimethyl phenylene oxide), polystyrene, and poly(vinyl chloride), was reported in a classic early study of the molecular dynamics by NMR (50). The results showed both how the relaxation times report on the molecular dynamics and how such relaxation times can be related to important macroscopic properties. Figure 5.41 shows an example of $T_{1\rho}^C$ data typically obtained for glassy polymers. A prominent feature of such relaxation rate measurements is the dispersion in relaxation rates that makes such semilog plots nonlinear. This dispersion in relaxation times is due to the inherent dynamic heterogeneity for polymers in the amorphous state, where some chains are less mobile than others because they are more constrained by other chains in the immediate environment. Factors that affect the dynamic heterogeneity, such as annealing, are also shown to affect the dispersion in relaxation rates. Information about polymer dynamics is obtained by comparing the initial relaxation rates. As shown in Fig. 5.42, a correlation has been reported between the dynamics measured by NMR and the notched impact strength.

Because of their commercial importance, the study of the molecular dynamics of polystyrene and polycarbonate has been the focus of much attention in the polymer community. The molecular dynamics of polystyrene have been investigated by deuterium NMR for main-chain- and side-chain-labeled samples (3). In the studies of ring-labeled polystyrene, at least two components were observed on the basis of the spin–lattice relaxation studies, a fast and a slowly relaxing component. The lineshapes for the two components can be separated due to these differences, and Fig. 5.43 shows the wide-line deuterium spectra of atactic polystyrene-d_5 below the glass transition

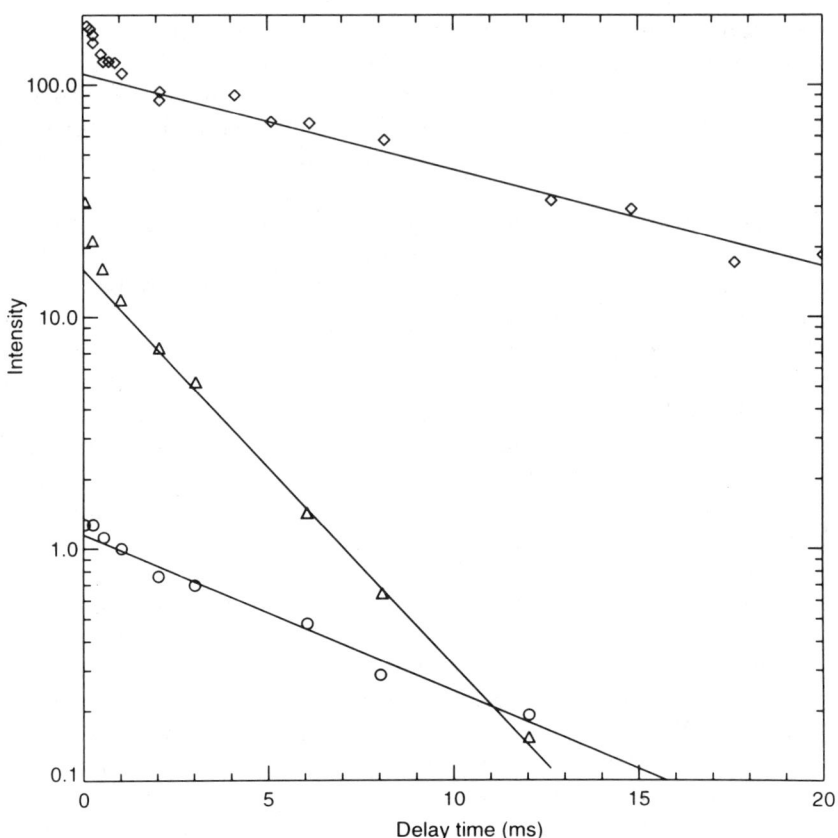

FIGURE 5.41 The $T_{1\rho}^C$ relaxation for (◇) the nonprotonated ring carbons of poly(ether sulfone) and the (△) protonated and (○) quaternary carbons of polycarbonate. Note that the decay is not linear. Adapted with permission (50).

temperature. Using a long relaxation delay a fully relaxed spectrum is observed that appears to have contributions from both rigid and motionally averaged materials. The lineshape of the motionally averaged material, which can be observed in the spectrum with a short relaxation delay (0.05 s), is typical of those observed for aromatic rings undergoing 180° flips, and the fact that it is possible to use the T_1 to discriminate between the two populations demonstrates that they are spatially separated from each other. It is estimated that the correlation time for ring flipping must be on the order of 10^{-7}–10^{-8} s to give the motionally averaged spectrum for the mobile material.

Solid-state ^{13}C NMR relaxation and rotational dipolar spin echo methods have also been used to study the dynamics of a variety of

5.3. Polymer Dynamics in the Solid State

ring- and main-chain-substituted polystyrene derivatives (72). Figure 5.44 shows the rotational dipolar spin echo spectra of the protonated aromatic and methine carbon resonances, in which the dipolar couplings appear as stick spectra tracing out the dipolar lineshape. The dipolar lineshapes are not those of a rigid solid, but indicate that the width and shape of the line are reduced by fast anisotropic molecular motion. The best fit for the aromatic peak was for a model in which 7% of the polystyrene aromatic rings undergo 180° flips. Further support for this model was obtained from the observation of multiexponential spin–lattice relaxation and the observation of the chemical

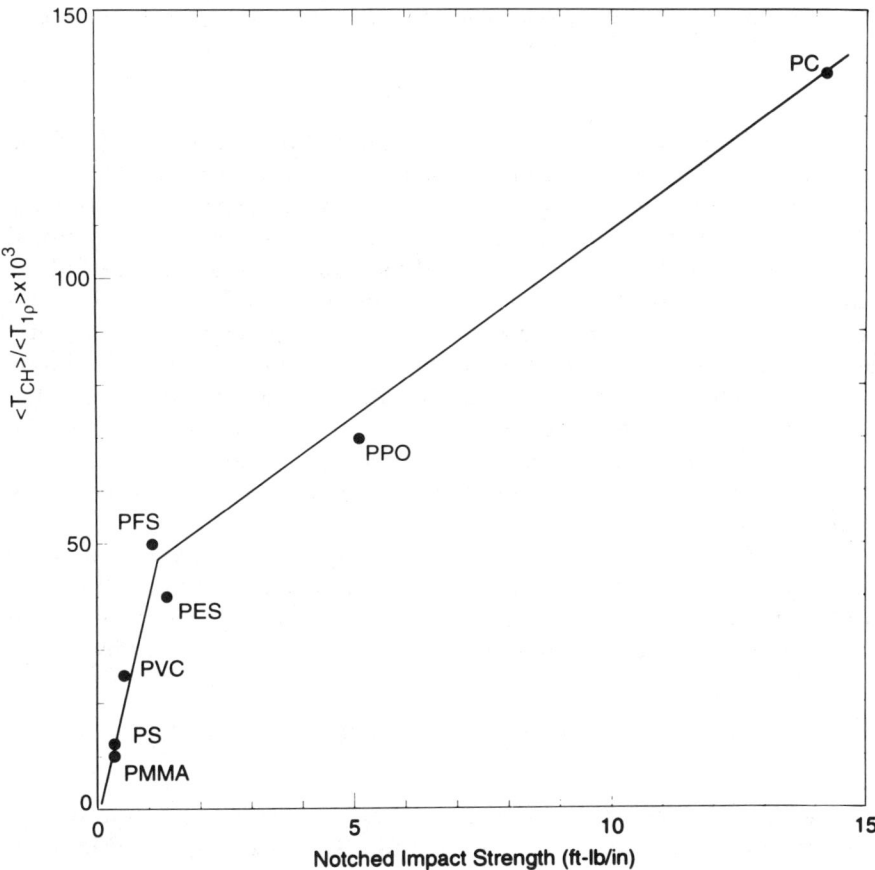

FIGURE 5.42 The correlation between the notched impact strength and the ratio of the ^{13}C NMR relaxation parameters. PC, polycarbonate; PPO, poly(phenylene oxide); PSF, polysulfone; PES, poly(ether ketone); PVC, poly(vinyl chloride); PS, polystyrene; PMMA, poly(methyl methacrylate). Adapted with permission (50).

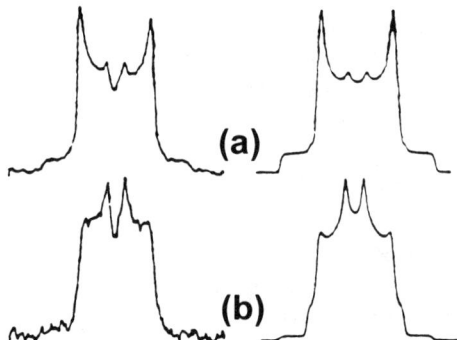

FIGURE 5.43 The experimental (left) and calculated (right) wide-line deuterium NMR spectra of polystyrene-d_5 with recycle times of 3 s (a) and 0.05 s (b) at 373 K. The lineshapes were simulated assuming that 20% of the polystyrene rings undergo 180° flips. Adapted with permission.

shift anisotropy tensor for the nonprotonated aromatic carbon. If the averaging of the dipolar interactions is due to ring flips about the 1,4 axis of the aromatic ring, these flips would average the dipolar interactions for the protonated carbons but would leave the chemical shift tensor unaffected. The ring flips appear to be unaffected by chain configuration defects, so it has been proposed that the flipping rings are at sites in the glass where high-frequency chain wiggling leads to a lower barrier for ring flipping, and that the ring flips can be associated with the weak γ transition in the mechanical loss spectra (72). Analysis of the data for the aliphatic carbons shows that the same fraction of main-chain carbons also experience motional averaging, suggesting that such motions are cooperative with the ring flipping.

In addition to this population of ring flipping sites, there appears to be some smaller amplitude low-frequency motions that affect $T_{1\rho}^C$ and lead to averaging of the dipolar lineshape. For the protonated aromatic carbons it is proposed that the relaxation is given by

$$\frac{1}{T_{1\rho}^C} = K^2 (\sin^2 \theta) J(\omega), \tag{5.48}$$

where K^2 is a constant, $\sin^2 \theta$ is the average dipolar fluctuation orthogonal to the applied RF field, and $J(\omega)$ is the spectral density associated with the ring motion at the frequency (37 kHz) of the spin-locking field. The spectral densities for ring motions were shown to be similar, and if it is assumed that ring rotation occurs about the 1,4 axis, then the $T_{1\rho}^C$ should be a simple function of θ, which can be

5.3. *Polymer Dynamics in the Solid State* 419

measured from the ratio of the second and first sidebands in the rotational dipolar spin echo spectra. This prediction is confirmed experimentally.

These same techniques, and others, have been used to investigate the molecular dynamics of polycarbonate. Figure 5.45 shows the wide-line deuterium NMR spectra of methyl- and ring-labeled polycarbonate at 293 and 380 K (3). At both temperatures the methyl deuterons give Pake spectra that can be fit to a model in which there is no molecular motion other than rotation of the methyl groups,

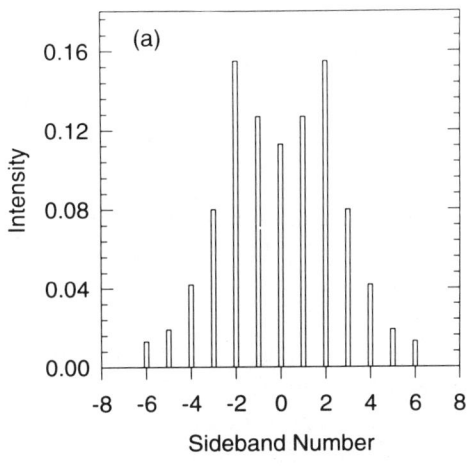

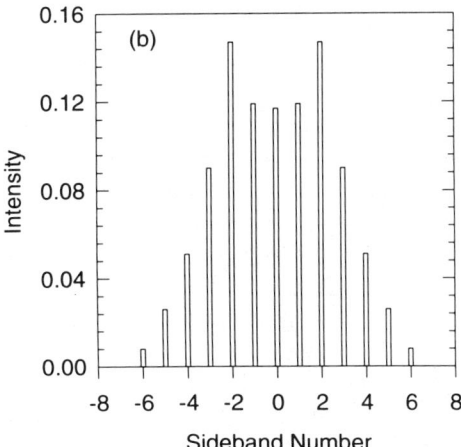

FIGURE 5.44 The dipolar sideband patterns for the protonated aromatic (a) and main-chain (b) resonances of polystyrene. Adapted with permission (72).

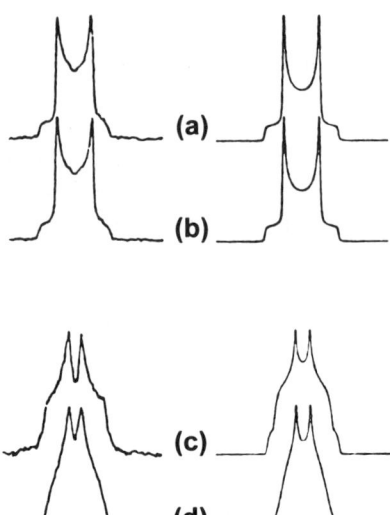

FIGURE 5.45 The experimental (left) and calculated (right) wide-line deuterium spectra for (a and b) methyl deuterated and (c and d) ring deuterated polycarbonate at 293 and 380 K. Adapted with permission (3).

demonstrating that the motion of the C–CH$_3$ bond vector is in the slow motion limit on the deuterium NMR time scale in the glassy state. The ^{2}H spectra for the ring-labeled polycarbonate show the characteristic lineshape observed for aromatic rings undergoing 180° flips. These spectra are best described by a model in which the rings undergo small-amplitude oscillations in addition to the ring flips. The amplitude of the oscillations increases from $\pm 15°$ at room temperature to $\pm 35°$ at 380 K with a Gaussian distribution of amplitudes with a variance of 10°.

The chemical shift anisotropy lineshape of polycarbonate labeled with ^{13}C at the ring position *ortho* to the carbonate group has also been used to characterize the molecular dynamics of polycarbonate over a range of temperatures. A variety of models were tested to fit the chemical shift anisotropy lineshapes shown in Fig. 5.46 (73,74). Among the models that did not give a good fit to the data were 180° flips and restricted rotations of various amplitudes. However, a good fit was obtained for models combining ring flips with restricted rotations, similar to that reported from the deuterium NMR data presented in Fig. 5.45. Combined with these data were the measurements of the proton T_1 and $T_{1\rho}$ relaxation times in an attempt to relate the NMR relaxation with the relaxation peaks measured by dynamic mechanical and dielectric spectroscopy. The results are

5.3. Polymer Dynamics in the Solid State

shown in Fig. 5.47 (74). The correlation between the data measured by NMR and other methods shows that such NMR experiments can be used to obtain a molecular-level understanding of the relaxation peaks observed by other spectroscopies.

In addition to the above studies, rotational dipolar spin echo has also been used to investigate the molecular dynamics of bisphenol polycarbonate and derivatives modified on the links between the aromatic rings and for ring-substituted polycarbonates. Figure 5.48 shows the rotational dipolar spin echo dipolar lineshapes for several of the carbons in polycarbonate (75). Again, information about the molecular dynamics is obtained by comparing the experimental data with that simulated with models for the molecular dynamics. The best fit for the aromatic rings is for 180° ring flips superimposed on 30° ring oscillations around the phenyl 1,4 axis. Unlike the ^{2}H data reported for polycarbonate (3), it is found that 20° amplitude oscillations of the methyl groups are required for the best fit of the experimental data. It is also reported that chlorine substitution on the

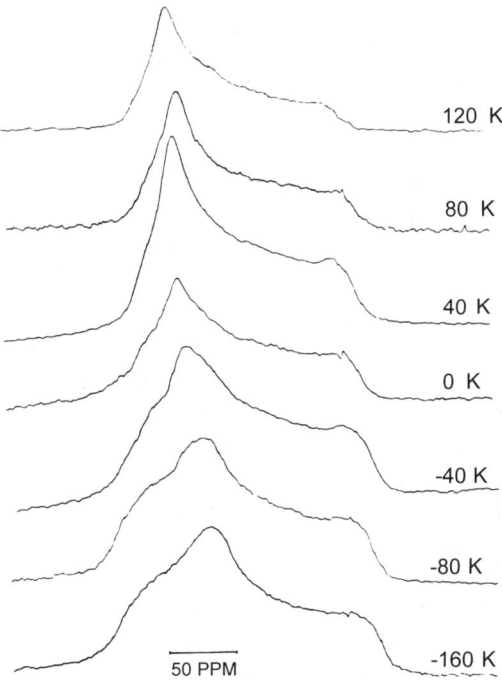

FIGURE 5.46 The ^{13}C chemical shift anisotropy lineshape as a function of temperature for bisphenol polycarbonate labeled with ^{13}C at the two phenylene carbons *ortho* to the carbonate group. Adapted with permission (74).

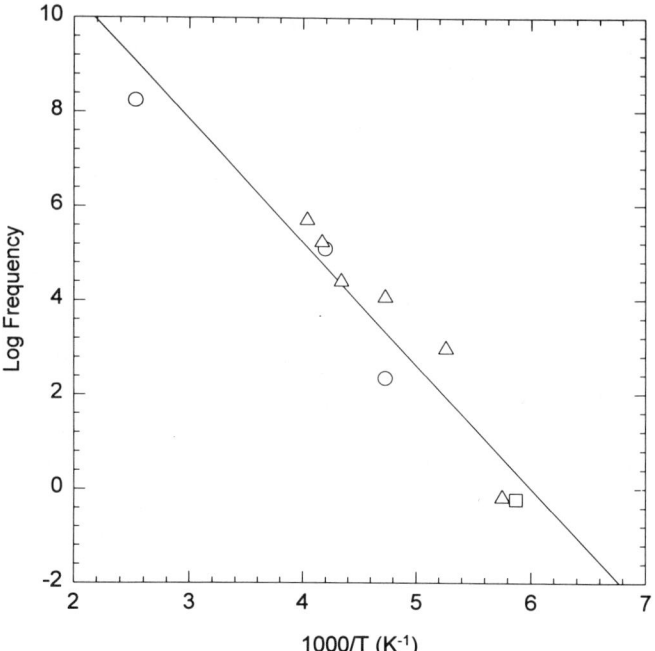

FIGURE 5.47 A relaxation map for polycarbonate combining (○) NMR, (□) dynamic mechanical, and (△) dielectric data. Adapted with permission (74).

aromatic rings quenches ring flips, and substitutions at the linking groups between the rings can also inhibit chain motion.

The molecular dynamics of the carbonate linkage have been examined by ^{13}C NMR using polycarbonate labeled at the carbonyl carbon, and Fig. 5.49 shows the nonspinning spectra of this material near ambient temperature along with the best-fit powder pattern (76). Only small changes are observed as the temperature is increased from -177 to $100°C$. The spectral changes over this temperature range are ascribed to oscillations of the carbonyl with an amplitude of about $40°$.

The molecular dynamics of polymers have also been investigated as a function of carbon dioxide and hydrostatic pressure and aging. The effect of carbon dioxide on the dynamics of polymers is of interest since the molecular-level processes leading to gas permeability in polymers are not well understood. The pressure of carbon dioxide (up to 35 atm) has only a small effect on the deuterium lineshape of deuterated polycarbonate, polystyrene, or poly(methyl methacrylate), but some of the T_1's are sensitive to the pressure (77). The T_1's for polystyrene showed the greatest temperature sensitivity, while poly-

5.3. Polymer Dynamics in the Solid State

carbonate was intermediate and poly(methyl methacrylate) showed no effect. The effect of pressure on the T_1's for polycarbonate and polystyrene can be correlated with the pressure dependence of the carbon dioxide diffusivity. These data suggest that either the polymer dynamics measured by deuterium NMR are altered by the gas or both

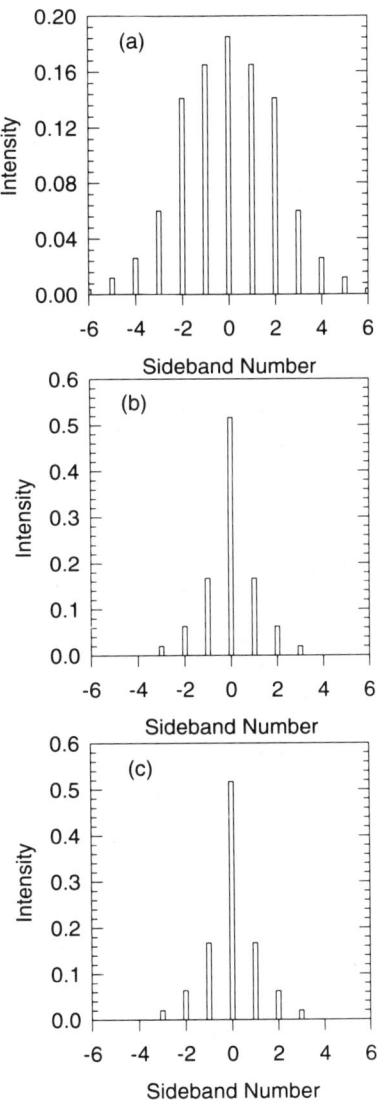

FIGURE 5.48 The experimental dipolar lineshapes obtained by 2D rotational dipolar spin echo for the (a) protonated aromatic, (b) methyl, and (c) quaternary carbons of polycarbonate. Adapted with permission (75).

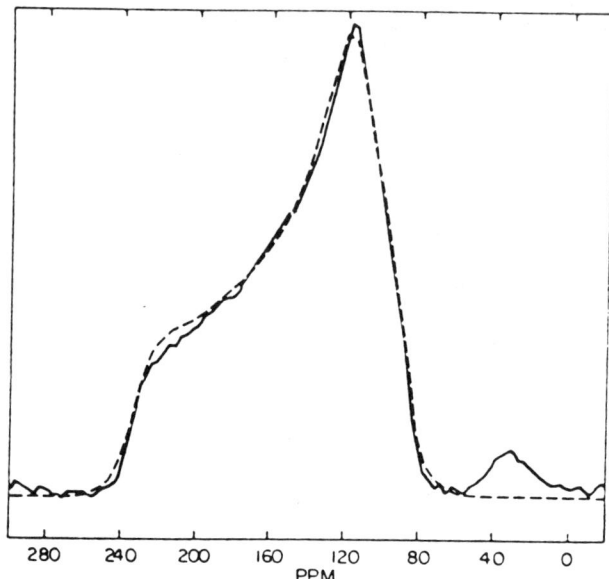

FIGURE 5.49 The ^{13}C NMR spectrum of carbonyl-labeled polycarbonate obtained at 12°C without magic angle spinning. The dotted line shows the simulated spectrum. Adapted with permission (76).

the NMR parameters and the diffusivity have the same dependence on some other process, such as swelling or plasticization. Hydrostatic pressure has been shown to affect ring flips in polycarbonate (78) and polystyrene (79), but has only a small effect on the main-chain dynamics of polystyrene (80).

The effect of aging on the dynamics of polycarbonate and poly(ester carbonates) derived from bisphenol-A polycarbonate and tere- or isophthalic acids has also been studied. Physical aging has little effect on bisphenol-A polycarbonate, but leads to small lineshape changes in the terephthalate-labeled rings in poly(ester carbonates) (81).

Deuterium NMR has also been used to study the sub-T_g processes in the amorphous phase of poly(phenylene sulfide). The data show that this polymer experiences ring flips like those in polycarbonates, and small-amplitude oscillations accompany the ring flips (82). There is a large distribution in the flipping rates for poly(phenylene sulfides) that is presumably related to a distribution in packing environments.

The low-frequency molecular dynamics of amorphous polymers have been measured using real-time exchange methods such as those used to measure molecular dynamics in crystals (4). The major differences between the study of crystalline and amorphous materials is that the chain conformation and dynamics are less well defined, and while the molecular dynamics in a crystalline lattice can often be described by a two-site jump model, where the monomer is translated and rotated to a neighboring site in the lattice, the initial and final conformations in the amorphous phase are much more poorly defined. The molecular dynamics in amorphous polymers can be better modeled with a distribution of jump angles and a wide distribution in the correlation times for the motions that arise from the statistical nature of the intermolecular interactions in the amorphous phase.

In some of the early studies of polymers in the solid state, differences were observed between the spectra in the solid state and those in solution. In poly(2,6-dimethyl phenylene oxide), for example, two aromatic signals are observed in the solid, while only one is observed in solution (50). This is due to the difference in dynamics between solids and solutions, where the chemical shift in solution is averaged by rapid molecular motion. Similar observations have been made for atactic polypropylene, as shown in Fig. 5.50, where the methylene peak is split into a doublet at the lower temperature (83). As the temperature increases above 259 K the peaks coalesce into a single broad peak. The splitting of the methylene signals at low temperatures is due to a freezing out of the *trans–gauche* interconversion. The peaks are assigned via the γ-*gauche* effect (65) to the *tt* and *tg* conformations at 48.6 and 44.6 ppm. Interconversion of these conformations can be observed in the solid-state 2D exchange spectra shown in Fig. 5.51, where cross peaks are observed connecting the methylene peaks at long mixing times. Analysis of the correlation times for the motions leading to cross peaks in the 2D spectra demonstrates that *gauche–trans* is isomerization is tightly coupled to the α relaxation in atactic polypropylene (83).

Two-dimensional wide-line exchange NMR has been used to investigate the molecular dynamics of several amorphous polymers, including polystyrene, atactic polypropylene, and polyisoprene, at temperatures above T_g, where the chain dynamics are in the range that can be measured by 2D NMR. Figure 5.52 shows the deuterium 2D experimental and simulated 2D exchange spectra for main-chain-labeled polystyrene at 391 K, 10 K above the T_g (84). Instead of the well-defined ridge pattern observed for crystalline polymers, a more diffuse pattern is observed. The key features of this pattern include extended ridges in the 2D plane at the frequency of the singularities and a decay and broadening of the center of the powder pattern along the diagonal. The pattern of cross-peak intensity in the 2D spectra as a function of mixing time and temperature cannot be described by a model with a single correlation time, but can be described by a model

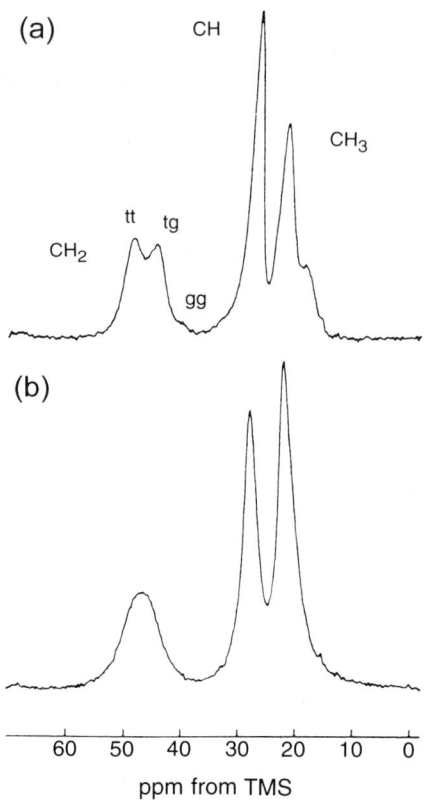

FIGURE 5.50 The ^{13}C cross-polarization and magic angle spinning spectra of atactic polypropylene at (a) 250 K and (b) 262 K. Adapted with permission (83).

with isotropic rotational diffusion with a broad distribution of correlation times. For polystyrene at this temperature the mean correlation time is 0.006 s and the width of the distribution is 3 decades. The inset plot in the simulation shows the final angular distribution of C–D bond vectors after the mixing time. Note that the differences in the simulated and experimental data at the edges of the spectra that arise from pulse imperfections.

The 2D NMR results for polystyrene can be combined with other data measured on the same polymer for a more global understanding of structure–property relationships in amorphous polymers. Figure 5.53 shows a plot of the correlation times for polystyrene acquired by 2D NMR, solid echo, broad-line NMR, and T_1 relaxation rate measurements that spans 12 orders of magnitude (85). It is found that the data for the 2D exchange, solid echo, and broad-line NMR experi-

5.3. Polymer Dynamics in the Solid State

ments lie on a single line that can be fit to the Williams–Landel–Ferry equation (86) describing the temperature dependence of polymers above T_g. This shows that these data are a measure of the dynamics associated with the α transition. Conversely, the T_1 data deviate from this temperature dependence and are associated with the β transition. Similar behavior is observed for atactic polypropylene, where a single curve over many orders of magnitude was constructed from the 2D exchange data just above T_g and the spin–lattice relaxation measurements at higher temperatures (67).

The molecular dynamics of 1,4-cis-polyisoprene above the glass transition temperature have been studied in considerable detail using the polymer with deuterated methyl groups (87). Better spectra are obtained for this polymer since the width of the spectra is reduced for methyl deuterons compared to that of the main-chain deuterons in

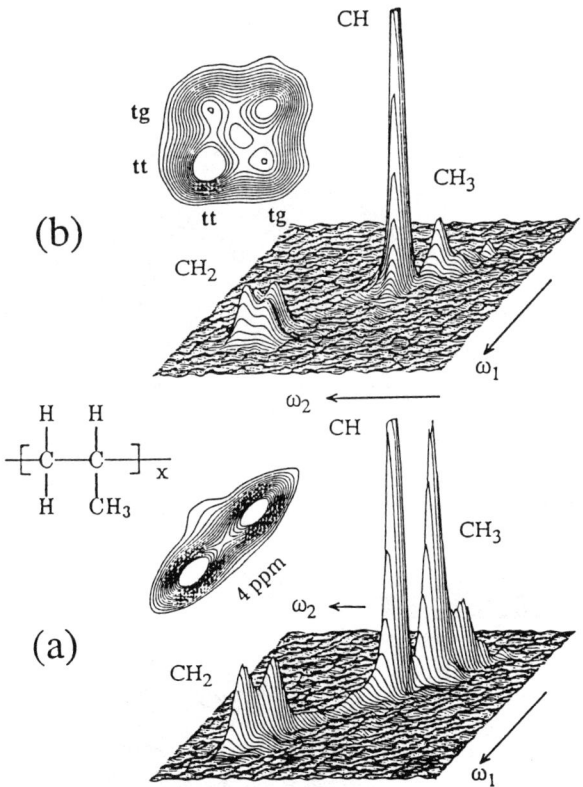

FIGURE 5.51 The 2D ^{13}C exchange spectra for atactic polypropylene at 250 K with mixing times (a) 0.005 and (b) 0.5 s. Contour plots of the methylene peaks are shown in the insets. Adapted with permission (83).

polystyrene. In this study there is a close correlation between the dynamics measured by 2D NMR and those measured by dielectric spectroscopy. In addition, the width of the distribution in correlation times is reported to decrease from 2.2 decades at 5 K above the T_g to 0.5 decades at 55 K above the T_g.

Multidimensional NMR has been used to investigate the β relaxation of poly(methyl methacrylate) via the molecular dynamics of the carbonyl carbon and the methyl ester groups. Evidence for the molecular dynamics of the carbonyl groups was obtained from the lineshape as a function of temperature as shown in Fig. 5.54 (88). At ambient temperature a regular chemical shift anisotropy pattern is observed with $\omega_{11} = 268$ ppm, $\omega_{22} = 150$ ppm, and $\omega_{33} = 112$ ppm. As the temperature is increased, changes are observed in the spectra

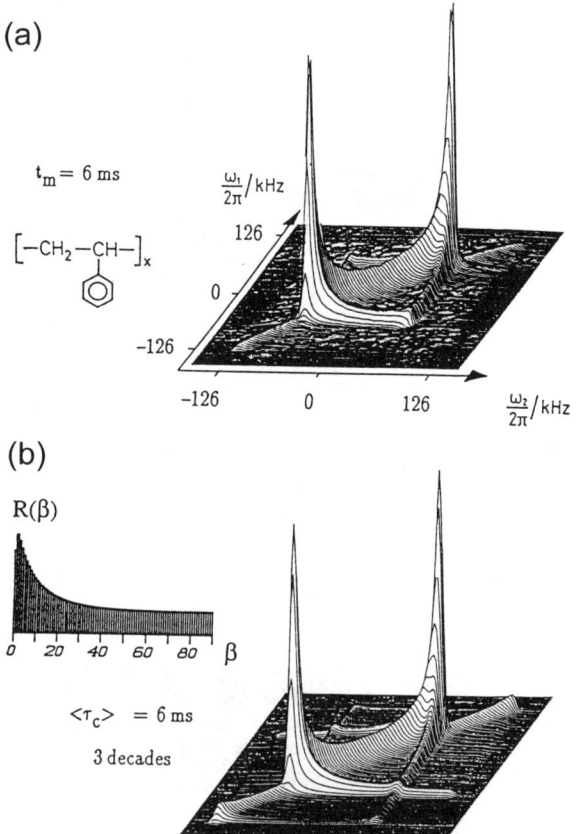

FIGURE 5.52 The (a) experimental and (b) simulated 2D ^{2}H exchange spectra for chain deuterated polystyrene at 391 K. Adapted with permission (84).

5.3. Polymer Dynamics in the Solid State

that are indicative of large-amplitude reorientations that are fast on the chemical shift anisotropy time scale (10 kHz). More detailed information about the molecular dynamics can be obtained by 2D exchange experiments and consideration of the chemical shift anisotropy tensor for the carbonyl group in poly(methyl methacrylate) shown in Fig. 5.55. The important features to note are that the ω_{33} axis is perpendicular to the plane of the carbonyl and, because of steric constraints, the plane of the carbonyl group is perpendicular to the chain axis.

The 2D exchange spectrum of poly(methyl methacrylate) containing 20% ^{13}C-labeled carbonyl groups is shown in Fig. 5.56 along with three spectra simulated for different models for the molecular dynamics of the carbonyl group (88). One important feature to note about the experimental spectrum is that the off-diagonal intensity that arises from the change in orientation of the carbonyl group is observed only between the ω_{11} and the ω_{22} chemical shifts, while the ω_{33} magnetization remains along the diagonal. Such an observation is consistent with a 180° flip of the carbonyl group which interchanges

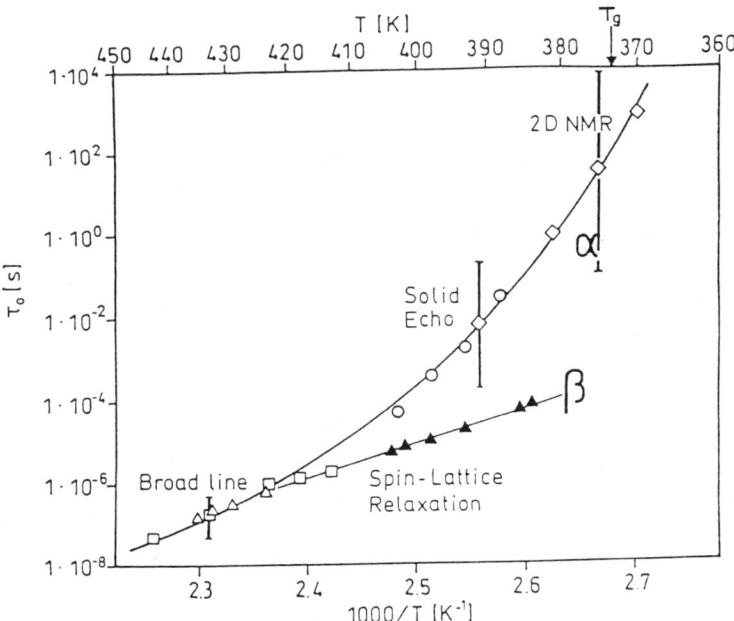

FIGURE 5.53 An Arrhenius plot of the correlation times for the main-chain dynamics of polystyrene measured by several NMR methods, including 2D exchange NMR, the solid echo, T_1 measurements, and broad-line NMR. Adapted with permission (85).

the ω_{11} and ω_{22} chemical shifts but leaves the ω_{33} chemical shift unaffected. However, 180° flips alone cannot account for the 2D spectra, as shown by the simulation in Fig. 5.55(b). However, if the 180° flips are combined with a concomitant rotation about the local-chain axis with an amplitude of $\pm 20°$, excellent agreement is observed between the experimental and the simulated spectra. The simulations also show that the local-chain motion without the 180° flips is a poor model for the local dynamics. The correlation between the dynamics required for simulation of the experimental spectra and those measured by dielectric and dynamic mechanical spectroscopy demonstrates that these motions are associated with the β relaxation peak.

Further evidence for the flipping motion of the carbonyl group was obtained from the selective excitation 3D exchange spectra acquired

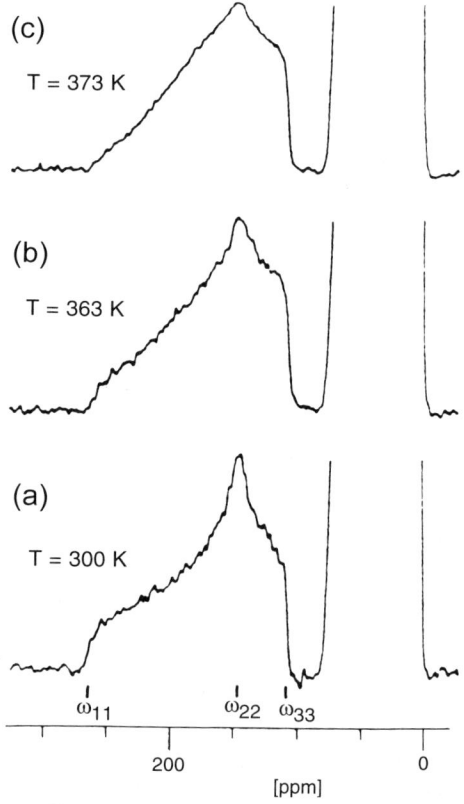

FIGURE 5.54 The ^{13}C NMR powder pattern of the carboxyl group of poly(methyl methacrylate) at (a) 300 K, (b) 363 K, and (c) 373 K. Adapted with permission (88).

5.3. Polymer Dynamics in the Solid State

using the pulse sequence shown in Fig. 5.57 (88). This experiment differs from the normal 2D exchange experiment in that a narrow band of chemical shifts is selected prior to the exchange experiment with a selective excitation pulse sequence (89). The final result is a pseudo-3D NMR spectrum corresponding to a plane in a true 3D spectrum at the frequency of the chemical shift that was selectively excited. This experiment can be used to distinguish between diffusional and jump models since magnetization appears diffusely in the 2D plane for diffusional models and at well-defined frequencies in a two-site jump model. Such a pseudo-3D spectrum of poly(methyl methacrylate) is shown in Fig. 5.58. In addition to the intense diagonal resonance, well-defined peaks are observed along the $\omega_3 = \omega_1$

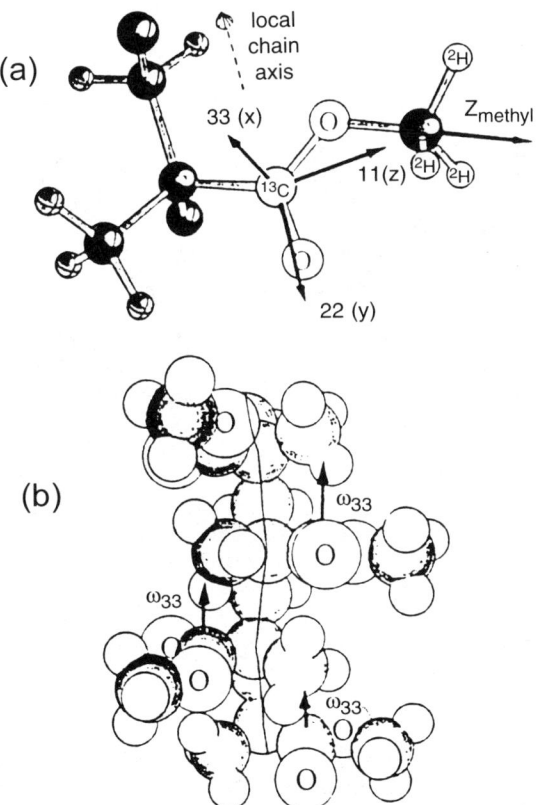

FIGURE 5.55 A drawing illustrating (a) the relative orientations of the chemical shift anisotropy tensor and (b) the local chain axis in poly(methyl methacrylate). In a typical chain segment the chain axis is perpendicular to the OCO plane. Adapted with permission (88).

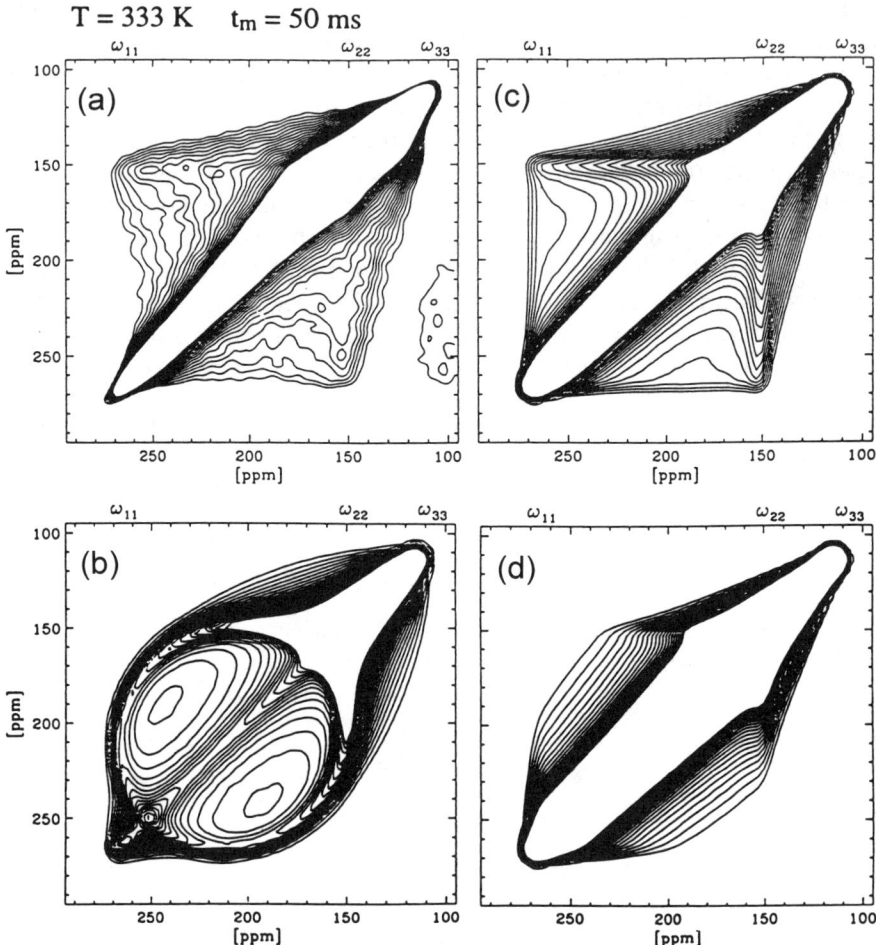

FIGURE 5.56 Comparison of the 2D exchange spectrum of poly(methyl methacrylate) with spectra simulated using several models for the molecular dynamics. The (a) experimental data are compared with spectra simulated assuming (b) a Gaussian distribution of side-group flip angles a $\pm 25°$ amplitude, (c) $180° \pm 10°$ flips with concomitant rotation about the local chain axis, and (d) rotations about the local chain axis. $T = 333$ K; $t_m = 50$ ms. Adapted with permission (88).

and $\omega_3 = \omega_2$ lines that arise from jumps between orientations and jumps back to the starting orientation. From these studies it is concluded that 50% of the groups are not flipping on the time scale of this experiment but experience low-amplitude rocking, 25% have undergone one or an odd number of flips, and 25% have undergone two or an even number of flips, and that main-chain rocking accompa-

5.3. Polymer Dynamics in the Solid State

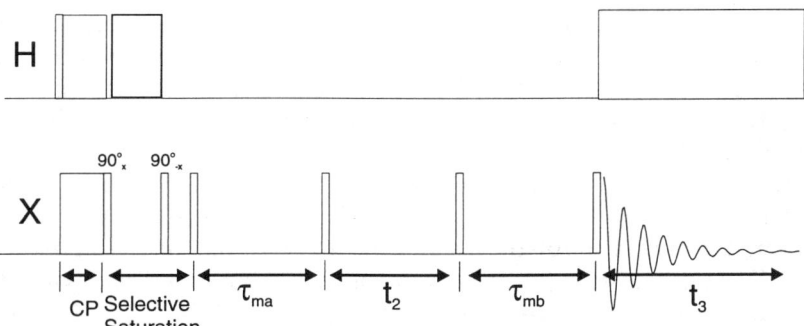

FIGURE 5.57 The pulse sequence diagram for the pseudo-3D exchange NMR experiment (88). In the pseudo-3D experiment the t_1 period is replaced with the selective excitation of a selected band of frequencies (89).

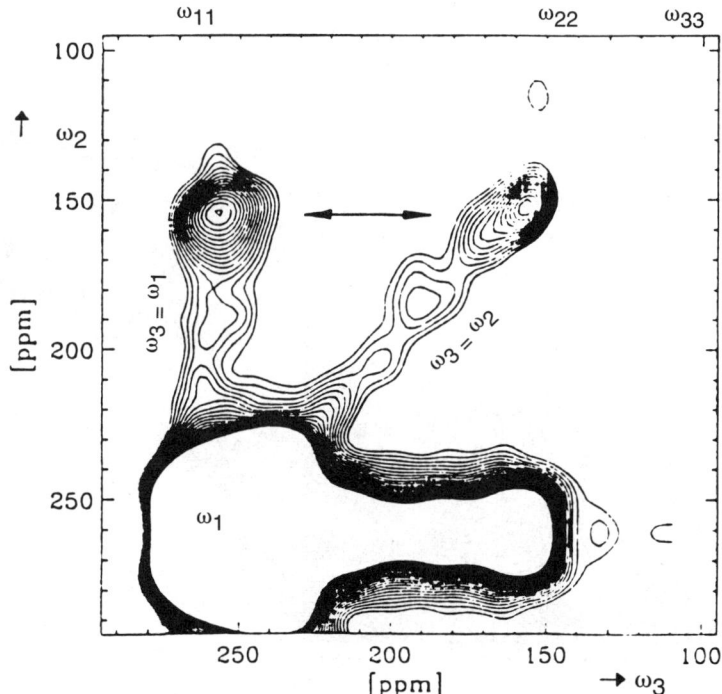

FIGURE 5.58 The selective excitation 3D exchange NMR spectrum of poly(methyl methacrylate) with ^{13}C-labeled carbonyl groups at 333 K with delays of 0.05 s for the first and second mixing times. Adapted with permission (88).

nies the flipping. The coupling of main-chain motion to carbonyl flipping is related to the asymmetry of the bulky side chain, and it was suggested that a flipped carbonyl can better fit into the space that it previously occupied if there is a small rotation of the main chain (88).

Similar methods have been used to study the side-chain dynamics in poly(ethyl acrylate), where it was also observed that flipping of the carbonyl group is coupled to ±20° rocking motions of the main chain (90). However, in this study it was observed that the carbonyl motion remains anisotropic even above the T_g. At 365 K (T_g + 27 K) the motion can be described as a combination of carbonyl flipping with an increased amplitude (±50°) of rocking. Such anisotropic motion above T_g is contrary to that observed in other amorphous polymers above T_g.

Two-dimensional wide-line deuterium exchange NMR has also been used to study the ring flips in polycarbonate and how they are affected by tensile stress, as shown by the spectrum in Fig. 5.59 (91). From these data the mean reorientation angle for the ring deuterons was found to be 120°, exactly the value expected for flipping about the

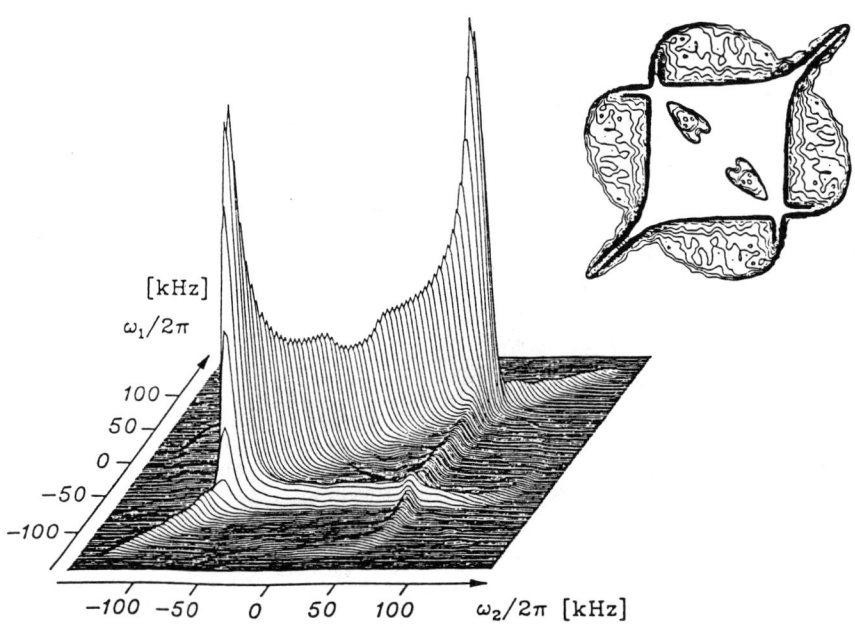

FIGURE 5.59 The 2D ^{2}H exchange spectrum of ring deuterated polycarbonate at 233 K and with a mixing time of 0.5 s. Note the well-defined ridge off the diagonal that arises from flipping rings. Adapted with permission (91).

5.3. Polymer Dynamics in the Solid State

1,4 axis of the aromatic ring. However, the best fit to the experimental data was obtained from a model with an asymmetric flip angle distribution, where jumps around 0° have a smaller distribution than those around 180°. It was reported that the distribution width for the flipping rings increased from 50° for the polymer without stress to 90° for the sample with 3% strain. Studies on carbonyl–labeled polycarbonate did not show evidence for large-amplitude motion at ambient temperature even with mixing times as long as 2 s (92).

The relaxation for glassy polymers above the T_g is often modeled using nonexponential correlation functions. Such behavior could arise either to a distribution of correlation times for the same process or due to different process. These two possibilities cannot be distinguished in a 2D NMR experiment that measures the initial and final orientations, but can be measured in 3D or 4D NMR experiments. The novel reduced 4D exchange experiment shown in Fig. 5.60 has been used to make this distinction (93). It is a reduced 4D experiment because fixed values of t_1 and t_2 are used while the decay period t_3 is incremented in the normal way and data are collected during the t_4 time period. The values for t_1 and t_2 are chosen to be equal to each other such that a stimulated echo is formed from those components that did not exchange during the first mixing time. In this way the exchange behavior of those components that did not exchange during the first mixing time can be probed during the following mixing times.

The reduced 4D exchange experiment has been used to probe the dynamics of poly(vinyl acetate) at 20 K above its glass transition temperature (93). Figure 5.61 shows the 2D exchange spectra for poly(vinyl acetate) for a mixing time of 10 ms and two spectra from the reduced 4D experiment obtained with different mixing times. In the 2D spectrum with a 10-ms mixing time some off-diagonal intensity is observed, but there is also a large diagonal peak, demonstrating that there is dynamical heterogeneity. Figure 5.61(b) shows the reduced 4D experiment with all three mixing times set to 10 ms. This spectrum has little off-diagonal intensity and demonstrates that those molecules that did not change orientations in the first mixing time

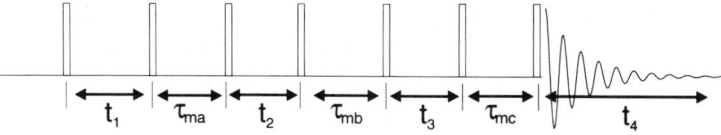

FIGURE 5.60 The pulse sequence diagram for the reduced 4D exchange NMR exchange exchange spectra. The reduced spectrum is recorded with fixed values of t_1 and t_2, and the value of t_3 is incremented in the normal manner (93).

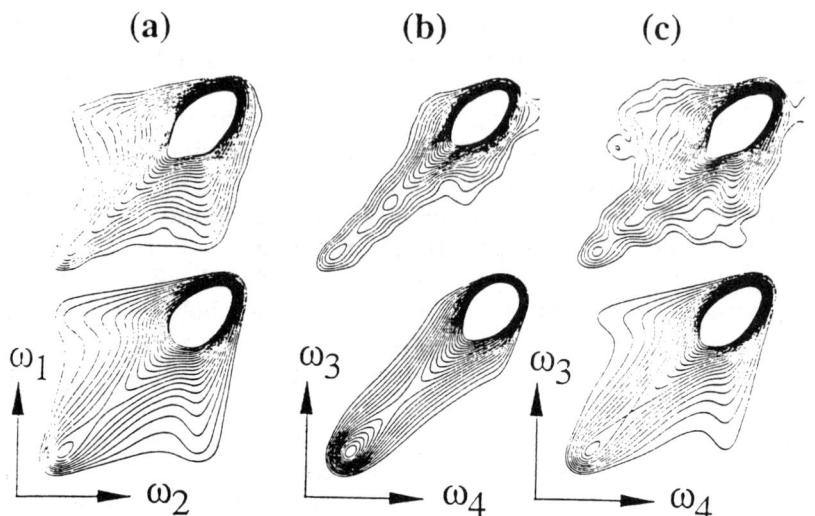

FIGURE 5.61 The experimental (top) and simulated (bottom) (a) 2D exchange spectra acquired with a mixing time of 0.01 s, (b) reduced 4D exchange spectra acquired with $t_{ma} = t_{mb} = t_{mc} = 0.01$ s, and (c) reduced 4D exchange spectra acquired with $t_{ma} = t_{mc} = 0.01$ s and $t_{mb} = 1$ s at 20 K above the T_g of poly(vinyl acetate). Adapted with permission (93).

remain static on the 10-ms time scale. However, off-diagonal intensity is observed if the second mixing time is increased to 1 s, demonstrating that some of the molecules that were initially slow have changed their correlation time.

5.3.4 THE DYNAMICS OF POLYMER BLENDS

Polymer blends are materials of interest due to their tunable properties and to the possibility that materials with desirable new properties can be obtained simply by mixing pairs of polymers together. The mixing of polymers is also known to affect their molecular dynamics, and it is frequently reported that the relaxation peaks observed in dielectric or dynamic mechanical spectroscopy are shifted or suppressed, or that new peaks appear that are not observed in the component polymers. NMR studies of the dynamics of polymer are often performed to attempt to gain a molecular-level understanding of the relaxation transitions in polymer blends. Many of the methods previously presented for the study of the dynamics of semicrystalline and amorphous polymers have also been applied to the study of polymer blends.

5.3. Polymer Dynamics in the Solid State   437

Blends of 1,4-polyisoprene and 1,2-polybutadiene are an example of polymers that are miscible over their entire composition range, as evidenced by the spontaneous interdiffusion of the two polymers (94). Unlike many other blends, however, the glass transition temperature is extremely broad, suggesting that the blend is inhomogeneous on some level. The dynamics of the blend have been investigated by magic angle spinning NMR, and Fig. 5.62 shows the portion of the spectra containing the resonances from the double-bonded carbons

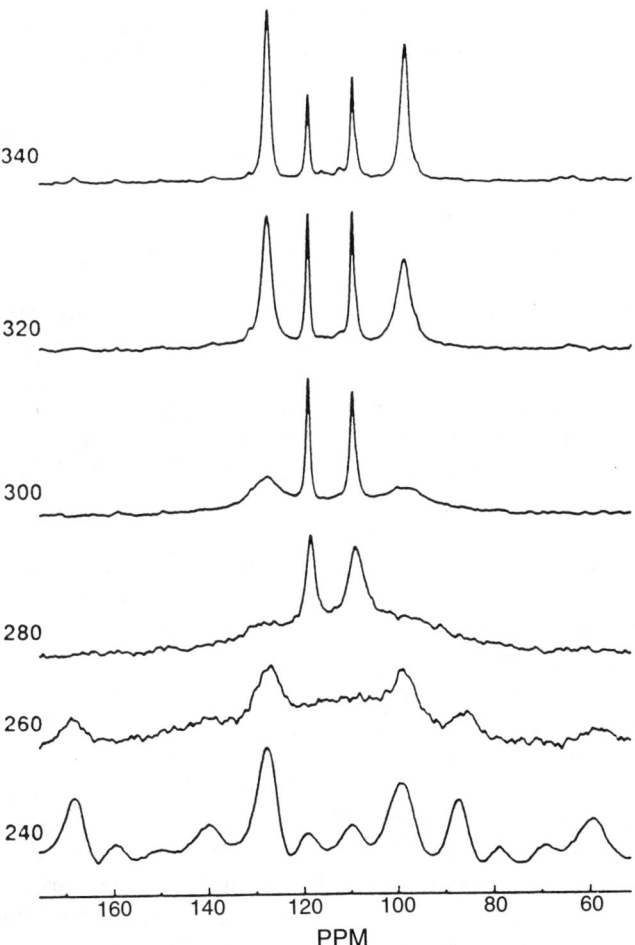

FIGURE 5.62 The ^{13}C magic angle spinning NMR spectra of the blend of 1,4-polyisoprene and 1,2-polybutadiene acquired between 240 and 340 K. This region of the spectrum contains the peaks from 1,2-polybutadiene at 130 and 100 ppm and from 1,4-polyisoprene at 120 and 110 ppm. Adapted with permission (95).

between 240 and 340 K (95). Relatively sharp spectra are obtained at high temperatures, and the outer lines are assigned to the 1,2-polybutadiene resonances while the inner ones are assigned to 1,4-polyisoprene. At intermediate temperatures it becomes obvious that the temperature dependence of the linewidth is remarkably different for the two polymers, even though they appear to be mixed on the molecular level. The maximum linewidths observed for the 1,2-polybutadiene and 1,4-polyisoprene are at 280 and 260 K. As noted in Section 5.3.1.1, such linewidth changes can arise from interference between the dipolar couplings and the magic angle spinning or dipolar decoupling or the chemical shift anisotropy (40). Such a temperature-dependent linewidth is also observed in the pure polymers, but the temperature at which the maximum linewidths are observed is shifted from 300 K for 1,2-polybutadiene and 240 K for 1,4-polyisoprene. Thus, even though the polymers are thermodynamically miscible, each polymer undergoes its own glass transition, and these NMR studies provide a molecular-level insight into the broad T_g observed by DSC.

Similar observations have been made for blends of polystyrene and poly(methyl vinyl ether), as shown in Fig. 5.63 (96). In this study it was observed that the broadening as a function of temperature depends on the ratio of the hard (polystyrene) and soft [poly(methyl vinyl ether)] polymer, as shown in Fig. 5.64. Similar results were reported in another study where the linewidths for both polymers were plotted as a function of $T-T_g$ to correct for the changes in T_g that accompany the changes in the composition (97). These studies showed that the maximum in the observed linewidth was composition dependent for poly(methyl vinyl ether) but not for polystyrene. This shows that the dynamics giving rise to the line broadening are intramolecular for polystyrene, while both intra- and intermolecular interactions contribute to the dynamics for poly(methyl vinyl ether) in the blends. As with the butadiene/isoprene blends noted above, different chain dynamics are associated with the two types of chains in the thermodynamically miscible blend.

Deuterium lineshape analysis has also been used to monitor polymer molecular dynamics in blends. Figure 5.65 shows the lineshapes from a study of ring deuterated polycarbonate blends with poly(methyl methacrylate) and poly(cyclohexylenedimethylene terephthalate) (98). The blend with poly(methyl methacrylate) can be prepared as a one-phase or two-phase system since this blend exhibits a lower critical solution temperature near 200°C. A distribution of ring flipping rates was observed in the pure polymer, in agreement with earlier studies of polycarbonate (3). Deuterium NMR and dynamic mechanical spectroscopy show that blend formation leads to a slowing

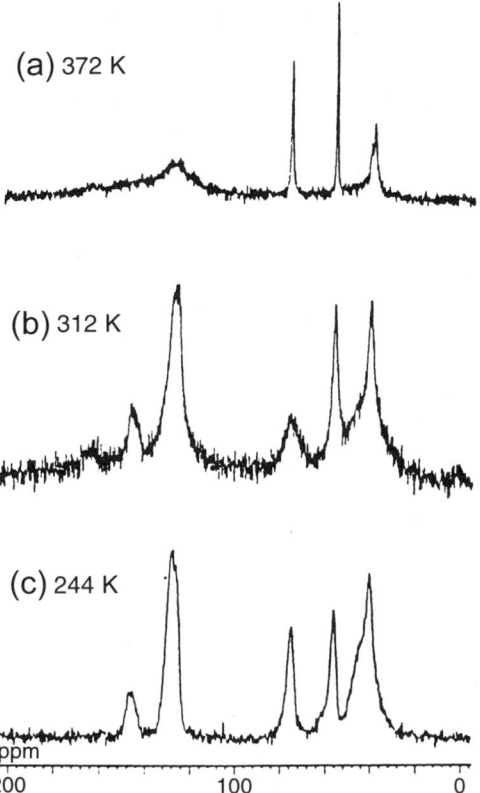

FIGURE 5.63 The ^{13}C magic angle spinning NMR spectra of polystyrene/poly(methyl vinyl ether) blends at (a) 372 K, (b) 312 K, and (c) 244 K. Adapted with permission (96).

of the main-chain motions in polycarbonate, while the deuterium NMR analysis of blends made with deuterated poly(methyl methacrylate) shows that the poly(methyl methacrylate) dynamics are not affected by blend formation (98).

Deuterium NMR relaxation has been used to investigate ring dynamics in blends of ring deuterated polystyrene and poly(2,6-dimethyl phenylene oxide) by using the observation that the mobile fraction of flipping rings has a faster relaxation time (99). A series of blends were prepared with low-molecular-weight poly(2,6-dimethyl phenylene oxide) polymers with the same T_g as polystyrene to remove the compositional dependence of the T_g. The results showed that there is no significant change in the fraction of rapidly flipping rings as a function of composition, demonstrating that blend formation does

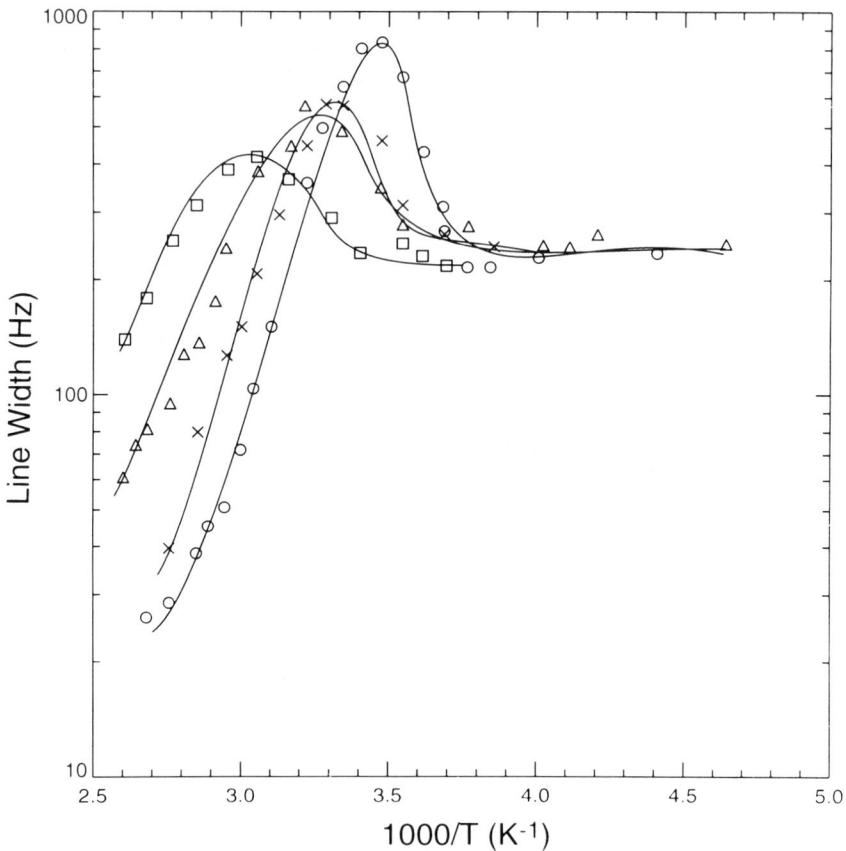

FIGURE 5.64 The temperature dependence of the linewidth of the CH resonance of (○) poly(methyl vinyl ether) (PVME) and blends with polystyrene (PS) at ratios of (×) PS/PVME = 20/80, (△) PS/PVME = 50/50, and (□) PS/PVME = 80/20. Adapted with permission (96).

not have a large effect on the polystyrene local dynamics as measured by deuterium NMR.

The proton relaxation rates T_1 and $T_{1\rho}$ in polymer blends have been extensively used to study the length scale of polymer mixing in blends (Section 4.3) but are not often useful for characterizing the molecular dynamics of polymer blends. For polymers mixed on the molecular level, rapid spin diffusion equilibrates the relaxation rates, and the measured relaxation time can usually be related to the rotating methyl groups. One strong piece of evidence for the molecular-level mixing of chains is the dependence of the proton relaxation rates on the weight fraction of the component polymers. However,

5.3. Polymer Dynamics in the Solid State   441

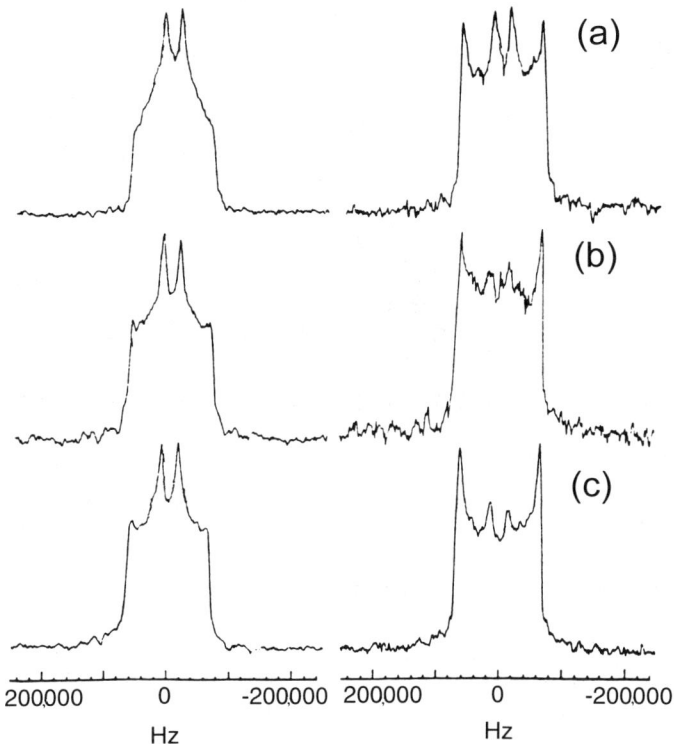

FIGURE 5.65 The wide-line ^{2}H NMR spectra for (a) ring deuterated bisphenol-A polycarbonate and blends with (b) poly(methyl methacrylate) and (c) poly(cyclohexylene-dimethylene terephthalate). The spectra on the left were acquired at 32°C, and those on the right were acquired at −23°C. Adapted with permission (98).

this analysis assumes that the relaxation of the components in the blends is the same as that in the pure polymers. It is sometimes observed that the proton relaxation rates for the blend are outside of the range of those of the component polymers, demonstrating that intermolecular interactions between the component polymers can affect the correlation times and change the observed relaxation rates (100,101).

Information about the molecular dynamics of blends can sometimes be obtained from the carbon relaxation rates, particularly $T_{1\rho}^C$. For example, we noted earlier in this section that the lineshape changes observed in magic angle spinning experiments can be used to identify those motions that interfere with either the chemical shift anisotropy or the dipolar decoupling of the blends of polystyrene and poly(methyl vinyl ether) (97). It was also reported that the tempera-

442   5. *The Dynamics of Macromolecules*

ture dependence of $T_{1\rho}^C$ for the protonated aromatic carbons of polystyrene paralleled the lineshape changes, and could be used to probe the molecular dynamics of blends. It has also been reported that the temperature dependence of the $T_{1\rho}^C$'s for the interacting groups in the blends of poly[2-(3,5-dinitrobenzoyl)oxy ethyl methacrylate] and poly[(N-ethycarbazole-3-yl)methyl methacrylate] differs between the component polymers and the blend (102). This is presumably a consequence of the strong interaction between the donor and the acceptor groups in the charge transfer complex.

The low-frequency molecular dynamics of polymer blends have been studied using the 2D spin exchange experiments developed by Spiess and co-workers (4). Figure 5.66 shows the results for one such blend, polystyrene and poly(2,6-dimethyl phenylene oxide) labeled with ^{13}C at the methyl group, as a function of temperature and mixing time (103). At ambient temperature and with a mixing time of 1 s, off-diagonal intensity is observed that is characteristic of rotational Brownian motion. Molecular motion can be distinguished from carbon–carbon spin diffusion, which has a weak temperature dependence, by measuring the spectrum at low temperature with the same

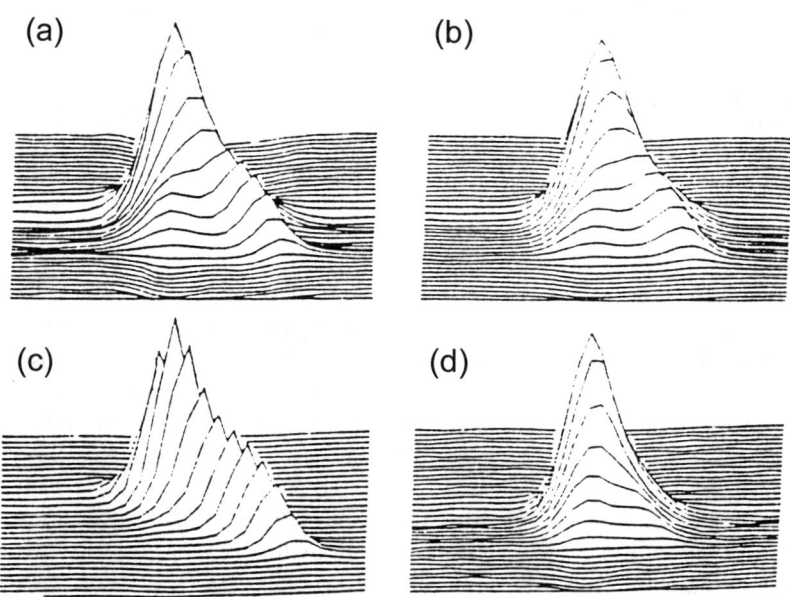

FIGURE 5.66 The ^{13}C 2D exchange spectra for blends of polystyrene with methyl-labeled poly(2,6-dimethyl phenylene oxide) at (a) τ_m = 1 s and T = −70°C, (b) τ_m = 1 s and T = 25°C, (c) τ_m = 0.05 s and T = 27°C, and (d) τ_m = 0.05 s and T = 156°C. Adapted with permission (103).

5.3. Polymer Dynamics in the Solid State

mixing time, as shown in Fig. 5.66. Figures 5.66(c) and 5.66(d) show the exchange spectra acquired with a 50-ms mixing time at ambient temperature and near the glass transition temperature. The dynamics of poly(2,6-dimethyl phenylene oxide) in the blend differ significantly from the behavior of the pure polymers in that motion is detected in the 2D exchange experiments at temperatures 10°C below the T_g, while motion is detected only for the pure polymers at temperatures well above the T_g. Modeling of the dynamics of the blend as a function of temperature and mixing time demonstrates that the motion is Brownian and is associated with a wide distribution of correlation times. The best fits between the experimental and the simulated data are obtained for models with a bimodal distribution of correlation times using a lattice model. It should be noted that the molecular dynamics measured by the 2D exchange experiments are much slower than those for the T_1 studies of polystyrene/poly(2,6-dimethyl phenylene oxide) noted above (99), and the difference in time scale probed by the two experiments may account for the different results.

This same blend has been studied by deuterium 2D exchange for blends with main-chain deuterated polystyrene, and some of the results are shown in Fig. 5.67 (104). Again, a bimodal distribution of correlation times is required to fit the experimental spectra. The relative weighting of the fast and slow modes corresponds to the local concentration fluctuations calculated using a statistical lattice model, where the monomers with the fastest dynamics correspond to those surrounded by polystyrene while the slowest monomers correspond to those surrounded by poly(2,6-dimethyl phenylene oxide). Such lattice models can be tested by fitting the 2D spectra as a function of polymer composition.

The dynamics of the individual components in blends of polyisoprene and 1,2-polybutadiene have been studied by deuterium 2D exchange NMR (105). As in the magic angle spinning studies of the linewidth as a function of temperature (95), the deuterium studies show that different transition temperatures are observed for the individual polymers in the blend. These data are summarized in Fig. 5.68, which shows the mean correlation times as a function of temperature obtained by simulating the 2D exchange spectra as a function of temperature for the component polymers and the blends (105). The broad T_g observed for the polyisoprene and 1,2-polybutadiene blends is ascribed to the difference in dynamics between the chains, the breadth of the distribution of correlation times, and the local compositional variations that affect both of these factors (106).

Two-dimensional ^{31}P NMR has been used to study the molecular dynamics of the diluent tetraxylyl hydroquinone diphosphate dis-

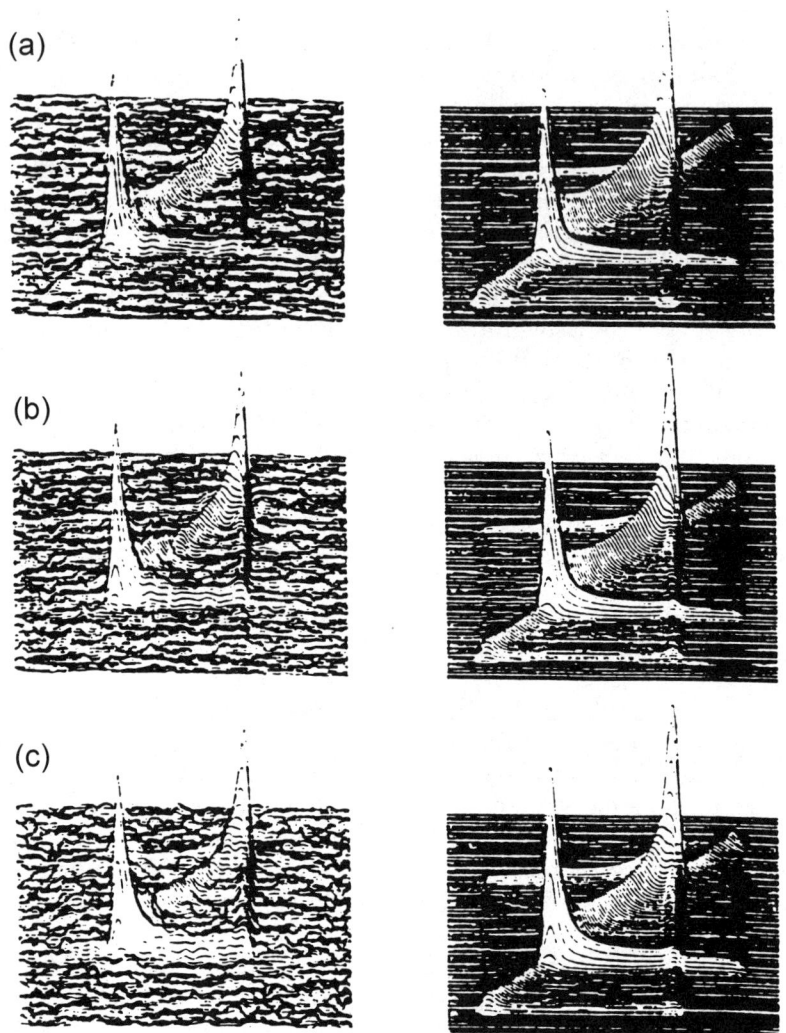

FIGURE 5.67 The experimental (left) and simulated (right) 2D ^{2}H spin exchange spectra in blends of polystyrene-d_3 and poly(2,6-dimethyl phenylene oxide) at 128°C with mixing times of (a) 0.006, (b) 0.018, and (c) 0.05 s. Adapted with permission (104).

solved in poly(2,6-dimethyl phenylene oxide) (107). Figure 5.69 shows the exchange spectra of the 50:50 mixture as a function of temperature and mixing time (107). The data show that the onset of Brownian rotational motion appears at the same temperature as the loss peak in the dynamic mechanical spectra. As with the blends, the best simulations of the experimental exchange spectra were obtained with a bimodal distribution of correlation times. In this case the most

5.3. Polymer Dynamics in the Solid State   445

mobile diluent molecules were assigned to those in contact with other diluent molecules, and the more rigid ones were those surrounded by the polymer.

$$\left(\underset{CH_3}{\overset{CH_3}{\bigcirc}}\!-\!O \right)_2 \!\!\overset{O}{\underset{\|}{P}}\!-\!O\!-\!\bigcirc\!-\!O\!-\!\overset{O}{\underset{\|}{P}}\!\left(O\!-\!\underset{H_3C}{\overset{H_3C}{\bigcirc}} \right)$$

Tetraxylyl hydroquinone diphosphate

5.3.5 THE DYNAMICS OF MULTIPHASE POLYMER SYSTEMS

Microphase-separated polymer systems, including diblock copolymers, immiscible blends, and multiblock thermoplastic elastomers, are commercially important materials that derive their function from the structure and dynamics of their phase-separated natures. They have been extensively studied by NMR in order to characterize the structure and dynamics of the phases and the interfaces between them. Among the important questions are how the dynamics changes

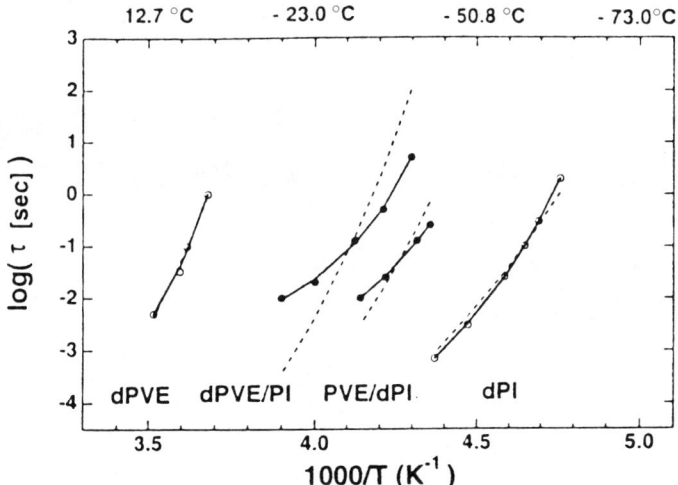

FIGURE 5.68 The log mean correlation times as a function of inverse temperature measured by simulation of the 2D ^{2}H exchange spectra as a function of temperature for deuterated polymers and blends. The dashed lines are estimated from mechanical measurements. Adapted with permission (105).

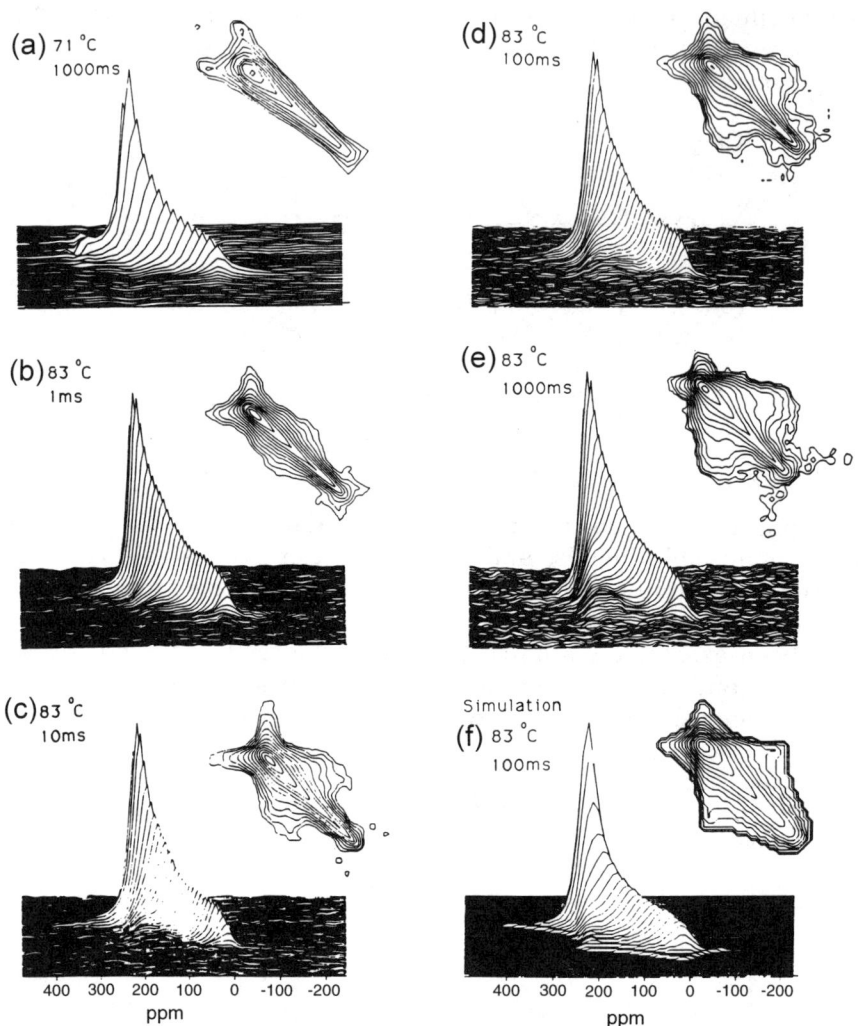

FIGURE 5.69 The ^{31}P 2D exchange spectra of tetraxylyl hydroquinone diphosphate dissolved in poly(2,6-dimethyl phenylene oxide) as a function of temperature and mixing time. The experimental data at (a) $\tau_m = 1$ s and $T = 71°C$, (b) $\tau_m = 0.001$ s and $T = 83°C$, (c) $\tau_m = 0.01$ s and $T = 83°C$, (d) $\tau_m = 0.1$ s and $T = 83°C$, and (e) $\tau_m = 1$ s and $T = 83°C$ are compared with the (f) simulated spectrum at $\tau_m = 0.1$ s and $T = 83°C$. Adapted with permission (107).

with temperature, how the dynamics are affected by the chemical structure of the monomers, and how the mechanical properties are related to the dynamics measured by NMR.

The structure and dynamics of polyurethanes and phase-separated segmented copolymers have been investigated by wide-line deuterium

5.3. Polymer Dynamics in the Solid State

NMR and magic angle spinning cross-polarization methods. In many of these materials there is a large difference in the molecular dynamics of the segments in "hard" and "soft" domains that results in large difference in the lineshapes and relaxation times. These differences have been used to characterize the length scale of phase separation (108) and the fraction of polymer in each phase in microphase-separated polymers, as shown in Section 4.3.2 (109).

The molecular dynamics of the hard segments of polyurethanes have been studied by wide-line deuterium NMR in polymers based on bis(4-isocyanophenyl) methane (MDI) and 2,4-toluenediyl di-isocyante (TDI) specifically labeled at the butanediol sites (110). Figure 5.70 shows deuterium spectra as a function of temperature and recycle delay time to separate the more rapidly and slowly relaxing components. These spectra differ from the normal Pake patterns observed for rigid polymers and demonstrate that the butanediol groups exhibit a wide distribution of correlation times. Further evidence for a wide distribution of correlation times is obtained from multiexponential spin–lattice relaxation behavior. From the studies of molecular dynamics as a function of temperature it is clear that not only are the molecular dynamics changing, but the fraction of polymer in the hard and soft domains is also changing with temperature. The lineshape for the more mobile fraction can be fit to a two-site jump model with a jump angle near 109°, corresponding to *gauche–trans* isomerizations, combined with small-amplitude librational motion, although this model is by no means the only one which will fit the data.

MDI

TDI

This same approach has been used to study the dynamics of the deuterium-labeled butanediol unit in poly(butylene terephthalate) and segmented copolymers, and some of the results are shown in Fig. 5.71 for poly(butylene terephthalate) and copolymers containing 96 and 87% hard segments (111). A motionally averaged lineshape is observed for poly(butylene terephthalate) at room temperature, which

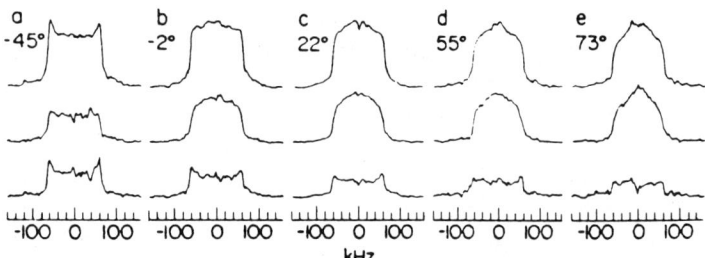

FIGURE 5.70 The solid-state ^{2}H NMR spectra of deuterium-labeled polyurethanes as a function of temperature and recycle delay time. The spectra were recorded at (a) −45, (b) −2, (c) 22, (d) 55, and (e) 73°C. The top row shows the spectra with a long delay (2–4 s), and the middle row those with a short delay (100–160 ms); the bottom row is the difference. Adapted with permission (110).

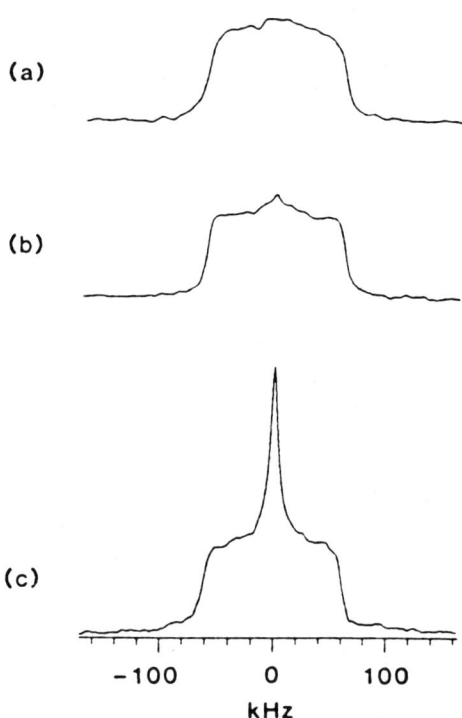

FIGURE 5.71 The wide-line ^{2}H NMR spectra of (a) poly(butylene terephthalate) and segmented copolyesters containing (b) 96% and (c) 87% hard segments. Adapted with permission (110).

can be fit to a two-site jump with an angle of 103° and a rate of 1.4×10^5 s^{-1}. Similar lineshapes are observed for the hard segments in the copolymers, but an additional resonance is also observed that is assigned to mobile deuterons in the soft segments that undergo nearly isotropic motion that is rapid on the deuterium NMR time scale. These same polymers were also studied by ^{13}C static and magic angle spinning NMR, where the chemical shift anisotropy lineshapes were reported (112). These data confirm the results from the deuterium NMR studies and show that the aromatic rings in the hard segments undergo 180° ring flips.

$$\left[\overset{O}{\underset{\parallel}{C}} - \bigcirc - \overset{O}{\underset{\parallel}{C}} OCH_2CD_2CD_2CH_2O \right]_m \left[\overset{O}{\underset{\parallel}{C}} - \bigcirc - \overset{O}{\underset{\parallel}{C}} O[CH_2CH_2CH_2CH_2O]_{12} \right]$$

Poly(butylene terephthalates)

The molecular dynamics of the hard segment have been studied in model polyurethane polymers containing five piperazine rings separated by carbonyloxytetramethyleneoxycarbonyl spacers. The piperazine rings are deuterium labeled at the outer (1,5), intermediate (2,4), or center position (3) to measure the dynamics at different parts of the hard-segment domains (113). Figure 5.72 shows, as in the other polyurethanes, that the lineshape is a composite of hard and soft domains. At sufficiently low temperatures all motion is frozen out and a Pake spectrum is observed, while at sufficiently high temperatures all of the domains are mobile and a motionally narrowed spectrum is observed. The data in Fig. 5.72 show that the relative fraction of hard and soft domains depends strongly on the position of the deuterium label, with more rigid domains observed for the polymer labeled at the 3 position relative to the polymers labeled at the 2,4 and 1,5

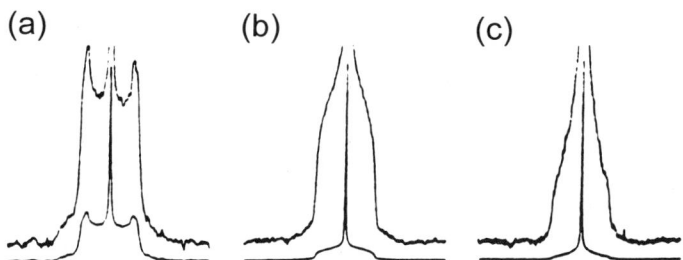

FIGURE 5.72 The wide-line ^{2}H NMR spectra for model polyurethanes with deuterium-labeled rings at the (a) center, (b) intermediate, or (c) outer positions. Adapted with permission (112).

positions. Thus there appears to be a range of molecular environments for the rings in the hard segments, ranging from the most rigid at the center to the more mobile at the surface. Further evidence for this interpretation is obtained from the temperature dependence of the lineshape for the rings labeled at the three positions.

$$\left[-N\underset{}{\bigcirc}N-\overset{O}{\overset{\|}{C}}OCH_2CH_2CH_2CH_2O\overset{O}{\overset{\|}{C}}-N\underset{}{\bigcirc}N-\overset{O}{\overset{\|}{C}}\right]_4\left[-OCH_2CH_2CH_2CH_2-\right]_{\sim 27.6}$$

A similar strategy has been used to study the dynamics of model polyurethanes with a soft–hard–soft architecture, where a five-ring piperazine ring hard segment is labeled at the center, intermediate, or outer positions (114). At temperatures below 300 K the soft phase solidifies and hinders the mobility of the hard segments at the interface. Between 310 and 350 K composite lineshapes are observed, indicating the coexistence of the hard and soft domains. Between 350 and 400 K there is a dynamic coupling between the two phases, and above 360 K the tetrahedral jumps occur at a sufficient rate that the lineshapes for the hard domains are motionally averaged.

REFERENCES

1. F. Heatley, *Prog. NMR Spectrosc.* **13**, 47 (1979).
2. J. R. Lyerla, "High Resolution NMR of Glassy Amorphous Polymers" (J. R. Lyerla, ed.), p. 63. VCH, Deerfield Beach, Florida, 1986.
3. H. Spiess, *Colloid Polym. Sci.* **261**, 193 (1983).
4. H. Spiess, *Chem. Rev.* **91**, 1321 (1991).
5. W. Gronski, *Makromol. Chem.* **177**, 3017 (1976).
6. A. Allerhand and R. K. Hailstone, *J. Chem. Phys.* **56**, 3718 (1972).
7. K. S. Cole and R. H. Cole, *J. Chem. Phys.* **9**, 341 (1941).
8. R. M. Fuoss and J. G. Kirkwood, *J. Am. Chem. Soc.* **63**, 385 (1941).
9. J. Schaefer, *Macromolecules* **6**, 882 (1973).
10. G. Lipari and A. Szabo, *J. Am. Chem. Soc.* **104**, 4546 (1982).
11. G. Lipari and A. Szabo, *J. Am. Chem. Soc.* **104**, 4559 (1982).
12. B. Valeur, J. P. Jarry, F. Genny, and L. Monnerie, *J. Polym. Sci. Polym. Phys. Ed.* **13**, 2251 (1975).
13. A. A. Jones and W. H. Stockmayer, *J. Polym. Sci. Polym. Phys. Ed.* **15**, 847 (1977).
14. J. L. Viovy, L. Monnerie, and J. C. Brochon, *Macromolecules* **16**, 1845 (1983).
15. C. K. Hall and E. Helfand, *J. Chem. Phys.* **77**, 3275 (1982).
16. R. Dejean de la Batie, F. Laupretre, and L. Monnerie, *Macromolecules* **21**, 2052 (1988).
17. R. Dejean de la Batie, F. Laupretre, and L. Monnerie, *Macromolecules* **21**, 2045 (1988).
18. W. O. Howarth, *Faraday Trans. 2* **75**, 863 (1979).

References

19. F. Heatley and A. Begum, *Polymer* **17**, 399 (1976).
20. G. C. Levy, D. E. Axelson, R. Schwartz, and J. Hochmann, *J. Am. Chem. Soc.* **100**, 410 (1978).
21. D. J. Gisser, S. Glowinkowski, and M. D. Ediger, *Macromolecules* **24**, 4270 (1991).
22. D. Ghesquiere, B. Ban and C. Chachaty, *Macromolecules* **10**, 743 (1977).
23. A. Guillermo, R. Dupeyre, and J. P. Cohen-Addad, *Macromolecules* **23**, 1291 (1990).
24. W. Zhu and M. D. Ediger, *Macromolecules* **28**, 7549 (1995).
25. A. Spyros, P. Dais, and F. Heatley, *Macromolecules* **27**, 5845 (1994).
26. M. Paci, *Polym. Commun.* **26**, 116 (1985).
27. E. Helfand, *J. Chem. Phys.* **54**, 4651 (1971).
28. K. J. Lui and J. E. Angerson, *Macromolecules* **3**, 163 (1970).
29. M. C. Lang, F. Laupretre, C. Noel, and L. Monnerie, *J. Chem. Soc. Faraday Trans.* **2**, 349 (1979).
30. S. Glowinkowski, D. J. Gisser, and M. D. Ediger, *Macromolecules* **23**, 3520 (1990).
31. G. E. Pake, *J. Chem. Phys.* **16**, 327 (1948).
32. J. Hirschinger and A. D. English, *J. Magn. Reson.* **85**, 542 (1989).
33. H. Spiess, *H. Chem. Phys.* **72**, 6755 (1980).
34. D. A. Torshia and A. Szabo, *J. Magn. Reson.* **49**, 107 (1982).
35. C. Schmidt, B. Blumich, and H. W. Spiess, *J. Magn. Reson.* **79**, 269 (1988).
36. W. T. Dixon, J. Schaefer, M. Sefcik, E. O. Stejskal, and R. A. McKay, *J. Magn. Reson.* **49**, 341 (1982).
37. Y. Yang, M. Schuster, B. Blumich, and H. W. Spiess, *Chem. Phys. Lett.* **139**, 239 (1987).
38. A. Kentgens, E. de Boer, and W. Veeman, *J. Chem. Phys.* **87**, 6859 (1987).
39. A. Hagemeyer, K. Schmidt-Rohr, and H. Spiess, *Adv. Magn. Reson.* **13**, 85 (1989).
40. W. P. Rothwell and J. S. Waugh, *J. Chem. Phys.* **74**, 2721 (1981).
41. J. Schaefer, R. McKay, and E. Setjskal, *J. Magn. Reson.* **52**, 123 (1983).
42. D. E. Axelson, "Carbon-13 Solid State NMR of Semicrystalline Polymers." VCH, Deerfield Beach, Florida, 1986.
43. J. R. Lyerla and C. S. Yannoni, *IBM J. Res. Dev.* **27**, 302 (1983).
44. V. J. McBrierty, D. C. Douglass, and D. R. Falcone, *J. Chem. Soc., Faraday Trans.* 2 **68**, 1051 (1972).
45. H. T. Edzes and W. S. Veeman, *Polym. Bull.* **5**, 255 (1981).
46. A. Naito, S. Ganapathy, K. Akasaka, and C. A. McDowell, *J. Magn. Reson.* **54**, 226 (1983).
47. J. L. White and P. A. Mirau, *Macromolecules* **26**, 3049 (1993).
48. R. A. Komoroski, "High Resolution NMR Spectroscopy of Synthetic Polymers in Bulk," Vol. 7. VCH, Deerfield Beach, 1986.
49. J. J. Marcinko and A. A. Parker, *Macromol. Symp.* **86**, 251 (1994).
50. J. Schaefer, E. O. Stejskal, and R. Buchdahl, *Macromolecules* **10**, 384 (1977).
51. W. S. Veeman, E. M. Menger, W. Ritchey, and E. de Boer, *Macromolecules* **12**, 924 (1979).
52. J. J. Dechter, *J. Polym. Sci. Poly. Lett. Ed.* **23**, 261 (1985).
53. A. Jonansson and J. Tegenfeldt, *Macromolecules* **25**, 4712 (1992).
54. H. Mirau and A. English, *Macromolecules* **21**, 1543 (1988).
55. J. Hirschinger, H. Miura, K. H. Gardner, and A. D. English, *Macromolecules* **23**, 2153 (1990).
56. J. Wendoloski, K. Gardner, J. Hirschinger, H. Miura, and A. English, *Science* **247**, 431 (1990).
57. H. Miura, J. Hirschinger, and A. D. English, *Macromolecules* **23**, 2169 (1990).

58. C. L. Jackson, R. J. Schadt, K. H. Gardner, D. B. Chase, S. R. Allen, V. Gabara, and A. D. English, *Polymer* **35**, 1123 (1994).
59. R. J. Schadt, E. J. Cain, K. H. Gardner, V. Gabara, S. R. Allen, and A. D. English, *Macromolecules* **26**, 6503 (1993).
60. R. J. Schadt, K. H. Gardner, V. Gabara, S. R. Allen, D. B. Chase, and A. D. English, *Macromolecules* **26**, 6509 (1993).
61. J. H. Simpson, D. M. Rice, and F. E. Karasz, *J. Polym. Sci. Part B, Polym. Phys.* **30**, 11 (1992).
62. J H. Simpson, D. M. Rice, and F. A. Karasz, *Polymer* **32**, 2340 (1991).
63. M. Poliks and J. Schaefer, *Macromolecules* **23**, 3426 (1990).
64. S. J. Opella and M. H. Frey, *J. Am. Chem. Soc.* **101**, 5854 (1979).
65. A. Tonelli, "NMR Spectroscopy and Polymer Microstructure: The Conformational Connection." VCH, New York, 1989.
66. K. Schmidt-Rohr and H. Spiess, *Macromolecules* **24**, 5288 (1991).
67. D. Schaefer, H. Spiess, U. Suter, and W. Flemming, *Macromolecules* **23**, 3431 (1990).
68. J. Hirschinger, D. Schaefer, H. Spiess, and A. Lovinger, *Macromolecules* **24**, 2428 (1991).
69. W. Maas, A. Kentgens, and W. Veeman, *J. Chem. Phys.* **87**, 6854 (1987).
70. N. McCrum, B. Read, and G. Williams, "Anelastic and Dielectric Effects in Polymeric Solids." Wiley, New York, 1967.
71. Y. Takahashi, H. Miyaji, and K. Asai, *Macromolecules* **16**, 1789 (1983).
72. J. Schaefer, M. D. Sefcik, E. O. Stejskal, R. A. McKay, W. T. Dixon, and R. E. Cais, *Macromolecules* **17**, 1107 (1984).
73. P. T. Inglefield, R. A. Amici, J. F. O'Gara, C. C. Jung, and A. A. Jones, *Macromolecules* **16**, 1552 (1983).
74. J. F. O'Gara, A. A. Jones, C. C. Hung, and P. T. Inglefield, *Macromolecules* **18**, 1117 (1985).
75. J. Schaefer, E. Stejskal, R. McKay, and W. Dixon, *Macromolecules* **17**, 1479 (1984).
76. P. Henrichs, M. Linder, J. Hewitt, D. Massa, and H. Isaacson, *Macromolecules* **17**, 2412 (1984).
77. P. B. Smith and D. J. Moll, *Macromolecules* **23**, 3250 (1990).
78. M. T. Hansen, A. S. Kulik, K. O. Prins, and H. W. Spiess, *Polymer* **33**, 2231 (1992).
79. A. S. Kulik and K. O. Prins, *Polymer* **34**, 4635 (1993).
80. A. S. Kulik and K. O. Prins, *Polymer* **34**, 4629 (1993).
81. P. B. Smith, R. A. Bubeck, and S. E. Bales, *Macromolecules* **21**, 2058 (1988).
82. P. M. Henrichs, V. A. Nicely, and D. R. Fagerburg, *Macromolecules* **24**, 4033 (1991).
83. K. Zemke, B. Chmelka, and H. Spiess, *Macromolecules* **24**, 6874 (1991).
84. S. Kaufmann, S. Wefing, D. Schaefer, and H. W. Spiess, *J. Chem. Phys.* **93**, 197 (1990).
85. U. Pschorn, E. Rossler, H. Sillescu, S. Kaufmann, D. Schaefer, and H. Spiess, *Macromolecules* **24**, 398 (1991).
86. M. L. Williams, R. F. Landel, and J. D. Ferry, *J. Am. Chem. Soc.* **77**, 3701 (1955).
87. D. Schaefer and H. W. Spiess, *J. Chem. Phys.* **97**, 7944 (1992).
88. K. Schmidt-Rohr, A. S. Kulik, H. W. Beckham, A. O. Ohlemacher, U. Pawelzik, C. Boeffel, and H. W. Spiess, *Macromolecules* **27**, 4733 (1994).
89. P. Tekely, J. Brondear, K. Elbayed, A. Retournard, and D. Canet, *J. Magn. Reson.* **80**, 509 (1988).
90. A. S. Kulik, H. W. Beckham, K. Schmidt-Rohr, D. Radloff, U. Pawelzik, C. Boeffel, and H. W. Spiess, *Macromolecules* **27**, 4746 (1994).

91. M. T. Hansen, B. Blumich, C. Boeffel, and H. W. Spiess, *Macromolecules* **25**, 5542 (1992).
92. B. Blumich and H. Spiess, *Angew. Chem., Int. Ed. Engl.* **27**, 1655 (1988).
93. K. Schmidt-Rohr and H. W. Spiess, *Phys. Rev. Lett.* **66**, 3020 (1991).
94. C. M. Roland, *Macromolecules* **20**, 2557 (1987).
95. J. Miller, K. McGrath, C. Roland, C. Trask, and A. Garroway, *Macromolecules* **23**, 4543 (1990).
96. K. Takegoshi and K. Hikichi, *J. Chem. Phys.* **94**, 3200 (1991).
97. C. Le Menestrel, A. M. Kenwright, P. Sergot, F. Laupretre, and L. Monnerie, *Macromolecules* **25**, 3020 (1992).
98. C. J. T. Landry and P. M. Henrichs, *Macromolecules* **22**, 2157 (1989).
99. M. A. de Araujo, D. Oelfin, R. Stadler, and M. Moller, *Makromol. Chem. Rapid Commun.* **10**, 259 (1989).
100. T. Suzuki, E. M. Pearce, and T. K. Kwei, *Polymer* **30**, 198 (1992).
101. A. Simmons and A. Natansohn, *Macromolecules* **24**, 3651 (1991).
102. A. Simmons and A. Natansohn, *Macromolecules* **25**, 3881 (1992).
103. Y. H. Chin, C. Zhang, P. Wang, P. T. Inglefield, A. A. Jones, R. P. Kambour, J. T. Bendler, and D. M. White, *Macromolecules* **25**, 3031 (1992).
104. Y. H. Chin, P. T. Inglefield, and A. A. Jones, *Macromolecules* **26**, 5372 (1993).
105. G. C. Chung, J. A. Kornfield, and S. D. Smith, *Macromolecules* **27**, 964 (1994).
106. G. C. Chung, J. A. Kornfield, and S. D. Smith, *Macromolecules* **27**, 5729 (1994).
107. C. Zhang, P. Wang, A. Jones, and P. Inglefield, *Macromolecules* **24**, 338 (1991).
108. R. Assink, *Macromolecules* **11**, 1233 (1978).
109. J. J. Dumais, L. W. Jelinski, L. M. Leung, I. Gancarz, A. Galambos, and J. T. Koberstein, *Macromolecules* **18**, 116 (1985).
110. A. Kintinar, I. Jelinski, I. Gancarz, J. T. Koberstein, *Macromolecules*, **19**, 1876 (1986).
111. L. W. Jelinski, J. J. Dumais, and A. J. Engel, *in* "Solid State 2H NMR Studies of Molecular Motion. Poly(butylene terephthalate) and poly(butylene terephthalate)-Containing Segmented Copolymers." (L. W. Jelinski, J. J. Dumais, A. K. Engel, eds.), Vol. 247, p. 55. American Chemical Society, Washington, D.C., 1984.
112. L. W. Jelinski, J. J. Dumais, and A. K. Engel, *Macromolecules* **16**, 403 (1983).
113. J. A. Kornfield, H. W. Spiess, H. Nefzger, H. Hayden, and C. D. Eisenbach, *Macromolecules* **24**, 4787 (1991).
114. A. D. Meltzer, H. W. Spiess, C. D. Eisenbach, and H. Hayen, *Macromolecules* **25**, 993 (1992).

Index

A
Allyl methacrylate, 204
Alternating copolymers, 131
Amorphous polymers, 225–260, 412–436
Angular momentum, 2
Anisotropic shielding, 38–42
Asymmetry parameter (η), 381
Atactic, 119

B
Bernoullian statistics, 133–151, 184, 187, 219
Bloch equations, 23–29
Block copolymers, 131, 142, 214, 214–217, 292–300
Block length, 150, 215, 216, 219
Boltzman spin temperature, 12
Branching, 117, 129–131, 199–208

C
Carbon–carbon couplings, 53
Carbon chemical shifts, 42–46
    α-effect, 43
    β-effect, 43
    γ-effect, 43, 44
Carbon–proton couplings, 53
Carbon–proton cross relaxation time (T_{CH}), 72, 80, 392, 415
Carbon rotating frame relaxation time ($T_{1\rho}^C$), 72, 82, 392–394, 418, 441
Carr–Purcell, 84
Chain conformation, 167, 220–231, 245–273
Chain packing, 248, 249
Charge-transfer complex, 301
Chemical shift, 30–47
Chemical shift anisotropy, 17, 35, 36, 65, 357, 386–388, 420, 428–434
Chemical shift filter, 278, 298
Chemical shift gradients, 278, 317
Chemical shifts referencing, 59–61
Cole–Cole distribution function, 361, 370, 371, 376
Coleman–Fox model, 145–153, 187
Combined rotation and multipulse NMR (CRAMPS), 256, 277, 278, 310–313, 317
Configurational statistics, 133–151
Continuous wave, 8
Copolymers, 131–133, 212–220
Copolymer polymerization, 153
Correlation time, 16, 74, 354, 358
COSY, 97, 165, 186, 195, 203, 211, 227
    double quantum filtered, 97, 165
Cross linking, 117, 129–131
Cross polarization, 69–72, 245, 318–322, 392
Cross polarization spin-lattice relaxation time (CPT_1), 79, 253, 286
Cross relaxation, 76

D
Decoupling, 61, 62, 65
DEPT, 107, 172, 175

Dendrimers, 206–208
Deuterated solvents, 58
Deuterium NMR, 288, 290, 333–335, 381–386, 395–405, 422–424, 434, 435, 438–440, 443
Diamagnetic susceptibility, 39
Diamond lattice model, 365, 366, 373, 374
Diffusion constants, 277, 279, 313
Dimethyl sulfone, 386
Dipolar broadening, 19
Dipolar dephasing, 108, 254
Dipolar filter, 278, 281, 294
Distributions of correlation times, 360–364, 413, 435, 443
Disyndiotactic, 120
Domain size, 275, 279–286, 304–306, 318
Draw ratio, 332

E

Electric quadrupole moment, 15
Empirical chemical shift rules, 161–165
End groups, 208–211
Epimerization, 159
Equivalent chemical shifts, 51
Erythro, 120, 121
Erythrodi-isotactic, 120
Evolution period, 89

F

Field gradients, 338–340
Fluorine chemical shifts, 46
Fluorine couplings, 53, 179, 180
Four-dimensional NMR, 435
Fourier transformation, 11
Free induction decay, 10
Frequency-domain spectrum, 11
Frequency labeling, 89
Frequency locking, 57
Fuoss–Kirkwood distribution function, 361, 362, 370, 371, 376

G

γ-*gauche* effect, 157, 161, 163, 172, 178, 182, 190, 246–250, 256, 261–264, 425
Geminal couplings, 51, 166, 226
Geometric isomerism, 117, 122–127, 196–199

Goldman–Shen, 274, 278, 279, 290
Graft copolymers, 131

H

Hall–Helfand model, 366, 367, 375, 376
Hartmann–Hahn condition, 69
Hemi-isotactic, 184
Heteronuclear correlation, 166, 185, 187, 192, 211, 322–330
Heteronuclear decoupling, 62, 65, 87
Heteronuclear multiple-quantum coherence (HMQC), 101, 204
Heterotactic, 129, 142
Hydrogen bonding, 303

I

Imaging, 337–349
INADEQUATE, 173, 174, 206
Inductive effects, 36
Inclusion compounds, 271–273
INEPT, 172, 193
Intermolecular association, 232–239
Intermolecular cross polarization, 303, 304
Internal chemical shift reference, 59
Inversion recovery, 77
Isoregic, 118
Isotactic, 119, 142
Isotropic rotational diffusion, 359, 369

J

J coupling, 47–55

K

Karplus relationship, 52, 220
Kramer's Theory, 377

L

Larmor frequency, 4, 5
Lattice, 12
Lineshapes, 380–389
Log χ^2 distribution, 363, 364, 370, 376

Index

M

Magic angle spinning (MAS), 65–68, 245, 268, 394
Magnetic field strength, 6
Magnetic moment, 2
Magnetic quantum numbers, 2
Magnetogyric ratio, 3, 72
Markov statistics, 133–151, 184, 187
Meiboom–Gill, 84
meso, 121
Microstructure, 117, 155
Model compounds, 158, 159
Model free relaxation, 365
Multiphase polymers, 289–300, 445–450

N

Natural abundance, 4
Neoprene, 123, 198
Nitrogen chemical shifts, 46
Nitrogen couplings, 54
Nitrogen NMR, 204, 252, 258, 259
NOESY, 102–105, 167, 169, 186, 214, 220, 223, 228, 229, 232–239, 283
Nonselective excitation, 75
Novolac, 205
Nuclear electric quadrupolar relaxation, 15, 29, 357
Nuclear Overhauser effects, 85–87, 210, 356
Nylon 6, 252
Nylon 66, 395–401

O

Optically active polymers, 128, 129
Orientational distribution function, 332–335
Oriented polymers, 331–337

P

Peak width at half height, 23, 27
Perhydrotriphenylene, 271, 272
Persistance ratio, 142
Phosphorus chemical shifts, 46
Phosphorus couplings, 54
Phosphorus NMR, 267, 283–286, 443–445
Polarization transfer, 107

Polybenzimidazole, 303, 316–318
Polybutadiene, 197, 261, 268, 271, 342–344, 369, 371–373, 437, 438, 443
Poly(1-butene), 174, 368, 369
Polycaprolactone, 94, 99, 100, 272, 327
Polycarbonate, 299, 324, 325, 327, 330, 347, 414, 415, 419–424, 434, 438
Poly[2-(3,5-dinitrobenzoyl)oxy ethyl methacrylate], 302, 442
Polyether epoxy resins, 258
Polyetherimide, 316–318
Poly[(N-ethylcarbazole-3-yl) methyl methacrylate], 442
Polyethylene, 200, 245, 254, 271, 279–283, 286, 287, 334, 368, 406–408
Poly(acrylic acid), 237
Poly(amic acid), 258
Poly(α-amino acids), 105, 129
Poly(benzyl-L-glutamate), 105, 228
Poly[bis(4-ethyl phenoxy) phosphazine)], 267
Poly[bis(3-methoxyphenoxy)phosphazine], 283–286
Poly[bis(phenoxy)phosphazine], 283–286
Poly(butylene terephthalate), 261–263, 330, 447–449
Poly(cyclohexylenedimethlene terephthalate), 438
Poly(diacetylenes), 264–266
Poly(diethyl oxetane), 248
Poly(dimethyl phenylene oxide), 110, 211, 257, 306, 325, 330, 415, 425, 439, 442–445
Poly(dimethyl siloxane), 250
Poly(ether ether ketone) (PEEK), 405, 406
Poly(ether sulfone), 415
Poly(ethyl acrylate), 434
Poly(ethylene-co-methacrylic acid), 296, 302
Poly(ethylene-co-propylene), 217
Poly(ethylene-co-vinyl acetate), 219
Poly(ethylene oxide), 237, 255, 288, 369, 373, 377, 395
Poly(ethylene terephthalate), 281, 330, 337
Poly(3-hexyl thiophene), 168
Polyimide, 303
Polyisobutylene, 369
Polyisoprene, 198, 373, 375, 377, 425, 427, 428, 437, 438, 443
Poly(lactic acid), 129
Polymer blends, 167, 300–331, 436–445

Poly(methyl acrylate), 326
Poly(methyl methacrylate), 109, 185–189, 227–228, 236, 287, 318–322, 369, 370, 391, 415, 422, 428–434, 438
Poly(methyl methacrylate-co-4-vinyl pyridine), 233
Poly(4-methyl styrene), 306
Poly(methyl thriiane), 369
Poly(n-alky acrylates), 377
Poly(n-butyl methacrylate), 370
Poly(n-hexyl methacrylate), 370
Poly(napthal acrylates), 371
Poly(oxymethylene), 255, 337, 394, 409–411
Poly(p-acetoxystyrene), 239
Poly(p-hydroxystyrene), 239
Poly(pentadiene), 271
Poly(phenyl thriiane), 369
Poly(phenylene terephthalamide), 401–403
Poly(phenylene vinylene), 335, 403–405
Polypropylene, 159, 171–174, 181–185, 247, 248, 368, 369, 391, 394, 409, 425
Poly(propylene oxide), 95, 129, 137, 138, 174–177, 209, 210, 221, 369, 370
Poly(propylene sulfide), 369, 424
Poly[(S)-4-methyl-1-hexane], 377
Polysilanes, 250, 266, 267
Polysiloxane, 344
Polystyrene, 234, 255, 256, 304, 307–316, 325, 330, 369, 370, 373, 415–419, 425–427, 439, 442
Poly(styrene-alt-methyl methacrylate), 213, 229
Poly(styrene-b-butadiene), 278
Poly(styrene-b-dimethyl siloxane), 293
Poly(styrene-b-methyl phenyl siloxane), 294–296
Poly(styrene-co-2-phenyl-1,1-dicyanoethene), 169
Poly(styrene-co-acrylonitrile), 307
Poly(styrene-co-methyl methacrylate), 298, 299
Poly(styrene-co-styrene sulfonic acid), 233
Poly(styrene-divinyl benzene), 271
Poly(styrene-peroxide), 369
Polysuccinimide, 203
Polysulfones, 129
Poly(t-butyl ethylene oxide-b-ethylene oxide-b-t-butyl ethylene oxide), 216
Poly(tetrafluoro ethylene), 333

Poly(tetramethyl-p-silphenylenesiloxane-b-dimethyl siloxane), 214
Polyurethanes, 289–292, 446–450
Poly(vinyl acetate), 348, 435
Poly(vinyl chloride), 163, 190–195, 201, 202, 236, 368
Poly(vinyl fluoride), 180, 195, 196
Poly(vinyl formal), 122
Poly(vinylidine chloride), 369, 415
Poly(vinylidine cyanide-alt-vinyl acetate), 226
Poly(vinylidine cyanide-co-isobutylene) 220
Poly(vinylidine fluoride), 177–180, 318–322, 411, 412
Poly(vinyl methyl ether), 234, 304, 307–316, 373, 375, 377
Poly(vinyl phenol), 325, 326
Poly(2-vinyl pyridine), 302
Poly(4-vinyl pyridine), 302
Product operator formalism, 92, 102, 107
Progressive saturation, 79
Propagation errors, 151–153
Propagation mechanism, 133–151
Proton chemical shifts, 37, 39, 41
Proton–proton couplings, 49–52
Proton rotating frame relaxation time ($T_{1\rho}^H$), 72, 81, 392
Pulsed NMR spectroscopy, 8, 9

Q

Quadrupolar echo, 385
Quadrupolar relaxation, 17, 29, 30, 357

R

racemic, 121
Random copolymers, 131
Rate phenomenon, 55
Regioisomerism, 117, 118, 165, 168–181
Relayed coherence transfer, 99, 165
Resolution, 155, 244, 255, 257, 274, 338
Resonance assignments, 155, 158–167, 192
Ring currents, 41
Ring flips, 381, 401–406, 416–420, 434, 435
ROESY, 105
Rotating frame spin-lattice relaxation, $T_{1\rho}$, 81, 415

Rotational dipolar spin echo, 389, 405, 406, 421
Rotational isomeric states (RIS), 157, 163, 172, 182, 190

S

Saturation, 12, 28
Screeing constant, 32–34
Segmental dynamics, 359
Selective excitation, 74, 274, 310–313, 430
Semicrystalline polymers, 245–255, 279–289, 394–412
Silicon NMR, 214, 250, 266, 267
Solid–solid phase transitions, 261–267
Solid-state NMR, 63
Solid-state NOEs, 313–316
Solution dynamics, 358–378
Spectral density, 74–77, 86, 354, 361–364, 383, 393
Spectral editing, 107–111
Spin diffusion, 76, 244, 274–300
Spin echoes, 83
Spin-lattice relaxation (T_1), 11–18, 72–81, 210, 356, 357, 380
Spin-locking, 70
Spin–spin relaxation time (T_2), 19–23, 82–85, 341, 356, 357, 380, 389, 415
Stereochemical configuration, 117–122, 181–196
Stereosequence assignments, 166
Superconducting magnet, 7
Syndioregic, 118
Syndiotactic, 119, 142

T

T_{1Q}, 382–385, 398
Tetramethylsilane (TMS), 59, 60
threo, 120, 121, 185
Threodi-isotactic, 120
Three-dimensional NMR, 222–227, 430
Time domain spectrum, 10
Tip angle, 9
TOCSY, 98, 105, 193
Total suppression of sidebands (TOSS), 110, 388
Two-dimensional NMR, 87–105, 157, 165–167
correlated, 88, 96–107
heteronuclear correlation, 102
J-resolved, 91–95, 221
resolved, 88, 91–95

V

Vicinyl couplings, 51, 166, 226

W

Wide line separation 2D NMR (WISE), 292, 296, 313
Windowless isotropic mixing (WIM), 322

Z

Zeeman, 5